Applied Genomics and Public Health

Applied Genomics and Public Health

Edited by

George P. Patrinos

Department of Pharmacy, School of Health Sciences, University of Patras, Patras, Greece

Department of Pathology, College of Medicine and Health Sciences, United Arab Emirates University, Al-Ain, United Arab Emirates

Zayed Center of Health Sciences, United Arab Emirates University, Al-Ain, United Arab Emirates

Department of Pathology - Bioinformatics Unit, Faculty of Medicine and Health Sciences, Erasmus University Medical Center, Rotterdam, The Netherlands

ELSEVIER

ACADEMIC PRESS

An imprint of Elsevier

Academic Press is an imprint of Elsevier
125 London Wall, London EC2Y 5AS, United Kingdom
525 B Street, Suite 1650, San Diego, CA 92101, United States
50 Hampshire Street, 5th Floor, Cambridge, MA 02139, United States
The Boulevard, Langford Lane, Kidlington, Oxford OX5 1GB, United Kingdom

British Library Cataloguing-in-Publication Data
A catalogue record for this book is available from the British Library

Library of Congress Cataloging-in-Publication Data
A catalog record for this book is available from the Library of Congress

ISBN: 978-0-12-813695-9

For Information on all Academic Press publications
visit our website at https://www.elsevier.com/books-and-journals

Publisher: Andre Gerhard Wolff
Acquisition Editor: Kattie Washington
Senior Editorial Project Manager: Kiristi Anderson
Production Project Manager: Kiruthika Govindaraju
Cover Designer: Mark Rogers

Typeset by MPS Limited, Chennai, India

Contents

List of Contributors

Asimina Andritsou
Department of Pharmacy, School of Health Sciences, University of Patras, Patras, Greece

Neeraj Arora
Department of Pathology, Tata Medical Centre Kolkata, Kolkata, India

Evaggelia Barba
Department of Pharmacy, School of Health Sciences, University of Patras, Patras, Greece

Adam Binder
Division of Hematologic Malignancies and Hematopoietic Stem Cell Transplantation, Department of Medical Oncology, Sidney Kimmel Cancer Center, Thomas Jefferson University, Philadelphia, PA, United States

Nitin Chakravarti
Division of Hematologic Malignancies and Hematopoietic Stem Cell Transplantation, Department of Medical Oncology, Sidney Kimmel Cancer Center, Thomas Jefferson University, Philadelphia, PA, United States

Konstantina Chalikiopoulou
Department of Pharmacy, School of Health Sciences, University of Patras, Patras, Greece

Catalina Lopez Correa
Genome British Columbia, Vancouver, BC, Canada

Adam Ertel
Department of Cancer Biology, Sidney Kimmel Cancer Center, Thomas Jefferson University, Philadelphia, PA, United States

Paolo Fortina
Department of Cancer Biology, Sidney Kimmel Cancer Center, Thomas Jefferson University, Philadelphia, PA, United States; Department of Translation and Precision Medicine, Sapienza University, Rome, Italy

Vassileios Fragoulakis
The Golden Helix Foundation, London, United Kingdom

Konstantinos Ghirtis
Human Medicines Assessment Section, Product Assessment Division, National Organization for Medicines, Athens, Greece; Department of Pharmaceutical Chemistry, Faculty of Pharmacy, National and Kapodistrian University of Athens, Zografou, Greece

Despina Giannouri
Department of Pharmacy, School of Health Sciences, University of Patras, Patras, Greece

Heidi Carmen Howard
Centre for Research Ethics and Bioethics, Uppsala, Sweden

Manousos E. Kambouris
Department of Pharmacy, School of Health Sciences, University of Patras, Patras, Greece

Kostas Kampourakis
Section of Biology and IUFE, University of Geneva, Geneva, Switzerland

Kariofyllis Karamperis
Department of Twin Research and Genetic Epidemiology, King's College London, London, United Kingdom

Zoe Kordou
Department of Pharmacy, School of Health Sciences, University of Patras, Patras, Greece

Maria Koromina
Department of Pharmacy, School of Health Sciences, University of Patras, Patras, Greece

Gaurav Kumar
Department of Cancer Biology, Sidney Kimmel Cancer Center, Thomas Jefferson University, Philadelphia, PA, United States

Yiannis Manoussopoulos
Plant Protection Division of Patras, Institute of Industrial and Forage Plants, Patras, Greece

Effrosyni Mendrinou
Department of Pharmacy, School of Health Sciences, University of Patras, Patras, Greece

Anjali Mishra
Division of Hematologic Malignancies and Hematopoietic Stem Cell Transplantation, Department of Medical Oncology, Sidney Kimmel Cancer Center, Thomas Jefferson University, Philadelphia, PA, United States

Christina Mitropoulou
The Golden Helix Foundation, London, United Kingdom

Emilia Niemiec
Centre for Research Ethics and Bioethics, Uppsala, Sweden

Vural Özdemir
Science Communication and Emerging Technology Governance, Toronto, ON, Canada

Konstantina Papaioannou
Independent Scholar, London, United Kingdom

George P. Patrinos
Department of Pharmacy, School of Health Sciences, University of Patras, Patras, Greece; Department of Pathology, College of Medicine and Health Sciences, United Arab Emirates University, Al-Ain, United Arab Emirates; Zayed Center of Health Sciences, United Arab Emirates University, Al-Ain, United Arab Emirates; Department of Pathology - Bioinformatics Unit, Faculty of Medicine and Health Sciences, Erasmus University Medical Center, Rotterdam, The Netherlands

Konstantina Politopoulou
Department of Pharmacy, School of Health Sciences, University of Patras, Patras, Greece

Pierluigi Porcu
Division of Hematologic Malignancies and Hematopoietic Stem Cell Transplantation, Department of Medical Oncology, Sidney Kimmel Cancer Center, Thomas Jefferson University, Philadelphia, PA, United States

Nicole Probst-Hensch
Department of Epidemiology and Public Health, Swiss Tropical and Public Health Institute, Swiss TPH, Basel, Switzerland University of Basel, Basel, Switzerland

Boris Rebolledo-Jaramillo
Rare Diseases Program, Bioinformatics Analysis Core, Center for Genetics and Genomics, Instituto de Ciencias e Innovación en Medicina, Facultad de Medicina, Clínica Alemana Universidad del Desarrollo, Santiago, Chile

Gabriela M. Repetto
Rare Diseases Program, Center for Genetics and Genomics, Instituto de Ciencias e Innovación en Medicina, Facultad de Medicina, Clínica Alemana Universidad del Desarrollo, Santiago, Chile

Stavroula Siamoglou
Department of Pharmacy, School of Health Sciences, University of Patras, Patras, Greece

Maria Skokou
Psychiatric Clinic, Patras General Hospital, Patras, Greece

Evangelia-Eirini Tsermpini
Department of Pharmacy, School of Health Sciences, University of Patras, Patras, Greece

Aristea Velegraki
Department of Microbiology, School of Medicine, National and Kapodistrian University of Athens, Athens, Greece

Athanassios Vozikis
Department of Economics, University of Piraeus, Piraeus, Greece

Sam Wadge
Department of Twin Research and Genetic Epidemiology, King's College London, London, United Kingdom

Janet L. Williams
Research Genetic Counselors, Genomic Medicine Institute, Geisinger, Danville, PA, United States

Preface

The *Translational and Applied Genomics* book series was launched in 2018, aiming to provide succinct and comprehensive textbooks pertaining to the various translational and applied genomic disciplines that constitute the cornerstone of personalized medicine.

Unfortunately until now, the significant advancements in the field of personalized medicine, in terms of both genomic technology as well as the findings from genomics discovery research, are not matched with reciprocal advances in the translation of these findings into the clinic. In particular, there are often significant hurdles that decelerate the smooth incorporation of genomics research findings in the daily medical practice. These obstacles are more related to public health genomics disciplines rather than genomics research itself but are of equal importance since they contribute greatly to the transition from genomics research to genomic and personalized medicine. These include ethical, legal, and societal aspects in genomics, genome informatics, improving and harmonizing the genetics education of healthcare professionals and biomedical scientists, raising genetics awareness among the general public, and health economic evaluation in relation to genomic medicine.

Given the fact that there was no textbook in the literature exclusively devoted to applied genomics and public health, and immediately upon the launch of the Translational and Applied Genomics book series, I have submitted a proposal to edit such textbook envisaging to fill in this important literature gap, which Elsevier/Academic Press has agreed to include in the book series.

The contents of this textbook are divided in two parts. The first part starts with genetic epidemiology and continues with genomics applications in key medical specialties, such as cancer genomics and psychiatric genomics. Also, dedicated chapters have been compiled for the genomics of rare diseases and pharmacogenomics; not only this, a chapter on microbial genomics is also included in the textbook. In all the above chapters, emphasis has been given on the impact of genomics in public health. Also, given the fact that all of the above areas are entirely dependent upon genome informatics, two chapters have been devoted to genome informatics pipelines and browsers, and the impact of translational tools and databases in personalized medicine. The last chapter of the first part is dedicated to molecular diagnostics and genetic testing, which constitutes a fundamental part of applied genomics.

The second part of the book addresses the various adjacent, to personalized medicine, disciplines that are important for the implementation of genomics discoveries in the clinic. As such, there are chapters dedicated to the means of assessing of the stakeholder environment and stance toward personalized medicine, the issue of genomics awareness, and education of the general public and healthcare professionals, respectively, the ethical, legal, and regulatory aspects of personalized medicine and the economics and the closely related issue of pricing and reimbursement in genomic and personalized medicine. Elements such as the Internet of Things, artificial intelligence, genetic counseling, and the various promotional strategies for genetic testing delivery are also covered in the contents of this textbook, some of which for the first time in a textbook of this kind. Lastly, the book ends with a chapter dedicated to the application of personalized medicine interventions in emerging economies and developing and low-resource countries.

As with all edited textbooks, my effort has been assisted by many internationally renowned experts in their field, who have kindly accepted my invitation to compile the 21 chapters of this book and share with me and the readers their expertise, experience, and results, making effort to formulate the contents of the book such that the notions described therein are explained in a simple language and terminology, so that the book is useful not only to experienced physicians and health-care specialists and academics but also to undergraduate medical and life sciences students. The numerous self-explanatory illustrations clearly contributes to this end, making this book an ideal reference material for courses related to applied genomics and public health. After all, the contents of this textbook are structured in such way so that it covers specialized courses in the field of public health genomics.

I am grateful to the colleagues who provided constructive comments and criticisms from the proposal stage and during the preparation of this textbook. I expect that some points in this book can be further improved and, therefore, I would welcome comments and constructive criticism from attentive readers, which will contribute to improve the contents of this book in its future editions. I am also indebted to the editors, Mr. Peter Linsley and Ms. Kattie Washington, Kristi Anderson, and Kiruthika Govindaraju at Elsevier, who helped in close collaboration to overcome encountered difficulties at the various stages of this project. I also express my gratitude to all contributors for delivering outstanding compilations that summarize their experience and many years of hard work in their field of research. I am also indebted to Mark Rogers who was responsible for the design and the cover of this book and to the copy editor, Jayaprakash, who has refined the final manuscript prior to publication. Furthermore, I owe my special thanks to the academic reviewers of the proposal for this textbook for their constructive criticisms on the chapters and their positive evaluation.

Last, but not least, I wish to cordially thank my wonderful family for their patience and continuous support over the years, who has significantly contributed to my scientific advancement and, most importantly, personal improvement and from whom I have taken considerable amount of time to devote to this project.

George P. Patrinos

Department of Pharmacy, Faculty of Health Sciences, University of Patras, Patras, Greece;
Department of Pathology, College of Medicine and Health Sciences, United Arab Emirates University, Al-Ain, United Arab Emirates;
Zayed Center of Health Sciences, United Arab Emirates University, Al-Ain, United Arab Emirates;
Department of Pathology - Bioinformatics Unit, Faculty of Medicine and Health Sciences, Erasmus University Medical Center, Rotterdam, The Netherlands

August 2019

CHAPTER

APPLIED GENOMICS AND PUBLIC HEALTH

1

George P. Patrinos[1,2,3]

[1]*Department of Pharmacy, School of Health Sciences, University of Patras, University Campus, Patras, Greece* [2]*Department of Pathology, College of Medicine and Health Sciences, United Arab Emirates University, Al Ain, United Arab Emirates* [3]*Zayed Center of Health Sciences, United Arab Emirates University, Al Ain, United Arab Emirates*

Doctors have always recognized that every patient is unique, and doctors have always tried to tailor their treatments as best they can to individuals. You can match a blood transfusion to a blood type—that was an important discovery. What if matching a cancer cure to our genetic code was just as easy, just as standard? What if figuring out the right dose of medicine was as simple as taking our temperature?

—**Former US President Barack H. Obama, January 30, 2015**

1.1 INTRODUCTION

Personalized medicine exploits an individual's genomic profile to support the clinical decision-making process and to individualize drug treatment modalities.[1] The concept of personalized medicine has gained momentum in the last decade; however, its founders have described the concept many centuries ago. Around 400 BCE, Hippocrates of Kos (460–370 BCE) stated, "... it is more important to know what kind of person suffers from a disease than to know the disease a person suffers." Also, Ibn Sina (c. CE 980–1037) mentioned, "... in the make-up of most people there is somewhere a natural tendency to get out of order, some congenital weakness in one particular organ, tissue or system." In particular, he termed this a "personal disposition" and put forward the view that each patient should be looked upon as a distinct and separate case. Interestingly, the Talmud (Yevamot 64b; 2nd century BCE) mentions that if a woman's first two children had died from blood loss after circumcision, the third son should be excused from circumcision, hence indicating that the abnormal bleeding tendency was hereditary. These ancient statements and examples, if put together, may encapsulate the essence of modern, personalized, genomic medicine.

In recent years, significant advances have been made in understanding the genetic etiology of a wide range of human-inherited diseases. These advances have been made possible thanks to the significant breakthroughs and rapid pace of development of the genomic technology, aiding clinicians in their task of estimating disease risk as well as individualizing treatment modalities.[2]

Although there have been major leaps in genomics research and discovery work, facilitated by the genomic technology revolution,[3] the pace of these discoveries has not met reciprocal advances in the translation of these findings into the clinic. In other words, there are often significant barriers

Applied Genomics and Public Health. DOI: https://doi.org/10.1016/B978-0-12-813695-9.00001-7

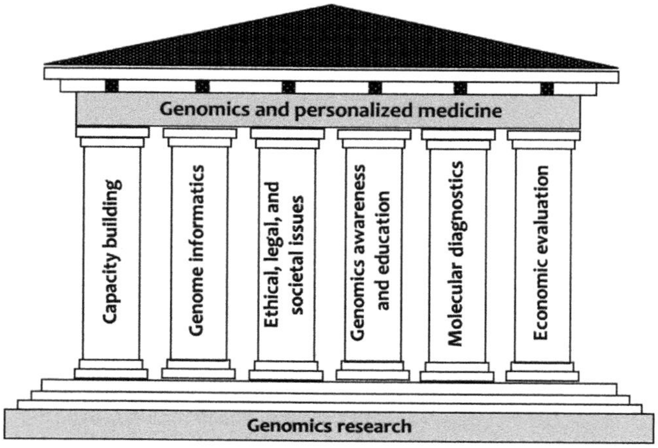

FIGURE 1.1

Illustrative depiction of the transition of genomics research findings into personalized medicine interventions. Genomics research is metaphorically shown as the foundations of an ancient Greek temple. The rooftop of the temple depicts the various genomic and personalized medicine interventions. For the rooftop to hold, strong pillars need to be erected, each one representing the various public health disciplines, each one described in detail in the various chapters of this textbook (see also the text for details).

that hamper the smooth incorporation of genomics research findings into daily medical practice, which have to do more with disciplines related to public health genomics rather than genomics research itself. These disciplines are of utmost importance since they contribute to the transition from genomics research to genomic and personalized medicine. These include ethical, legal, and societal aspects in genomics, also termed ELSI, genome informatics, improving and harmonizing the genetics education of health-care professionals and biomedical scientists, raising genetics awareness among the general public, and health economic evaluation in relation to genomic medicine. These disciplines can be exemplified as the supporting pillars that need to be erected, from the solid bedrock of genomics research, to firmly hold the superstructure of genomic and personalized medicine (Fig. 1.1). Presently, although the foundations of genomics research are becoming stronger with ever-increasing hopes and expectations, the pillars themselves are still largely under construction.[4]

This textbook is a collection of timely contributions related to the various disciplines touching upon the implementation of genomic and personalized medicine and are closely related to public health genomics.

1.2 GENOMICS IN HEALTH CARE

The contribution of germline gene variants to disease etiology is also known as genetic epidemiology. The genetic basis of several Mendelian diseases is elucidated through family-based studies and subsequent functional cloning. As highlighted in Chapter 2, From Genetic Epidemiology to Exposome and Systems Epidemiology, which describes this emerging discipline, although genome-wide association studies identified thousands of variants that are associated with complex genetic traits and conditions, the variants that are actually held responsible for these conditions largely remain undeciphered. As such, although genetic tests for high-penetrance gene variants have clinical utility for individuals, such as in preimplantation, prenatal, or postnatal diagnostics, and in preventive cascade screening of biological relatives (see also Chapter 10: Genetic Testing), genetic testing for complex phenotypes has utility only for research and public health rather than individual testing. Such complex phenotypes include rare diseases, cancer, cardiovascular and psychiatric diseases to name a few.

Rare diseases are defined by their prevalence and although individually infrequent, they affect almost half billion persons globally. Most of these rare diseases have a strong genetic component but till date, their underlying genetic etiology remains elusive. Chapter 3, Rare Diseases: Genomics and Public Health, summarizes the impact of genomic discoveries in accelerating clinical diagnosis, discoveries, and therapeutic interventions in rare diseases.

Of equal importance, cancer is one of the leading causes of death and a global public-health burden. Carcinogenesis originates from a cascade of tumor-promoting events, which result from a genomic-variant overload. Also, despite the emergence of new therapeutic modalities as the standard of care for several cancers, our ability to predict a patient's response to these therapies, or the duration of these responses, is still lagging behind. Chapter 4, Applied Genomics and Public Health Cancer Genomics, addresses our current understanding of the contribution of various genetic aberrations and different regulatory pathways, genomics, and epigenomics in carcinogenesis and their impact in the era of precision medicine. This chapter also attempts to highlight how various genomics and integrated bioinformatics tools are being used to address many of the crucial questions in cancer medicine.

Similarly, psychiatric illnesses are also characterized by vast phenotypic heterogeneity, resulting from their underlying complex genetic basis, which largely remains unknown. The same is true for the genetic basis of interindividual differences in psychiatric drug-treatment response and toxicity, namely, for antipsychotics, antidepressants, and mood stabilizers. Chapter 5, Genomic Basis of Psychiatric Illnesses and Response to Psychiatric Drug Treatment Modalities, overviews psychiatric genomics and its implications for public health and care. It also attempts to highlight differences in psychiatric drug-treatment response and the genetic basis for the development of adverse drug reactions in these patients.

Chapter 6, Pharmacogenomics in Clinical Care: Implications for Public Health, goes a step further and attempts to summarize, more broadly, our current knowledge on the genomic etiology of variable drug treatment, both in terms of efficacy and also toxicity, namely, the development of adverse drug reactions. This chapter outlines the genetic basis of interindividual drug response for different medical specialties, namely, cardiology, oncology, psychiatry, neurology, and antiinfectious agents, and focuses on these drugs that are approved by the major regulatory bodies. Lastly, the chapter attempts to expand on the various applications of pharmacogenomics, in public health–related disciplines, in line with the overall aims of the textbook.

Monitoring an extended spectrum of agents, which has the potential of becoming pathogenic and virulent, is of major concern in public health. Also, new microbiota has emerged, requiring them to be included in the mainstream monitoring and surveillance practice to prevent more aggressive outbreaks. Chapter 7, Microbial Genomics in Public Health: A Translational Risk-Response Aspect, touches upon microbiomics and related applications including genomic-related approaches for adaptive and massive testing.

All the aforementioned genomics approaches yield a huge amount of data that need to be processed for them to be correlated with the underlying phenotype. In particular, genome-wide association studies and next generation–sequencing approaches rely on big-data analysis and hence, bioinformatics plays a vital role in the analysis and interpretation of these genomic data.[5] Chapter 8, Genome Informatics Pipelines and Genome Browsers, attempts to summarize the various bioinformatics pipelines used in translational research and figure out how these data, from the genomic analysis at the level of individuals or populations, can be gradually integrated into the therapeutic and preventative guidelines of modern health-care systems. The chapter also addresses the various computational barriers and challenges, which arise from analyzing massive volumes of genetic data, such as annotation and quantitative data and read alignments. Similarly, genome informatics and genomic data analysis rely on the development and expert curation of genomic databases and translational tools that convert genomics data, which are often difficult to be understood by the treating physicians, to a clinically meaningful format. Chapter 9, Translational Tools and Databases in Genomic Medicine, provides an update on the main genomics databases that are developed to accommodate and curate the huge volume of data resulting from biomarker discovery and next generation–sequencing analysis. It also touches upon the translational tools that facilitate the practice of pharmacogenomics and personalized medicine for pharmacogenomic data interpretation.

As mentioned previously, there are several genetic tests with clinical utility in preimplantation, prenatal, postnatal, and even preventive molecular diagnostics, which resulted from genetic epidemiology and genome-wide association studies, and are developed for high-penetrance gene variants.[6] Chapter 10, Genetic Testing, provides an overview on the various aspects of molecular diagnostics, the main types of genetic tests, their advantages and limitations, and their usefulness in personalized medicine, clinical practice, and disease diagnosis.

1.3 PERSONALIZED MEDICINE AND PUBLIC HEALTH

As mentioned earlier, the smooth incorporation of the genomic discoveries into modern medical practices relies on addressing a number of obstacles, one of which is the incomplete mapping of the key stakeholders involved in this translation process, namely, the major players, their power of intervention and policy positions, their interests and networks, and coalitions that connect them. Also, such comprehensive mapping should be complemented with the proper understanding of the policies, opinions, and overall policy content. Chapter 11, Assessing the Stakeholder Landscape and Stance Point on Genomic and Personalized Medicine, aims to provide an overview of the process for assessing the views and opinions of stakeholders toward personalized medicine and an example of implementing such an approach in a health-care environment, so that adoption of genomics into the mainstream medical interventions is expedited.

Also, a crucial bottleneck that requires rectification is the poor genomics education and literacy among health-care professionals and biomedical scientists. Chapter 12, Health-Care Professionals'

Awareness and Understanding of Genomics, critically examines the level of genomics literacy among general physicians, specialists, and other relevant health-care professionals. In particular, this chapter provides a holistic, critical overview of the level of genomics education and the extent of health-care professionals' understanding toward genomics applications into their medical practices, and highlights a significant gap in the required genomics knowledge and understanding in order to explain the related technologies and their value to the general public. Also, in this chapter the conclusions from various related studies are critically evaluated in an attempt to raise awareness about future educational needs.

As previously indicated, apart from the various societal challenges, there are ethical and legal issues that are related to personalized medicine interventions. Chapter 13, Genethics and Public Health Genomics, and Chapter 14, Legal Aspects of Genomic and Personalized Medicine, address ethical and legal issues related to personalized medicine, respectively. Chapter 13, Genethics and Public Health Genomics, presents ethical and societal issues in genomics, focusing in particular on public health, and also discusses the ethical issues related to the actual implementation of genomic technologies with focus on genome sequencing used for diagnosis, screening, and in the context of reproductive technologies and genome editing. Most importantly, a fundamental issue, to ensure ethical use of genomics in health care, is the provision of a frank and objective view of the present limitations and risks of these modern technologies. Similarly, Chapter 14, Legal Aspects of Genomic and Personalized Medicine, attempts to summarize the existing legal framework that oversees personalized medicine–related issues in various countries worldwide, such as genetic testing, genome editing, and informed consent, to determine possible inconsistencies among different countries, and to highlight eventual legislative gaps, possibly allowing for future harmonization of these legal measures.

The Internet of Things is a new concept, which refers to a pervasive computing environment that is producing a digital replica of all living things and inanimate objects worldwide. This, together with artificial intelligence–assisted data analysis, may constitute an innovative approach pertaining to genomics and public health in particular. Chapter 15, Genomics, The Internet of Things, Artificial Intelligence and Society, discusses this modern concept and provides interesting examples resulting from the application of this innovative approach, such as remote phenotypic data capture, genotype–phenotype association, and multiomics data integration, while it also highlights the sociotechnical aspects of this digital connectivity and their implications for personalized medicine.

An important pillar in the personalized medicine superstructure (Fig. 1.1) is health economics in genomic medicine.[7] The progress we have witnessed in personalized medicine interventions is tightly linked with the economic viability of these innovative interventions. Chapter 16, Economic Evaluation of Genomic and Personalized Medicine Interventions: Implications in Public Health, describes in an illustrative manner the decision-making process within the context of personalized medicine. First, the chapter describes the methods employed for economic evaluation, together with the challenges for researchers, accompanied by practical examples of use. Also, the chapter describes new economic models to be used in evaluating personalized medicine interventions that incorporate public health aspects, such as economic affordability, innovation, social preferences, personal utility, and clinical ethics, hence providing further insights in the resource-allocation process of modern health-care systems. In the same vein, Chapter 17, Pricing, Budget Allocation, and Reimbursement of Personalized Medicine Interventions, refers to the concept of pricing and reimbursement of genomics technologies and personalized medicine interventions, and attempts to summarize the pricing and reimbursement policies as well as being closely related to pricing issues pertaining to, for example, the balance between price and access to innovative testing, monitoring, and evaluation for cost-effectiveness and safety, and the development of research capacity.

Genetic counseling is a multiskill process, which employs genomics knowledge and communication and psychological skills and expertise to support patients, caregivers, and their families in understanding and coping with genetic diseases. Chapter 18, Genetic Counseling, touches upon the crucial role of genetic counselors in communicating genomic information through consultation with individual patients and their families. This chapter also highlights the key role played by genetic counselors in the translation of the findings derived from genomic sequencing technology and emphasizes that the number of the existing genetic-counseling graduate programs is inadequate to support the role that genetic counselors are expected to play in promoting the benefit of public health genomics.

As summarized in Chapter 10, Genetic Testing, a considerable number of genetic-testing laboratories have emerged, which offer a plethora of different prenatal and postnatal genetic tests. However, the means of communicating and advertising the genetic testing services to the interested parties, namely health-care professionals and the general public, is still poorly developed, which also poses a critical ethical and legal challenge to overcome toward provision of personalized medicine interventions. Chapter 19, Defining Genetic Testing Delivery and Promotional Strategies for Personalized Medicine, discusses the existing models for the delivery of genetic testing services, presents the steps to be undertaken when defining the optimal promotional strategy for genetic testing services, and the various means to promote these services to the interested stakeholders.

As with every public health system, regulation is needed to ensure the highest possible level of confidence between health-care providers and patients/consumers for the provision of health services to function properly. The accelerated pace of discoveries in the field of personalized medicine dictates that acquisition and appraisal of genomic information has to be suitably incorporated in new and/or revised legislation, regulatory guidance, and medical practice. Chapter 20, Regulatory Aspects of Genomic Medicine and Pharmacogenomics, addresses the regulatory aspects of personalized medicine interventions and discusses issues, such as incentives, conditional marketing, and authorization measures, and a balanced mix of incentives and sanctions to encourage corporate and other entities to pursue potential new uses of their approved products along with the evolution of personalized medicine.

Lastly, implementation of personalized medicine often lags behind in developing countries and low-resource environments compared to developed countries, such as those in Northwestern Europe and the United States.[8] Chapter 21, Genomic Medicine in Emerging Economies, explores the various advances made in genomics and the impact of these technologies in clinical settings across resource-limited countries and emerging economies. In particular, this chapter outlines some of the challenges related to the clinical implementation of genomics in emerging economies, with examples from Latin America and Africa, such as capacity building, lack of genomics education, and qualified laboratory personnel, and discusses opportunities that arise in emerging economies, such as the "fast-second winner" model,[9] where emerging economies have the potential to implement genomics faster. Lastly, the chapter provides a list of international organizations, such as the Golden Helix Foundation (www.goldenhelix.org) and the Global Genomic Medicine Collaborative (www.g2mc.org), which are developing standards for the global implementation of personalized medicine.

1.4 CONCLUSION

As outlined in the previous sections, with this content composition, this textbook aims to highlight the importance of the closely related, to genomics research, public health genomics disciplines toward maximizing the usefulness of the genomics research findings for the benefit of the patients. It is anticipated that the indicative aspects of public health, described in this textbook, and their role in expediting implementation of personalized medicine interventions, could stimulate large-scale collaborative studies and encourage related multicenter efforts, as a part of large, international public health genomics–research consortia, to catalyze integration of applied genomics into health care and public health.

ACKNOWLEDGMENTS

I wish to thank all the authors of the chapters of this textbook for their contributions, highlighting the importance of implementing public health genomics approaches to facilitate and expedite the implementation of personalized medicine interventions in modern medicine.

REFERENCES

1. Manolio TA, Chisholm RL, Ozenberger B, et al. Implementing genomic medicine in the clinic: the future is here. *Genet Med*. 2013;15(4):258–267.
2. Manolio TA, Abramowicz M, Al-Mulla F, et al. Global implementation of genomic medicine: we are not alone. *Sci Transl Med*. 2015;7:290ps13.
3. Gullapalli RR, Lyons-Weiler M, Petrosko P, Dhir R, Becich MJ, LaFramboise WA, et al. Clinical integration of next-generation sequencing technology. *Clin Lab Med*. 2012;32(4):585–599.
4. Cooper DN, Brand A, Dolzan V, et al. Bridging genomics research between developed and developing countries: the Genomic Medicine Alliance. *Per Med*. 2014;11(7):615–623.
5. Lambert C, Baker D, Patrinos GP, eds. *Human Genome Informatics: Translating Genes Into Health*. Burlington, CA: Elsevier/Academic Press; 2018.
6. Patrinos GP, Danielson P, Ansorge W, eds. *Molecular Diagnostics*. 3rd ed. Burlington, CA: Elsevier/Academic Press; 2016.
7. Fragoulakis V, Mitropoulou C, Williams MS, et al. *Economic Evaluation in Genomic Medicine*. Burlington, CA: Elsevier/Academic Press; 2015.
8. Correa-Lopez C, Patrinos GP, eds. *Genomic Medicine in Developing Countries*. Burlington, CA: Elsevier/Academic Press; 2018.
9. Mitropoulos K, Cooper DN, Mitropoulou C, et al. Genomic medicine without borders: which strategies should developing countries employ to invest in precision medicine? A new "fast-second winner" strategy. *OMICS*. 2017;21(11):647–657.

PART

I

GENOMICS IN HEALTHCARE

CHAPTER

2 FROM GENETIC EPIDEMIOLOGY TO EXPOSOME AND SYSTEMS EPIDEMIOLOGY

Nicole Probst-Hensch

Department of Epidemiology and Public Health, Swiss Tropical and Public Health Institute, Swiss TPH, Basel, Switzerland University of Basel, Basel, Switzerland

2.1 GENETIC VARIATION

The haploid human genome is the complete DNA sequence consisting of approximately 3.3 billion base pairs. Many different types of genetic variation exist. They can be classified in different ways, for example, by their physical nature, by their biological consequence, or by their effect on disease risk. Common physical types of sequence variation are differences in the number of sequence repeats of different length and single-nucleotide polymorphisms (SNPs). The biological relevance of these genetic variants ranges from functionally irrelevant or silent to the abolishment of protein production, with differences in the penetrance and clinical relevance associated with respective variants.[1,2] Assessing the functional relevance of a genetic variant and assigning it a causal role in the etiology of a specific phenotype or disease remains the biggest challenge of genetic epidemiology, particularly with regard to common complex diseases. Two major aspects are at the heart of this challenge. First, only a very minor part of the DNA sequence is coding for proteins.[3–6] Elucidating the biological relevance of the DNA sequence, which is not protein-coding, is the focus of intense international research efforts. Second, DNA variants in proximity of each other on the same chromosome are not transmitted to subsequent generations independently of each other. Crossover of homologous chromosome segments during meiosis, when haploid gametes are formed, is less likely to separate genetic variants lying close to each other. As a result, genetic variants can be in complete or partial linkage disequilibrium (LD) over many subsequent generations. It is therefore not possible to differentiate with family-based or genetic association studies alone, which, of several disease-associated genetic variants, is causally responsible for the disease association.

2.2 WHAT IS GENETIC EPIDEMIOLOGY?

Genetic epidemiology focuses on the effect of genetic variation on health and disease.[1,2] To maximize causal inference and understanding of disease etiology, it ideally integrates the current understanding of biology as well as nongenetic disease determinants. Genetic epidemiology has evolved in parallel to technological advances from studying patterns of disease aggregation and distribution

Applied Genomics and Public Health. DOI: https://doi.org/10.1016/B978-0-12-813695-9.00002-9

in families to assessing the health effects of single and combined genetic variants across the whole genome.

Genetic epidemiology has successfully identified the cause of many monogenetic disorders, for example, disease exhibiting a clear Mendelian inheritance pattern.[1] These mostly rare genetic syndromes are linked to the inheritance of high-penetrance gene variants. The penetrance of a gene variant reflects the likelihood that its carrier will develop the specific disease. Carriers of such variants face a very high, sometimes up to 100%, likelihood of developing the specific disease before a certain age, irrespective of environment, behavior, or other genetic variants. An example for a genetic disease with 100% penetrance is Huntington disease.

The identification of the genetic variants contributing to common, noncommunicable diseases (NCDs), such as cardiovascular, respiratory, metabolic, or neurological disorders, is more challenging.[1,7,8] These diseases develop against a background of complex risk patterns consisting of several genetic variants and biological pathways as well as exogenous factors to which a person is exposed, that is, lifestyle, physical environment, social environment. The contribution to disease risk for each factor separately is usually small, and each new diagnosis develops from a personalized risk profile.

2.3 APPROACHES TOWARD IDENTIFYING GENETIC VARIANTS

The following four questions are at the heart of genetic epidemiology[9]: Does a phenotypic trait show familial aggregation? Does it show genetic segregation? Does it show cosegregation in families with a genetic marker? Does it show association in the population with a genetic marker? Answering the last two questions in the context of linkage studies and genetic association studies, respectively, requires access to DNA and genotyping for appropriate genetic markers. The advances in molecular genotyping technology combined with progress made in computational efficiency have considerably improved the power of linkage and association studies at a reasonable cost over the recent decades.

The approach toward identifying disease-causing gene variants depends importantly on their penetrance (Table 2.1). While high-penetrance gene variants, which tend to be rare, are usually first studied with the help of family studies, the common family study approaches are often not efficient for identifying low-penetrance gene variants, which are often prevalent. On the contrary the two approaches can also be combined to benefit from the advantages of both methods.[10]

2.4 IMPLYING GENETIC LOCI IN DISEASE ETIOLOGY IN THE CONTEXT OF FAMILY STUDIES—SEGREGATION AND LINKAGE ANALYSIS

The first indication for the contribution of a highly penetrant gene variant to a disease risk comes for the aggregation of specific traits in families as well as from a higher observed trait concordance in monozygotic compared to dizygotic twins.[1] Yet, disease aggregation in families or monozygotic twins is not sufficient proof of a genetic disease cause. Family members and monozygotic twins are also more likely to share environment and lifestyle compared to unrelated persons.

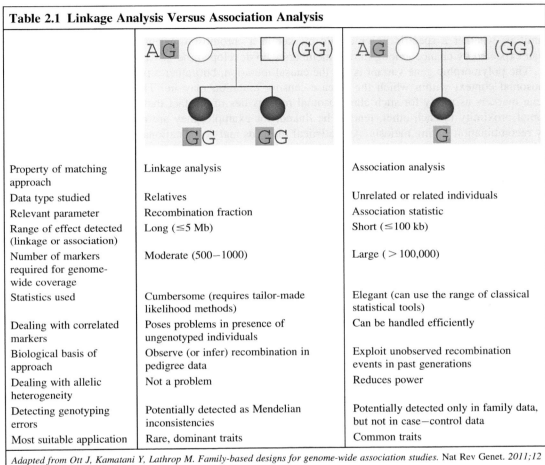

Table 2.1 Linkage Analysis Versus Association Analysis

	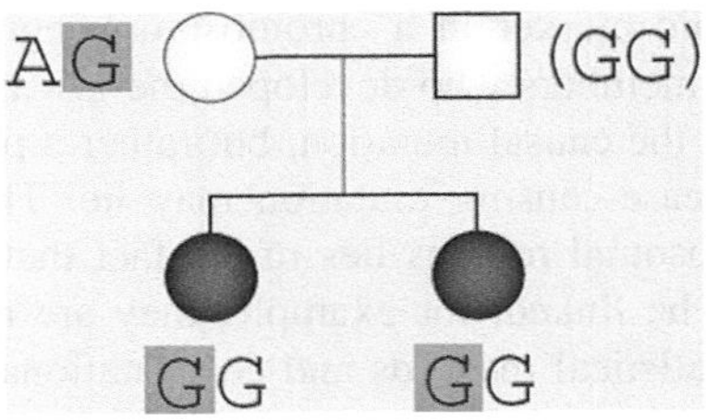	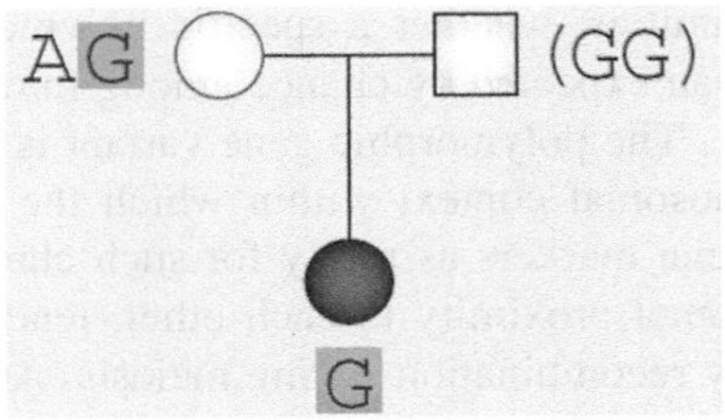
Property of matching approach	Linkage analysis	Association analysis
Data type studied	Relatives	Unrelated or related individuals
Relevant parameter	Recombination fraction	Association statistic
Range of effect detected (linkage or association)	Long (≤5 Mb)	Short (≤100 kb)
Number of markers required for genome-wide coverage	Moderate (500–1000)	Large (>100,000)
Statistics used	Cumbersome (requires tailor-made likelihood methods)	Elegant (can use the range of classical statistical tools)
Dealing with correlated markers	Poses problems in presence of ungenotyped individuals	Can be handled efficiently
Biological basis of approach	Observe (or infer) recombination in pedigree data	Exploit unobserved recombination events in past generations
Dealing with allelic heterogeneity	Not a problem	Reduces power
Detecting genotyping errors	Potentially detected as Mendelian inconsistencies	Potentially detected only in family data, but not in case–control data
Most suitable application	Rare, dominant traits	Common traits

Adapted from Ott J, Kamatani Y, Lathrop M. Family-based designs for genome-wide association studies. Nat Rev Genet. *2011;12 (7):465–474.*

Nevertheless, if familial aggregation is observed, it justifies investigating the role of genetics further and searching for disease genes.

In a next step, the patterns of disease distribution within families can be compared to patterns expected, assuming one or more biologically rational models, for example, additive, dominant, recessive, and polygenic genetic effects. This approach of statistically estimating the mode of inheritance of a disease trait based on disease distributions within families is called segregation analysis, which does not require access to genetic material.[1,9,11]

When DNA is available from several family members across different generations, the cosegregation of genetic variants with disease in these families allows identifying the chromosomal regions within which the disease-causing genetic variant(s) are likely to lie.[1,9] In linkage studies, highly polymorphic genetic markers, spread across the chromosomes, are measured in the DNA collected from family members from different generations. Linkage analysis investigates, in many different families, whether a specific polymorphic marker in a chromosomal region is present more often than expected by chance among family members who developed the disease (Fig. 2.1A).

The polymorphic gene variant is not the causal mutation, but rather a proxy indicator for a chromosomal context within which the disease-causing mutation may lie. The utility of polymorphic gene markers as proxy for such chromosomal regions lies in the fact that genetic loci, in chromosomal proximity to each other, tend to be linked, for example, they are not likely to be separated by recombination during meiosis. As statistical methods and computational power for linkage analysis evolved, it became possible to handle complex parametric and nonparametric models and to take into consideration anticipation (the earlier onset of a disease sometimes seen in succeeding generations), imprinting (the phenotype that is observed depends on the sex or the parent transmitting the disease-causing allele), admixture, and missing information.

Linkage analysis approaches can be differentiated into model-based and model-free linkage analysis. Model-based linkage analyses need to assume a genetic model for the phenotype of interest. Model misspecification poses a challenge to model-based linkage analysis. Despite this challenge, model-based linkage analyses successfully led to the eventual cloning of monogenic disease genes.[9,10] Model-free linkage analysis methods in contrast, which do not need to specify the model of inheritance, have been developed toward identifying genetic variants involved in more complex disease etiology.

2.5 IMPLYING GENETIC LOCI IN DISEASE ETIOLOGY IN THE CONTEXT OF GENETIC ASSOCIATION STUDIES

A limitation of family-based studies is the fact that given the few generations studied in each family, the likelihood for recombination events remains very limited over large chromosomal regions. As a result, genetic variants over extended chromosomal regions are mostly coinherited and their cosegregation with the disease of interest is therefore identical. Zooming in on the disease-causing variant, in order to ultimately clone it, can in part be achieved in the context of association studies of indirectly related individuals. Association studies allow finer mapping of disease-causing variants than family-based linkage analysis. The correlation of genetic variants lying next to each other on the same chromosome is present over much shorter chromosome stretches, because the likelihood of recombination events between polymorphic markers, in proximity to each other, increases over numerous ancestral generations. On the one hand, these shorter distances of linked markers allow narrowing down the chromosomal region within which a disease-causing variant may lie. On the other hand, the fact that even after numerous generations, a disease-causing variant is still present in a specific chromosomal context of other genetic variants, for example, the disease-causing variants is in LD with other genetic variants, supports the efficient search for disease-causing loci (Fig. 2.1B).

While genetic association studies, which assess the overrepresentation of specific gene variants in persons expressing the phenotype trait of interest, were initially limited to testing few variants in candidate genes related to complex diseases, advances in high-throughput genotyping methods have

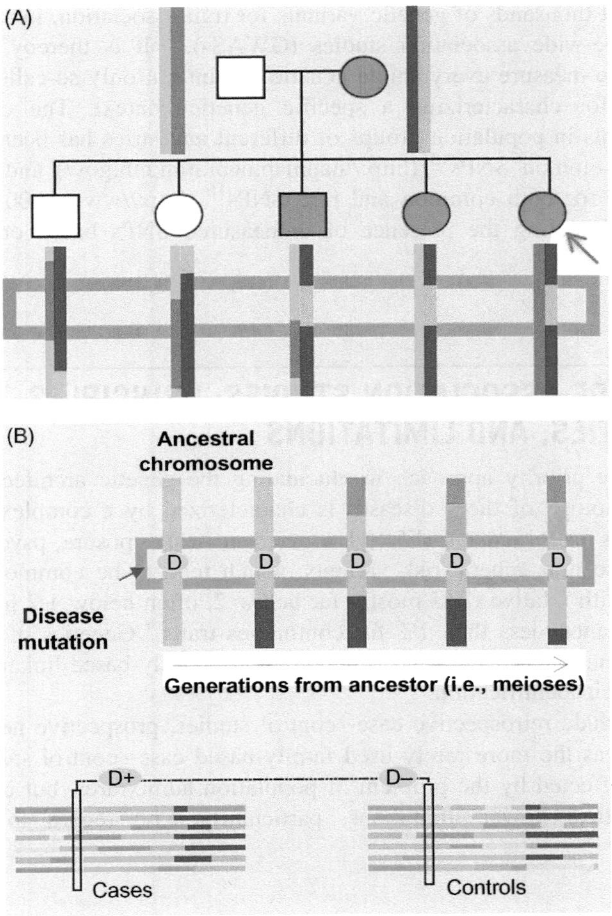

FIGURE 2.1

(A) Linkage analysis for the discovery of disease susceptibility loci. Linkage reflects the tendency for chromosomal segments to be inherited intact from parent to offspring. Here, all affected offspring share a chromosome segment (blue) inherited from the mother, which is not shared with the unaffected offspring, suggesting a susceptibility locus within the shared chromosome segment. (B) Association analysis for the discovery of disease susceptibility loci. LD is the nonrandom association of alleles at two or more loci. Here, association testing would identify a correlation between SNPs in the block of high LD (blue) and case status, which would suggest the presence of an unobserved disease variant (D). *LD*, Linkage disequilibrium; *SNPs*, single-nucleotide polymorphisms.

Adapted from Mathias RA. Introduction to genetics and genomics in asthma: genetics of asthma. Adv Exp Med Biol. *2014;795:125–155.*[12]

made it possible to test thousands of genetic variants for trait association, in an agnostic manner, in the context of genome-wide association studies (GWASs).[14] It is thereby due to this LD that GWASs do not need to measure every single genetic variant but only so-called tag variants. These variants are proxies for characterizing a specific genetic context. The correlational structure between genetic variants in population groups of different ancestries has been characterized by the HapMap Project for common SNPs[15] (http://hapmap.ncbi.nlm.nih.gov/) and subsequently by the 1000 Genome Project for both common and rare SNPs[16] (http://www.1000genomes.org/). These reference data allow imputing the presence of unmeasured SNPs based on a set of genotyped SNPs.[17]

2.6 GENOME-WIDE ASSOCIATION STUDIES: PRINCIPLES, OPPORTUNITIES, AND LIMITATIONS

GWASs have been the priority approach in elucidating the genetic architecture of complex diseases.[7,14,18–23] The etiology of these diseases is characterized by a complex interplay of genetic variants and exogenous factors, that is, lifestyle, environmental exposure, psychosocial factors. The influence of each of several genetic risk variants, which tend to be common rather than rare, is comparatively small with relative risks mostly far below 2, often below 1.2 for dichotomous traits, or with explained variances less than 1% for continuous traits.[6] Genetics of complex diseases do not follow-up Mendelian inheritance patterns and common family-based linkage approaches are not usually efficient for their identification.

GWAS designs include retrospective case–control studies, prospective nested case–control or cohort studies as well as the more rarely used family-based case–control studies, which have the benefit of being less affected by the problem of population admixtures, but come at the disadvantage of facing statistical power limitations, particularly with regard to polygenic complex disorders.[6,10,24]

2.7 QUALITY CONTROL IN GENOME-WIDE ASSOCIATION STUDIES

A vital step of GWAS relates to the application of quality control (QC) protocols as previously published and of great relevance for minimizing bias in observed associations.[6,25,26] Reasons for data errors include, for example, poor DNA quality, poor DNA hybridization to the array, poorly performing genotype probes, and sample mix-ups or contamination. Seven essential QC steps, which can all be executed by the PLINK software,[27] consist of filtering out SNPs or individuals based on (1) individual and SNP missingness (particular care must be taken if missingness differs between cases and controls), (2) sex discrepancy between self-reported and genetic sex, (3) minor allele frequency (MAF) (commonly used MAFs are ≥ 0.01 for large sample size with $N \geq 100{,}000$ and ≥ 0.05 for smaller sample sizes), (4) deviations from Hardy–Weinberg equilibrium (HWE) as the indicator for genotyping error; since small sample size or disease-related factors may also lead to HWE deviations, HWE is either estimated in control subjects only to avoid excluding potentially meaningful HWE deviations, (5) heterozygosity rate, which can indicate sample contamination and also inbreeding, (6) relatedness, and (7) ethnic outliers (population stratification).[6,26]

2.8 ASSOCIATION TESTING, SIGNIFICANCE LEVELS, AND VISUALIZING ASSOCIATIONS IN GENOME-WIDE ASSOCIATION STUDIES

In GWAS, thousands of measured and imputed genetic variants, mostly SNPs, are each agnostically investigated for association with dichotomized (i.e., diseased vs not diseased) or continuous traits in the context of logistic and linear regressions, respectively.

Before applying an association test the genotype data needs to be coded according to a presumed genetic model. It is common practice to run the GWAS under the assumption of a codominant genetic model with genotype data coded as 0 (homozygous for major allele), 1 (heterozygous), and 2 (homozygous for minor allele), indicating that the presence of each copy of the minor allele increases disease risk or phenotype trait in an additive manner on the scale of the regression function.[6]

Population stratification is the single most important source of bias in GWAS.[6,26] It refers to the presence of population subgroups of different ethnic backgrounds in the study sample. As both allele frequencies, including LD as well as phenotype traits, differ by ethnic groups, not accounting for this important source of confounding can lead to masked or false-positive associations. Removing individuals, who are clear ethnic outliers, in comparison to the majority of the study population is not sufficient as a measure to protect against confounding due to ethnic admixture. Subtle degrees of population stratification can even exist in a presumably homogenous ethnic group. Therefore additional adjustment for population stratification in regression models is important.[26] As population stratification leads to inflated variances and increased spurious associations, the presence of population stratification can be examined with the help of quantile–quantile plots [observed test statistics (or some function of the *P* values) vs values expected from a theoretical distribution; deviation from the diagonal line can be indicative of population stratification]. Population stratification can be adjusted by either the genomic control method (correcting test statistic for genomic inflation) or with the help of adjusting for principle components, which summarize the genotypic variation due to population structure in orthogonal vectors.

As conducting a GWAS implies running thousands of statistical tests, stringent *P* values (standard *P* value for GWAS: $P < 5 \times 10^{-8}$) have been commonly applied to protect against false-positive findings as a result of multiple testing. This stringent *P* value has certainly led to ignoring relevant associations. Associations of SNPs at higher *P* values seem to still be of great relevance toward studying gene–gene interactions, gene networks, and gene pathways.[7,23]

The association of each genetic variant with the phenotypic trait of interest is generally visualized in a Manhattan plot (Fig. 2.2).

Each dot in the figure represents an SNP. Its position in the plot is characterized by the − log 10 (*P* value) of its association with the phenotype (*y*-axis) and by the chromosomal position (*x*-axis). Stacks seen in signals are reflecting the fact that neighboring loci are in high LD.

A regional Manhattan plot zooming into a specific genomic region is called a locus zoom plot, which provides additional information regarding chromosome position, genes, recombination rate, and LD levels (Fig. 2.3).

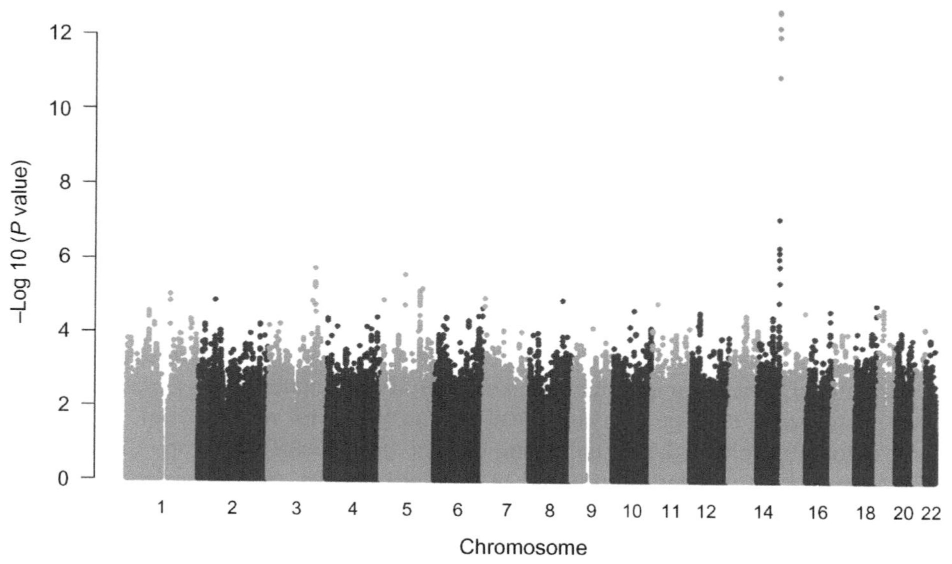

FIGURE 2.2

Manhattan plot of a genome-wide association study for identifying single-nucleotide polymorphisms associated with circulating alpha-1 antitrypsin.[13]

In Fig. 2.2 is given an example of a Manhattan plot derived from a GWAS to search for genetic determinants of circulating alpha-1 antitrypsin (AAT).[13] Rare, genetically determined severe AAT deficiency is a strong risk factor for emphysema in smokers and represents a model for the relevance of gene–environment interactions.[28,29] The major signal of the GWAS reflects the association of common SNPs in the chromosomal region that contains the *SERPINA1* gene, which codes for the AAT protein. Fig. 2.3 shows the locus zoom plot of the associated chromosomal region.

2.9 META-ANALYSIS, REPLICATION, VALIDATION, AND THE VALUE OF IMPUTATION IN GENOME-WIDE ASSOCIATION STUDIES

The small effect sizes of SNPs as well as the stringent cutoffs for statistical significance needed to avoid chance findings imply the need of large sample sizes to unequivocally establish SNP–phenotype associations. Research consortia focusing on specific phenotypic traits have been established, which bring together all relevant epidemiological studies able to contribute as either a discovery or replication study to phenotype-specific GWAS meta-analyses. For example, the SpiroMeta and CHARGE consortia bring together researchers and cohorts with a focus on the genetic background of lung function and chronic obstructive pulmonary disease (COPD).[30–32]

Discovery meta-analysis and replication efforts of SNP–phenotype associations should be conducted in populations of similar ethnic background and environmental context, as both population

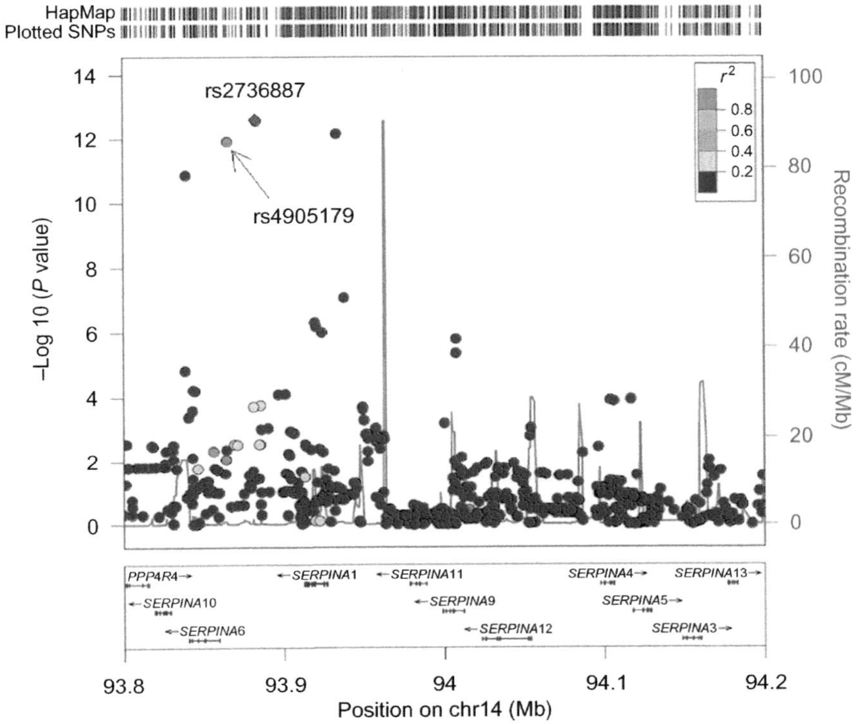

FIGURE 2.3

Regional plot for the *SERPINA* gene cluster (93.8–94.2 Mb on chromosome 14q32.13, reference panel: NCBI build 36.3).[13]

stratification and exogenous factors can modify the SNP–phenotype association.[6] Furthermore, phenotyping should be harmonized as best possible. Interstudy differences in the single-study endpoints, for example, different molecular subtypes of cancer, can mask true SNP associations only present for subtypes of a disease.

Replicated SNP–phenotype association should show effects in the same direction. Joint meta-analysis of associations from discovery and replication samples should lead statistically more significant *P* values compared to results from just the discovery sample. While the purpose of replication studies is to firmly establish SNP–phenotype associations in a given ethnic and exposure context, validation studies in contrast evaluate the generalizability of reported associations to different population contexts such as to different ethnic groups. Validation in different populations is of particular relevance when evaluating the clinical utility of genetic tests. In the absence of a firm understanding of the causality of genetic variants, it is important to keep in mind that the LD structure of the genome varies between different ethnic groups.

Discovery, replication, and validation of GWAS do not need to be restricted to measured genotype data. In fact, the genotypes of different studies may well have been obtained from different arrays. Harmonization of genotype data for GWAS meta-analysis can be achieved through imputation, which at the same time leads to the fine mapping of genetic regions. Imputation is the method to predict unmeasured SNPs and genotypes by using information from a densely assayed reference panel.[6] In other words, it predicts indirectly measured information using measured genotypes as proxy and benefiting from LD. Imputation can expand directly measured genotype data containing 300,000–2500,000 SNPs to nearly 40 million variants. The reference panels for imputation evolved from the International HapMap project,[33] to the 1000 Genomes Project,[16] to the Haplotype Reference Consortium that provides a combined panel from several studies and populations including the 1000 Genome Project phase 3 and the UK10K project.[34] Given the broad availability of ethnically diverse reference panels, investigators can select the ethnically best matched population sample for imputation.

2.10 INTERPRETATION AND FOLLOW-UP OF GENOME-WIDE ASSOCIATION STUDIES FINDINGS—CHALLENGES AND LIMITATIONS

An observed and replicated SNP–phenotype association does not imply a causal effect of the respective SNP.[3] A positive SNP–phenotype association can be the result of several explanations: (1) in extremely rare cases the SNP may have a causal effect on disease; (2) much more likely, the associated SNP is in LD with a yet-unknown causal SNP in chromosomal proximity; (3) the association may be a chance finding, although in the light of the large sample sizes of GWAS meta-analysis, this explanation is increasingly less likely. In fact the bigger problem is that due to the stringent *P* value cutoffs, true SNP–phenotype associations are being missed by not meeting these stringent *P* values; (4) the association may be due to bias such as residual confounding by unrecognized population stratification or unaddressed quality issues; (5) effect of an SNP on survival with the disease past the reproductive phase.

In that sense, GWAS results can be viewed as hypothesis generating. They help establish the genetic architecture of a disease or phenotype. The hypothesis raised is that a causal genetic variant may be located in a specific chromosomal region to which the GWAS result points. This region is considerably narrower than a typical chromosomal region identified by family-based linkage studies. Nevertheless, additional research zooming into the chromosomal region of interest needs to follow any GWAS in order to identify the causal genetic variant and to understand its biological–functional effect. Numerous reviews and approaches on the follow-up of GWAS findings have been published. The following are possible but not exclusive steps toward identifying the causal variant underlying an SNP–phenotype association[3,6,35]:

1. Identifying all common and rare variants in LD with the SNP found to be associated with the trait of interest and narrowing down the region of interest through fine mapping. Transethnic fine mapping is one approach that can be used to refine the region of association.[35,36] The lower LD and shorter haplotype blocks, particularly in African populations, may reduce the number of candidate causal variants.[3] The advent of next-generation sequencing data is expected to improve the exhaustive identification of all relevant genetic variants in the GWAS-identified chromosomal region. Yet, as recently evidenced for breast cancer, the combination of next-generation sequencing and a multiethnic approach for fine mapping may still not be

sufficient for identifying causal variants, in part, because extremely large sample sizes are needed.[37] Once all common and rare gene variants in the region of interest and in LD have been identified, conditional analyses can be performed to determine which of the variants remain independently associated with the phenotype. A causal effect would be much more likely for variants with independent and strong effect; this can be the case for one or several variants in fact.[3] This approach was chosen to follow up on a published association between a common SNP rs4905179 located in the *SERPINA* gene cluster on 14q32.13 with emphysema and lung function. In general the relevance of common *SERPINA1* SNPs for lung-function variation in the population is poorly understood, despite the fact that rare *SERPINA1* mutations causing severe AAT deficiency are an established emphysema risk factor among smokers. A GWAS was therefore conducted on circulating AAT measured in over 5000 subjects of the SAPALDIA cohort.[13] This GWAS identified the common SNP rs4905179 as the top ranking signal, which was previously associated with emphysema and lung function. Yet, fine mapping of this genetic region and subsequent conditional analyses to unravel which of the AAT-associated variants contribute independently to the phenotype's variability showed that the association of AAT with common variants was entirely attributable to rare and functionally well-established variants, most important to the so-called Z-allele (rs28929474). This finding was replicated in an independent population-based Danish cohort. The study is a textbook example of how a large part of a trait's heritability can be hidden in infrequent genetic polymorphisms. Subsequent follow-up analysis with lung function as an endpoint furthermore suggested that common variant association may also be explained by the rare variant in individuals with poor pulmonary health.

2. Identifying genetic sequences of possible relevance in the region of interest, for example, genes or regulatory elements. The vast majority of disease-associated variants reside in noncoding regions of the DNA. This suggests that these SNPs might, for example, affect gene expression through effects on transcription, splicing, or mRNA stability.[3] Functional follow-up studies of GWAS SNP–phenotype associations found enrichment for SNPs that reside in *cis*-regulatory elements, which are active in the respective target tissue of interest.[3] Once disease-associated SNPs have been found to reside in regulatory elements, experimental studies may follow up on the finding, such as reporter assays or genome editing in cell lines or animal models. High-throughput genome and epigenome editing screens have been developed that facilitate efficient testing of candidate *cis*-regulatory regions and their associated variants in parallel. In addition, it is has become possible to investigate all long-range chromosomal contacts of regulatory regions of interest, recognizing that chromosomal loops also play an important role in regulating gene expression (e.g., 3C, 4C techniques). For example, consideration of the three-dimensional organization of chromatin facilitated the functional understanding of intronic genetic variants in the obesity-associated *FTO* gene, in the absence of their effect on *FTO* gene expression. 4C analysis combined with transgenic models demonstrated that the regulatory SNPs were rather interacting with *the IRX3* promoter, several hundred of kilobases away from the variants and the *FTO* gene. This approach identified the region containing the obesity-associated regulatory SNPs as the *FTO* enhancer region for regulating *IRX3* expression. The role of *IRX3* in body-weight maintenance was confirmed in mouse models.[3,38]
3. Examining the effect of SNPs on functional outcomes, for example, examination of tissue banks for gene expression effects in different target tissues. According to recent evidence that SNPs in

regulatory regions that control the level of gene expression are overrepresented in complex disease,[4,5] testing of SNPs for expression quantitative trait loci (eQTLs) is of great relevance.[3] The evidence from published functional follow-up studies, in fact, suggest that the pathways by which many GWAS causal variants influence disease risk, whether by *cis*- or *trans*-acting mechanisms, converge on alteration of expression levels of a target gene. The altered gene or gene network expressions are then thought to be responsible for the pathophysiological effects, although it remains poorly understood how the relatively small, usually not more than twofold changes in gene expression can lead to disease development.[3] As an example for a post-GWAS approach based on tissue expression, the follow-up of GWAS meta-analysis results for lung function, produced by the SpiroMeta/CHARGE consortia, is being presented.[30] As gene regulation is often tissue-specific, it is important to discover relevant eQTLs by investigating tissue expression in the target tissue of interest.[3,6] In following up on reported SNP—respiratory health associations, results from the lung eQTL consortium were of particular interest. In a large-scale eQTL mapping study including 1111 human lung tissues, 468,300 *cis*-acting (affecting expression of genes within 1 Mb of the transcript start site) and 16,677 *trans*-acting (further than 1 Mb away or on a different chromosome) eQTLs out of 2598,263 SNPs have been identified at a false discovery rate of 10%.[39] The goal of the systems genetics follow-up of lung-function GWAS results was to identify the SNP—mRNA—lung-function correlations and thereby genes and molecular mechanisms that would in part explain interindividual differences in lung function, which are predictive of morbidity and mortality (Fig. 2.4).

First, lung-function SNPs were found to be enriched for lung eQTLs and vice versa. SNPs associated with both a lung-function phenotype and with lung gene expression are more likely involved in a causal pathway. Second, in many cases the eQTL-regulated genes were different from genes assigned to an SNP—lung-function association in the GWAS meta-analysis papers. In fact, in many cases, several eQTL-regulated genes were associated with a single SNP. This points to SNP effects on regulatory networks rather than single-gene expressions and underlines the potential of systems genetics to unravel these pathways. Third, testing of the lung-function expression quantitative trait loci (eSNP)-regulated genes for enrichment in gene-ontology biological processes and pathways with the web-based Gene Set Analysis Toolkit (WebGestalt) found that eSNP-regulated genes are involved in development and inflammatory pathways. A comparison of eSNP-regulated genes in blood with those in lung tissue showed that lung-function eSNP-regulated genes restricted to blood were enriched solely for inflammatory processes, whereas eSNP-regulated genes in lung tissue and those shared between lung and blood were enriched for both developmental and inflammatory processes. Lung-function eSNP-regulated genes were also overrepresented among genes differentially expressed during human fetal lung development. Fourth, the expression of several of the lung eSNP-regulated genes was shown to be associated with lung function in the tissue donors. Fifth, ChIP-Seq data from the ENCODE project demonstrated that lung-function eSNP-regulated genes were enriched in binding sites for several transcription factors, that is, c-Myc, NF-κb, P300, Pol2, Ebf1, and Sin3a. At the SNP level, lung-function eSNPs were associated with several ENCODE functional annotations consistent with their role as regulatory SNPs. Sixth, the study also generated hypotheses about potential therapeutic agents for COPD. First, the mRNA levels of lung-function eSNP-regulated genes were tested for association

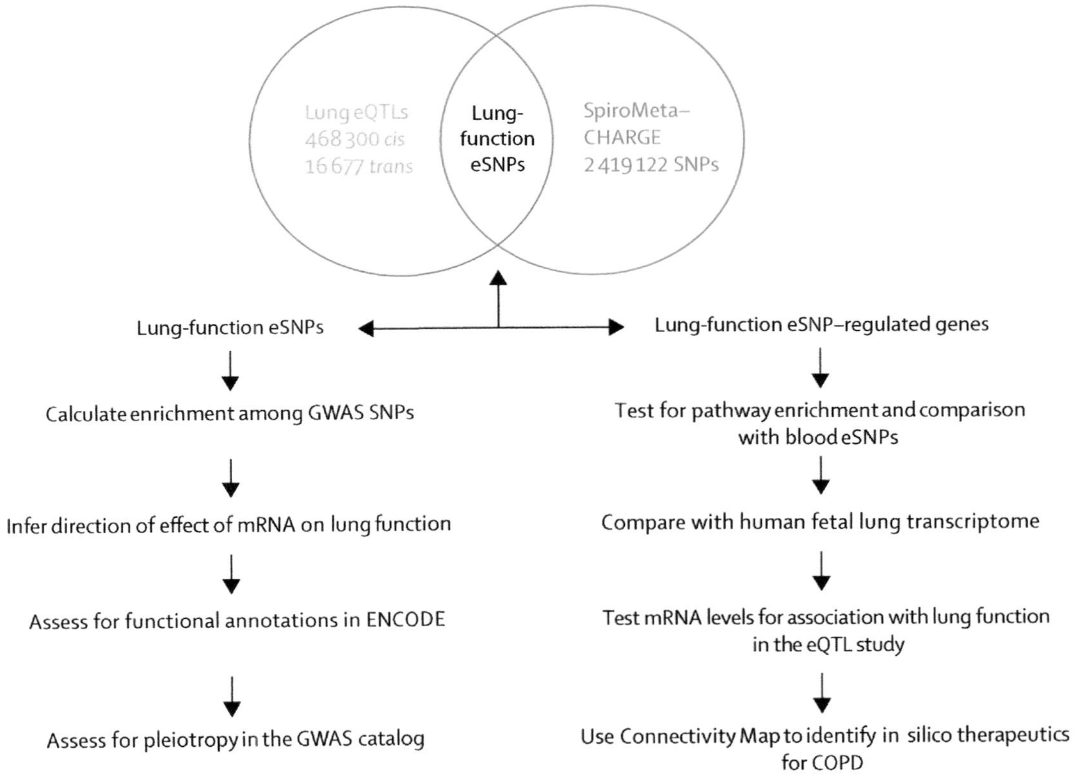

FIGURE 2.4

Study design for systems genetics analysis to assess molecular mechanisms underlying variations in lung function.

Adapted from Obeidat M, Hao K, Bosse Y, et al. Molecular mechanisms underlying variations in lung function: a systems genetics analysis. Lancet Respir Med. 2015;3(10):782–795.[30]

with COPD. The resulting *COPD* gene signature was then used to query the Connectivity Map database, which hosts a publicly available database of transcriptional profiles produced by existing drugs. Several agents had a negative enrichment score suggesting that they are predicted to reverse the COPD gene signature. Finally, and of great relevance, several of the eSNPs did not meet the stringent statistical significance criteria of the GWAS and would have missed by only focusing on GWAS main findings. These results underline the relevance of studying SNPs and their downstream functional consequences in networks and pathways to fully capitalize from GWAS meta-analysis results.

It is important to point out that there are several other ways than alteration of gene expression by which disease-associated variants could have a functional role. For example, a variant could affect a protein's amino acid sequence for SNPs lying in an exon or protein levels through effects on translation or protein stability without an effect on mRNA levels.[3]

2.11 GENETIC EPIDEMIOLOGY—WHERE DO WE STAND—WHERE DO WE GO

The genetic variants underlying many Mendelian diseases are known. In many cases, researchers have also determined how the mutation affects protein function and therefore pathophysiology. It needs to be acknowledged, though, the variants of unknown significance (VUS) remain a challenging result in genetic clinical practice, one that will become increasingly relevant with the implementation of genetic sequencing into medical practice. Accordingly, the epidemiological and functional characterization of VUS is a research priority in diseases involving high-penetrance gene variants.[3,40]

Elucidating the etiology including causal genetic variants of complex diseases remains a major challenge. These diseases seem to mostly lack a necessary cause and often develop a background of "personalized" risk patterns.[41] GWAS meta-analysis of considerable size and statistical power have over the past years led to the identification of thousands of firmly established and well-replicated SNP–phenotype associations.[18,42] The numbers of studies that have investigated the mechanisms underlying SNP–phenotype associations remains considerably smaller compared to GWAS publications.[3] But functional follow-up of GWAS findings has been greatly facilitated over the past years by consortia efforts to systematically characterize the human genome and epigenome sequence for its functional relevance. Data generated by the ENCODE Project,[43] the NIH Roadmap EpiGenome Project,[44] the FANTOM consortium,[45] and others provide extensive information in numerous human cell types and tissues. A wealth of expression quantitative trait loci (eQTL) and methylation quantitative trait loci (mQTL) and other high-throughput data is now available for various cell and tissue types and allow to efficiently study the functionality of causative SNPs and genes. Nevertheless it remains true that only a small fraction of GWAS findings, which in essence represent statistical associations, have been followed up to determine[1] which variant or variants are causal,[2] what their molecular functions are,[3] what genes the causal variants affect, and[4] how changes in the function or regulation of the causal genes ultimately alter disease risk.[3]

Rather than conducting ever larger GWAS studies to confirm statistical association for increasingly small SNP effects, it is now important to make use and sense of the data and associations produced by research consortia over the past decades and to address aspects still poorly understood or investigated, as summarized next.

2.12 HERITABILITY GAP OF GENOME-WIDE ASSOCIATION STUDIES FOR COMPLEX DISEASES

For most complex phenotypic traits the variance explained by the SNPs found to be firmly associated lags far behind those predicted by heritability estimates. This fact has been termed the heritability gap.[7] Only for some rare complex conditions, notably age-related macular degeneration, the contributions of several variants of large effect (increasing disease risk by twofold or more) have been observed. Most of the common variants associated with complex phenotypes individually or in combination confer relatively small increments in risk (1.1- to 1.5-fold) and explain only a small proportion of heritability, for example, phenotypic variance in a population.[7]

Causes for the heritability gap include[7] (1) GWAS design aspects, for example, unprecise and nonharmonized phenotyping in GWAS meta-analysis, selection bias related to control groups, population stratification, (2) inaccurate heritability estimates, as they depend on the study setting including the environmental context of families and participants from whom heritability estimates were derived; (3) the focus of GWAS studies on common variants and the SNP that was statistically most significant. This is unlikely the causal SNP in most cases. In fact, rare variants in LD with the common noncausal SNP may explain a considerably larger part of variance in the subgroup carrying these rare variants. Still the fact that most complex diseases have seen an increase with the Westernization of lifestyle within a single generation suggests that it is unlikely that a combination of personalized rare variants alone would explain a considerable percentage of disease risk variation at the population level.[46] (4) Complex diseases arise from complex gene–gene–environment interactions. These interactions were poorly considered in GWASs.[47] (5) Some authors have put forward that ignoring nonadditive gene variant effects in GWAS may explain in part the heritability gap. But an important role of dominant gene effects is not supported by evidence.[48–50] (6) Epigenetic variation has been put forward as another source of the heritability gap. But existing evidence support the notion that genetic and epigenetic impacts on disease are mostly independent.[48] (7) structural variants, such as copy number variants, may account for some of the missing heritability.

These aspects are now being addressed in numerous post-GWAS research efforts through various approaches to further improve understanding of the genetic architecture of complex phenotypes. Of particular relevance, also from the perspective of clinical utility, is study of the combined disease effect of several gene variants.

2.13 BEYOND A SINGLE LOCUS—EPISTASIS AND POLYGENIC RISK SCORES

As GWAS meta-analysis implied several to numerous independent SNPs in most complex phenotypes, this provides an opportunity to systematically examine the quantitative impact of these common genetic variants as part of a genetic risk score (GRS). GRS can be of interest to identify individuals at extreme ends of a genetic risk distribution, for example, persons how have inherited mostly low-risk alleles versus individuals who have inherited mostly high-risk alleles and be of value for individual risk prediction. GRS are also investigated for closing the heritability gap. Yet, a recent investigation of GRS derived for 32 complex disease traits in the Dutch Lifelines Cohort Study showed that the highest explained variance by the weighted GRS was for height (15.5%) and otherwise varied between 0.02% and 6.7%.[48]

2.14 CLINICAL AND PUBLIC HEALTH UTILITY OF GENETIC EPIDEMIOLOGY

2.14.1 CLINICAL UTILITY

Population-based heritability estimates provide a valuable metric for completeness of available genetic risk information. But individualized disease prevention and treatment will ultimately require identifying the variants accounting for risk in a given individual rather than on a population basis.[7] The small explained variance reported for complex trait–weighted GRS[48] suggests that polygenic risk scores are not yet ready for personalized risk prediction in preventive and clinical practice.[51] It has to be considered, though, that screening recommendations are also made in the light of comparatively low and average population risks. For example, mammography screening is recommended to start at age 50 in European countries. It has been demonstrated that this average breast cancer risk is reached by some women much earlier and by some women never in their lifetime depending on their polygenic breast cancer score. The benefit of a genetically targeted, personalized screening approach is currently under investigation.[52,53]

Even if the sensitivity and specificity for predicting risk of complex phenotypes could be improved or even if personalized screening recommendations were feasible, the question still arises about the actual utility of the test results. A genetic test for predicting susceptibility to traits, such as obesity, diabetes, lung cancer, or nonfamilial breast and colorectal cancer, is only of value, if it motivates for behavioral adaptation, such as persons at high risk of lung cancer to stop smoking; persons at high risk of diabetes to decrease energy intake, particularly from sweets including soft drinks, and to become more physically active; women at the extreme end of at-risk breast cancer alleles to undergo mammography screening and possibly at an earlier age then recommended. There is currently neither sufficient evidence for the impact of genetic at-risk tests for behavioral adaption nor indication that low genetic risk would decrease the motivation to live a healthy life. A healthy lifestyle is in fact desirable from a healthy aging perspective irrespective of individual disease risks.[54] These same considerations regarding clinical utility not only apply to common SNPs and derived scores, but also for findings of rare gene variants predicting disease risk. While these rare variants may have improved test characteristics over more common lower penetrance variants for individual risk prediction, testing is only relevant if the health outcome of the tested person can ultimately be improved.

In contrast to the situation of common complex diseases, evidence of the clinical utility of genetic testing clearly exists in many cases of high-penetrance variants in the setting of rare, Mendelian disorders. Being able to test for causal genetic variants in these settings allows for preventive interventions such as preimplantation diagnostics, prenatal screening, or neonatal screening in the setting of severe, early onset syndromes such as for example cystic fibrosis.[55] In adult-onset diseases such as for example familial breast cancer or familial hypercholesterolemia (FH), confirmation of the genetic variation in the index case helps initiate testing of biological family members in the form of cascade screening to offer preventive options such as regular screening, mastectomy, oophorectomy, or preventive chemotherapy in the case of breast cancer[56] or regular screening or statin treatment in the case of FH.[57]

2.14.2 PUBLIC HEALTH UTILITY

While clinical utility refers the potential of genetic tests to improve the health of an individual, public health utility refers to the utility of genetic epidemiology findings to improve the health of populations or subgroups thereof.

The public health utility of genetic testing for rare Mendelian disorders can in principle be interpreted as offering preventive and treatment options for a subgroup of the population much in need. While from a cost-effectiveness perspective, it may be considered beneficial being able to offer diagnostics that prevent severe rare syndromes that cause high direct and indirect costs from occurring, from an ethical and societal perspective this obviously raises serious issues and concerns.[58] The increased understanding of the genetic architecture of rare genetic diseases leads to the development of personalized treatments. As evidenced for the case of cystic fibrosis, where novel drugs target genetically characterized subgroups of an already rare disease that comes at a cost of a very high price and against limited evidence for long-term benefits,[59,60] discussion about the public health utility of genetic testing and personalized treatments need to include discussions about priorities in drug development and drug pricing.

The public health utility of genetic testing for complex diseases is much higher than their clinical utility and is evidenced in different domains. Findings from GWAS meta-analysis and follow-up studies continue to improve etiological disease understanding and thereby contribute to the strengthening of primary prevention, which is a global priority in the fight against the NCD epidemic.[46,61]

Identifying exogenous modifiable risk factors for complex diseases through observational epidemiology remains a challenge. Confounding, selection bias, exposure and disease misclassification, as well as reverse causation are among the factors that limit causal inference, particularly for risk factors with small relative effects which are mostly in the focus of modern chronic disease epidemiology. Many examples exist, where randomized trials were unable to confirm previously reported firm observational associations between exposures and disease.[62] Genetic variants as well as other biomarkers (see next) can be of help in strengthening causal inference.

In particular, genetic stratification of disease phenotypes can contribute to an improved identification of modifiable risk factors, which are unlikely to have an equivalent effect on all disease subphenotypes. Phenotype stratification may prevent diluting true exposure effects in one disease subtype with no exposure effects in another disease subtype.

The interaction of genetic variants with exogenous exposures can improve mechanistic and therefore causal understanding of modifiable environmental and lifestyle risk. For example, observing an interaction between GRSs related to insulin resistance with air pollution exposure strengthens the belief that air pollution effects on diabetes risk may be causal.[63] Challenges related to gene–environment interactions are severalfold and include the need for large sample size, particularly in the context of genome-wide interaction studies, which agnostically identify relevant interactions. But even with highly powered studies, an important limitation is the likelihood that the measured environments may be correlated with unmeasured genetic variation. Genetic variation identified by trait GWASs partially captures environmental risk or protective factors. Therefore some of the same genetic variation underlies both traits and environments. An environmental factor may then be acting as proxy for a gene-by-gene interaction instead of a $G \times E$ interaction, and this limits the value of gene–environment interactions in causal inference.[64,65]

Finally, Mendelian randomization (MR) offers a unique opportunity for testing the causality of exposure–phenotype associations with the help of genetic variants as proxies for the exposure of interest[66,67] (Fig. 2.5). Genetic proxies for exposures are less likely to be subject to confounding and reverse causation, and they are measured with little error. Genetic variants are randomized to persons at the time of conception and parental variants are randomly assorted during meiosis. Germline genetic variation clearly precedes the onset of the phenotypic trait of interest and is not influenced by them. Results from GWASs are now providing abundant genetic instruments to investigate causal effects of biological traits. It is for example possible to interrogate the observed

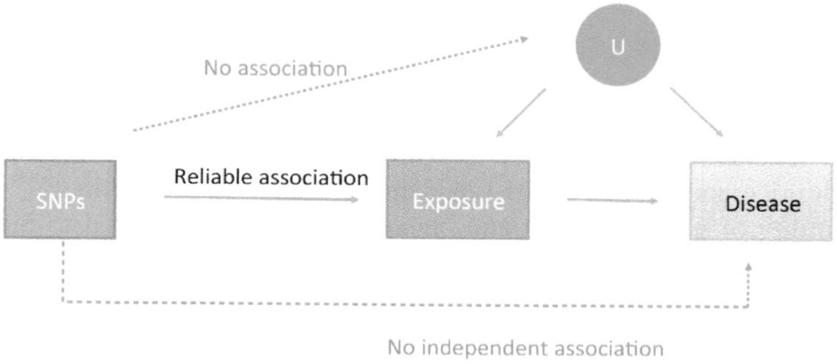

FIGURE 2.5

Instrumental variable analysis to generate causal estimates through Mendelian randomization.[67]
The three principles of instrumental variable analysis are the instrumental variable (in this case a genetic variant either in isolation or in combination with other variants) must associate with the exposure; the instrumental variable must not associate with confounders that are either known or unknown (U); and there is no pathway from the SNP to disease that does not include the exposure of interest. This figure is a schematic representation and should not be interpreted as a formal directed acyclic graph. *SNP*, Single-nucleotide polymorphism.

association of body mass index (BMI) with cardiovascular outcomes by replacing measured BMI with its genetic proxy, namely, a polygenic score for BMI. This approach is the equivalent of a randomized trial, where the long-term effect of a weight reduction intervention on cardiovascular phenotypes is being studied.[68] It is important to stress the fact that it is not required that the causal SNPs for a biological trait is known, it is sufficient to use as instrument SNPs in strong LD with the causal variants. In some instances, it is possible to apply an instrumental variable approach to estimate the magnitude of the causal effect. For MR to make valid causal conclusions a set of conditions must be met as summarized in recent publication,[66,69] and the results of an MR study must still be interpreted in the light of additional evidence and biological understanding.[66]

An important, but rarely discussed limitation of the MR approach is the fact that for major environmental risk factors, such as, for example, air pollution, no genetic exposure proxies exist that would allow interrogating causality of their health effects. It is important that this fact does not lead to a shift of research focus away from environmental exposures, particularly those that are not related to individual behavior but rather to policy and built environment.[61]

2.15 OUTLOOK: EXPOSOME—SYSTEMS EPIDEMIOLOGY—CITIZEN COHORTS AND BIOBANKS

Over recent decades, research and funding for disentangling complex disease understanding have been strongly focused on genetics research. Yet, despite broad investments into the main effect of common genetic variants in the context of GWAS, we can still explain only a small percentage of the interindividual variability of NCD risk. While genetic research continues to make an impact on understanding the pathophysiology of diseases, it is now a priority to pay equal attention and contribute more funds to studying and integrating the modifiable risk and protective factors influencing NCDs. This will most likely also close the gap of missing heritability.[7] The global NCD epidemic with its steep increase in NCD incidence and prevalence that occurred over a very short period of time is clear proof of the important impact of modifiable risks and the potential for primary prevention.[41,46]

The long-term health effects of environmental risks, for example, nongenetic risks related behavior as well as physical and social environment can mostly neither be studied in the context of randomized intervention trial nor with the help of MR approaches in the absence of genetic proxies. Observational epidemiology, which is ideally conducted in the context of longitudinal cohorts, is facing severe challenges in assigning causality to risks.[41] These challenges include (1) *Exposure misclassification*: It is in part related to the challenge of assessing exposure during critical windows of exposure. These windows of exposure can date back to early childhood and even in utero and in some cases event to exposures in the generation of parents and grandparents. Exposures can be challenging to measure, even in the context of human biomonitoring, where a short half-life of chemical and metabolites pose a limitation. Furthermore, exposures often reflect complex mixtures that are in part correlated. The components of mixtures may have not additive effects on health. (2) *Phenotype misclassification*: Exposures are not expected to impact broad phenotype traits but rather on pathophysiological and molecular subtypes of disease. (3) *Susceptibility*: The effects of specific exposures on phenotypic traits depend on susceptibility of individuals, which is in part genetically determined. Yet, gene–environment interactions, particularly conducted in an agnostic manner require very large sample sizes. Nevertheless examples of replicated agnostic genome-wide interaction studies that are based on a more limited, but very well characterized study sample exist in the domain of air pollution for example.[70,71] (4) *Sample size*: Considering the long-latency and life course history of many NCDs and that several of them are quite rare, such as for example certain types of cancer, it requires large and long-running cohorts to accumulate sufficient incident diagnoses that allow studying exposure effects in a prospective and therefore less likely biased manner.[72] (5) *Low exposures and relative risks*: The strong NCD risk factors such as smoking or human papilloma virus have been identified. The current focus of chronic disease and environmental epidemiology is on low levels of exposures each associated with a small relative risk increase. These low levels of exposure are difficult to measure, and the low relative risks are susceptible to bias and confounding. Yet, these exposures which affect large percentages of the population can nevertheless have strong public health impacts.

The exposome concept has been developed to parallel the GWAS paradigm and to benefit from recent technological developments that will help improve the credibility of causal chains. The exposome has been defined as the totality of exposures to which a human organism is exposed from the time of conception to death.[41,73–78] Much like genetic variants are studied in an integrated and agnostic manner for disease association, technological advances these days make it feasible to study nongenetic exogenous and endogenous factors in an integrated manner and over the course of life.[74,75] Today's options for capturing lifestyle and environmental exposures go far beyond answering of a questionnaire. Study participants can be equipped with wearables to capture their physical activity or exposures such air pollutants, noise, or electromagnetic fields; participants can provide detailed nutrient information by filling in apps over their smartphone or by taking pictures of what they eat. GPS tracking of participants' movement in space and time and the collection of detailed residential and occupational address histories form the basis for modeling participants short- and long-term exposures with the help of geographic information systems. Finally, different biospecimens donated by participants can these days be analyzed for far more than genetic sequences. With the help of various -omics platforms and approaches, it is possible to measure thousands of molecules, for example, transcriptome, methylome, proteome, adductome, microbiome, or various chemicals from different sources in different types of biospecimens. These biomarkers can reflect exposure, susceptibility, intermediate, and clinical phenotypes.

The EU-funded Exposomics project (http://exposomics-project.eu/) for example has taken a bottom-up approach to identify pathways and networks by which air pollution, an exposure of great policy and public health relevance, exerts its observed broad health effects.[41,74] It combined data from short-term exposure studies in children and adults with existing data and biobanks from adult and children cohorts in order to identify -omics markers that reflect specific air pollution exposures in a more precise manner and molecular pathways that may mediate the association of air pollutants with chronic disease outcomes such as cardiovascular phenotypes and asthma. The former could help calibrate external air pollution exposures, while the latter would lend mechanistic understanding and credibility for causality of air pollution effects. According to the "meet-in-the-middle," concept intermediate biomarkers that are associated both with the exposure (retrospectively) and with the disease (prospectively) strongly support a causal relationship.[74] As an example of projects embedded into Exposomics, a case–control study nested into the EPIC cohort was able to strengthen the belief with the help of protein markers and DNA methylation measured in blood up to 17 years before disease incidence that air pollution exposure in the long-term impacts cardiovascular disease etiology through oxidative and proinflammatory pathways.[79] In a separate approach based on prospectively collected biospecimens in two adults cohorts, SAPALDA and EPIC, a broader phenome perspective was applied, and it was shown that perturbations of metabolic pathways mediating air pollution exposures were shared between asthma and cardiovascular diseases.[80]

Complex diseases mostly lack a necessary cause in contrast to infections. This means that each new diagnosis arises against an individualized risk and protective factor profile that evolved over the course of life. The accumulation of molecular damage and the downstream health effect of this long-term risk profiles certainly depends on a person's genetic background. But primary prevention of NCDs can only be strengthened by tackling risks that are modifiable through individual behavior and more importantly through policy.[61] Understanding these modifiable risks in the context of observational epidemiology and systems epidemiology necessitates access to nationally and internationally harmonized cohorts and biobanks. These cohorts and biobanks allow studying various exposures and various phenotypes. They allow studying public health relevant pathways such as systemic low-grade inflammation, which may integrate the effect of numerous prevalent exposures on a broad aging phenome, in as such be very attractive as a target for primary prevention[81,82] (Fig. 2.6).

In fact, a success of GWAS is their advocacy for the setup of large, nationally, and internationally harmonized cohorts and research consortia. But the study of genetic markers on health does not depend on prospectively sample biospecimens and is less prone to confounding than any other exposure of interest. In contrast, all other biomarkers that are not fixed over time need to be studied in a prospective manner when investigating disease etiology and associations must be very careful in considering short- and long-term confounders. Therefore large, internationally harmonized population-based citizen cohorts with associated biobanks are needed to fully capitalize on the technological opportunities arising from exposure and biomarker measurements as well as from

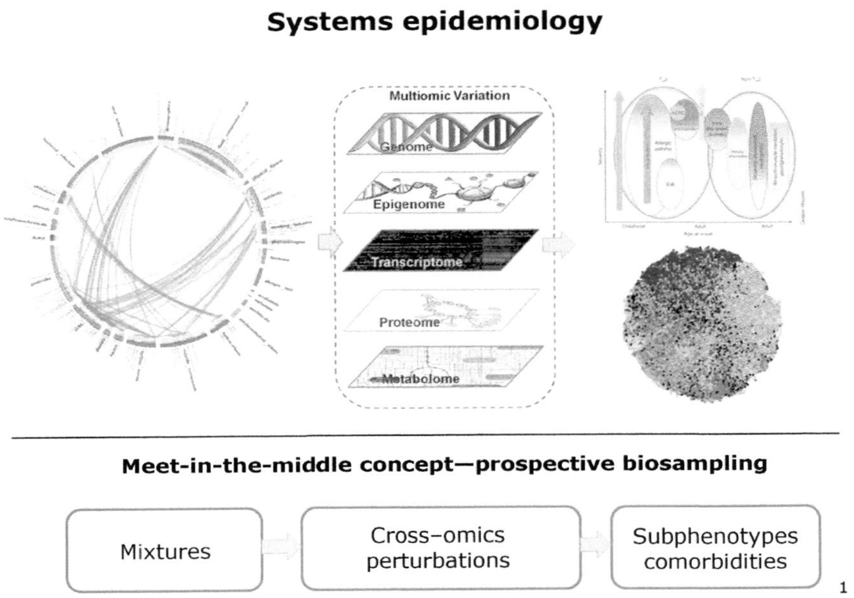

FIGURE 2.6

Systems epidemiology from exposome to phenome, integrated figure from references.[82–86] Longitudinal cohorts with associated biobanks offer the opportunity for holistic approaches toward studying exposure mixtures and the correlation between its components as well as the cross-omics fingerprints mediating their effects on single phenotypes, subphenotypes, and comorbidities in a clear time-resolved manner.

imaging to improve the prevention of common complex disease. As such citizen cohorts are a central pillar of personalized health research and will be essential to assess the predictive value and utility of novel biomarkers including genetic markers evolving from this research.[72,87]

REFERENCES

1. Burton PR, Tobin MD, Hopper JL. Key concepts in genetic epidemiology. *Lancet*. 2005;366 (9489):941–951.
2. Feero WG, Guttmacher AE, Collins FS. Genomic medicine—an updated primer. *N Engl J Med*. 2010;362 (21):2001–2011.
3. Gallagher MD, Chen-Plotkin AS. The post-GWAS era: from association to function. *Am J Hum Genet*. 2018;102(5):717–730.
4. Maurano MT, Humbert R, Rynes E, et al. Systematic localization of common disease-associated variation in regulatory DNA. *Science*. 2012;337(6099):1190–1195.
5. Nicolae DL, Gamazon E, Zhang W, Duan S, Dolan ME, Cox NJ. Trait-associated SNPs are more likely to be eQTLs: annotation to enhance discovery from GWAS. *PLoS Genet*. 2010;6(4):e1000888.

6. Wang MH, Cordell HJ, Van Steen K. Statistical methods for genome-wide association studies. *Semin Cancer Biol*. 2019;55:53–60.
7. Manolio TA, Collins FS, Cox NJ, et al. Finding the missing heritability of complex diseases. *Nature*. 2009;461(7265):747–753.
8. Hindorff LA, Gillanders EM, Manolio TA. Genetic architecture of cancer and other complex diseases: lessons learned and future directions. *Carcinogenesis*. 2011;32(7):945–954.
9. Elston RC, Anne Spence M. Advances in statistical human genetics over the last 25 years. *Stat Med*. 2006;25(18):3049–3080.
10. Ott J, Kamatani Y, Lathrop M. Family-based designs for genome-wide association studies. *Nat Rev Genet*. 2011;12(7):465–474.
11. Elston RC. Segregation analysis. *Adv Hum Genet*. 1981;11(63–120):372–373.
12. Mathias RA. Introduction to genetics and genomics in asthma: genetics of asthma. *Adv Exp Med Biol*. 2014;795:125–155.
13. Thun GA, Imboden M, Ferrarotti I, et al. Causal and synthetic associations of variants in the SERPINA gene cluster with alpha1-antitrypsin serum levels. *PLoS Genet*. 2013;9(8):e1003585.
14. Dube JB, Hegele RA. Genetics 100 for cardiologists: basics of genome-wide association studies. *Can J Cardiol*. 2013;29(1):10–17.
15. International HapMap Consortium. The International HapMap Project. *Nature*. 2003;426(6968):789–796.
16. Genomes Project Consortium, Abecasis GR, Auton A, et al. An integrated map of genetic variation from 1,092 human genomes. *Nature*. 2012;491(7422):56–65.
17. van Leeuwen EM, Kanterakis A, Deelen P, Kattenberg MV, Genome of the Netherlands C, Slagboom PE, et al. Population-specific genotype imputations using minimac or IMPUTE2. *Nat Protoc*. 2015;10(9):1285–1296.
18. Welter D, MacArthur J, Morales J, et al. The NHGRI GWAS catalog, a curated resource of SNP-trait associations. *Nucleic Acids Res*. 2014;42(Database issue):D1001–D1006.
19. Hakonarson H, Grant SF. Genome-wide association studies (GWAS): impact on elucidating the aetiology of diabetes. *Diabetes Metab Res Rev*. 2011;27(7):685–696.
20. Kottgen A. Genome-wide association studies in nephrology research. *Am J Kidney Dis*. 2010;56(4):743–758.
21. Willis-Owen SA, Cookson WO, Moffatt MF. Genome-wide association studies in the genetics of asthma. *Curr Allergy Asthma Rep*. 2009;9(1):3–9.
22. Seng KC, Seng CK. The success of the genome-wide association approach: a brief story of a long struggle. *Eur J Hum Genet*. 2008;16(5):554–564.
23. McCarthy MI, Abecasis GR, Cardon LR, et al. Genome-wide association studies for complex traits: consensus, uncertainty and challenges. *Nat Rev Genet*. 2008;9(5):356–369.
24. Laird NM, Lange C. Family-based designs in the age of large-scale gene-association studies. *Nat Rev Genet*. 2006;7(5):385–394.
25. Turner S, Armstrong LL, Bradford Y, et al. Quality control procedures for genome-wide association studies. *Curr Protoc Hum Genet*. 2011;. Chapter 1: Unit119.
26. Marees AT, de Kluiver H, Stringer S, et al. A tutorial on conducting genome-wide association studies: quality control and statistical analysis. *Int J Methods Psychiatr Res*. 2018;27(2):e1608.
27. Purcell S, Neale B, Todd-Brown K, et al. PLINK: a tool set for whole-genome association and population-based linkage analyses. *Am J Hum Genet*. 2007;81(3):559–575.
28. Ferrarotti I, Thun GA, Zorzetto M, et al. Serum levels and genotype distribution of alpha1-antitrypsin in the general population. *Thorax*. 2012;67(8):669–674.
29. Banauch GI, Brantly M, Izbicki G, et al. Accelerated spirometric decline in New York City firefighters with alpha(1)-antitrypsin deficiency. *Chest*. 2010;138(5):1116–1124.

30. Obeidat M, Hao K, Bosse Y, et al. Molecular mechanisms underlying variations in lung function: a systems genetics analysis. *Lancet Respir Med.* 2015;3(10):782–795.
31. Jackson VE, Latourelle JC, Wain LV, et al. Meta-analysis of exome array data identifies six novel genetic loci for lung function. *Wellcome Open Res.* 2018;3:4.
32. Wain LV, Shrine N, Artigas MS, et al. Genome-wide association analyses for lung function and chronic obstructive pulmonary disease identify new loci and potential druggable targets. *Nat Genet.* 2017;49(3):416–425.
33. International HapMap Consortium. A haplotype map of the human genome. *Nature.* 2005;437 (7063):1299–1320.
34. McCarthy S, Das S, Kretzschmar W, et al. A reference panel of 64,976 haplotypes for genotype imputation. *Nat Genet.* 2016;48(10):1279–1283.
35. Schaid DJ, Chen W, Larson NB. From genome-wide associations to candidate causal variants by statistical fine-mapping. *Nat Rev Genet.* 2018;19(8):491–504.
36. Sabater-Lleal M, Huffman JE, de Vries PS, et al. Genome-wide association transethnic meta-analyses identifies novel associations regulating coagulation factor VIII and von Willebrand factor plasma levels. *Circulation.* 2019;139:620–635.
37. Lindstrom S, Ablorh A, Chapman B, et al. Deep targeted sequencing of 12 breast cancer susceptibility regions in 4611 women across four different ethnicities. *Breast Cancer Res.* 2016;18(1):109.
38. Smemo S, Tena JJ, Kim KH, et al. Obesity-associated variants within FTO form long-range functional connections with IRX3. *Nature.* 2014;507(7492):371–375.
39. Hao K, Bosse Y, Nickle DC, et al. Lung eQTLs to help reveal the molecular underpinnings of asthma. *PLoS Genet.* 2012;8(11):e1003029.
40. Ricks-Santi L, McDonald JT, Gold B, et al. Next generation sequencing reveals high prevalence of BRCA1 and BRCA2 variants of unknown significance in early-onset breast cancer in African American women. *Ethn Dis.* 2017;27(2):169–178.
41. Vineis P, Fecht D. Environment, cancer and inequalities—the urgent need for prevention. *Eur J Cancer.* 2018;103:317–326.
42. Hindorff LA, Sethupathy P, Junkins HA, et al. Potential etiologic and functional implications of genome-wide association loci for human diseases and traits. *Proc Natl Acad Sci USA.* 2009;106(23):9362–9367.
43. Consortium EP. An integrated encyclopedia of DNA elements in the human genome. *Nature.* 2012;489 (7414):57–74.
44. Roadmap Epigenomics Consortium, Kundaje A, Meuleman W, et al. Integrative analysis of 111 reference human epigenomes. *Nature.* 2015;518(7539):317–330.
45. Andersson R, Gebhard C, Miguel-Escalada I, et al. An atlas of active enhancers across human cell types and tissues. *Nature.* 2014;507(7493):455–461.
46. Probst-Hensch N, Tanner M, Kessler C, Burri C, Kunzli N. Prevention—a cost-effective way to fight the non-communicable disease epidemic: an academic perspective of the United Nations High-level NCD Meeting. *Swiss Med Wkly.* 2011;141:w13266.
47. Kaprio J. Twins and the mystery of missing heritability: the contribution of gene-environment interactions. *J Intern Med.* 2012;272(5):440–448.
48. Nolte IM, van der Most PJ, Alizadeh BZ, et al. Missing heritability: is the gap closing? An analysis of 32 complex traits in the Lifelines Cohort Study. *Eur J Hum Genet.* 2017;25(7):877–885.
49. Zhu Z, Bakshi A, Vinkhuyzen AA, et al. Dominance genetic variation contributes little to the missing heritability for human complex traits. *Am J Hum Genet.* 2015;96(3):377–385.
50. Zaitlen N, Kraft P, Patterson N, et al. Using extended genealogy to estimate components of heritability for 23 quantitative and dichotomous traits. *PLoS Genet.* 2013;9(5):e1003520.
51. Cornel MC, van El CG, Borry P. The challenge of implementing genetic tests with clinical utility while avoiding unsound applications. *J Community Genet.* 2014;5(1):7–12.

52. Shieh Y, Eklund M, Madlensky L, et al. Breast cancer screening in the precision medicine era: risk-based screening in a population-based trial. *J Natl Cancer Inst*. 2017;109(5).
53. Naslund-Koch C, Nordestgaard BG, Bojesen SE. Common breast cancer risk alleles and risk assessment: a study on 35 441 individuals from the Danish general population. *Ann Oncol*. 2017;28(1):175–181.
54. Hollands GJ, French DP, Griffin SJ, et al. The impact of communicating genetic risks of disease on risk-reducing health behaviour: systematic review with meta-analysis. *BMJ*. 2016;352:i1102.
55. Castellani C, Massie J, Sontag M, Southern KW. Newborn screening for cystic fibrosis. *Lancet Respir Med*. 2016;4(8):653–661.
56. D'Andrea E, Marzuillo C, De Vito C, et al. Which BRCA genetic testing programs are ready for implementation in health care? A systematic review of economic evaluations. *Genet Med*. 2016;18(12):1171–1180.
57. Knowles JW, Rader DJ, Khoury MJ. Cascade screening for familial hypercholesterolemia and the use of genetic testing. *JAMA*. 2017;318(4):381–382.
58. Deem MJ. Whole-genome sequencing and disability in the NICU: exploring practical and ethical challenges. *Pediatrics*. 2016;137(suppl 1):S47–S55.
59. Harutyunyan M, Huang Y, Mun KS, Yang F, Arora K, Naren AP. Personalized medicine in CF: from modulator development to therapy for cystic fibrosis patients with rare CFTR mutations. *Am J Physiol Lung Cell Mol Physiol*. 2018;314(4):L529–L543.
60. Gulland A. Cystic fibrosis drug is not cost effective, says NICE. *BMJ*. 2016;353:i3409.
61. Pearce N, Ebrahim S, McKee M, et al. The road to 25 × 25: how can the five-target strategy reach its goal? *Lancet Glob Health*. 2014;2(3):e126–e128.
62. Davey Smith G, Ebrahim S. Epidemiology—is it time to call it a day? *Int J Epidemiol*. 2001;30(1):1–11.
63. Eze IC, Imboden M, Kumar A, et al. Air pollution and diabetes association: modification by type 2 diabetes genetic risk score. *Environ Int*. 2016;94:263–271.
64. Fletcher JM, Conley D. The challenge of causal inference in gene-environment interaction research: leveraging research designs from the social sciences. *Am J Public Health*. 2013;103(suppl 1):S42–S45.
65. Krapohl E, Hannigan LJ, Pingault JB, et al. Widespread covariation of early environmental exposures and trait-associated polygenic variation. *Proc Natl Acad Sci USA*. 2017;114(44):11727–11732.
66. Davey Smith G, Hemani G. Mendelian randomization: genetic anchors for causal inference in epidemiological studies. *Hum Mol Genet*. 2014;23(R1):R89–R98.
67. Holmes MV, Ala-Korpela M, Smith GD. Mendelian randomization in cardiometabolic disease: challenges in evaluating causality. *Nat Rev Cardiol*. 2017;14(10):577–590.
68. Lyall DM, Celis-Morales C, Ward J, et al. Association of body mass index with cardiometabolic disease in the UK Biobank: a Mendelian randomization study. *JAMA Cardiol*. 2017;2(8):882–889.
69. Yarmolinsky J, Wade KH, Richmond RC, et al. Causal inference in cancer epidemiology: what is the role of Mendelian randomization? *Cancer Epidemiol Biomarkers Prev*. 2018;27(9):995–1010.
70. Imboden M, Kumar A, Curjuric I, et al. Modification of the association between PM10 and lung function decline by cadherin 13 polymorphisms in the SAPALDIA cohort: a genome-wide interaction analysis. *Environ Health Perspect*. 2015;123(1):72–79.
71. Kim HJ, Min JY, Min KB, et al. CDH13 gene-by-PM10 interaction effect on lung function decline in Korean men. *Chemosphere*. 2017;168:583–589.
72. Burton PR, Hansell AL, Fortier I, et al. Size matters: just how big is BIG?: Quantifying realistic sample size requirements for human genome epidemiology. *Int J Epidemiol*. 2009;38(1):263–273.
73. Wild CP. Complementing the genome with an "exposome": the outstanding challenge of environmental exposure measurement in molecular epidemiology. *Cancer Epidemiol Biomarkers Prev*. 2005;14 (8):1847–1850.
74. Vineis P, Chadeau-Hyam M, Gmuender H, et al. The exposome in practice: design of the EXPOsOMICS project. *Int J Hyg Environ Health*. 2017;220(2 Pt A):142–151.

75. Maitre L, de Bont J, Casas M, et al. Human Early Life Exposome (HELIX) study: a European population-based exposome cohort. *BMJ Open*. 2018;8(9):e021311.
76. Siroux V, Agier L, Slama R. The exposome concept: a challenge and a potential driver for environmental health research. *Eur Respir Rev*. 2016;25(140):124–129.
77. Rappaport SM. Implications of the exposome for exposure science. *J Expo Sci Environ Epidemiol*. 2011;21(1):5–9.
78. Rappaport SM, Barupal DK, Wishart D, Vineis P, Scalbert A. The blood exposome and its role in discovering causes of disease. *Environ Health Perspect*. 2014;122(8):769–774.
79. Fiorito G, Vlaanderen J, Polidoro S, et al. Oxidative stress and inflammation mediate the effect of air pollution on cardio- and cerebrovascular disease: a prospective study in nonsmokers. *Environ Mol Mutagen*. 2018;59(3):234–246.
80. Jeong A, Fiorito G, Keski-Rahkonen P, et al. Perturbation of metabolic pathways mediates the association of air pollutants with asthma and cardiovascular diseases. *Environ Int*. 2018;119:334–345.
81. Probst-Hensch NM. Chronic age-related diseases share risk factors: do they share pathophysiological mechanisms and why does that matter? *Swiss Med Wkly*. 2010;140:w13072.
82. Sun YV, Hu YJ. Integrative analysis of multi-omics data for discovery and functional studies of complex human diseases. *Adv Genet*. 2016;93:147–190.
83. Patel CJ, Manrai AK. Development of exposome correlation globes to map out environment-wide associations. *Pac Symp Biocomput*. 2015;231–242.
84. Vineis P, van Veldhoven K, Chadeau-Hyam M, Athersuch TJ. Advancing the application of omics-based biomarkers in environmental epidemiology. *Environ Mol Mutagen*. 2013;54(7):461–467.
85. Wenzel SE. Asthma phenotypes: the evolution from clinical to molecular approaches. *Nat Med*. 2012;18 (5):716–725.
86. Hoehndorf R, Schofield PN, Gkoutos GV. Analysis of the human diseasome using phenotype similarity between common, genetic, and infectious diseases. *Sci Rep*. 2015;5:10888.
87. Wijmenga C, Zhernakova A. The importance of cohort studies in the post-GWAS era. *Nat Genet*. 2018;50(3):322–328.

CHAPTER

RARE DISEASES: GENOMICS AND PUBLIC HEALTH

3

Gabriela M. Repetto[1] and Boris Rebolledo-Jaramillo[1,2]

[1]*Rare Diseases Program, Center for Genetics and Genomics, Instituto de Ciencias e Innovación en Medicina, Facultad de Medicina, Clínica Alemana Universidad del Desarrollo, Santiago, Chile* [2]*Bioinformatics Analysis Core, Center for Genetics and Genomics, Instituto de Ciencias e Innovación en Medicina, Facultad de Medicina, Clínica Alemana Universidad del Desarrollo, Santiago, Chile*

3.1 RARE DISEASES ARE A RELEVANT PUBLIC HEALTH PROBLEM WORLDWIDE

Rare diseases (RDs) or orphan diseases, by definition, are conditions that affect a small number of individuals. The exact number of RDs is unknown, but it is estimated to be approximately 6000–8000.[1] Although each individual disease is rare, their collective prevalence is estimated at around 6%–8% of the population.[2] This implies that over 450 million persons worldwide may have an RD. In addition, although heterogeneous in causes and manifestations, most RDs are chronic and debilitating and are a substantial cause for disability and early death. They manifest as congenital anomalies, intellectual disabilities, epilepsy, and many other severe signs and symptoms. Most lack curative therapies. They constitute a substantial cause of health-care spending for individuals and society and also have indirect economic impact due to loss of productivity.[3] For these reasons, RDs constitute a relevant and global public health-care problem.

A comprehensive catalog of RDs can be found in Orphanet,[1] an initiative funded by the European Commission, that gathers, curates, and disseminates knowledge on RDs. Over 6000 diseases are listed and described there, as well as information on expert services, orphan drugs, patient organizations, and other relevant materials. Based on Orphanet's disease inventory, it is evident that the majority of RDs are of genetic etiology, and a smaller percentage is autoimmune or infectious disorders, in addition to some rare cancers.

Given the substantial role of genetic variation as a cause of RD, the development and introduction of high-throughput sequencing (HTS), in particular whole-exome sequencing (WES) and whole-genome sequencing (WGS) in research and clinical settings, have had a substantial impact on the diagnosis, discovery, and understanding of RDs. Other relevant fundamentals that have contributed to advances in the field of RDs in this decade are other -omics (transcriptomics, proteomics, metabolomics, and microbiome data), potent bioinformatics analysis tools, patient participation, incentives for therapeutic developments, and initiatives, which foster data sharing and international collaboration.

Applied Genomics and Public Health. DOI: https://doi.org/10.1016/B978-0-12-813695-9.00003-0

Among the many areas that are developing under the broad precision medicine umbrella, RDs are among the health-care and research fields where genomics has grown at the fastest pace, along with oncology and pharmacogenomics.[4] These fast-paced developments are resulting in many advances that have impacted patients' lives. Nevertheless, many challenges still remain in discovery, understanding of mechanisms, drug development, and access to timely diagnosis and therapies. Elements of these advances and challenges will be described further in this chapter.

3.2 DIAGNOSTIC STRATEGIES

Patients with RD face many challenges. One of them is what has been called the "diagnostic odyssey" or "diagnostic labyrinth," where patients are evaluated by many clinicians and have many tests performed without achieving a specific diagnosis. Not having a diagnosis hinders patients from having access to therapies, assessing prognosis, and estimating recurrence risks, among other benefits. A survey by the European Organization for Rare Disorders on eight RDs representing 12,000 patients found that for 25% of patients, it took between 5 and 30 years to achieve a correct diagnosis and that almost half of them received an erroneous diagnosis initially.[5] A similar study in Canada showed that 60% of patients visited between 3 and 20 specialists before arriving to a diagnosis.[2]

The development of HTS tools and bioinformatics analysis and their deployment into the clinical space have been crucial to facilitate hypothesis-free approaches. Altogether, they have dramatically increased the rate of discovery and diagnosis over the past decade. Conventional genetic approaches, such as positional cloning, linkage analysis, and candidate genes, led to an average of 100 new disease-related gene discovery per year between years 2000 and 2010. The incorporation of WES and WGS has tripled the discovery rate in the past 10 years, although this rate seems to be decreasing.[6]

DNA-sequencing technologies have greatly advanced over the past decade. Sanger sequencing can only be used to analyze a single-DNA molecule in a labor intensive and time-consuming way, whereas HTS allows for the analysis of thousands of genomic regions simultaneously, which is desirable in a hypothesis-free clinical context.[7] Analysis by HTS can include single genes (e.g., *CFTR* for cystic fibrosis), gene panels for conditions with genetic heterogeneity (e.g., hearing loss, epilepsy, or intellectual disability that include hundreds or thousands of genes), capture of nearly all exons plus a limited number of base pairs of exon–intron borders in the 20,000 or so protein-coding genes in WES, or untargeted sequencing of coding, regulatory, intronic, and intergenic regions in WGS, the latter two representing hypothesis-free diagnostic approaches.[8] Diagnostic yield for WES and WGS is increased if performed as a trio analysis, with samples from the patient and parents. This helps in the interpretation of de novo genetic variants, parent of origin, and rare familial variants.[9]

Although HTS is time efficient and ever lowering in costs, the huge amount of data generated when either WES or WGS is performed poses a new challenge, converting raw sequencing data into clinically relevant information.[10] Fortunately, thanks to individual researchers worldwide and large consortia, such as the Broad Institute and the International Genome Sample Resource (IGSR or the 1000 Genomes Project), as sequencing technologies progressed, so have the tools needed to analyze it. Many researchers have led the way generating or improving the bioinformatics tools used in clinical research today (see also Chapter 8), especially in the field of RDs. They also have helped define the current guidelines to manage HTS data efficiently.[11,12]

In a very succinct way the bioinformatics pipeline used to identify clinically relevant genetic variants from a HTS sample can be summarized in three major steps: (1) *alignment*: first, data from sequencers, stored in FASTQ format (a plain text format defining a DNA fragment sequence along with their corresponding base-wise probability of error), is assigned a location within the human genome using tools called "aligners," (2) *variant calling*: using these locations, another tool called "variant caller" defines the probabilistic presence of a variant at a genomic location and its genotype, given the nucleotides observed at that position, and (3) *variant annotation*: finally, to help defining clinically relevant variation, genetic variants are given biological context using a myriad of databases, with tools such as ANNOVAR[13] that describe a variant's features such as class [i.e., single-nucleotide variant (SNV) or insertion deletion (indel)], function (i.e., exonic and intronic), impact (i.e., synonymous and nonsynonymous), frequency in known human populations (1000 Genomes database,[12] gnomAD,[14] and others), or even clinical information that can be obtained from repositories such as the Online Mendelian Inheritance in Man[15] and ClinVar[16] databases. In the clinical setting, these pieces of information serve as input to classify variants into the categories defined by the American College of Medical Genetics (ACMG) and the Association for Molecular Pathology (AMP): (likely) pathogenic, (likely) benign, and variant of unknown significance (VUS).[17]

In a fourth stage, annotated and classified variants are analyzed by a multidisciplinary team of experts in order to extract those variants, plausibly explaining a patient's phenotype.[18] Detailed phenotypic characterization is a relevant component of the diagnostic process. Harmonized clinical ontologies facilitate the analysis and prioritization of genomic variant findings, as well as database interoperability and sharing. The Human Phenotype Ontology is a resource for annotation, stratification, and differential diagnosis, which is currently being used by many programs worldwide.[19]

This, at first glance, straightforward process aimed at ending a patient's diagnostic odyssey, is challenged at several levels. For instance, there are multiple different aligners and variants calling tools available. A systematic comparison of 13 different pipelines using four different aligners and four different variant callers showed that concordance on Illumina HTS data was on average $91.7\% \pm 3.9\%$ and that individual pipelines showed between 1% and 3% private variants calls each.[20] Another study compared concordance and speed among six different pipelines and concluded that some pipelines can be up to 20 times faster than others and reinforced the observation that variants can be reliably called in only 70% of the genome and that there still exists at least a 10% of the genome where it is unfeasible to call SNVs and indels.[21] The "pipeline of choice" trade-off can be somewhat alleviated if multiple pipelines are used on the same dataset and statistically merge the results to generate a set of high-quality variant calls.[22] However, this could increase processing time and, consequently, time to diagnosis.

Another source of uncertainty is the speed at which databases used for the annotation of variants are updated. For instance, ClinVar is a public archive of relationships between genetic variants and phenotypes hosted by the National Institutes of Health (NIH) in the United States. It includes user-submitted data with different levels of evidence: clinical testing, research, or literature.[16] The website's database (https://www.ncbi.nlm.nih.gov/clinvar) is updated weekly as user submissions become processed; however, the whole database VCF is updated and made available monthly.

Another example is the 1000 Genomes Project. It launched in 2008 and ended in 2015. It is now a static resource maintained by the IGSR, and it is primarily used to extract global or population-specific allele frequencies during a variant annotation. The IGSR is committed to maintain access to the resource and expand it increasing population diversity. So, in the future, allele frequencies may vary, or a patient whose population of origin has not been represented in public databases might receive an improved annotation.[12]

Variant annotation and classification have gone a long way toward perfecting standards. Current guidelines such as those proposed by the ACMG/AMP ensure that clinicians and patients are provided with evidence-based information, but evidence is dynamic, and it can change over time. There are no guidelines governing VUS reinterpretation. In fact, there is an ongoing discussion on how to approach this, both technical and ethically, and results show that a variant classification could change its clinical status as information becomes available (i.e., unknown significance to likely benign or likely pathogenic), leading to complex genetic counseling scenarios but improved clinical care.[23–25]

Despite the possible impact on the credibility of genetic testing (although evidence shows mostly positive or neutral reactions in upper class patients after informing them changes to their original diagnosis[23]), variant reclassification, and more in general reanalysis of WES or WGS data on the light of new information, might actually be a necessity in the race for diagnosis. Recently, this has been demonstrated, for example, with a cohort of 416 patients, where 156 WESs without a diagnosis were reanalyzed with the same pipeline a year after, obtaining 24 (15.4%) additional diagnoses.[26] A similar increase in diagnosis rate was observed in a cohort of patients with epilepsy. Forty eight out of 124 (38.7%) patients with undiagnosed conditions had their results reclassified with information from the past 2 years,[27] suggesting that reinterpretation of genomic tests should be performed routinely, especially since reanalysis does not imply resequencing, so costs are considerably lower on the provider and patient's end.

3.3 THERAPEUTIC DEVELOPMENTS

RDs are a highly heterogeneous group of disorders, with marked differences in clinical manifestations, age of onset, course, and prognosis. Although substantial advances have been made in the past two to three decades in the development and approval of targeted therapies for RDs, less than 5% of the 7000 or so diseases have specific treatments. Inborn errors of metabolism, immunodeficiencies, and cancers are among the areas with the fastest growing therapies. Major breakthroughs have occurred recently also in neuromuscular diseases such as Duchenne muscular dystrophy and spinal muscular atrophy.[28,29] Disorders that affect the central nervous system remain a challenging area.

Although the majority of RDs still lack specific treatments, many benefit from symptomatic therapies that are also used for common diseases. Having a specific diagnosis can certainly help guide decisions. Broad categories of "common" therapeutic approaches that are beneficial to persons with RD, for example, are early intervention and school support programs for infants and children with developmental delays, nutritional support, surgical repair of congenital anomalies, or ventilatory assistance for respiratory insufficiency, among many others. Genetic counseling, education, and support are also crucial in management.

Some specific interventions have been used for decades, such as dietary management of inborn errors of metabolism in phenylketonuria, organic acidemias, and many others, where specific offending substrates are reduced from the diet and/or deficient products are supplemented.

The development of new targeted therapies for RD has many challenges: complex biological mechanisms, difficulties in conducting clinical trials in small populations, and lack of financial incentives, among others. To overcome these barriers, some countries have established specific regulations. The first public policy on RD was the Orphan Drug Act (ODA), passed in 1983 in the United States.[30] This initiative was largely led by the National Organization for Rare Disorders (NORD), a patient association formed to advocate and disseminate knowledge on RD. Work between families, legislators, and public figures led to increased public awareness, writing, and passing of the law. The ODA generated incentives for the development of orphan drugs, such as tax incentives, extended market exclusivity, fee waivers, and clinical research subsidies.[30] Similar legislations were passed subsequently in Japan, Singapore, Australia, and the European Union. These laws dramatically changed the landscape for therapies for RD: from approximately 10 drugs developed during the decade before the ODA to more than 500 new developments approved for clinical use in the three decades after.[2] Nevertheless, concerns have been raised recently on the impact of these laws on high medication prices and misuse for nonorphan disease drug development.[31,32]

The advances in molecular understanding of mechanisms of disease have facilitated the development and approval of specific targeted therapies for patients with RD. Therapeutic strategies can be broadly categorized into two groups: biologicals and small molecules. A third is drug repurposing or repositioning of existing approved therapies for new uses.[33]

Biologicals or biologic products are complex products that can be isolated from natural sources or produced using advanced biotechnology methods. They are usually administered intravenously or to target organs. One major pitfall is that they usually do not cross the blood–brain barrier and require intrathecal administration when the central nervous system is the target. Biologicals approved for RD include as follows:

1. *Recombinant proteins*, such as enzymes, peptides, and antibodies. Examples include recombinant factors VIII and IX for hemophilia types A and B, respectively, and enzyme replacement therapies that were among the first orphan drugs to be developed and approved and are currently available for lysosomal storage diseases, such as Gaucher and Fabry diseases, and mucopolysaccharidosis types I, II, IVA, VI, and VII.
2. *Immunology-based therapies* for certain cancers, immunodeficiencies, autoimmune disorders, and others that involved dysregulation of the immune system. This category includes monoclonal antibodies, such as canakinumab that targets interleukin 1β and is approved for active juvenile idiopathic arthritis and cryopyrin-associated periodic syndromes, or eculizumab that targets complement component 5 and is approved for paroxysmal nocturnal hemoglobinuria and atypical hemolytic uremic syndrome, among others.[34]
3. *Stem cell-based therapies*, where cord blood or bone marrow is used to differentiate hematopoietic cells into specific subpopulations. These approaches have been used for some lysosomal storage and peroxisomal diseases, as well as congenital disorders of hematopoiesis. As in many conditions, patients at presymptomatic or early stages of the disease tend to have better outcomes than patients in later stages.[35]

4. *Gene therapy*, in which normal copies of genes are introduced into cells using different delivery systems. This is particularly useful for diseases caused by loss of function alleles. There have been a large number of clinical trials for different RDs, but to date, only one in vivo gene therapy product, voretigene neparvovec, has been approved by the FDA in December 2017. It is used for the treatment of Leber congenital amaurosis. It consists of an adeno-associated viral vector containing human RPE65 cDNA, and it is delivered via intraretinal injection.[33,36]

Other biological strategies currently at the research or clinical trial level are induced pluripotent stem cells (iPSCs) and gene editing. iPSCs can be differentiated into specific cell types and be used as an autologous transplant. Gene editing can be considered a highly specific form of gene therapy, where a DNA sequence is changed within cells with single base-pair precision. At the time of this writing, there were no approved therapies using gene editing for RD, but there are several ongoing clinical trials targeting relatively common cancers and some monogenic diseases, such sickle cell disease, hemophilia, and several primary immunodeficiencies (reviewed in Ref. [37]).

Small molecules have been a cornerstone of therapies for decades; common examples are antibiotics, analgesics, and beta blockers. Characterization of precise genetic defects in disease has accelerated the recognition of novel potential therapeutic targets. Development of these drugs is strategically different than development of the biologicals described above. Therapeutic targets are initially identified through high-throughput screening, for example, of large databases of genomic variants. Subsequently, assays (biochemical or cell based) are conducted to determine therapeutic activities of potential candidates and define later screening of small molecule libraries usually using fast automated systems. These steps lead to the identification of drug–target "hits" that would need further confirmation in secondary, more specific assays before actual drug development and preclinical evaluation. This model allows to search for candidates not only of diseases with known mechanisms but can be expanded to conditions where the pathophysiology is unknown or unclear. Examples of small molecule developments include potentiators and correctors of the different alterations in cystic fibrosis, ivacaftor and lumacaftor, and several novel chemotherapeutic agents.[33,38]

RNA-based therapies share elements of both small molecules and biologicals. Synthetic RNAs can be produced to modulate RNA function. Their aim is to specifically bind to their targets and trigger different effects, such as enzymatic destruction of mRNA or changes in pre-mRNA splicing. Clinically approved examples of the former are mipomersen and inclisiran for hyperlipidemias, and of the latter, eteplirsen that causes skipping of exon 51 in Duchenne muscular dystrophy and nusinersen for exons 7 and 8 inclusion in spinal muscular atrophy.[39]

Drug repurposing is yet another strategy, which consists of using already approved drugs to treat new diseases. It can involve target and effect screening of known drugs, as described in the small molecule approach. Given that they are already approved, and their safety profile is known, the evaluation and approval process for new indications can be accelerated. Examples of the nonsteroidal antiinflammatory drug in RD are celecoxib for familial adenomatous polyposis and the antifungal ketoconazole for Cushing disease.[40]

3.4 NATIONAL PUBLIC POLICIES AND PROGRAMS

The complex problems that affect patients with RD, their families, and the health-care systems require comprehensive and complex solutions. Several countries have developed public policies to address these burdens, but many barriers remain still. The patient community has been critical in raising awareness of the needs of persons with RD and promoting the development of legislation. As mentioned previously, the ODA in the United States was the first public policy specifically developed for RD.[30]

Other national RD programs and policies have different emphasis on access to diagnosis and therapies, coordination of care, research, and/or patient involvement. Newborn screening (NBS) programs in different countries are an example of timely and universal diagnosis, therapeutic implementation, and follow-up. Composition of diseases included in NBS panels varies widely, partly due to differences in prevalence and access to technologies and treatments in different regions.

There are examples of national policies for a broader range of rare and unknown disorders. France led the development of a national plan for RD that was implemented in 2004. Among other initiatives, it fostered the creation Centers of Excellence designed to provide timely, equitable, and evidence-guided care, as well as a curated information system, Orphanet, currently in use worldwide. The third version of the plan for the period 2018–20 was recently published.[41] The Centers of Excellence model was subsequently adopted and expanded by the European Union through the formation of the European Reference Networks, generating virtual networks of health-care providers to facilitate discussion about patients with RDs and complex conditions that require highly specialized treatment and concentrated knowledge and resources [42,43] The work of the networks is facilitated through a dedicated IT platform and telemedicine tools.

Another large program in Europe is the 100,000 UK Genomes project, announced in 2012, to sequence whole genomes from patients from the National Health Service (NHS) with RD and some types of cancer and infectious diseases while generating a transformation of clinical care in the NHS to include WGS as routine clinical practice. An organization, Genomics England, was created to deliver the program. Contracts with industry were made for sequencing, bioinformatics analysis, data storage, and other related services. The NHS formed 13 genomic medicine centers and a comprehensive Genomics Education Programme for staff in the health-care system. The 100,000 genomes milestone was reached in December 2018, and results are being delivered to the first participants.[44,45]

The Undiagnosed Diseases Network (UDN) in the United States is funded by the NIH. It was designed to develop a paradigm for the diagnosis, management of, and research on rare and unknown disorders. It consists currently of seven clinical sites, sequencing, metabolomics, and animal model cores and a biorepository.[46] Patients or their referring physicians apply online, their clinical data is reviewed by a committee, and are accepted for evaluation if no diagnosis has been previously reached through extensive evaluations or when manifestations suggest a novel genetic disorder. In its first 20 months, more than 1500 patients were referred, close to 600 accepted for evaluation, and close to 400 had completed evaluation. Of the latter, a diagnosis was achieved in 35%, and a fifth of these led to changes in therapies. More than 30 new syndromes were identified.[47] These initial results illustrate the power of the collaborative programs.

China has the largest RD population in the world. In 2013, it launched a pilot project to optimize prevention, diagnosis, and therapies within the existing medical system. A national network was created, involving more than 100 medical centers, covering a population of over half a billion individuals. Clinical pathways and registries were developed for 20 "example" RD; their implementation process and outcomes are currently being evaluated. Support for selected NGS testing is also included in the pilot projects. It is expected that this initiative will improve care for patients with these disorders and that it will be scaled to 200–300 diseases in the near future.[48] Japan launched the Initiative on Rare and Undiagnosed Disease in 2015, integrating Nan-Byo ("difficult illness") patients, researchers, and the Japanese universal health-care system as a nationwide consortium, focused on facilitating diagnosis and care for persons with yet undiagnosed conditions.[49]

Similarly, the Australian Genomics Health Alliance was created as a clinical and research collaboration to address the challenges of genomic medicine implementation and integration in health-care system and includes flagship projects on rare disorders.[50,51]

Some Latin American countries, such as Brazil, Peru, Colombia, and Costa Rica, have recently developed policies related to care of individuals with RDs. These policies recognize the needs of the patients and provide some coverage. However, substantial barriers to care exist, due, among other reasons, to the scarcity of resources, the lack of certified clinical and laboratory geneticists, absence of funding and reimbursement of genetic testing and therapies, and difficulties in access to centers in countries that have large populations spread across vast landscapes (see also Chapter 20: Regulatory Aspects of Genomic Medicine and Pharmacogenomics). Peru and Colombia have legislation to protect the rights of all patients with RDs. Chile has implemented a more targeted approach, providing coverage for diagnosis and high-cost treatments for 18 RDs, including certain lysosomal storage disease and hemophilia.[52,53] Many other disorders are not covered, and patients are resorting to the supreme courts to obtain coverage for therapies, a trend that is growing in Latin America.[54] These situations call for the development of integral policies.

3.5 INTERNATIONAL COLLABORATIVE INITIATIVES

Given their low incidence, it is very likely that most clinicians seeing a patient with an RD will have difficulty in recognizing a diagnosis and in implementing best care practices. Similarly, research for these complex conditions constitutes a challenge for individual research groups. The recognition of these barriers has led to a culture of international collaboration and data sharing, aimed at maximizing the outcomes for patients. These initiatives have contributed substantially to harmonization of nomenclature, protocols, and interpretation criteria for genomic data. Among these, large collaborative alliances are as follows:

- *UDN International* was created in Europe in 2015, modeled in part after the UDN in the United States, with the aims of improving patient diagnosis and therapies, facilitate research into the causes and mechanisms of RDs, and to foster a collaborative clinical and research community, with common evaluation and data sharing standards. It currently has members from 18 countries, and among other activities, it organizes yearly international meetings on RDs.[55]
- *International Rare Diseases Research Consortium (IRDiRC)* was created in 2011, bringing together research organizations from Asia, Australasia, Europe, the Middle East, and North America and patient advocacy groups such as the NORD and Genetic Alliance from the United States and Rare Diseases Europe (EURORDIS). Their goals are to enable the diagnosis of all RDs and to deliver 200 new therapies for RDs. IRDiRC comprises three scientific committees (diagnostics, interdisciplinary, and therapies) and a consortium assembly of public and public funders from more than 40 member institutions.[6]

3.6 THE ROLE OF PATIENT ORGANIZATIONS

Patient organizations have been crucial to raise awareness about the reality and needs of persons with RDs. Support and advocacy groups have a strong role in the development of legislation and policies, as was described earlier for the ODA and other national policies. They also provide education, and several also promote or support research.[56] Thousands of patient organizations exist worldwide, and many converge into country or continental networks, thus becoming a common voice and stronger advocate for patients. Examples of continental networks are EURORDIS,[57] the Iberoamerican Alliance for Rare Diseases (ALIBER) that encompasses organizations in the Iberian Peninsula and Latin America,[58] and the African Alliance for Rare Diseases.[59]

Rare Disease Day is a yearly global awareness and educational campaign that takes place on the last day of February (and falls, every 4 year on the rare date of February 29) since 2008 and is organized by EURORDIS. Among other resources, videos with patient stories and other educational materials are made available and translated to over 20 languages. For the 2019 version, activities occurred in more than 100 countries in all continents.[60]

3.7 DATA SHARING

Large cohort studies interrogate hundreds to thousands of people independently, generating a vast collection of hypothesis-specific knowledge. In a world eager to see the completion of the promise of personalized medicine, where the Internet connects us all, it makes no sense for disconnected knowledge to exist. Combining multiple-cohort genomic studies to assist genetic testing might sound as simple as joining datasets, but in practice, it comprises a myriad of technical, social, and ethical issues. For example, as mentioned earlier, different groups can choose from a variety of analysis pipelines, and experiments not necessarily were performed under the same experimental conditions, so heterogeneity complicates this task.

Pioneering the "data aggregation" initiative, the ExAC and later the gnomAD projects made available data from 125,000 exomes and 15,000 genomes from different population and disease-specific studies (14 in ExAC and expanded to 50 in gnomAD), processed them through a common analytical pipeline, and generated an invaluable resource set to serve as an allele frequencies reference.[14] All the data is available under an ODC Open Database License, where "you are free to share and modify the gnomAD data so long as you attribute any public use of the database, or works produced from the database; keep the resulting datasets open; and offer your shared or adapted version of the dataset under the same ODbL license" (https://gnomad.broadinstitute.org/about). The authors stated that "ExAC was made possible by the willingness of multiple large disease-focused consortia to share their raw data," and that modern computational resources made it possible today to harmonize the analysis pipeline efficiently. They finalized their main communication of results stating that a larger dataset would only be possible if the emphasis continues on the value of genomic data sharing.[14] ExAC/gnomAD are now integrated in clinical analysis pipelines to facilitate variant prioritization based on their occurrence in the human population, something that previous databases such as the 1000 Genomes Project was not best suitable to do given its smaller sample size.

Concomitantly to the ExAC initiative, the Global Alliance for Genomics and Health (GA4GH) was launched in 2013 to "accelerate the potential of research and medicine to advance human health."[61] Their mission is to generate a framework of standards to share genomic and health-related data sharing responsibly. They are grouped into two main work streams: (1) foundational, in charge of generating standards for data security and regulatory and ethical aspects of data sharing, and (2) technical, including guidelines for clinical and phenotypical data capture. Unlike ExAC and gnomAD, GA4GH is not a data repository nor aims to be, it instead offers a set of country-free guidelines to guarantee responsible data sharing among participants. Similarly, the "Managing Ethico-social, Technical and Administrative issues in Data ACcess" (METADAC) initiative of the United Kingdom developed and maintains the administrative, technical activities, and policies to optimize the access and use of scientific resources of the United Kingdom, such as genomic data from large genomic cohorts [62].

Phenotype and patient data sharing among colleagues and collaborators locally or at conferences, something intuitively done by clinicians and researchers across the globe, has now been proven to be a valuable mean to accelerate genomic medicine. The database of genomic variation and phenotype in humans using Ensembl resources took this idea and converted it into a public resource where researchers and clinicians can interact securely.[63] Another initiative is Matchmaker Exchange aimed at facilitating the communication between phenotypic and genomic databases and search for patients with similar phenotypes and variants.[64]

Considering all initiatives, there is no doubt that this culture of connecting clinicians, researchers, and patients through secure frameworks has definitely helped increasing diagnostic rate in patients who would otherwise be considered unique in isolation.[26] However, stricter regulations in personal data protection, such as the EU General Data Protection Regulation,[65,66] might deaccelerate data exchange between different parts of the world, but discussions and framework proposals generated by policy makers' international genomic consortia and the general public on how to best adapt to the digital world can only ultimately benefit us all.

3.8 RARE DISEASE IN DEVELOPING COUNTRIES

The high prevalence of infectious and malnutrition diseases, and the low incidence of individual RD, the high costs involved in diagnosis and therapies, and the limited resources for clinical research pose significant difficulties to the RD communities in low- and middle-income countries (LMICs; see also Chapter 20: Regulatory Aspects of Genomic Medicine and Pharmacogenomics). In fact, countries in Africa and Latin America are not current members of the large international collaborations mentioned above. This contributes to the health-care disparities between regions in the world. Nevertheless, efforts are gradually being made: as described in Section 3.4, some countries are gradually implementing policies and programs, many of them as a result of the advocacy of patient associations.

HTS technologies are scarce and, if available, expensive in LMICs; the price of WGS in developed countries is frequently quoted at US$1000 or less,[4] but this is far from reality in LMICs, even when technology is available, due to lower volumes and high import, tax, and shipping costs. One alternative for countries in this scenario is the purchase of sequencing services abroad, which substantially decreases costs but impede the development of local capacities and developments. Others have developed more targeted, but still broad approaches to diagnoses, such as using proband-only clinical exome sequencing (the 6000 or so protein-coding genes) instead of trio WES, resulting in similar diagnostic yield but lower cost than the latter option.[67] Another alternative would be the creation of regional networks of RD to optimize use of resources, share capacities, and facilitate bulk purchases. One successful example in the area of common disease prevention and treatment is the PanAmerican Health Organization Revolving Fund for immunizations where "Member States pool their national resources to procure high-quality life-saving vaccines and related products at the lowest price."[68] This model is also being used for therapies for cervical cancer in the region and could be applied to diagnostic and treatment tools for persons with RDs.

3.9 CONCLUDING REMARKS

RDs represent a substantial global public health problem, both in frequency (despite their name) and in severity. Genomic and bioinformatic technologies are changing the diagnostic landscape,

Table 3.1 Selected List of Useful Online Rare disease Resources for Families, Clinicians, and Researchers

Resource	URL
1000 Genomes	http://www.1000genomes.org
African Alliance for Rare Diseases	http://africa-rare.org
ClinGen	http://www.clinicalgenome.org
ClinVar	http://www.ncbi.nl.nih.gov/clinvar
DECIPHER	https://decipher.sanger.ac.uk
EURORDIS	http://www.eurordis.org
Exome Aggregation Consortium	http://exac.broadinstitute.org
GeneMatcher	http://genematcher.org
GeneReviews	https://www.ncbi.nlm.nih.gov/books/NBK1116/
Genomics England	https://www.genomicsengland.co.uk
Global Alliance for Genomics and Health	http://genomicsandhealth.org
gnomAD	http://gnomAD.broadinstitute.org
Human Phenotype Ontology	http://www.human-phenotype-ontology.org
Iberoamerican Alliance for Rare Disease	http://aliber.org
International Rare Diseases Research Consortium	http://www.irdirc.org
Matchmaker Exchange	http://www.matchmakerexchange.org
National Organization for Rare Disorders	https://rarediseases.org
Online Mendelian Inheritance in Man	http://omim.org
Orphanet	http://www.orpha.net
PhenoDB	http://phenodb.org
POSSUM	http://www.possum.net.au
Rare Disease Day	http://www.rarediseaseday.org
Undiagnosed Diseases Network	http://undiagnosed.hms.harvard.edu
Undiagnosed Diseases Network International	http://www.udninternational.org
UNIQUE	https://www.rarechromo.org/

and novel therapies are being developed. International collaboration and data sharing have also been fundamental to advance the field. Some countries have more or less comprehensive public policies. Nevertheless, substantial barriers remain for the majority of patients, and greater efforts are required reach the World Health Organization's universal goals of equity in access, quality care, and reduction of financial risk for persons with RDs. A selected list of useful online RD resources for families, clinicians, and researchers is provided in Table 3.1. From the genomics perspective, RDs are a cornerstone of personalized medicine; understanding of diseases mechanisms will continue to shed light on fundamental processes in biology, with the hope that knowledge will transform care for patients and their families.[6]

FUNDING

Fondecyt-Chile grants 1171014 (GMR) and 3170290 (BR-J).

REFERENCES

1. *Orphanet*. <http://www.orpha.net> Accessed 02.03.19.
2. Dharssi S, Wong-Rieger D, Harold M, Terry S. Review of 11 national policies for rare diseases in the context of key patient needs. *Orphanet J Rare Dis*. 2017;12(1):63. Available from: https://doi.org/10.1186/s13023-017-0618-0.
3. Angelis A, Tordrup D, Kanavos P. Socio-economic burden of rare diseases: a systematic review of cost of illness evidence. *Health Policy (New York)*. 2015;119(7):964–979. Available from: https://doi.org/10.1016/j.healthpol.2014.12.016.
4. Shendure J, Findlay GM, Snyder MW. Genomic medicine–progress, pitfalls, and promise. *Cell*. 2019;177(1):45–57. Available from: https://doi.org/10.1016/j.cell.2019.02.003.
5. *EURORDIS*. Survey of the delay in diagnosis for 8 rare diseases in Europe ('*EURORDISCARE 2*'). <www.eurordis.org> Accessed 04.03.19.
6. Boycott KM, Rath A, Chong JX, et al. International cooperation to enable the diagnosis of all rare genetic diseases. *Am J Hum Genet*. 2017;8(25). Available from: https://doi.org/10.1016/j.ajhg.2017.04.003.
7. Jamuar SS, Tan E-C. Clinical application of next-generation sequencing for Mendelian diseases. *Hum Genomics*. 2015;9(1):10. Available from: https://doi.org/10.1186/s40246-015-0031-5.
8. Adams DR, Eng CM. Next-generation sequencing to diagnose suspected genetic disorders. *N Engl J Med*. 2018;379(14):1353–1362. Available from: https://doi.org/10.1056/NEJMra1711801.
9. Retterer K, Juusola J, Cho MT, et al. Clinical application of whole-exome sequencing across clinical indications. *Genet Med*. 2016;18(7):696–704. Available from: https://doi.org/10.1038/gim.2015.148.
10. Reuter JA, Spacek DV, Snyder MP. High-throughput sequencing technologies. *Mol Cell*. 2015;58(4):586–597. Available from: https://doi.org/10.1016/j.molcel.2015.05.004.
11. Van der Auwera GA, Carneiro MO, Hartl C, et al. From FastQ data to high confidence variant calls: the Genome Analysis Toolkit best practices pipeline. *Curr Protoc Bioinf*. 2013;43:11.10.1–11.10.33. Available from: https://doi.org/10.1002/0471250953.bi1110s43.
12. 1000 Genomes Project Consortium RA, Auton A, Brooks LD, et al. A global reference for human genetic variation. *Nature*. 2015;526(7571):68–74. Available from: https://doi.org/10.1038/nature15393.
13. Yang H, Wang K. Genomic variant annotation and prioritization with ANNOVAR and wANNOVAR. *Nat Protoc*. 2015;10(10):1556–1566. Available from: https://doi.org/10.1038/nprot.2015.105.

14. Lek M, Karczewski KJ, Minikel EV, et al. Analysis of protein-coding genetic variation in 60,706 humans. *Nature*. 2016;536(7616):285–291. Available from: https://doi.org/10.1038/nature19057.
15. McKusick-Nathans Institute of Genetic Medicine, (Johns Hopkins University). OMIM – Online Mendelian Inheritance in Man. <https://www.omim.org/>; 2016 Accessed 09.09.16.
16. Landrum MJ, Lee JM, Benson M, et al. ClinVar: improving access to variant interpretations and supporting evidence. *Nucleic Acids Res*. 2018;46(D1):D1062–D1067. Available from: https://doi.org/10.1093/nar/gkx1153.
17. Richards S, Aziz N, Bale S, et al. Standards and guidelines for the interpretation of sequence variants: a joint consensus recommendation of the American College of Medical Genetics and Genomics and the Association for Molecular Pathology. *Genet Med*. 2015;17(5):405–424. Available from: https://doi.org/10.1038/gim.2015.30.
18. Mestek-Boukhibar L, Clement E, Jones WD, et al. Rapid Paediatric Sequencing (RaPS): comprehensive real-life workflow for rapid diagnosis of critically ill children. *J Med Genet*. 2018;55(11):721–728. Available from: https://doi.org/10.1136/jmedgenet-2018-105396.
19. Köhler S, Vasilevsky NA, Engelstad M, et al. The human phenotype ontology in 2017. *Nucleic Acids Res*. 2017;45(D1):D865–D876. Available from: https://doi.org/10.1093/nar/gkw1039.
20. Hwang S, Kim E, Lee I, Marcotte EM. Systematic comparison of variant calling pipelines using gold standard personal exome variants. *Sci Rep*. 2015;5(1):17875. Available from: https://doi.org/10.1038/srep17875.
21. Laurie S, Fernandez-Callejo M, Marco-Sola S, et al. From wet-lab to variations: concordance and speed of bioinformatics pipelines for whole genome and whole exome sequencing. *Hum Mutat*. 2016;37(12):1263–1271. Available from: https://doi.org/10.1002/humu.23114.
22. Gézsi A, Bolgár B, Marx P, Sarkozy P, Szalai C, Antal P. VariantMetaCaller: automated fusion of variant calling pipelines for quantitative, precision-based filtering. *BMC Genomics*. 2015;16(1):875. Available from: https://doi.org/10.1186/s12864-015-2050-y.
23. Taber JM, Klein WMP, Lewis KL, Johnston JJ, Biesecker LG, Biesecker BB. Reactions to clinical reinterpretation of a gene variant by participants in a sequencing study. *Genet Med*. 2018;20(3):337–345. Available from: https://doi.org/10.1038/gim.2017.88.
24. Chisholm C, Daoud H, Ghani M, et al. Reinterpretation of sequence variants: one diagnostic laboratory's experience, and the need for standard guidelines. *Genet Med*. 2018;20(3):365–368. Available from: https://doi.org/10.1038/gim.2017.191.
25. Vears DF, Niemiec E, Howard HC, Borry P. Analysis of VUS reporting, variant reinterpretation and recontact policies in clinical genomic sequencing consent forms. *Eur J Hum Genet*. 2018;26(12):1743–1751. Available from: https://doi.org/10.1038/s41431-018-0239-7.
26. Nambot S, Thevenon J, Kuentz P, et al. Clinical whole-exome sequencing for the diagnosis of rare disorders with congenital anomalies and/or intellectual disability: substantial interest of prospective annual reanalysis. *Genet Med*. 2018;20(6):645–654. Available from: https://doi.org/10.1038/gim.2017.162.
27. SoRelle JA, Thodeson DM, Arnold S, Gotway G, Park JY. Clinical utility of reinterpreting previously reported genomic epilepsy test results for pediatric patients. *JAMA Pediatr*. 2018;173(1):e182302. Available from: https://doi.org/10.1001/jamapediatrics.2018.2302.
28. Farrar MA, Park SB, Vucic S, et al. Emerging therapies and challenges in spinal muscular atrophy. *Ann Neurol*. 2017;81(3):355–368. Available from: https://doi.org/10.1002/ana.24864.
29. Mah JK. An overview of recent therapeutics advances for Duchenne muscular dystrophy. *Methods Mol Biol (Clifton, N.J.)*. 2018;1687:3–17. Available from: https://doi.org/10.1007/978-1-4939-7374-3_1.
30. *Orphan Drug Act*. <https://www.govinfo.gov/content/pkg/STATUTE-96/pdf/STATUTE-96-Pg2049.pdf>; 1983.
31. The balancing act of orphan drug pricing. *Lancet*. 2017;390(10113):2606. <https://doi.org/10.1016/S0140-6736(17)33305-6>.
32. Thomas S, Caplan A. The Orphan Drug Act Revisited. *JAMA*. 2019. Available from: https://doi.org/10.1001/jama.2019.0290.

33. Sun W, Zheng W, Simeonov A. Drug discovery and development for rare genetic disorders. *Am J Med Genet Part A*. 2017;173(9):2307–2322. Available from: https://doi.org/10.1002/ajmg.a.38326.
34. Park T, Griggs SK, Suh D-C. Cost effectiveness of monoclonal antibody therapy for rare diseases: a systematic review. *BioDrugs*. 2015;29(4):259–274. Available from: https://doi.org/10.1007/s40259-015-0135-4.
35. Chiesa R, Wynn RF, Veys P. Haematopoietic stem cell transplantation in inborn errors of metabolism. *Curr Opin Hematol*. 2016;23(6):530–535. Available from: https://doi.org/10.1097/MOH.0000000000000289.
36. Sinclair A, Islam S, Jones S. Gene therapy: an overview of approved and pipeline technologies. <http://www.ncbi.nlm.nih.gov/pubmed/30855777>; 2016 Accessed 18.03.19.
37. Porteus MH. A new class of medicines through DNA editing. *N Engl J Med*. 2019;380(10):947–959. Available from: https://doi.org/10.1056/NEJMra1800729.
38. Mijnders M, Kleizen B, Braakman I. Correcting CFTR folding defects by small-molecule correctors to cure cystic fibrosis. *Curr Opin Pharmacol*. 2017;34:83–90. Available from: https://doi.org/10.1016/j.coph.2017.09.014.
39. Levin AA. Treating disease at the RNA level with oligonucleotides. *N Engl J Med*. 2019;380(1):57–70. Available from: https://doi.org/10.1056/NEJMra1705346.
40. Pushpakom S, Iorio F, Eyers PA, et al. Drug repurposing: progress, challenges and recommendations. *Nat Rev Drug Discov*. 2018;18(1):41–58. Available from: https://doi.org/10.1038/nrd.2018.168.
41. Ministry for Solidarity and Health and Ministry for Higher Education, Research and Innovation. *French National Plan for Rare Diseases 2018-2022*. <https://solidarites-sante.gouv.fr/IMG/pdf/pnmr3_-_en.pdf>; 2018 Accessed 10.03.19.
42. Ferrelli RM, De Santis M, Egle Gentile A, Taruscio D. Health systems sustainability and rare diseases. *Adv Exp Med Biol*. 2017;1031:629–640. Available from: https://doi.org/10.1007/978-3-319-67144-4_33.
43. *European Commission*. Rare diseases. European Commission. <https://ec.europa.eu/health/non_communicable_diseases/rare_diseases_en> Accessed 28.01.19.
44. *Genomics England*. 100,000 Genomes Project. <https://www.genomicsengland.co.uk/> Accessed 19.03.17.
45. Genomics Education Programme. <https://www.genomicseducation.hee.nhs.uk/> Accessed 28.01.19.
46. Gahl WA, Mulvihill JJ, Toro C, et al. The NIH Undiagnosed Diseases Program and Network: applications to modern medicine. *Mol Genet Metab*. 2016;117(4):393–400. Available from: https://doi.org/10.1016/j.ymgme.2016.01.007.
47. Splinter K, Adams DR, Bacino CA, et al. Effect of genetic diagnosis on patients with previously undiagnosed disease. *N Engl J Med*. 2018;379(22):2131–2139. Available from: https://doi.org/10.1056/NEJMoa1714458.
48. Cui Y, Zhou X, Han J. China launched a pilot project to improve its rare disease healthcare levels. *Orphanet J Rare Dis*. 2014;9:14. Available from: https://doi.org/10.1186/1750-1172-9-14.
49. Adachi T, Kawamura K, Furusawa Y, et al. Japan's initiative on rare and undiagnosed diseases (IRUD): towards an end to the diagnostic odyssey. *Eur J Hum Genet*. 2017;25(9):1025–1028. Available from: https://doi.org/10.1038/ejhg.2017.106.
50. *Australian Genomics*. Australian Genomics Health Alliance | Home. <https://www.australiangenomics.org.au/> Accessed 28.01.19.
51. Long JC, Pomare C, Best S, et al. Building a learning community of Australian clinical genomics: a social network study of the Australian Genomic Health Alliance. *BMC Med*. 2019;17(1):44. Available from: https://doi.org/10.1186/s12916-019-1274-0.
52. Ricarte Soto – Ley 20.850 - Ministerio de Salud - Gobierno de Chile. <https://www.minsal.cl/leyricarte/> Accessed 30.01.19.
53. FONASA – Ley Ricarte Soto. <http://leyricartesoto.fonasa.cl/> Accessed 03.04.17.

54. El Mercurio.com – Blogs: Decenas de pacientes se alistan para exigir en tribunales cobertura de fármacos de alto costo. <http://www.elmercurio.com/blogs/2018/11/09/64719/Decenas-de-pacientes-se-alistan-para-exigir-en-tribunales-cobertura-de-farmacos-de-alto-costo.aspx> Accessed 20.03.19.
55. Taruscio D, Groft SC, Cederroth H, et al. Undiagnosed diseases network international (UDNI): white paper for global actions to meet patient needs. *Mol Genet Metab*. 2015;116(4):223–225. Available from: https://doi.org/10.1016/j.ymgme.2015.11.003.
56. Pinto D, Martin D, Chenhall R. The involvement of patient organisations in rare disease research: a mixed methods study in Australia. *Orphanet J Rare Dis*. 2016;11(1):2. Available from: https://doi.org/10.1186/s13023-016-0382-6.
57. *European Organisation for Rare Disorders*. Eurordis Rare Disease Europe. <http://www.eurordis.org/sites/default/files/publications/princeps_document-EN.pdf. www.eurodis.org>; 2005.
58. ALIBER – Iberoamerican Alliance for Rare Diseases. <https://aliber.org/web/en/> Accessed 20.03.19.
59. Africa-Rare – Moving Rare Diseases Forward in Africa. <http://africa-rare.org/> Accessed 21.03.19.
60. Rare Disease Day ® 2019. <https://www.rarediseaseday.org/> Accessed 06.03.19.
61. Global Alliance for Genomics and Health. <https://www.ga4gh.org/>.
62. Murtagh MJ, Blell MT, Butters OW, et al. Better governance, better access: practising responsible data sharing in the METADAC governance infrastructure. *Hum Genomics*. 2018;12(1):24. Available from: https://doi.org/10.1186/s40246-018-0154-6.
63. Firth HV, Richards SM, Bevan AP, et al. DECIPHER: database of chromosomal imbalance and phenotype in humans using Ensembl resources. *Am J Hum Genet*. 2009;84(4):524–533. Available from: https://doi.org/10.1016/j.ajhg.2009.03.010.
64. Sobreira NLM, Arachchi H, Buske OJ, et al. *Matchmaker exchange. Current Protocols in Human Genetics*. vol. 95. Hoboken, NJ: John Wiley & Sons, Inc; 2017:9.31.1–9.31.15. Available from: http://doi.org/10.1002/cphg.50.
65. EUGDPR – Information Portal. <https://eugdpr.org/> Accessed 19.03.19.
66. Phillips M. International data-sharing norms: from the OECD to the General Data Protection Regulation (GDPR). *Hum Genet*. 2018;137(8):575–582. Available from: https://doi.org/10.1007/s00439-018-1919-7.
67. Hu X, Li N, Xu Y, et al. Proband-only medical exome sequencing as a cost-effective first-tier genetic diagnostic test for patients without prior molecular tests and clinical diagnosis in a developing country: the China experience. *Genet Med*. 2018;20(9):1045–1053. Available from: https://doi.org/10.1038/gim.2017.195.
68. PAHO/WHO | PAHO Revolving Fund. <https://www.paho.org/hq/index.php?option = com_content&view = article&id = 1864:paho-revolving-fund&Itemid = 4135&lang = en> Accessed 22.03.19.

CHAPTER

APPLIED GENOMICS AND PUBLIC HEALTH CANCER GENOMICS

4

Pierluigi Porcu[1], Gaurav Kumar[2], Nitin Chakravarti[1], Neeraj Arora[3], Anjali Mishra[1], Adam Binder[1], Adam Ertel[2] and Paolo Fortina[2,4]

[1]Division of Hematologic Malignancies and Hematopoietic Stem Cell Transplantation, Department of Medical Oncology, Sidney Kimmel Cancer Center, Thomas Jefferson University, Philadelphia, PA, United States [2]Department of Cancer Biology, Sidney Kimmel Cancer Center, Thomas Jefferson University, Philadelphia, PA, United States [3]Department of Pathology, Tata Medical Centre Kolkata, Kolkata, India [4]Department of Translation and Precision Medicine, Sapienza University, Rome, Italy

4.1 INTRODUCTION

Cancer is now well understood as a "disease of the genome," arising and evolving in the context of the cumulative accrual of somatically inherited, primary (driver) and secondary (passenger) DNA mutations in cancer stem or progenitor cells.[1] Driver genomic events in one or more cells lead to a sequence of radical changes in cell function, such as impaired differentiation, resistance to apoptosis, loss of response to inhibitory signals, acquisition of metastatic potential, metabolic reprogramming, and immune evasion, which in aggregate define the neoplastic phenotype and have therefore been defined as the "hallmarks of cancer."[2–4] The contributing role of inherited germ-line variants in directly causing or increasing the risk of malignancy, long recognized in pediatric cancers, is also gradually being revealed in adults.[5] Finally, inheritable epigenetic aberrations, leading to a disruption of the posttranslational histone modification code and reorganization of chromatin structure and architecture, are extremely common in human tumor cells, and transcriptional dysregulation, often due to enhancer dysfunction and epigenetic reprogramming, has been identified as an additional functional hallmark of cancer.[6–9]

According to the World Health Organization (WHO), cancer is responsible for one in six deaths globally, making it the second leading cause of deaths.[10] In 2018 alone, ~18 million new cancer cases were diagnosed worldwide, and 9.6 million cancer deaths were reported. It is now well established that even though the high incidence and mortality of cancer depend in significant part on lifestyle, socio-economic status, environmental factors, and access to care, racial and ethnic backgrounds are also very important elements that impact overall cancer risk and outcomes. Differences in germ-line and somatic mutations in different races, sexes, and ethnic groups are directly linked to mutation frequencies, which are often associated with cancer subtype, prognosis, and treatment response.[11]

The development of massively parallel sequencing technologies, also referred to as next-generation sequencing (NGS), combined with advanced computational data analysis, allowing

Applied Genomics and Public Health. DOI: https://doi.org/10.1016/B978-0-12-813695-9.00004-2

high-throughput base-level DNA mapping of whole tumor genomes, now also possible at the single cell level,[12,13] has ushered in a new era of "cancer genomics," which can be defined as the characterization of the differences in the totality of DNA sequence and gene expression in tumor cells, as compared to normal cells. Cancer genomics has transformed our understanding of the genetic causes of tumorigenesis[14] enabling unbiased discovery of oncogenes and tumor suppressors and also vastly expanded our view of tumor heterogeneity and evolution[15–17] and is increasingly being applied to individual patient's care, under the heading "precision or personalized medicine."[18] However, the application of cancer genomics to the field of public health, which is concerned with the translation of research into health benefits at the population level, has not yet been broadly adopted.[19] Furthermore, the study of cancer genomics across ethnic and racial groups, including vulnerable populations, which is an important goal of public health, remains in its infancy.[11,20–22]

The present chapter provides a brief overview of the national (United States) and global burden of cancer, with an emphasis on hematologic malignancies, including geographical, racial, and ethnic disparities, which are the focus of ongoing public health efforts worldwide. We then provide a concise summary of the genetic and epigenetic foundations of cancer that are relevant to the utilization of genomic analysis tools to the field of public health and describe specific applications of NGS for the characterization of cancer's mutational and epigenetic landscapes, including functional genomics studies. We also discuss details of the NGS-based assays employed to study different cancer "omics" dimensions, such as genomics, epigenomics, and transcriptomics, along with the computational workflow of their data analysis. Finally, we discuss newer "omics" platforms, including strategies for their integration, in the clinic and relevance to public health.

4.2 PUBLIC HEALTH AND THE GLOBAL BURDEN OF CANCER

Over the last 30 years, there has been a significant progress in medicine such that communicable, maternal, neonatal, and nutritional-related mortality has declined and as of 2017, noncommunicable diseases are two of the three leading causes of death worldwide.[23–25] In addition to technological and pharmaceutical advances, health-care delivery and logistical solutions have also improved, providing better access to medical care for remote populations.[26,27] Average global life expectancy is rising and, consequently, modifiable risk factors for noncommunicable diseases (i.e., smoking) are being adopted at increasing rates in various parts of the world. As a result, the incidence of noncommunicable, chronic diseases, such as chronic obstructive pulmonary disease (COPD), diabetes, chronic kidney disease, and cancer, is increasing.[28] This is most concerning in low- and middle-income countries (LMICs) where cancer incidence is rising most rapidly, but where the infrastructure and financial resources available to treat patients are less established. Presently, 46% of all new cancers are diagnosed in China, Russia, and India, and these three countries which combined 40% of the world's population experience 52% of all cancer-related deaths.[29] In 2017 the WHO updated its cancer resolution, which recognizes that cancer is a leading cause of morbidity and mortality globally. This resolution provides a unifying framework through which incremental steps may be followed to reduce the burden of the disease.[30]

To address the global burden of cancer from a public health perspective, including the potential application and impact of genomic studies, a general roadmap is vital to guide governments and institutions. A large integrative network is necessary to develop robust cancer registry tools that can be used to drive local, regional, national, and international policies that will allow for optimal care of the patient throughout the cancer continuum; from prevention strategies to end-of-life care. While governmental support is crucial to a "cancer control" plan,[31] local health-care delivery strategies must be unique and personalized to the patient population due to cultural/social interpretation of disease and regional variation of cancer incidence due to environmental exposures, endemic viruses, and racial, ethnic, and gender diversity. As a result, a roadmap detailing broad goals, such as increasing screening programs, early and effective treatment, high value low-cost therapeutic interventions, and reducing modifiable risk factors, is important, but it may differ at the national, regional, and local levels. This variation will be one of the largest hurdles to overcome the rising burden of cancer globally as interventions will be limited by their scalability. It will require the effort of individual nations to understand the epidemiology of cancer and the social barriers to health-care implementation within their borders in order to effectively design and implement effective treatment strategies.[32,33]

Before the development of any intervention program, such as genetic screening for risk-assessment purposes, countries must ensure that they will be able to reach a large portion of the population. The screening technique should be safe, cost-effective, and equitable. In addition, resources must exist to address the increase in demand for services. There must be the appropriate expertise to diagnose, stage, and treat any new findings from the screening program, including those derived from genomic assays. Without appropriate treatment capabilities, and the counseling infrastructure and resources needed to communicate to patients the meaning and impact of mutations, screening programs will not add the intended value of early interventions and improve outcomes. While some screening programs will require a global effort, such as breast cancer, cervical cancer, and colon cancer screening, others will require a clear understanding of regional cancer epidemiology. Screening for gastric cancer, for example, has shown to be effective in Japan; however, it has not been proven to be broadly applicable in the United States.[34,35] Similarly, the incidence of lung cancer varies greatly globally. About one-third of all cases of lung cancer occur in China and 50% of the cases occur in LMICs.[36,37] However, some regions with many LMICs, such as sub-Saharan Africa, have some of the lowest rates of lung cancer worldwide. Therefore implementing a lung cancer screening program with low-dose computerized axial tomography (CAT) scans, while reducing mortality by 20% in high-risk populations,[38] must be carefully considered, given the cost and technological requirements. While not standard of care, genetic risk scores are also being studied in order to further risk stratify patient populations and provide increasingly high value care.[39,40]

Similarly, prevention strategies will combine both global and also local interventions. Vaccination programs for hepatitis B and human papilloma viruses can be successful at reducing various malignancies including hepatocellular carcinoma, cervical cancer, as well as head and neck cancer. However, the implementation of vaccine programs requires complex integration of governmental support, health-care infrastructure, cold chain delivery, robust logistical services, and overcoming social barriers or preconceived falsehoods of vaccination programs.[41,42] In addition, even in LMICs, there are clear health disparities based on income and success of vaccination strategies that must be addressed and overcome for a successful cancer prevention program.[42]

Local prevention strategies should be geared toward cancers related to high-risk activities of the community all the while considering social and cultural influences that drive these practices.[43] For example, various strategies have been implemented to reduce the use of betel quid, which is mainly used in the Asia-Pacific region and is a class 1 carcinogen and results in high incidences of oral cancer.[44,45] Another local approach is to coordinate efforts with legislative systems to change behavior and reduce modifiable risk factors. All these interventions coordinated public health strategies focusing on decreasing cancer incidence.

The rising global cancer burden will be a public health challenge globally both in high- and low-income countries. While the cost of care will likely be more limiting in LMICs, implementation and acceptance of equitable, safe, effective prevention, and screening initiatives are challenging in any setting. There should be clear unifying themes and goals, while structuring programs to the specific populations' racial, ethnic, cultural, and genetic diversity. Only through this complex integrative strategy will nations successfully reduce the overall burden of disease. To be effective, the global introduction and integration of genomic technologies in public health, in addition to cost and infrastructure considerations, will require a detailed working knowledge of each country and each population's social and cultural landscapes.

4.3 CANCER AS A GENETIC AND EPIGENETIC DISEASE

More than a century of investigative work supports the conclusion that cancer is primarily a genetic disease of somatic cells. The risk of developing cancer increases steadily with age. Exposure to mutagens, such as ultraviolet light, radiation, and DNA-damaging physical or chemical agents, significantly adds to cancer risk,[46] and individuals with inherited or acquired defects in DNA repair have an extremely high risk of developing cancer.[47] These observations all support the conclusion that the somatic mutational burden is a key driver of oncogenesis. The clonal evolution theory of cancer, according to which cancer is an evolutionary process driven by stepwise somatic cell mutations, with sequential subclonal selection, has been criticized, revised, and updated,[48–50] after its initial formulation by Nowell in 1976.[51] However, its fundamental tenet that cancer is rooted in the lifelong accrual of mutations in somatic cells, only some of which are evolutionarily advantageous (*driver mutations*), remains a cornerstone of cancer biology and is now increasingly being accepted as a basis for therapeutic failure.[52] The acquired cellular traits that are most competitive from an evolutionary standpoint include enhanced cell proliferation, escape from apoptosis, immune evasion, and acquisition of stem cell–like properties. However, advantageous traits may vary depending on the phases of the natural history of the disease (early development, progression, acceleration and transformation, posttreatment relapse, etc.). This principle also continues to inform studies aimed at defining and interpreting the molecular landscape of human cancer for diagnostic and therapeutic purposes and is increasingly used to study cancer at the population level. The prediction by Nowell in 1976,[51] "One may ultimately have to consider each advanced malignancy as an individual therapeutic problem, after as many cells as possible have been eliminated through the nonspecific modalities of surgery, radiation and chemotherapy," may only need to be revised to affirm that an individualized approach to cancer therapy and patient care is in fact desirable from the beginning.

A recent example of the impact of the somatic evolution theory of cancer on our investigation and understanding of hematologic disorders, with a major public health impact, is the discovery of the phenomenon called clonal hematopoiesis of indeterminate potential (CHIP),[52–55] in which hematopoietic stem cells (HSCs) or early blood cell progenitors in otherwise healthy individuals contribute to the formation of genetically distinct subclones, characterized by one or more shared mutations, most commonly affecting *DNMT3a*, *TET2*, *ASXL1*, *JAK2*, *SF3B1/2*, and *TP53*. Some of these mutations are *bona fide* driver mutations and likely contribute directly to the development of the clone. Others may be selected via mechanisms, such as neutral drift, in the stem cell population.[56] Clones most typically have mutations in a single gene, but in a significant number of cases, mutations in two or more genes are found. Clones can vary in size from 2% to 100% of peripheral blood mononuclear cells. The prevalence of CHIP increased dramatically with age: it is detected in less than 1% of the population under age 40 but ~10%–20% of the population over age 70. CHIP is associated with a 10-fold increased risk of developing myelodysplastic syndromes (MDS) or acute myeloid leukemia (AML), and of significant impact for public health, and also leads to an increased risk of cardiovascular disease.[57]

NGS techniques have revolutionized in human cancer genomics, offering the capability to comprehensively analyze the profiles (landscapes) of thousands of clinical tumor samples,[58] characterize genome-wide alterations, detect mutation signatures, and discover genetic and genomic mechanisms that lead to drug resistance.[59] The establishment of vast collaborative research infrastructures, such as The Cancer Genome Atlas (TCGA) in 2005 and the International Cancer Genome Consortium (ICGC) in 2008, has accelerated the comprehensive understanding of the genetics of cancer and helped in generating new cancer therapies and diagnostic methods.

Despite these advances, many unsolved questions in cancer genomics remain, posing a major challenge to the application of this knowledge to patient care, both at the individual (personalized medicine) and at the population (public health) levels.[60] For cancer prevention, which hinges on accurate risk assessment, we still do not have good estimates of the number of mutations required to induce and drive most cancers and how the diversity of the spectrum and rates of mutations across tumor types may impact risk.[61] The discovery of a number of mutation-rate modifiers, such as transcription-induced strand bias[62] and epigenetic changes, including variations in chromatin structure,[63,64] adds to this complexity. Another limitation is the uncertainty on how to define driver mutations, and what the relative importance of distinct driver mutations in the same cancer might be. Age-incidence curves have been used to estimate the number of rate-limiting steps required for a cancer to develop.[65] However, this model assumes a one-to-one correspondence between rate-limiting steps and driver mutations, and this is not always true. Some driver mutations are not rate limiting[66] and not all rate-limiting genetic events are driver mutations.[61] The identity and number of driver mutations have often been inferred from the rate of mutations affecting known cancer genes, with typical mutational landscapes showing a handful of high-frequency (>50%) mutations, followed by a "long tail" of low-frequency (<10%) mutations that are presumed to be passengers, and not drivers. This approach, however, is limited by the fact that our knowledge of cancer driver genes is far from complete, passenger mutations can affect known cancer genes, and not all passenger mutations may be functionally irrelevant.

Another critically important area of genomic investigation, from the public health standpoint, is how therapeutic interventions shape the cancer genome. Antineoplastic therapies apply negative and positive selection pressures to variable fractions of the tumor cell population. For example, Campbell et al.[67] have shown that in many cancers, prior alkylating agent therapy with temozolomide leads to a late "hypermutation" signature, with a dramatic increase in mutation burden. This may have both a positive and a negative impact on patient outcomes. On one hand, the presence of these mutations may accelerate the development of resistant clones, particularly in the presence of a selective environment characterized by repeated exposure to the same chemotherapy.[68] On the other hand, a high mutational burden has been shown to enhance response to checkpoint blockade[69,70] and may potentiate the clinical benefit of immunotherapeutic approaches.

Thus even though the fact that major advances in high-throughput technologies have led to the sequencing of thousands of cancer genomes, therefore producing cancer landscape maps of unprecedented definition, we still have not identified the key driver mutations for most cancers, and this remains the primary, sometimes crippling, limiting factor in our efforts to develop truly "targeted" and personalized therapies. In this sense, while the introduction of NGS has transformed our understanding of the genomic "anatomy" of cancer, the real clinical impact will come only when we understand the genomic "pathophysiology" of cancer. Functional cancer genomics, defined as the use of NGS-based techniques to characterize not only the DNA sequence of the tumor genome but also the full transcriptional landscape (transcriptome) of the same tumor, often in association with chromatin accessibility and occupancy studies,[71,72] will therefore become increasingly important for cancer genomics to productively and efficiently move into clinical practice, where therapeutic decisions for individual patients critically depend on the accurate identification of the specific mutations driving each patient's cancer. These approaches can now be enhanced by methodological advances that have accelerated the functional interrogation of individual cancer-associated alterations within in vivo models. For example, the emergence of CRISPR-Cas9-based strategies to rapidly generate increasingly complex somatic alterations and the development of multiplexed and quantitative approaches to ascertain gene function in vivo will certainly lead to a better understanding of the relative in vivo impact of each aberration.

In that regard, a key part of understanding the functional genomics of cancer is the characterization of the role of epigenetic dysregulation in promoting oncogenesis and driving diversity within tumors. Changes in DNA sequence, such as point mutations and indels, translocations, and copy number variations (CNVs), result in the activation of oncogenic genes and in the silencing of tumor-suppressor genes (TSGs). Genomic instability, driven by deficiencies in DNA repair, DNA replication stress, telomere dysfunction, and mitotic aberrations, may cause large-scale abnormal chromosomal aberrations. However, epigenetic alterations, such as DNA methylation, histone modifications, chromatin remodeling, and regulatory loops orchestrated by aberrantly expressed noncoding RNAs [microRNA (miRNA)], can also contribute to the initiation and the progression of cancer.[73] Defects in chromatin modifiers and remodelers have been described in various malignancies, highlighting that epigenetic dysregulation acts in concert with genetic abnormalities in the etiology of both solid and hematological cancer.[74]

4.4 NEXT-GENERATION SEQUENCING TECHNOLOGIES

"Omics" technologies are aimed at the universal detection of DNA structural variations (genomics), regulatory changes (epigenomics), and gene expression (transcriptomics) in an unsupervised manner. These technologies are now playing an important role in screening, diagnosis, and prognosis as well as aiding our understanding of the etiology of diseases. These whole-genome approaches rely on massive parallel sequencing (NGS) methodologies, using a number of different platforms with different technologies. At present, there are five major commercially available NGS technologies. They include Illumina (San Diego, California, www.illumina.com), Pacific Biosciences (Menlo Park, California, www.pacb.com), Oxford Nanopore (Oxford Science Park, United Kingdom, www.nanoporetech.com), Qiagen (Germantown, Maryland, www.qiagen.com/us), and BGI (Shenzhen, PRC, www.bgi.com). All these different technologies aim to sequence the genome accurately and efficiently to facilitate the understanding of health and disease. By implementing any of these technologies with a suitable analytics platform, one can study the many different aspects of "omics."

As discussed, genomics is the systematic study of organisms' whole genomes and is performed for both research and clinical applications. At the research level, the genome is studied for different kinds of variations, such as single-nucleotide variants, indels, structural variants (SVs), and CNVs, using whole-genome sequencing (WGS) and whole-exome sequencing (WES) technologies. The only difference between WGS and WES is that the latter examines only the coding regions of DNA ($\sim$2% of the genome) which are transcribed into mRNA and translated into proteins, while the former examines entire genomes, hence covering all coding and noncoding variants. For a typical research-based sequencing experiment, biological sample (tissue and/or blood) is obtained and processed to prepare sequencing libraries—collections of nucleotide fragments ligated with platform-specific adapter molecules and/or barcodes, which are then sequenced. After sequencing, data are analyzed to obtain the genetic variants followed by annotation with respect to genes and clinical significance using specialized software.

In the field of cancer genomics, WGS and WES facilitate the discovery of cancer risk mutations and prognostic markers that can be developed into targeted NGS assays that, in a manner similar to WES, specifically target only those regions of interest. Several targeted gene panel platforms are commercially available which can provide sequence information for an entire genes or partial sequences of relevant "hot spots" for high-risk variants in genes, such as *ATM*, *BARD1*, *BRCA1*, *BRCA2*, *BRIP1*, *CDH1*, *CHEK2*, *EPCAM*, *MLH1*, *MSH2*, *MSH6*, *NBN*, *PALB2*, *PMS2*, *PTEN*, *RAD51C*, *RAD51D*, *STK11*, *TP53*, and so on. Targeted panels typically use amplicon-based library generation, where gene-specific polymerase chain reaction (PCR) primers enrich the regions of interest, followed by standard sequencing. While this is a reliable approach to identify well-established cancer biomarkers, it can only identify pathogenic or likely pathogenic mutations in 8%−15% of cases, while remaining cases harbor rare or uncharacterized variants that can only be detected by WGS or WES.

Although research applications implementing epigenomics are under development, understanding the impact on chromatin accessibility and gene regulation is a high-interest research area that utilizes NGS technologies. These technologies facilitate better understanding of DNA methylation and nucleosome histone protein modifications by utilizing several techniques including whole-genome bisulfite sequencing, bisulfite-pyrosequencing (methyl-seq), chromatin immunoprecipitation sequencing, Hi-C seq, and assay for transposase accessible chromatin. All these techniques are being utilized to identify gene networks dysregulated by hyper- or hypomethylation of gene promoter regions of DNA in different cancers.

Transcriptomics is a significant tool for understanding the molecular mechanisms of complex diseases, such as cancer. On the research level, it is utilized to identify and quantify expressed mRNA transcripts, alternative transcriptional start sites, splicing variants, and gene fusions. Multigene mRNA signature–based assays are being increasingly incorporated into clinical applications. Like genetic variant targeted gene panels, these assays validate gene expression of several genes to study their implication in cancer. In addition, recently, gene fusions have also been associated with cancer prognosis and are making their way into clinical testing, for example, AML1-ETO fusion testing in AML and TMPRSS2-ERG fusion in prostate cancer.[75]

4.5 FUTURE DIRECTIONS OF NEXT-GENERATION SEQUENCING

Long-read sequencing—The present sequencing technologies can provide massive information from different biological dimensions to understand the molecular basis of disease. However, these current technologies are based on the principle of short-read sequencing and pose computational challenges in covering the complex repetitive regions and large SVs in the genome. To overcome this, long-read sequencing may alleviate computational challenges surrounding genome assembly and transcript reconstruction. This could be useful in identifying variations in the genome that are hard to study like coinherited alleles, haplotype information, and phasing of de novo mutations using short-read sequencing.

Genomic instability is one of the hallmarks of cancer, leading to SVs including CNVs, chromosomal fusions, insertions, deletions, duplications, inversions, or translocations of at least 50 bp in size. These variations can promote cancer initiation and development by creating gene fusions, amplifying oncogenes, or deleting TSGs. Long-read sequencing will facilitate the discovery of new isoforms and novel gene fusions to understand better evolution of the cancer genome.

Expansion from bulk tissues to individual cells—Single-cell technologies are becoming increasingly important tools in biological analysis. Complementing average measurements from whole tissue, single-cell technologies provide a finer-grained picture of the complex biology and may help unmask tissue heterogeneity. It can also reveal complex and rare cell populations, uncover regulatory relationships between genes, and track the trajectories of distinct cell lineages in development. In cancer, it is helping to investigate clonal evolution and intratumor diversity and understanding the role of rare cells in tumor progression. It is also facilitating the understanding of circulating tumor cells (CTCs).

4.6 GENOMIC FINDINGS IN HEMATOLOGIC MALIGNANCIES AND PUBLIC HEALTH

The recently published Global Burden of Disease study shows that the global incidence and mortality figures for hematologic malignancies are increasing.[76] In 2016 there were 467,000 new cases of leukemia and 461,000 new cases of non-Hodgkin's lymphoma (NHL) worldwide, with 310,000 and 240,000 leukemia deaths, respectively. Between 2006 and 2016, the global leukemia incidence increased by 26%, from 370,482 to 466,802 cases, and the global NHL increased by 45%, from 319,078 to 461,164 cases. The factors responsible for this increase included not only an aging and growing population but also rising age-specific incidence rates. In the United States, ~174,000 cases of hematologic malignancies were diagnosed in 2017, incidence trends are also increasing and leukemia, lymphoma, and myeloma in aggregate are the third most common cancer, and the second most common cause of cancer-related death. Even though the fact that leukemia and lymphoma were among the first malignancies to be studied at the genetic level, genomics and personalized medicine have had a relatively low impact on clinical practice in the United States and Europe and have been introduced only in a handful of selected centers in other countries. Thus very little information is available on the comparative molecular landscapes of leukemia, lymphoma, and myeloma from geographically and ethnically distinct parts of the world.

AML was one of the first cancers to be extensively studied and sequenced at the whole-genome level using novel high-throughput microarray and sequencing technologies.[77] Subsequent studies led to the identification of numerous novel recurrent somatic disease alleles with biologic, prognostic, and therapeutic relevances. Most AML NGS panels include a core set of 25–30 genes that are frequently mutated in myeloid neoplasms.[78] Some of these genes are used primarily to classify AML (*NPM1*, *CEBPA*, and *RUNX1*), others are used for risk stratification (*DNMT3A*, *TET2*, *FLT3*, *KIT*, *ASXL1*, *WT1*, and *TP53*); some genes, such as *JAK2*, *MPL*, *CSF3R*, and *CALR*, are used to diagnose myeloproliferative disorders, others (*IDH1*, *IDH2*, and *FLT3*) are used to identify patients for targeted therapy. Genomic variants in the *TET2* and *DNMT3A* genes are used to assess for the presence of CHIP; *SF3B1*, *SRSF2*, *U2AF1*, *ZRSR2*, *ASXL1*, *EZH2*, *BCOR*, and *STAG2* are used to assess risk for myelodysplastic syndrome. Recently, RNA methylation, in addition to DNA methylation, has been shown to play a role in AML pathogenesis. METTL3–METTL14 RNA methyltransferase complex has been shown to add N^6-methyladenosine (m6A) modification in mRNA resulting in enhanced translation of several oncoproteins, such as c-MYC and BCL-2.[78]

Chronic myeloid leukemia (CML) arises due to the t(9;22) translocation (the Philadelphia chromosome) in an HSC, resulting in constitutive expression of the fusion tyrosine kinase BCR-ABL1.[79,80] Along with these mutations, a number of histone-modifying systems (writer, eraser, or reader proteins) are dysregulated in CML, affecting numerous cell–signaling pathways that leukemic cells utilize for survival. Several studies have demonstrated dysregulation of polycomb repressive complex 2 (PRC2) in CML[81] and its expression level may alter in response to BCR-ABL1 inhibitors. PRC2 is a multiprotein complex that contains either the histone methyltransferases (HMTs) EZH1 or EZH2 and lays down histone H3 lysine 27 trimethylation, a hallmark of PRC2-EZH2 repressive activity. Loss of EZH2 activity in chronic phase (CP) patient samples, or in murine model of CML, significantly impairs the survival of leukemic stem cell (LSC).[82] EZH2 associates with β-arrestin to drive H4 lysine acetylation and BCR-ABL1 expression,[83] or with PRAME to repress the function of the death receptor TNFSF10.[84] DNA methylation represses gene transcription and occurs at cytosine residues in CpG dinucleotides where it is catalyzed by DNA methyltransferases (DNMTs). In CP-CML, CD34 + cells are shown to be overexpressing SIRT1, a NAD-dependent deacetylase, likely to be epigenetically driven by aberrant DNA methylation. Inhibition of SIRT1 has been shown to significantly increase TP53 acetylation, resulting in upregulation of the apoptotic response and a reduction of LSC and progenitor cell survival in vitro and in murine models.[85] Another emerging epigenetic process in tumorigenesis is the miRNAs, small or short noncoding RNAs of ≈20–23 nucleotides, that are processed from longer precursor RNAs and either prevent translation or induce cleavage of the mRNA. The miR-17/92 cluster which encodes six miRNAs processed from a single transcript is overexpressed in CML cell lines and CD34 + cells from chronic and blast phase.[86]

NGS-based technologies have revealed a remarkable genetic and epigenetic heterogeneity in B-cell chronic lymphocytic leukemia (CLL). These efforts have unearthed numerous somatic alterations in critical components of many cellular pathways. These include DNA damage and cell-cycle control (*TP53*, *ATM*, *POT1*, *BIRC3*), mRNA processing (*XPO1*, *SF3B1*), NOTCH signaling (*NOTCH1*), inflammatory pathways (*MYD88*), and chromatin modification (*CHD2*). Recurrent mutations in the splicing machinery cofactor *SF3B1*[87] as well as several lower frequency mutations in *NRAS*, *KRAS*, *HIST1H1E*, *SAMHD1*, and *MED12* have been reported.[88] In addition to genomic aberrations or genetic mutations, the epigenetic landscape adds another layer of complexity to the understanding of this clinically and biologically heterogeneous disease. The analysis of global DNA methylation profile has shown that aberrant methylation is an early leukemogenic event.[89] Hypermethylation of the promoter region of human telomerase reverse transcriptase (*hTERT*) gene, a negative regulator of p53 (*TWIST2*), a SYK protein tyrosine kinase—Zeta-chain-associated protein kinase 70 (*ZAP70*), homeobox protein (*HOX4A*), and the death-associated protein kinase 1 (*DAPK1*) gene, a mediator of apoptosis, has been linked to downregulation of these genes in virtually all CLL.[90]

Multiple myeloma (MM), a plasma cell malignancy, is a genetically highly complex and heterogeneous disease, reflected by the presence of a high percentage of nonrecurrent genetic mutations. Based on the karyotype, MM patients can be hyperdiploid or nonhyperdiploid. Majority of MM patients (50%–60%) display a hyperdiploid karyotype, characterized by trisomies involving chromosomes 3, 5, 7, 9, 11, 15, 19, and 21. Common nonhyperdiploid defects include monosomy 13, gains of 1p, or recurrent translocations involving the immunoglobulin heavy chain locus. The most-frequent translocations are t(11;14)(q13;q32) and t(4;14)(p16;q32) dysregulating *CCND1*, and *FGFR3* and *MMSET* genes, respectively.[91] In addition, the epigenetic machinery is also linked with MM progression, driving the differentiation of the malignant cells to a less mature and drug-resistant state.[92] Several mutations have been found in HMTs and histone demethylases.[93] In MM patients the repetitive elements LINE-1, Alu, and SAT-α are hypomethylated compared to healthy controls. MM is also characterized by the silencing of several cancer-related genes (*p73*, *p53*, *p15*, *p16*, *E-CAD*, *DAPK1*, *BNIP3*, *RB1*, *DIS3*, *CDKN2A*, and *CDKN2C*) through hypermethylation, correlating with genomic instability, disease progression, and poor prognosis.[94] Genome-wide analysis of DNA methylation patterns revealed that these patterns change during MM progression—hypomethylation is already present in the early stages of MM development and the methylation levels further decrease during disease progression.[95] Furthermore, the expression of several histone deacetylases is upregulated in MM patients, correlating with a poor prognosis.[96] Also, mutations in the histone acetyltransferases—*EP300* and *CREBBP*—were identified in MM patients; in relapsed patients, frequency of *CREBBP* mutations was higher suggesting its role in drug resistance in MM.

Genetic analysis in peripheral T-cell lymphoma[97] has identified frequent mutations in *TET2*, *DNMT3A*, *IDH2*, and *RHOA* in nodal T-cell lymphomas, including angioimmunoblastic T-cell lymphoma and peripheral T-cell lymphoma, not otherwise specified. Adult T-cell leukemia/lymphoma has frequent gain-of-function mutations in *PLCG1*, *PRKCB*, *CARD11*, and *VAV1* genes that are essential components of T-cell receptor signaling.[98] WES of cutaneous T-cell lymphoma patients displayed recurrent gains in chromosome 7, 8q, and 17q, as well as recurrent deletions involving TSGs in *TP53*, *RB1*, *PTEN*, and *CDKN1B*.[99] Extranodal NK/T-cell lymphoma shows constitutive activation of JAK/STAT pathway caused by increased frequency of *JAK3*, *STAT3*, and *STAT5B* mutations.[100]

4.7 APPLICATION OF GENOMICS TO CANCER IN THE CONTEXT OF PUBLIC HEALTH

Somatic mutations—The use of genetic or DNA-based assays in the context of public health is not new. However, particularly in the clinical oncology space, there has been a significant growth in their application and clinical utility over the past 10–15 years. Historically, genetic risk was assessed through the lens of Mendelian genetics, starting with a detailed family history. From a public health perspective, this knowledge was mostly applied in the setting of newborn screening for genetic disorders. While statutes vary from state to state, since its inception in 1963, there has been universal adoption of genetic screening for inheritable metabolic disorders across all states as an important public health initiative.[101] Momentous impact and action followed the recognition that tobacco is a potent DNA-damaging carcinogen, primarily though not exclusively, for the aerodigestive and genitourinary tracts, and that lifelong exposure to tobacco, including indirect exposure, is the primary cause of cancer deaths in the world. A distinct tobacco "mutational signature"[102] was identified, affecting entire anatomical regions exposed to tobacco carcinogens, providing molecular evidence supporting the concept of field cancerization. Genetic aberrations often found in premalignant field lesions associated with head and neck cancer, lung cancer, colorectal cancer, Barrett's esophagus, and bladder cancer include defects in DNA repair genes (*MLH1*, *MSH2*, *ERCC1*, and *XPF*), which in turn increase the risk of developing additional mutations. Field cancerization has implications for the design of optimal strategies for cancer surveillance, since entire organ systems are at risk, and multiple synchronous and metachronous cancers may develop in the same patient. Field cancerization has been the target of multiple chemoprevention strategies in high-risk individuals, but the clinical impact of these approaches has been so far modest.

With respect to cancer, initial sequencing efforts focused on single well-known cancer-promoting genes, such as *HRAS*, *KRAS2*, *NRAS*, *TP53*, and *BRAF*, known to be mutated in a broad spectrum or tumors. This led to the development in 2004 of the Catalogue of Somatic Mutations in Cancer (COSMIC),[103] an online database of somatically acquired mutations found in human cancer, based on highly curated data from papers in the scientific literature and large-scale experimental screens from the Cancer Genome Project at the Sanger Institute in the United Kingdom. The COSMIC database currently contains genome-wide screen data on over 32,000 genomes, from multiple databases, including TCGA and ICGC.

The identification of *BRCA1* and *BRCA2* genes through linkage analysis[104,105] led to the implementation of screening programs for at-risk women with a new diagnosis of breast and ovarian cancer. Subsequently, investigators identified other inheritable mutations in breast cancer, GI, and hematologic malignancies[106,107] leading to prophylactic interventions, family counseling, and outreach programs for a broader patient population. In addition, technological advances now allow for rapid genome-wide association studies (GWAS), WES, and functional genomic analysis in order to understand individual, gender, ethnic, racial, and population-based genomic risk factors.

Many of the initial findings that elucidated genomic risk in cancer were the result of close observation of a small group of individuals, and once the genes were discovered, they were more broadly applied to a larger population. With the development of GWAS, WES, and management of large data sets, researchers are now able to apply genomic research to large populations in order to identify relevant genomic variations that can then be applied in the context of public health initiatives. The challenge will be ensuring that these genomic applications have been validated and are ready for clinical use. A framework has been proposed and should be considered before the implementation of any public health program.[108] This framework requires that the genomic application has analytic validity, clinical validity, clinical utility, and evidence-based guidelines encouraging its use.[109]

To date, one of the largest non-newborn screening genetic programs in place exists for hereditary breast and ovarian cancer. In 2005 the US Preventive Service Task Force made its initial recommendations for screening patients at high risk for hereditary breast and ovarian cancer.[110] Since then, other gene mutations have been identified that carry increased susceptibility to these cancers and the gene panels have expanded to include them. With the development of NGS, at-risk populations can undergo genetic testing,[111] which provide detailed information on genetic risk. These recommendations have had broad public health implications. Women who tested positive for known pathogenic mutations will subsequently require family counseling, discussion of risk-reducing surgeries, and increased surveillance. While some of these strategies have been proven to be cost-effective (i.e., risk-reducing salpingo-oophorectomy and mastectomy), others still require further validation.[111,112]

While modern sequencing techniques have expanded our ability to identify at-risk patients, they have also introduced uncertainty into the field. Even though many genes are known to be pathogenic, mutational patterns within those genes can be quite diverse. The detection and disclosure of variants of unknown significance (VUS), abnormal genetic findings whose clinical significance is yet unknown, may create public health quagmires. In one recent meta-analysis, women with a family history of breast cancer who underwent germ-line testing had a significantly higher likelihood of detecting VUS as opposed to a pathogenic variant.[113] This finding illustrates our current limitations and need for improved classifications of genetic variation within populations so that we can truly understand how to determine at-risk patient subsets and adequately apply this vast amount of information in a meaningful way.

As we continue to expand our knowledge of disease at the genomic level, we will inevitably discover additional diversity within diseases that we once thought were more uniform. This diversity will exist based on genetic variation by gender, race, and ethnicity.[114–118] Researchers will need to synthesize data from large data sets using bioinformatics and robust algorithms to understand the significance of rare mutations, SV deletions, and epigenetic signatures that are found in various diseases. Once these findings are validated and the clinical utility is well delineated, institutional effort will be required to change policy, implement interventions, and support changes in the health-care system to support ongoing genomic-based public health initiatives.

4.7.1 GERM-LINE GENOMICS

While classical Mendelian inheritance of cancer is rare and typically observed in children, and past estimates of the role on inheritance in cancer were low, it is now becoming clear that cancer in adults has a sizable heritable component. A large twin study[119] estimated that heritable factors may explain between 20% and 40% of the variance in cancer risk. While relative risks (RRs) are highest in first-degree relatives of patients with early-onset cancers, the RRs for the common, nonearly onset cancers in first-degree relatives are only two- to threefold, with the exception of CLL, thyroid, and testicular cancer which have a RR of ~4–8. Nonetheless, RR as low as 5 can be imposed by well-defined cancer susceptibility genes (CSGs), with moderate penetrance, but clear Mendelian inheritance, such as *ATM*, *CHEK2*, *BRIP1*, and *PALB2*. To date, mutations in ~70 CSGs, associated with moderate-to high-penetrance cancer syndromes, have been described, with RR between 5 and 100. High-penetrance mutations, including those in *BRCA1* and *BRCA2*, and *APC* and DNA mismatch-repair genes, are estimated to account for less than 5% of all cases.[120,121] As for many other common complex diseases, it is likely that much of the inherited susceptibility to cancer can be explained by common low-penetrance alleles, which require very large cohorts to be identified.[122,123] Common genetic variants account for a large proportion of cancer incidence, even though they do not individually lead to strong clustering within families. Moreover, the combinations of effects from genetic and environmental factors may account for substantial differences in cancer susceptibility within and among populations.

Over the past decade, GWAS of cancer have discovered multiple low-penetrance loci. Given that the sizes of the effect are generally weak (RRs per allele <1.3), increasing the sample size has become crucial in identifying and characterizing true genetic associations. Genetic signatures of cancer etiology indicated novel influences in cancer development, thereby providing new insights into etiologic mechanisms that suggest interventions. By identifying many new loci influencing cancer development, genomic research has identified pathways that influence cancer development. Once the loci are identified, fine-mapping studies are a critical next step in finding functional variant(s) and in the discovery of nearby, independent, secondary signals which may increase the heritable fraction explained by each region. More than 90% of risk alleles lie in nonprotein coding DNA, and there is now unequivocal evidence that risk regions are enriched with regulatory elements including enhancers, promoters, insulators, and silencers.[124] In general, genome-wide estimates in humans indicate that about 500,000 enhancers may alter regulation of expression and thus alter risk by controlling expression of target susceptibility genes.[125] Analyses to date indicate that several regions harbor multiple distinct susceptibility variants for different cancer types, often falling into regions containing enhancers or super-enhancers, suggesting common mechanisms but tissue-specific regulation.[8] Thus fine-mapping of multiple cancer types using a common array is likely to be an effective strategy for finding new alleles influencing common cancers and for unraveling mechanisms in their etiology.

4.8 SPECIAL APPROACHES

Exosomes are extracellular vesicles carrying a myriad of cellular material including proteins, lipids, and nucleic acids—between cells to exchange information. This passage of material between cells is implicated in disease development—including cancer. As a result, exosome, specifically those that are released from tumor cells, can be collected and characterized through assays to determine cancer diagnosis. Recent evidence shows that exosomes contain abundant nucleic acids including single- and double-stranded genomic DNA as well as mitochondrial DNA. Transcriptomic analyses have cataloged both protein-coding messengers and noncoding RNAs packaged within exosomes, namely, miRNAs, piwi-interacting RNAs, ribosomal RNAs, small nuclear RNAs, small nucleolar RNAs, long ncRNAs, long intergenic RNAs, and circular RNAs. Among the small ncRNAs, miRNAs have been the focus of most exosome biomarker research and functional studies.

Exosomal RNA sequencing can be performed in various cell types, as well as in body fluids including plasma, urine, and serum.[126]

Finally, NGS technologies are also looking to expand on methods for cell-free tumor DNA.[127] This DNA is found throughout the body after being released from the primary tumor or from CTCs. These free-floating DNA fragments are found in the plasma of patients during the early stages of their disease—making them good candidates for a liquid biopsy biomarker. NGS can be utilized to identify the variations in sequence and epigenetic tags to indicate these fragments originated from tumor cells. As these methods of analysis advance, this noninvasive practice of liquid biopsy will allow for an advanced understanding of how to detect the presence of cancer without adding more additional stress to the patient. Also, as the nuances of the cell-free tumor DNA are understood over the course of the disease, the changes that occur in the tumor DNA as the cancer develops and is treated can be studied. This observation can provide a stronger knowledge base on the impact of disease type and development on the tumor DNA.

4.9 CONCLUSION

Cancer is a disease characterized by DNA aberrations conferring oncogenic traits, influencing tumoral transcriptomic landscape, and potentially affecting the clinical course of the disease. In recent years, impressive stride has been made in the field of cancer genomics and personalized medicine, churning out a large set of biological data containing information about genetic mutations, epigenetic modifications, proteome, and drug response. Integrating these complementary data repositories will enhance the overall understanding of this complex disease.

Several genomic and bioinformatics approaches have been developed to further enhance our current knowledge of the intricate interplay between different yet intertwined nodes of cancer. This chapter highlights different new and commonly used approaches currently applied in cancer genomics and precision medicine along with computational strategies to integrate biological data generated. In addition, we provide an overview of cancer as a global public health crisis requiring intervention and prevention programs both at the global and grassroot levels.

The true potential of a multifaceted genomic approach by successful application of these various integrative methods using omics data will help in answering some of the burning questions in the field of personalized cancer medicine. It also highlights data integration as a way forward in understanding complex tumor machinery to better target it. We expect that in the future further development of genomic techniques and utilization of data integration techniques will pave the way to address current problems aptly.

REFERENCES

1. Prager BC, Xie Q, Bao S, Rich JN. Cancer stem cells: the architects of the tumor ecosystem. *Cell Stem Cell*. 2019;24(1):41–53.
2. Bakhoum SF, Cantley LC. The multifaceted role of chromosomal instability in cancer and its microenvironment. *Cell*. 2018;174(6):1347–1360.
3. Hanahan D, Weinberg RA. Hallmarks of cancer: the next generation. *Cell*. 2011;144(5):646–674.

4. Pavlova NN, Thompson CB. The emerging hallmarks of cancer metabolism. *Cell Metab.* 2016;23(1):27–47.
5. Tawana K, Drazer MW, Churpek JE. Universal genetic testing for inherited susceptibility in children and adults with myelodysplastic syndrome and acute myeloid leukemia: are we there yet? *Leukemia.* 2018;32(7):1482–1492.
6. Bhagwat AS, Lu B, Vakoc CR. Enhancer dysfunction in leukemia. *Blood.* 2018;131(16):1795–1804.
7. Flavahan WA, Gaskell E, Bernstein BE. Epigenetic plasticity and the hallmarks of cancer. *Science.* 2017;357(6348).
8. Sengupta S, George RE. Super-enhancer-driven transcriptional dependencies in cancer. *Trends Cancer.* 2017;3(4):269–281.
9. Kim J, Zaret KS. Reprogramming of human cancer cells to pluripotency for models of cancer progression. *EMBO J.* 2015;34(6):739–747.
10. Bray F, Ferlay J, Soerjomataram I, Siegel RL, Torre LA, Jemal A. Global cancer statistics 2018: GLOBOCAN estimates of incidence and mortality worldwide for 36 cancers in 185 countries. *CA Cancer J Clin.* 2018;68(6):394–424.
11. Tan DS, Mok TS, Rebbeck TR. Cancer genomics: diversity and disparity across ethnicity and geography. *J Clin Oncol.* 2016;34(1):91–101.
12. Baslan T, Hicks J. Unravelling biology and shifting paradigms in cancer with single-cell sequencing. *Nat Rev Cancer.* 2017;17(9):557–569.
13. Krzywinski M. Visualizing clonal evolution in cancer. *Mol Cell.* 2016;62(5):652–656.
14. Garraway LA, Lander ES. Lessons from the cancer genome. *Cell.* 2013;153(1):17–37.
15. Amirouchene-Angelozzi N, Swanton C, Bardelli A. Tumor evolution as a therapeutic target. *Cancer Discov.* 2017.
16. Bowman RL, Busque L, Levine RL. Clonal hematopoiesis and evolution to hematopoietic malignancies. *Cell Stem Cell.* 2018;22(2):157–170.
17. Ferrando AA, Lopez-Otin C. Clonal evolution in leukemia. *Nat Med.* 2017;23(10):1135–1145.
18. Berger MF, Mardis ER. The emerging clinical relevance of genomics in cancer medicine. *Nat Rev Clin Oncol.* 2018;15(6):353–365.
19. Khoury MJ, Bowen MS, Clyne M, et al. From public health genomics to precision public health: a 20-year journey. *Genet Med.* 2018;20(6):574–582.
20. Heath EI, Lynce F, Xiu J, et al. Racial disparities in the molecular landscape of cancer. *Anticancer Res.* 2018;38(4):2235–2240.
21. Kader F, Ghai M. DNA methylation-based variation between human populations. *Mol Genet Genomics.* 2017;292(1):5–35.
22. Spratt DE, Chan T, Waldron L, et al. Racial/ethnic disparities in genomic sequencing. *JAMA Oncol.* 2016;2(8):1070–1074.
23. http://www.healthdata.org/sites/default/files/files/policy_report/2019/GBD_2017_Booklet.pdf (accessed 2019).
24. https://www.unfpa.org/sites/default/files/pub-pdf/9789241565141_eng.pdf (accessed 2019).
25. Misganaw A, Haregu TN, Deribe K, et al. National mortality burden due to communicable, non-communicable, and other diseases in Ethiopia, 1990-2015: findings from the Global Burden of Disease Study 2015. *Popul Health Metrics.* 2017;15:29.
26. https://www.un.org/africarenewal/magazine/december-2016-march-2017/taking-health-services-remote-areas (accessed 2019).
27. Weinberg J, Kaddu S, Gabler G, Kovarik C. The African Teledermatology Project: providing access to dermatologic care and education in sub-Saharan Africa. *Pan Afr Med J.* 2009;3:16.
28. https://www.cfr.org/report/emerging-global-health-crisis (accessed 2019).

29. Goss PE, Strasser-Weippl K, Lee-Bychkovsky BL, et al. Challenges to effective cancer control in China, India, and Russia. *Lancet Oncol*. 2014;15(5):489–538.
30. http://apps.who.int/gb/ebwha/pdf_files/WHA70/A70_ACONF9-en.pdf?ua = 1 (accessed 2019).
31. Shah SC, Kayamba V, Peek Jr. RM, Heimburger D. Cancer control in low- and middle-income countries: is it time to consider screening? *J Glob Oncol*. 2019;5:1–8.
32. Kihn-Alarcon AJ, Toledo-Ponce MF, Velarde A, Xu X. Liver cancer in Guatemala: an analysis of mortality and incidence trends from 2012 to 2016. *J Glob Oncol*. 2019;5:1–8.
33. Subramanian S, Gakunga R, Jones MD, et al. Establishing cohorts to generate the evidence base to reduce the burden of breast cancer in sub-Saharan Africa: results from a feasibility study in Kenya. *J Glob Oncol*. 2019;5:1–10.
34. Hamashima C. Systematic Review Group and Guideline Development Group for Gastric Cancer Screening Guidelines. Update version of the Japanese Guidelines for Gastric Cancer Screening. *Jpn J Clin Oncol*. 2018;48(7):673–683.
35. Kim GH, Liang PS, Bang SJ, Hwang JH. Screening and surveillance for gastric cancer in the United States: is it needed? *Gastrointest Endosc*. 2016;84(1):18–28.
36. Cheng TY, Cramb SM, Baade PD, Youlden DR, Nwogu C, Reid ME. The international epidemiology of lung cancer: latest trends, disparities, and tumor characteristics. *J Thorac Oncol*. 2016;11(10):1653–1671.
37. Torre LA, Siegel RL, Ward EM, Jemal A. International variation in lung cancer mortality rates and trends among women. *Cancer Epidemiol Biomarkers Prev*. 2014;23(6):1025–1036.
38. National Lung Screening Trial Research Team, Aberle DR, Adams AM, et al. Reduced lung-cancer mortality with low-dose computed tomographic screening. *N Engl J Med*. 2011;365(5):395–409.
39. Darabi H, Czene K, Zhao W, Liu J, Hall P, Humphreys K. Breast cancer risk prediction and individualised screening based on common genetic variation and breast density measurement. *Breast Cancer Res*. 2012;14(1):R25.
40. Weigl K, Thomsen H, Balavarca Y, Hellwege JN, Shrubsole MJ, Brenner H. Genetic risk score is associated with prevalence of advanced neoplasms in a colorectal cancer screening population. *Gastroenterology*. 2018;155(1):88–98.e10.
41. Ruiz JB, Bell RA. Understanding vaccination resistance: vaccine search term selection bias and the valence of retrieved information. *Vaccine*. 2014;32(44):5776–5780.
42. Shen AK, Fields R, McQuestion M. The future of routine immunization in the developing world: challenges and opportunities. *Glob Health Sci Pract*. 2014;2(4):381–394.
43. Hussain A, Zaheer S, Shafique K. Reasons for betel quid chewing amongst dependent and non-dependent betel quid chewing adolescents: a school-based cross-sectional survey. *Subst Abuse Treat Prev Policy*. 2018;13(1):16.
44. Chen G, Hsieh MY, Chen AW, Kao NH, Chen MK. The effectiveness of school educating program for betel quid chewing: a pilot study in Papua New Guinea. *J Chin Med Assoc*. 2018;81(4):352–357.
45. Moss J, Kawamoto C, Pokhrel P, Paulino Y, Herzog T. Developing a betel quid cessation program on the Island of Guam. *Pac Asia Inq*. 2015;6(1):144–150.
46. Vineis P, Schatzkin A, Potter JD. Models of carcinogenesis: an overview. *Carcinogenesis*. 2010;31(10):1703–1709.
47. Romero-Laorden N, Castro E. Inherited mutations in DNA repair genes and cancer risk. *Curr Probl Cancer*. 2017;41(4):251–264.
48. Baker SG. A cancer theory kerfuffle can lead to new lines of research. *J Natl Cancer Inst*. 2015;107(2).
49. Brucher BL, Jamall IS. Somatic mutation theory – why it's wrong for most cancers. *Cell Physiol Biochem*. 2016;38(5):1663–1680.

50. Gerlinger M, McGranahan N, Dewhurst SM, Burrell RA, Tomlinson I, Swanton C. Cancer: evolution within a lifetime. *Annu Rev Genet*. 2014;48:215–236.
51. Nowell PC. The clonal evolution of tumor cell populations. *Science*. 1976;194(4260):23–28.
52. Greaves M, Maley CC. Clonal evolution in cancer. *Nature*. 2012;481(7381):306–313.
53. Arends CM, Galan-Sousa J, Hoyer K, et al. Hematopoietic lineage distribution and evolutionary dynamics of clonal hematopoiesis. *Leukemia*. 2018;32(9):1908–1919.
54. Boettcher S, Ebert BL. Clonal hematopoiesis of indeterminate potential. *J Clin Oncol*. 2019;37 (5):419–422.
55. Xie M, Lu C, Wang J, et al. Age-related mutations associated with clonal hematopoietic expansion and malignancies. *Nat Med*. 2014;20(12):1472–1478.
56. Zink F, Stacey SN, Norddahl GL, et al. Clonal hematopoiesis, with and without candidate driver mutations, is common in the elderly. *Blood*. 2017;130(6):742–752.
57. Jaiswal S, Natarajan P, Silver AJ, et al. Clonal hematopoiesis and risk of atherosclerotic cardiovascular disease. *N Engl J Med*. 2017;377(2):111–121.
58. Vogelstein B, Papadopoulos N, Velculescu VE, Zhou S, Diaz Jr. LA, Kinzler KW. Cancer genome landscapes. *Science*. 2013;339(6127):1546–1558.
59. Hu X, Zhang Z. Understanding the genetic mechanisms of cancer drug resistance using genomic approaches. *Trends Genet*. 2016;32(2):127–137.
60. Yankeelov TE, Quaranta V, Evans KJ, Rericha EC. Toward a science of tumor forecasting for clinical oncology. *Cancer Res*. 2015;75(6):918–923.
61. Martincorena I, Campbell PJ. Somatic mutation in cancer and normal cells. *Science*. 2015;349 (6255):1483–1489.
62. Mugal CF, von Grunberg HH, Peifer M. Transcription-induced mutational strand bias and its effect on substitution rates in human genes. *Mol Biol Evol*. 2009;26(1):131–142.
63. Nair N, Shoaib M, Sorensen CS. Chromatin dynamics in genome stability: roles in suppressing endogenous DNA damage and facilitating DNA repair. *Int J Mol Sci*. 2017;18(7).
64. Schuster-Bockler B, Lehner B. Chromatin organization is a major influence on regional mutation rates in human cancer cells. *Nature*. 2012;488(7412):504–507.
65. Tomasetti C, Marchionni L, Nowak MA, Parmigiani G, Vogelstein B. Only three driver gene mutations are required for the development of lung and colorectal cancers. *Proc Natl Acad Sci USA*. 2015;112 (1):118–123.
66. Yates LR, Gerstung M, Knappskog S, et al. Subclonal diversification of primary breast cancer revealed by multiregion sequencing. *Nat Med*. 2015;21(7):751–759.
67. Campbell BB, Light N, Fabrizio D, et al. Comprehensive analysis of hypermutation in human cancer. *Cell*. 2017;171(5):1042–1056.e10.
68. Rebucci M, Michiels C. Molecular aspects of cancer cell resistance to chemotherapy. *Biochem Pharmacol*. 2013;85(9):1219–1226.
69. Conway JR, Kofman E, Mo SS, Elmarakeby H, Van Allen E. Genomics of response to immune checkpoint therapies for cancer: implications for precision medicine. *Genome Med*. 2018;10(1):93.
70. Rizvi NA, Hellmann MD, Snyder A, et al. Cancer immunology. Mutational landscape determines sensitivity to PD-1 blockade in non-small cell lung cancer. *Science*. 2015;348(6230):124–128.
71. Gerstung M, Pellagatti A, Malcovati L, et al. Combining gene mutation with gene expression data improves outcome prediction in myelodysplastic syndromes. *Nat Commun*. 2015;6:5901.
72. Winters IP, Murray CW, Winslow MM. Towards quantitative and multiplexed in vivo functional cancer genomics. *Nat Rev Genet*. 2018;19(12):741–755.
73. Taby R, Issa JP. Cancer epigenetics. *CA Cancer J Clin*. 2010;60(6):376–392.

74. You JS, Jones PA. Cancer genetics and epigenetics: two sides of the same coin? *Cancer Cell*. 2012;22 (1):9–20.
75. Arora K, Barbieri CE. Molecular subtypes of prostate cancer. *Curr Oncol Rep*. 2018;20(8):58.
76. Global Burden of Disease Cancer Collaboration, Fitzmaurice C, Akinyemiju TF, et al. Global, regional, and national cancer incidence, mortality, years of life lost, years lived with disability, and disability-adjusted life-years for 29 cancer groups, 1990 to 2016: a systematic analysis for the Global Burden of Disease Study. *JAMA Oncol*. 2018;4(11):1553–1568.
77. Shivarov V, Bullinger L. Expression profiling of leukemia patients: key lessons and future directions. *Exp Hematol*. 2014;42(8):651–660.
78. Cancer Genome Atlas Research Network, Ley TJ, Miller C, et al. Genomic and epigenomic landscapes of adult de novo acute myeloid leukemia. *N Engl J Med*. 2013;368(22):2059–2074.
79. Li Z, Weng H, Su R, et al. FTO plays an oncogenic role in acute myeloid leukemia as a *N*(6)-methyladenosine RNA demethylase. *Cancer Cell*. 2017;31(1):127–141.
80. Nowell PC, Hungerford DA. Chromosome studies in human leukemia. II. Chronic granulocytic leukemia. *J Natl Cancer Inst*. 1961;27:1013–1035.
81. Scott MT, Korfi K, Saffrey P, et al. Epigenetic reprogramming sensitizes CML stem cells to combined EZH2 and tyrosine kinase inhibition. *Cancer Discov*. 2016;6(11):1248–1257.
82. Xie H, Peng C, Huang J, et al. Chronic myelogenous leukemia – initiating cells require polycomb group protein EZH2. *Cancer Discov*. 2016;6(11):1237–1247.
83. Qin R, Li K, Qi X, et al. Beta-arrestin1 promotes the progression of chronic myeloid leukaemia by regulating BCR/ABL H4 acetylation. *Br J Cancer*. 2014;111(3):568–576.
84. De Carvalho DD, Binato R, Pereira WO, et al. BCR-ABL-mediated upregulation of PRAME is responsible for knocking down TRAIL in CML patients. *Oncogene*. 2011;30(2):223–233.
85. Li L, Wang L, Li L, et al. Activation of p53 by SIRT1 inhibition enhances elimination of CML leukemia stem cells in combination with imatinib. *Cancer Cell*. 2012;21(2):266–281.
86. Venturini L, Battmer K, Castoldi M, et al. Expression of the miR-17-92 polycistron in chronic myeloid leukemia (CML) CD34 + cells. *Blood*. 2007;109(10):4399–4405.
87. Wang L, Lawrence MS, Wan Y, et al. SF3B1 and other novel cancer genes in chronic lymphocytic leukemia. *N Engl J Med*. 2011;365(26):2497–2506.
88. Gruber M, Wu CJ. Evolving understanding of the CLL genome. *Semin Hematol*. 2014;51(3):177–187.
89. Cahill N, Bergh AC, Kanduri M, et al. 450K-array analysis of chronic lymphocytic leukemia cells reveals global DNA methylation to be relatively stable over time and similar in resting and proliferative compartments. *Leukemia*. 2013;27(1):150–158.
90. Mansouri L, Wierzbinska JA, Plass C, Rosenquist R. Epigenetic deregulation in chronic lymphocytic leukemia: clinical and biological impact. *Semin Cancer Biol*. 2018;51:1–11.
91. Corre J, Munshi N, Avet-Loiseau H. Genetics of multiple myeloma: another heterogeneity level? *Blood*. 2015;125(12):1870–1876.
92. Alzrigat M, Parraga AA, Jernberg-Wiklund H. Epigenetics in multiple myeloma: from mechanisms to therapy. *Semin Cancer Biol*. 2018;51:101–115.
93. Dupere-Richer D, Licht JD. Epigenetic regulatory mutations and epigenetic therapy for multiple myeloma. *Curr Opin Hematol*. 2017;24(4):336–344.
94. Wong KY, Chim CS. DNA methylation of tumor suppressor protein-coding and non-coding genes in multiple myeloma. *Epigenomics*. 2015;7(6):985–1001.
95. Walker BA, Wardell CP, Chiecchio L, et al. Aberrant global methylation patterns affect the molecular pathogenesis and prognosis of multiple myeloma. *Blood*. 2011;117(2):553–562.
96. Mithraprabhu S, Kalff A, Chow A, Khong T, Spencer A. Dysregulated class I histone deacetylases are indicators of poor prognosis in multiple myeloma. *Epigenetics*. 2014;9(11):1511–1520.

97. Van Arnam JS, Lim MS, Elenitoba-Johnson KSJ. Novel insights into the pathogenesis of T-cell lymphomas. *Blood.* 2018;131(21):2320–2330.
98. Watanabe T. Adult T-cell leukemia: molecular basis for clonal expansion and transformation of HTLV-1-infected T cells. *Blood.* 2017;129(9):1071–1081.
99. Mishra A, Porcu P. Expanding and expounding the genomic map of CTCL. *Blood.* 2017;130(12):1389–1390.
100. Haverkos BM, Coleman C, Gru AA, et al. Emerging insights on the pathogenesis and treatment of extranodal NK/T cell lymphomas (ENKTL). *Discov Med.* 2017;23(126):189–199.
101. Therrell BL, Johnson A, Williams D. Status of newborn screening programs in the United States. *Pediatrics.* 2006;117(5 Pt 2):S212–252.
102. Petljak M, Alexandrov LB. Understanding mutagenesis through delineation of mutational signatures in human cancer. *Carcinogenesis.* 2016;37(6):531–540.
103. https://cancer.sanger.ac.uk/cosmic (accessed 2019).
104. Hall JM, Friedman L, Guenther C, et al. Closing in on a breast cancer gene on chromosome 17q. *Am J Hum Genet.* 1992;50(6):1235–1242.
105. Kent P, O'Donoghue JM, O'Hanlon DM, Kerin MJ, Maher DJ, Given HF. Linkage analysis and the susceptibility gene (BRCA-1) in familial breast cancer. *Eur J Surg Oncol.* 1995;21(3):240–241.
106. Godley LA, Shimamura A. Genetic predisposition to hematologic malignancies: management and surveillance. *Blood.* 2017;130(4):424–432.
107. Shah PD, Nathanson KL. Application of panel-based tests for inherited risk of cancer. *Annu Rev Genomics Hum Genet.* 2017;18:201–227.
108. Khoury MJ, Bowen MS, Burke W, et al. Current priorities for public health practice in addressing the role of human genomics in improving population health. *Am J Prev Med.* 2011;40(4):486–493.
109. Bowen MS, Kolor K, Dotson WD, Ned RM, Khoury MJ. Public health action in genomics is now needed beyond newborn screening. *Public Health Genomics.* 2012;15(6):327–334.
110. Nelson HD, Huffman LH, Fu R, Harris EL. Force USPST. Genetic risk assessment and BRCA mutation testing for breast and ovarian cancer susceptibility: systematic evidence review for the U.S. Preventive Services Task Force. *Ann Intern Med.* 2005;143(5):362–379.
111. Ruiz A, Llort G, Yague C, et al. Genetic testing in hereditary breast and ovarian cancer using massive parallel sequencing. *Biomed Res Int.* 2014;2014:542541.
112. Castera L, Krieger S, Rousselin A, et al. Next-generation sequencing for the diagnosis of hereditary breast and ovarian cancer using genomic capture targeting multiple candidate genes. *Eur J Hum Genet.* 2014;22(11):1305–1313.
113. Balmana J, Sanz J, Bonfill X, et al. Genetic counseling program in familial breast cancer: analysis of its effectiveness, cost and cost-effectiveness ratio. *Int J Cancer.* 2004;112(4):647–652.
114. Anderson K, Jacobson JS, Heitjan DF, et al. Cost-effectiveness of preventive strategies for women with a BRCA1 or a BRCA2 mutation. *Ann Intern Med.* 2006;144(6):397–406.
115. Shan J, Mahfoudh W, Dsouza SP, et al. Genome-wide association studies (GWAS) breast cancer susceptibility loci in Arabs: susceptibility and prognostic implications in Tunisians. *Breast Cancer Res Treat.* 2012;135(3):715–724.
116. van Marcke C, Collard A, Vikkula M, Duhoux FP. Prevalence of pathogenic variants and variants of unknown significance in patients at high risk of breast cancer: A systematic review and meta-analysis of gene-panel data. *Crit Rev Oncol Hematol.* 2018;132:138–144.
117. Yoon HH, Shi Q, Alberts SR, et al. Racial differences in BRAF/KRAS mutation rates and survival in stage III colon cancer patients. *J Natl Cancer Inst.* 2015;107(10).
118. Zhang J, Yan S, Liu X, et al. Gender-related prognostic value and genomic pattern of intra-tumor heterogeneity in colorectal cancer. *Carcinogenesis.* 2017;38(8):837–846.

119. Lichtenstein P, Holm NV, Verkasalo PK, et al. Environmental and heritable factors in the causation of cancer—analyses of cohorts of twins from Sweden, Denmark, and Finland. *N Engl J Med*. 2000;343(2):78–85.
120. Lander ES, Schork NJ. Genetic dissection of complex traits. *Science*. 1994;265(5181):2037–2048.
121. Reich DE, Lander ES. On the allelic spectrum of human disease. *Trends Genet*. 2001;17(9):502–510.
122. Ponder BA. Cancer genetics. *Nature*. 2001;411(6835):336–341.
123. Peto J. Cancer epidemiology in the last century and the next decade. *Nature*. 2001;411(6835):390–395.
124. Maurano MT, Humbert R, Rynes E, et al. Systematic localization of common disease-associated variation in regulatory DNA. *Science*. 2012;337(6099):1190–1195.
125. Goldstein I, Hager GL. Dynamic enhancer function in the chromatin context. *Wiley Interdiscip Rev Syst Biol Med*. 2018;10(1).
126. Li X, Corbett AL, Taatizadeh E, et al. Challenges and opportunities in exosome research—perspectives from biology, engineering, and cancer therapy. *APL Bioeng*. 2019;3(1):011503.
127. Sole C, Arnaiz E, Manterola L, Otaegui D, Lawrie CH. The circulating transcriptome as a source of cancer liquid biopsy biomarkers. *Semin Cancer Biol*. 2019.

WEB REFERENCES

<http://apps.who.int/gb/ebwha/pdf_files/WHA70/A70_ACONF9-en.pdf?ua = 1> Accessed 2019.

<http://www.healthdata.org/sites/default/files/files/policy_report/2019/GBD_2017_Booklet.pdf> Accessed 2019.

<https://cancer.sanger.ac.uk/cosmic> Accessed 2019.

<https://www.bgi.com> Accessed 2019.

<https://www.cfr.org/report/emerging-global-health-crisis> Accessed 2019.

<https://www.nanoporetech.com> Accessed 2019.

<https://www.pacb.com/> Accessed 2019.

<https://www.qiagen.com/us/> Accessed 2019.

<https://www.un.org/africarenewal/magazine/december-2016-march-2017/taking-health-services-remote-areas> Accessed 2019.

<https://www.unfpa.org/sites/default/files/pub-pdf/9789241565141_eng.pdf> Accessed 2019.

<https://cancer.sanger.ac.uk/cosmic> Accessed 2019.

CHAPTER

GENOMIC BASIS OF PSYCHIATRIC ILLNESSES AND RESPONSE TO PSYCHIATRIC DRUG TREATMENT MODALITIES

5

Evangelia-Eirini Tsermpini[1], Maria Skokou[2], Zoe Kordou[1] and George P. Patrinos[1,3,4]

[1]*Department of Pharmacy, School of Health Sciences, University of Patras, Patras, Greece* [2]*Psychiatric Clinic, Patras General Hospital, Patras, Greece* [3]*Department of Pathology, College of Medicine and Health Sciences, United Arab Emirates University, Al Ain, United Arab Emirates* [4]*Zayed Center of Health Sciences, United Arab Emirates University, Al Ain, United Arab Emirates*

5.1 INTRODUCTION

The biological foundations of mental illness have been asserted since the era of Hippocrates, and observations of the heredity of these disorders are equally old. Researchers had exerted laborious efforts in the attempt to unravel the molecular and genetic foundations of psychiatric disease, but it is only recently that the field has witnessed an outstanding progress, with the new tools for genetic discovery, and a perspective for meaningful deepening of insights, translation of findings and integration to clinic, setting the foundation for precision medicine, is closer than ever.[1]

Progress in the field of genetic research has moved from seminal work on twin and family studies, to linkage analysis, studying of candidate genes, to the groundbreaking use of genome-wide association studies (GWASs), and whole-exome and whole-genome sequencing (WES and WGS, respectively).[2] From the early twin and family studies, it was evident that most major psychiatric illnesses are highly heritable, which ranges between 0.4 and 0.8.[3] The now emerging architecture of psychiatric illness is polygenic, with many common variants, hundreds or thousands, conferring each a slight increase in risk, with absolute risks around 1.1%–1.2% versus 1.0% population risk.[3] Rare variance is also important, resulting from rare and de novo copy number variants (CNVs), and rare and de novo variants, with large effect sizes, ranging from 2 to over 20.[2]

GWASs, assessing common variants throughout the genome, having set significance level to at least 5×10^{-8} would not have been fruitful if not a huge number of samples were available,[4,5] and this was only feasible by combining samples from many research centers, in a collaborative sense. The Psychiatric Genomics Consortium (PGC) is the largest of these efforts, beginning at 2007, and has yielded an astounding number of results.[1] The most cited of its works to date is the 2014 publication of 108 risk loci for schizophrenia (SZ), on a sample of 37,000 individuals of various ancestries.[6] On the other hand, the discovery of genetic variants resulting in differential pharmacodynamics and pharmacokinetics of drugs at the individual level has opened the way of addressing the large

Applied Genomics and Public Health. DOI: https://doi.org/10.1016/B978-0-12-813695-9.00005-4

interindividual variation in drug efficacy and tolerance, and consortia focusing on pharmacogenomics aim to validate the clinical use of pharmacokinetics genotyping, in order to optimize drug choice and dosing.[7] The expanding elucidation of the genetic background of psychiatric disorders is reasonably expected to better inform pathophysiological models of diseases and identify targets for design of novel pharmaceutical agents for treatment. The fact that most psychiatric disorders are comorbid with others, and risk genotypes cross diagnostic boundaries, is expected to better inform nosology. Taken together, advances in psychiatric genomics set the foundations for the applying of precision medicine, and moving to a new era in the diagnosis and treatment of the psychiatric patient. In this chapter a comprehensive up-to-date overview of the genomic findings in psychiatric illnesses is presented.

5.2 GENETICS OF SCHIZOPHRENIA

SZ is a severe psychiatric disorder that affects about 1% of the general population, regardless of populations and races.[8,9] The symptoms that characterize SZ include positive, negative, and cognitive symptoms, which involve hallucinations, anhedonia, decreased interest in social interactions, problems in concentration, memory, and learning, and various others.[8–10] These symptoms do not allow the patient to live independently and perform daily activities, such as education and work, and also reduce his or her ability to interact with other people, resulting in problems at the interpersonal level within the society.[10]

It is worth noting that the risk of developing the disease increases when it comes to first-, second-, or even third-degree family members of the patient. The reason for this association is due to the percentage of patients and relatives common genes; particularly in the case of monozygotic twins, the incidence increases by 50%.[11]

Genomic analysis studies, such as GWAS, WES, WGS, and candidate gene, are performed to identify the genetic variants associated with SZ and to evaluate their contribution to disease development. GWASs include a genomic comparison of unrelated individuals in order to find genetic biomarkers associated with a particular phenotype or a disease and develop new prevention and treatment strategies.[12] In WES studies, only exons, that is, gene regions encoding for proteins, are analyzed in order to unveil the genetic basis of the studied disease.[13] In contrast, in WGS studies, the entire genome, that is, coding, noncoding, and intergenic regions, is studied. Candidate-gene studies aim to confirm or not the correlation of specific genes with a disease. Individuals participating in these studies are divided into two groups, patients and healthy individuals. A candidate is considered a gene that is related to disease from previous studies. In candidate-gene studies the incidence of pathogenic variant(s) in the patient group is observed relative to that of healthy subjects, and then its correlation with the prognosis or diagnosis of the disease and its possible future use as a biomarker is evaluated.[14]

SZ is an exciting field of research. Numerous studies on identifying genes and polymorphisms associated with the disease have been conducted in both Caucasian and non-Caucasian populations.

GWASs have indicated many genes that are involved in the emergence of SZ, such as *NLGN1*,[15] *ACSM1 ACSM2 ANK3 GLUDP5*, *PHACTR1 OTOR* C3orf39/*SNRK*, *PSD3*, *GPHN*, *PCDH15*, *CTAGE1/RBBP8*, *SLC35F2*, *NPEPL1/GNAS*.[16] Other GWASs indicated significant associations between genomic variants in the *MSI* gene,[17] as well as duplications in 1p36.32 and 10p12.1 with SZ in a group of Han Chinese.[18]

According to studies that contain both patients with SZ and healthy individuals of Caucasian and non-Caucasian populations indicate a statistically significant association of specific genes with the development of the disease. Dopamine receptors have been widely studied in many research groups, but results seem to be incontinent. However, in most studies, variants in the *DRD1*,[19] *DRD2*,[19–26] *DRD3*,[27] as well *DRD4* genes,[28] were associated with the development of SZ. Other vital genes that are related to the phenotype of SZ include *DISC1*,[29–31] *GSK3B*,[32] *BDNF*,[20,22,33–40] *5-HT*$_{5A}$,[41] *NOTCH4*,[20,42] *COMT*,[22,27,43] *STT3A*, rs *ZNF804A*,[44,45] *GRM8*, *GRM7*,[46,47] and *NVL*[48] (Table 5.1). Furthermore, significant associations have been indicated for genes *DAOA*,[49] *PIK3C3*,[50] *DAT1*,[26] *ATXN2*,[56] *SLC18A2*,[27] *LMAN2L*,[51] *ITIH*,[52] *ZEB2*,[57] *HAGH*,[53] *MSH5-SAPCD1*, *SLC44A4*,[54] *MAD1L1*,[55] and MKL1.[58]

SZ is a very complex and heterogeneous psychiatric disease. It is certain that the risk of developing SZ is highly associated with the presence of the disease in the family members. This underlies the strong impact that genetic background has in the emergence of SZ. However, the exact mechanism and the etiology of SZ still remain unclear. In this part of the chapter, we focused on genes and genetic variants correlated with the development of SZ, regardless of race and population. According to GWAS and candidate-gene strategies, some genes seem to be associated with the development of SZ. The significant number of studies that exist indicates that there is not only one gene, which leads to SZ but rather the combination of a variety of genetic variants in different genes; hence, individual genes seem to have a low impact on SZ etiology. Large sample cohorts are needed, as well as studies that focus on different races and populations to get a better picture on the genetic basis of SZ.

5.3 GENETICS OF BIPOLAR DISORDER

Bipolar disorder (BD) affects around 1% of the general population worldwide regardless of nationality and economic and social status. Changes in the mood characterize the disease at regular intervals. According to the World Health Organization, BD is among the top 10 most significant diseases affecting the daily life of people aged 15–40.[70] BD is characterized by periods of depression and abnormal euphoria, during which, patients experience increased mobility of different severity and duration and have limited or no functionality. Sometimes the appearance of psychosis, that is, hallucinations, might occur as well as the need for admission to the hospital.[71]

BD is a complex and multifactorial disease, in which the interaction of genes with environmental factors plays a predominant role. Family and twin studies have shown that BD has a significant genetic component. The inheritance of the disease from parents to children is estimated at 70%–85%.[70,72] Also, according to twin studies, a high percentage of 70% of monozygotic twins have a similar clinical picture concerning BD.[72]

Initially, the investigations focused on finding a correlation between BD and specific candidate genes. This specific approach concerns the investigation of specific genes for possible involvement in the emergence and pathogenesis of the BD. Given the fact that the selection of these genes is based on their function, studies have focused primarily on genes related to systems, such as

Table 5.1 Genomic Variants Associated With the Development of Schizophrenia

Gene	Variant	Ethnic Group	Number of Samples	Study	References
NLGN1	rs13074723, rs1488547, rs2861598, rs4280663, rs4399918, rs4513478, rs9835385 and rs6792822	Han Chinese	746 cases/1599 controls (discovery GWAS) 1814 cases/1487 controls (replication)	GWAS and replication	[15]
PLAA	rs7045881, rs7035755	Norwegians and Europeans	201 cases/305 controls (discovery GWAS) 2663 cases/13,780 controls (replication)	GWAS and replication	[16]
ACSM2	rs433598, rs163274, rs163234, rs163275				
ACSM1	rs10761482, rs4984410				
ANK3	rs12324972, rs6497501				
RYR3	rs4313476 of GLUDP5, rs951442				
PHACTR1	rs9296494, rs9296494				
OTOR	rs4813250, rs6111277, rs6080400				
C3orf39/ SNRK	rs2372342				
PSD3	rs10106755, rs13273158				
GPHN	rs6573695, rs17247749, rs6573719, rs17836572, rs1885198, rs6573706, rs7154017, rs7157740, rs9783601, rs1952070				
PCDH15	rs4316429, rs1539318, rs7092779				
CTAGE1/ RBBP8	rs4264496, rs11662668				
SLC35F2	rs500063				
NPEPL1/ GNAS	rs6100223, rs6015375				
MSI	rs9892791, rs11657292, rs1822381	Han Chinese	921 cases/1244 controls	GWAS	[17]
DAOA	rs3916965, rs2391191	Han Chinese	537 cases/538 controls	Case–control	[49]
PIK3C3	rs3813065	Chinese	556 cases/563 controls	Case–control	[50]
SULT6B1	rs11895771	Japanese	1990 cases/5389 controls	Case–control	[42]
IL3RA	rs6603272, rs6645249	Han Chinese	NA	Case–control	[59]
DAT1	rs2455391	Han Chinese	368 cases/420 controls	Case–control	[26]

Table 5.1 Genomic Variants Associated With the Development of Schizophrenia *Continued*

Gene	Variant	Ethnic Group	Number of Samples	Study	References
ATXN2	rs7969300	Han Chinese	2725 cases/2857 controls	Case–control	[60]
SLC18A2	rs363399, rs10082463	Indians	118 families	Family	[27]
LMAN2L	rs2271893	Malaysians, Chinese, and Indians	471 cases/593 controls	Case–control	[51]
ITIH	rs2710322	Han Chinese	1235 cases/1235 controls	Case–control	[52]
ZEB2	rs6755392	Han Chinese	1248 cases/1248 controls	Case–control	[61]
HAGH	rs11859266	Japanese	2012 cases/2170 controls	Case–control	[53]
GSK3B	rs2199503, rs6771023	Han Chinese	NA	Case–control	[32]
GRM7	rs13353402, rs1531939	Han Chinese	1034 cases/1235 controls	Case–control	[62]
NVL	rs10916544	Han Chinese	1235 cases/1235 controls	Case–control	[48]
MSH5-SAPCD1	rs707937	Japanese	1518 cases/1784 controls	Case–control	[54]
SLC44A4	rs494620				
MAD1L1	rs12666575	Chinese	700 cases/700 controls	Case–control	[55]
MAD1L1	rs12666575	Europeans	2096 cases/2744 controls	Case–control	[55]
HOMER2	rs2306428	Europeans	NA	Case–control	[63]
ANKK1, DRD2	rs1800497, rs2242592	French	NA	Case–control	[64]
IL1B	rs16944, rs1143634	Caucasians	2330 cases, 3332 controls, 221 parent–offspring trios	Meta-analysis	[65]
IL1B, IL6R	rs4848306, rs4251961, rs4537545, rs2228145	Polish	621 cases/531 controls	Case–control	[66]
FOXP1	rs2684380	Irish	15,071 cases/33,675 controls	Case–control	[67]
IMMP2L	rs4730488				
PAK6	rs12595425				
MHC	rs2523722, rs204999				
SNAP25	rs1051312	Scottish	192 cases/213 controls	Case–control	[68]
PRKAR1A	rs8066131, rs8076465	Swedish	486 cases/514 controls	Case–control	[69]

GWAS, Genome-wide association study.

serotonergic, noradrenergic, and dopaminergic, as well as genes associated with the circadian rhythm.[73] The problem with these studies is that, while initially some genes seem to have positive effects and hence a promising role in the disease onset, the results are not replicated in other studies or appear to have completely different associations in other populations. It is noteworthy that many of the genes that are associated with BD also contribute to other psychiatric illnesses, such as SZ.[74]

Over the last years, GWASs have been extensively performed to investigate BD. GWASs include analysis of the entire genome of cohorts that include both patients and healthy individuals to discover genetic markers associated with the disorder. The emerged single-nucleotide polymorphisms (SNPs) are significant, as they gradually build up the profile of the pathogenesis of complex psychiatric illnesses, such as BD. Meta-analyses also play an essential role in BD genomics research, which process and statistically analyze the studies that address the same genetic variants, so that the results have higher statistical power. In Table 5.2 we summarized the findings of many studies that indicated an association between genetic variants in specific genes with the

Table 5.2 Genomic Variants Associated With the Development of Bipolar Disorder

Gene	Variant	Ethnic Group	Number of Samples	Study	References
ANK3	rs10994336	Pakistanis	120 cases/120 controls	Case–control	[75]
ANK3	rs10994415	Europeans, Canadians, and Australians	9747 cases/14,278 controls	GWAS	[76]
ANK3	rs10994336	British	87 cases/46 controls	Case–control	[77]
ANK3	rs10994336	British	47 cases/75 relatives/67 controls	Case–control	[78]
CACNA1C	rs1006737	Pakistani	120 cases/120 controls	Case–control	[75]
CACNA1C	rs1006737	Caucasians	2000 cases/3000 controls	Case–control	[79]
CACNA1C	rs1006737	Caucasians	4387 cases/6209 controls	and GWAS	
CACNA1C	rs1006737	British	1218 cases/2913 controls	GWAS	
CACNA1C	rs2159100	NA	7481 cases/9250 controls	GWAS	[19]
CACNA1C	rs1006737	Caucasians (Greeks)	530 cases	–	[87]
CACNA1C	rs1006737	Caucasians (Germans)	110 controls	–	[88]
CACNA1C	rs4765913	Caucasians	11,974 cases/51,792 controls	GWAS	[79]
CACNA1C	rs723672	Koreans	287 cases/340 controls	Case–control	[80]
CACNA1C	rs1051375				
MAOA	rs6323	Caucasians	1093 cases/1164 controls	Meta-analysis	[89]
MAOA	MAOA-CA	Caucasians	1259 cases/1309 controls		
MAOA	MAOA-VNTR	Iranians	156 cases/173 controls	Case–control	[90]
BDNF	rs6265	NA	4248 cases/7080 controls	Meta-analysis	[79]
BDNF	rs6265	Europeans	621 cases/998 controls	Case–control	[81]
BDNF	rs6265	Europeans, Asians,	283 families	Family-based association	[91]

Table 5.2 Genomic Variants Associated With the Development of Bipolar Disorder *Continued*

Gene	Variant	Ethnic Group	Number of Samples	Study	References
		African-Americans			
BDNF	rs6265	NA	111 cases/160 controls	Case–control	[82]
COMT	rs4680	Asians	2256 controls	Meta-analysis	[83]
COMT	rs4680	British	87 cases/50 controls	Case–control	[92]
DISC1	rs3738401	Finnish	227 cases /361 non-BP relatives/171 controls	–	[93]
DISC1	rs2492367	British	515 cases/1376 controls	Case–control	[94]
DISC1	rs7546310				
DTNBP1	rs2619522	Koreans	163 cases /350 controls	Case–control	[95]
DTNBP1	rs760761				
DTNBP1	rs1064395	Europeans	2411 cases/3613 controls	GWAS	[96]
DTNBP1	rs1064395	Europeans, Americans, Australians	6030 cases/31,749 controls	GWAS	
TRANK1	rs9834970	Europeans, Asians	14,000 cases and controls (meta-analysis phase I) and 17,700 (14,000 of phase I) + 3800 cases and controls (meta-analysis phase II)	Meta-analysis of GWAS	[84]
PERIOD3	rs707467	NA	209 cases/213 controls	Case–control	[97]
PERIOD3	rs10462020	NA			
FADS1	rs174576	Chinese	1146 cases/1956 controls	Case–control	[85]
MAD1L1	rs4236274				
MAD1L1	rs4332037				
SCN2A	rs17183814				
P2RX7	rs208294	Finnish	178 cases/1322 controls	Case–control	[86]

GWAS, Genome-wide association study.

development of BD. Many studies have focused on genomic variants located in *ANK3* gene,[75,76,98] which appears to regularly contribute not only to the buildup of voltage-dependent sodium channels in the early part of the neuron, but also to the generation of energy potential.

CACNA1C gene has been associated with BD.[99] The *CACNA1C* gene is located on the small arm of chromosome 12 and is a member of the family of genes encoding voltage-dependent calcium channels. It is expressed on a large scale in the human brain and participates in the regulation of signaling, neurotransmitter release, the plasticity of synapses, and stimulation of neurons.

MAOA gene is found on chromosome X. Its coding product is the enzyme monoamine oxidase that is located in the outer mitochondrial membrane and catalyzes the oxidation of monoamines.[89,90]

Another vital gene, *BDNF*, is located on chromosome 11 and encodes the brain-derived neurotrophic factor. This protein is involved in brain development and mainly in catecholamine's system. SNPs and gene haplotypes have been associated with the occurrence of BD.[72,91,100]

The *COMT* (catechol-*O*-methyltransferase) gene is located in the large arm of chromosome 22 and encodes enzymes involved in the degradation of epinephrine, norepinephrine, and dopamine. *COMT* is also crucial in the regulation of dopaminergic transmission in the prefrontal cortex and has been associated with cognitive functions such as short-term memory. Based on the previously mentioned data, *the COMT* gene is an exciting candidate in BD research.[72,92,101]

DISC1 gene has been associated not only with BD but also with SZ, and hence the name disrupted in SZ 1.[102] Gene expression occurs at high levels in the central nervous system, but mainly in the hippocampus and cerebral cortex.[103] Several studies have been studied this gene in conjunction with the *TSNAX* gene because of their proximity.[72,94]

The *DTNBP1* gene (dystrobrevin-binding protein 1) is located on the sixth chromosome, encodes a protein expressed in all hippocampal nerve cells, and plays a role in the signaling pathway glutamatergic.[72,95]

NCAN gene is located on chromosome 19 (19p13.11) and encodes a secreted protein located in the extracellular space of Golgi's organelle and lysosomal cavities. The role of this protein is mainly the regulation of cell migration, cell adhesion, and nerve cell axis guidance.[104] It has been observed that gene expression in mice is increased in regions of the cerebral cortex and hippocampus, two areas which in previous work have been associated with the occurrence of BD.[96] For the abovementioned reasons, it is one of the possible genes responsible for the onset of the disorder.[96,104]

TRANK1 gene is located on chromosome 3 (3p22) and can be found in the literature under the name LBA1.[104] The protein encoded by this gene is called TPR and ankyrin repeat-containing protein 1 but can be found as lupus brain antigen 1, and we do not know precisely how it works. However, we know that autoantibodies to this protein are related to the appearance of systemic lupus erythematosus, a disorder that often presents neuropsychiatric symptoms.[98,104,105]

PERIOD3 is a critical member of the PERIOD gene family associated with circadian rhythm. It is located on the first chromosome (1p36.23) and expresses with high intensity in the superheated nucleus. Genetic linkage studies have shown positive results for the region of chromosome 1, where the particular gene is located.[97]

The *DAOA* gene, found on chromosome 13, appears to be associated with the occurrence of BD.[72]

The *NRG1* gene is located on the small arm of chromosome 8 and encodes the NRG1 protein. It interacts with the ErbB4 receptor and thus regulates signaling via glutamate[72] (Table 5.2).

BD, like all psychiatric illnesses, is a complex and multifactorial disorder, making it quite challenging to determine its pathogenicity. Taking into account the results of studies in families, which show the continuous occurrence of BD among individuals from generation to generation, as well as twin studies, the genetic basis of the disease is unquestionable. Over the last decade, research has been carried out to find genes related to the onset of the disease. Several studies have shown the positive association of several genes and polymorphisms with BD. In contrast to positive correlations, some studies have shown a discrepancy in results with a negative correlation of disease for the same genes, leading the scientific community to a virtual impasse. The inability to replicate the results of some studies is one of the most important hurdles in the investigation of BD because it is not feasible to establish a solid genetic background. It is also imperative that future studies must include a very large number of patients.

A possible explanation of the inconsistent results is that they mainly reveal the high heterogeneity of patients with BD. This heterogeneity may relate variables, such as race, age, and the age of onset of the disorder. For this reason, it is proposed to carry out further investigations, which will examine more homogeneous groups of patients and to correlate the results with the specific characteristics of the group.

Despite the difficulties, we hope that the determination of the genes and the role played by each one of these variants in the manifestation of BD will gradually be achieved. Having this knowledge, as far as the unique genetic profile of patient groups, diagnosis and treatment of the BD is expected to be more successful.

5.4 GENETICS OF MAJOR DEPRESSION

Depression has been a problem for individuals and societies since antiquity, but public and professional awareness has markedly increased over the past half-century. In the DSM-5 classification the central condition of depression is major depressive disorder (MDD), which is experienced as a single or recurrent episode. This classification also has several specifiers relating to severity, remission status, the pattern of illness (single or recurrent episodes), onset (specifying a perinatal onset), and clinical features, as well as describing subtypes of depression. Symptoms of MDD must (1) be present for at least 2 weeks and, during that time, experienced for almost the entire day, and (2) represent a change from the person's usual presentation, causing significant distress and negatively impacting on the person's function.

MDD is modestly heritable (30%–40%), and as such is genetically complex and likely heterogeneous, complicating researchers' efforts to identify risk loci. The successful detection and interpretation of genetic associations require both large sample cohorts and empirically driven efforts to reduce phenotypic heterogeneity.[106] All the genetic variations that are associated with MDD are shown in Table 5.3. Here are probably many more genes involved in MDD, but (1) there is insufficient evidence for those already studied because of too few or inadequately powered studies or (2) they have yet not been identified as probable candidates for study because of our limited understanding of the pathophysiology of MDD.[108]

One of the most extensively studied genomic variants is the 44-base pair insertion/deletion (5-HTTLPR) occurring in the promoter region of the *SLC6A4* gene. Moreover, there are three more genetic variations on the same gene, a variable number tandem repeat (VNTR) in intron 2 and two SNPs rs7224199 and rs3813034. Furthermore, based on new studies together with the previous ones, there is statistical evidence supporting the role of the *MTHFR* gene and especially the genetic variation c. C677T, as well as a 48 bp VNTR in Exon 3 of the *DRD4* gene. Of high statistical significance and possibly linked to MDD appear to be two SNPs rs6313 (T102C) and rs6311 (A1438G) found on the *HTR2A* gene.[116] Many meta-analyses attribute statistical importance first to the p. V158M neutral genomic variant that is founded in the *COMT* gene and second to the p. S9G variant in the *DRD3* gene.

Table 5.3 Genomic Variants Leading to Major Depressive Disorder

Gene	Variation	Ethnic Group	Number of Samples	Study	References
SLC6A3 (DAT1)	40 bp VNTR-31-UTR region	NA	423	Meta-analysis	[107]
SLC6A3 (DAT1)	rs8179029	NA			
SLC6A3 (DAT1)	rs2550936	NA			
SLC6A4 (SERT)	44 bp Ins/Del promoter	NA	21,739	Meta-analysis	[109]
SLC6A4 (SERT)	VNTR intron 2	NA			
SLC6A4 (SERT)	rs7224199	NA			
SLC6A4 (SERT)	rs3813034	NA			
MTHFR	c. C677T	NA	17,181	Meta-analysis	[110]
TPH1	rs1800532	NA	4035	Meta-analysis	[111]
DRD4	48 bp VNTR exon 3	NA	1132	Meta-analysis	[107]
DRD3	p. S9G	NA	1147		
GNB3	c. C825T	NA	867		
HTR2A	rs6313	NA	1727	Meta-analysis	[112]
HTR2A	rs6311	NA			
BDNF	p. V66M	NA	1019	Meta-analysis	[113]
				Gene-based analysis	
COMT	p. V158M	NA	NA	Meta-analysis	[107]
GABRA3	GA repeat intron 8	NA	NA	Meta-analysis	[107]
HTR1A	c. C1019G	Japanese	331 cases/804 controls	Association study and meta-analysis	[114]
HTR1B	c. G861C	NA	NA	Meta-analysis	[107]
HTR2C	rs6318				
SLC6A2	T182C	NA	3083	Meta-analysis	[115]

The association between the *GNB3* c. C825T variant and MDD was investigated, and the result is statistically significant.[107] Also, the 40 bp VNTR in the *SLC6A3* gene was investigated in three studies, and each of them showed an increased MDD risk for carriers. The same is true for the SNPs, rs8179029, and rs2550936, which are also found in the *SLC6A3* gene.

Of importance is also the insertion/deletion in intron 16 of the *ACE* gene as well as the rs3730275 variant in the *CREB1* gene. The abovementioned variants are subject to many studies, with statistically significant results to relate them to MDD. Finally, a GA repeat in intron 8 of *GABRA3* gene, in chromosome X, seems to be implicated in the emergence of MDD.[117]

Taking into consideration that there are many genes that could also be related to the emergence of MDD or with depressive symptoms, it is critical that the next generation of genetic studies of depression should remain focused on systematically querying DNA sequence variation, whether through whole-genome linkage and association studies, dense linkage disequilibrium mapping (LD) mapping of all genes in a candidate region or an extensive network of interacting genes, whole-genome or whole-transcriptome analyses, etc. Levinson[116] provided that sample cohorts gradually become larger, MDD GWAS will likely elucidate the genetic architecture of this debilitating disorder.[118]

5.5 GENETICS OF ANXIETY DISORDERS

Anxiety disorders are common and confer a substantial burden.[119] Heritability is moderate, ranging between 30% and 50%, and there is a genetic overlap between most anxiety disorders,[120,121] schematically depicted in Fig. 5.1.

The shared genetic component among panic disorder phobias and generalized anxiety disorder is larger than the disorder-specific genetic components,[122] and this is also mirrored in the high comorbidity rates between anxiety disorders, major depression, and related disorders.[123] Panic disorder has received most research attention.[121] Candidate genes, including the 5-HTTLPR variant of the *SLC6A4*, *COMT*, *GABA* receptor genes, neurotrophic genes, and others have been studied, but recent meta-analyses have not revealed any robust associations.[124,125] Turning to GWASs, most

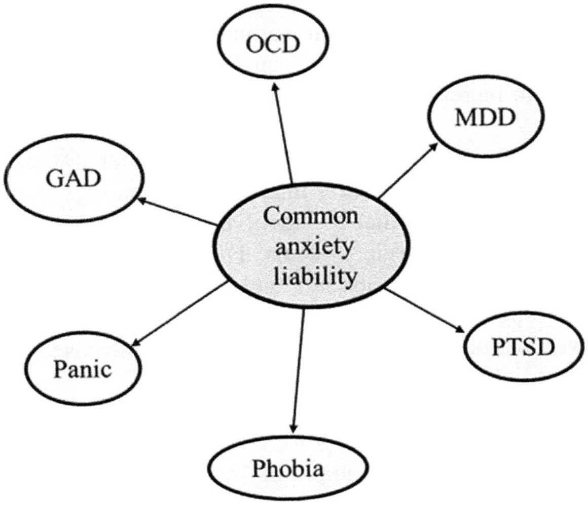

FIGURE 5.1

Schematic illustration of the shared genetic component among anxiety and stress. *GAD*, Generalized anxiety disorder; *MDD*, major depressive disorder; *OCD*, obsessive compulsive disorder; *PTSD*, posttraumatic stress disorder.

robust support has been reported for transmembrane protein 132D (*TMEM132D*), presumably having a role in threat processing.[126] Risk genotypes were rs7309727 and also an haplotype of this variant combined with rs11060369, resulting in higher expression of TMEM132D mRNA in the frontal cortex, which were found to be significantly associated with primary panic disorder in European ancestry cases.[127] There also have been studies of anxiety-related temperament and personality traits, namely, behavioral inhibition, neuroticism, harm avoidance, and introversion. Genetic variants in the *RGS2* gene have been shown to be associated with a range of anxiety- and stress-related disorders that is panic disorder, generalized anxiety disorder, and posttraumatic stress disorder (PTSD), as well as behavioral inhibition and introversion, which also predispose to anxiety disorders. Still, GWASs have not yet documented any significantly associated loci.[128–130] Estimates of SNP heritability are about 6% for neuroticism, 1%–19% for anxiety-related behaviors, and 17% for phobic anxiety, confirming the polygenic nature of anxiety personality traits, a modest effect of common variants, and a substantial contribution of environmental factors.[121,128]

5.6 GENETICS OF POSTTRAUMATIC STRESS DISORDER

PTSD is included in the section of "Trauma- and Stressor-Related Disorders in the DSM-5."[119] Diagnosis requires trauma exposure and subsequent development of a constellation of symptoms, extending to four categories, that is intrusion, avoidance, persistent negative alterations in cognition and mood, dysregulated arousal and reactivity. Trauma exposure is defined as direct or indirect (witnessing or learning for the event) exposure to actual or threatened death, serious injury, and sexual violence. Core symptoms are reexperiencing the event, persistent avoidance of related stimuli and chronic hyperarousal. It is a disorder often leading to long-lasting disability with large societal impact, related with suicide, substance use, and hospitalizations. Prevalence of PTSD is estimated about 7%–8%, with increased rates in populations exposed to violence, for example, battle combat.[131]

Heritability of PTSD is estimated between 24% and 72%, after trauma exposure, with women presenting two or three times higher heritability than men.[132] Findings from earlier candidate-gene studies have suggested genetic variants that are involved in the hypothalamic–pituitary–adrenal, adrenergic, serotonergic, and dopaminergic systems. Four SNPs in the FK506-binding protein 5 *9FKBP5* gene (rs9296158, rs3800373, rs1360780, and rs9470080; $P = .0004$), which is an important regulator of the stress response system and of the glucocorticoid receptor sensitivity, have been shown to be associated with childhood abuse and are related to increased reactivity of amygdala to threat stimuli and severity of PTSD symptoms in adulthood.[133–135] Pituitary adenylate cyclase-activating polypeptide is involved in cellular stress responses, and an SNP (rs2267735) in the *ADCYAP1R1* gene, encoding for the receptor PAC1 of this neuropeptide, was found to increase risk for PTSD in females ($P < .00002$ in a sample of 763 female subjects). Interestingly, this variant was not significant in male PTSD patients ($n = 474$).[136] This SNP, replicated in the study of Uddin et al., seems to disrupt an estrogen responsive element within ADCYAP1R1 and is associated with increased reactivity of amygdala and reduced functional connectivity between amygdala and hippocampus.[136,137] Other SNPs in genes shown to increase the risk for PTSD in adulthood have been found in the promoter region of the β2-adrenergic receptor gene (*ADRB2*),[138] within the gene encoding the serotonin transporter protein (SLC6A4),[139–141] also in DAT, a gene encoding the dopamine transporter,[142–144] in the dopamine receptor D2 gene in some[145,146] but not in other studies,[147,148] and in the COMT gene, in a patient sample from Brazil[149] and Rwanda,[150] possibly through decreased ability to reduce fear[151] (Table 5.4).

Turning to GWASs, initial efforts in relatively small samples have yielded some interesting findings. SNPs that were positively associated with PTSD were rs8042149 in the intronic region of the retinoid-related orphan receptor alpha gene (*RORA*), on chromosome 15, in a sample of 295 patients and 196 controls ($P = 2.5 \times 10^{-8}$). The product of the *RORA* gene is involved in the circadian rhythm, brain development, and neuroprotection.[152] Another variant, namely, the rs406001 on chromosome 7p12, closest to the *COBL* gene, was identified in a European American sample ($n = 1578$, $P = 3.97 \times 10^{-8}$). The product of *COBL* gene is probably involved in actin polymerization and neuronal development and function.[153] Another marker, rs10170218, located on *AC067818.1*, a long intergenic noncoding RNA gene on chromosome 2 was found to be statistically significant ($P = 5.1 \times 10^{-8}$).[154] Also, rs6482463 in the intronic region of the phosphoribosyl transferase domain containing the *PRTFDC1* gene was significantly associated with PTSD

Table 5.4 Genomic Variants Found to Increase Risk for Posttraumatic Stress Disorder, by Candidate-Gene Studies

Gene	SNPs	Function	References
9FKBP5	rs9296158, rs3800373 rs1360780, rs9470080	Encoding FK506-binding protein 5, regulator of the stress response system	[133–135]
ADCYAP1R1	rs2267735	Encoding receptor PAC1	[136,137]
ADRB2	rs2400707 (promotor)	Encoding B2 adrenergic receptor	[138]
SLC6A4	Short ("s") allele in the polymorphic region 5-HTTLPR L_G polymorphism in the long ("l") allele of the polymorphic region 5-HTTLPR, rs25531(G)	Encoding serotonin transporter	[139] [141] [140]
DAT (also known as *SLC6A3*)	40 bp VNTR in the 3′ noncoding region of the gene	Encoding dopamine transporter	[142–144,149]
D2 receptor gene	rs1800497	Encoding D2 receptor	[146] [145] [147][a] [148][a]
COMT	rs4680 (p. V58M)	Encoding catechol-*O*-methyltransferase, catabolizing enzyme of epinephrine, norepinephrine, dopamine	[143,149,150]

SNP, Single-nucleotide polymorphisms.
[a]*No association found.*

($P = 2.04 \times 10^{-9}$) ($n = 3494$, European Americans, $n = 2179$, African-Americans, $n = 205$, Hispanic and Native Americans, $n = 640$, and others, $n = 470$); however, it was not replicated in a sample of 491 white non-Hispanic US veterans.[155] The product of this gene is a small protein highly expressed in the brain.[155] The PGC PTSD Working Group, currently conducting a GWAS on civilian and veteran cohorts of over 40,000 subjects of multiethnic origin,[156] has recently published results on a sample of 20,070 subjects.[157] In this study, no variant reached genome-wide significance. Heritability estimate from SNP data was 29% for females, and 7% for males, but the latter was not distinguishable from 0.[157]

5.7 GENETICS OF SUBSTANCE USE DISORDERS

Among psychiatric disorders, substance use disorders (SUD) require an exposure to an environmental factor that is the substance, to manifest, similarly to the PTSD as nosological entity, requiring an exposure to a traumatic event. As such, the environmental component of SUD is substantial, and genetic factors account for nearly 50%–60% of SUD variability.[158] Most research has focused on alcohol and nicotine dependence. Candidate genes that have been shown to influence risk for alcohol abuse and dependence are *ADH1B*, *ADLDH2*, and *ADH4*, in earlier studies, but more recently GWASs are employed. Regarding nicotine dependence, the most robust findings implicate *CHRNA5*, *CHRNA3*, and *CHRNB4* on chromosome 15q25, which encode neuronal subunits of the nicotinic acetylcholine receptor (nAChr). The risk allele rs16969968 conferred 30% increased risk for nicotine dependence, and those carrying two copies have more than twice the risk to develop nicotine dependence.[159,160] GWASs targeting alcohol dependence have been less fruitful, and the largest studies so far have failed to identify SNPs that are significantly associated with risk of developing SUD. Given the (1) fact that exposure to the substance is necessary for the condition to evolve, and (2) phenotypic variability of the samples, for example, quantity of the substance that is consumed or severity of dependence is vast, and studies including larger sample sizes must be awaited.

5.8 GENETICS OF EATING DISORDERS

Heritability of the most studied eating disorders, namely, anorexia nervosa, bulimia nervosa, and binge eating disorder, derived from twin studies, averages between 40% and 65%.[161] Earlier studies of candidate genes have proved unfruitful, and as it is common in complex, polygenic psychiatric disorders, probably hundreds to even thousands of genes contribute to the disorder, each one of which contributing with a small risk percentage.[162,163] The Eating Disorders Working Group of the PGC was established in 2013 and yielded a genome-wide significant locus for anorexia nervosa on chromosome 12, close to genes involved in type I diabetes and other autoimmune conditions by combining data sets from previous GWASs.[164–166] SNP-based heritability for anorexia nervosa is estimated to be 20%, that is 20% of the liability to manifest this disorder is accounted for by common genetic variants.[167]

5.9 GENETICS OF AUTISM SPECTRUM DISORDER

Autism spectrum disorder (ASD) comprises a heterogeneous group of neurodevelopmental conditions, recognizable in early childhood, and characterized by defective social cognition and communication, restricted interests, and repetitive or compulsive behaviors, as well as impaired verbal communication. Previous diagnostic classifications included phenotypic presentations such as classical autism, Rett syndrome, Asperger syndrome, childhood disintegrative disorder, and pervasive developmental disorder not otherwise specified, which, with the exception of Rett disorder, are all subsumed under the diagnosis of ASD, according to DSM-5.[119] In this last classification, ASD is defined as a disorder with a history of deficits in two core domains: (1) social-communication ability and (2) restricted interest and repetitive behaviors. Median rate of ASD is 0.62%, ranging from 0.01% to 1.89% globally.[168] Lifetime prevalence is about 1%.[169]

A significant genetic component has been well documented in ASDs and was first suggested by twin and family studies. It has been found that ASD is developed in up to 26% of siblings of proband suffering from autism.[170,171] Monozygotic concordance rates range between 40% and 90%, whereas for dizygotic twins concordance rate is found to be 0%–30%. Estimates of heritability are highly variable, ranging from 36% to 95%,[172,173] and this is partly due to the disorder's heterogeneity. To express the disorder, female siblings seem to need an extended genetic variant load, as they are less likely to have ASD compared to their male counterparts.[170] Overall these findings support the view that the genetic component is strong; however, there also exists a significant environmental influence interacting with the genomic background of the subject, as with every psychiatric disorder mentioned earlier.

Moving to GWASs, a positive association was shown for a genetic variant in the *MACROD2* gene, which however was not replicated in a study by the Autism Genome Project.[174,175] In general, although polygenic risk due to a large number of variants is suggested for autism, search for common variants contribution has not yet been fruitful, and study samples are still underpowered.

On the other hand, rare genetic variants contribute substantially to patient's risk of developing ASD and include gross chromosomal rearrangements, submicroscopic deletions or duplications, such as CNV, or even smaller structural changes, such as small indels and single-nucleotide variants. Rare variants can be inherited from unaffected parents or arise de novo during meiosis of the gametes. Paternal and maternal ages at conception have been associated with ASD risk. Advanced maternal age is correlated with chromosomal aberrations, and advanced paternal age has been shown to correlate with increased rate of de novo single-nucleotide variants and CNV.[176]

Rare de novo CNVs have more robustly been found to increase risk for ASD.[177,178] The rate of de novo CNVs in patients with ASD is 5%–10%, which is significantly higher than that observed in control subjects or unaffected siblings (<1%–2%).[177,179]

5.10 GENETICS OF ATTENTION DEFICIT HYPERACTIVITY DISORDER

Attention deficit hyperactivity disorder (ADHD) is a neurodevelopmental condition of high heritability, estimated at 60%–90%, and highly polygenic, meaning that there are thousands of common and rare variants, each contributing a small increase in the overall risk of developing ADHD.[180]

There is familial overlap with ASD and intellectual disability, owing to the presumed genetic overlap.[181,182] Recent GWAS results on a sample comprising 20,183 individuals diagnosed with ADHD and 35,191 controls have yielded 304 genetic variants in 12 genetic loci at highly significance level ($P < 5 \times 10^{-8}$).[183] Implicated genes, for example, *FOXP2* and *SORCS3*, are involved in fundamental processes, such as synapse formation, development of speech, neuronal development, and plasticity.[184,185] CNVs have also been associated with ADHD, and there is a significant overlap with CNVs found in other neurodevelopmental disorders, such as ASD and SZ.[186,187]

5.11 PHARMACOGENOMICS OF PSYCHIATRIC TREATMENT MODALITIES

Since the discovery of genetic variations that are associated with the development of specific psychiatric disorders, such as SZ, BD, MDD, and others, researchers focused on genetic variations that are associated with patients' response to antipsychotics, antidepressants, and mood stabilizers. The following sections summarize recent innovations and perspectives in antipsychotic, antidepressant, and mood stabilizer pharmacogenomics.

5.11.1 PHARMACOGENOMICS OF ANTIPSYCHOTIC DRUGS

Antipsychotic drugs are only effective against clinical symptoms of psychiatric disorders, such as hallucinations, while about one-thirds of the patients have little or no benefit from the treatment.[188] Consequently, current pharmacotherapy strategies do not adequately address cognitive deficits of psychiatric illnesses, underlining the urgent need for prognostic therapies to develop.[189–191]

Pharmacogenomics research focuses on genomic markers that are involved in the metabolism of antipsychotic drugs through liver enzyme pathways, mainly cytochrome P450 (CYP) enzymes. Genetic changes in drug-metabolizing enzymes may lead to altered metabolic activity, which may affect the pharmacokinetic parameters of antipsychotics.[192]

Genomic variants in drug-metabolizing genes that may affect their function include SNPs, gene multiplications or insertions/deletions, small insertions/deletions, or more significantly CNVs that affect gene expression or protein conformation. Individuals that bear these variants can be categorized as intermediate metabolizers (IMs) or poor metabolizers (PMs), if they carry genomic variants in heterozygous or homozygous state, respectively, expensive metabolizers (EMs) if they have no deleterious genomics variants in these genes (see also Chapter 6) and lastly ultrarapid metabolizers (UMs) if they bear multiplications of the *CYP2D6* gene. EMs indicate regular enzyme activity, whereas IMs partially reduced enzyme activity, PMs indicate little or no enzyme activity, and UMs increased enzyme activity.[192]

Fifteen approved antipsychotics, among them aripiprazole, brexpiprazole, iloperidone, and pimozide are major or minor substrates of CYP2D6 that is an important hepatic enzyme, activity of which significantly affects the biotransformation of its substrates and, therefore, pharmacokinetics. Until today, hundreds of genetic variants have been identified in the *CYP2D6* gene. *CYP2D6**2, *33, and *35 (*1 being the wild-type allele) are classified as active alleles, whereas *CYP2D6**9, *10, *17, *29, *36, and *41 alleles cause reduced enzyme activity as a result of reduced gene expression (*9, *41) or modified protein conformation (*36).[192,193] According to many studies,

*CYP2D6**10 homozygosity indicates PMs and is associated with increased serum concentration risperidone levels and higher risk of adverse drug reactions.[194–201] CYP2D6 is also associated with the appearance of extrapyramidal symptoms and tardive dyskinesia (TD) in patients administered antipsychotic drugs.[202]

CYP1A2 represents another essential enzyme responsible for the biotransformation of certain antipsychotics.[203] Its most common variants are *CYP1A2**1F, *1K, *4, *5, *6, *7, *8, *11, *15, and *16.[203] *CYP1A2**1F is associated with high rates of TD in a cohort of Chinese patients that are treated with antipsychotic drugs but not in a cohort of Caucasian patients.[204,205] Notably, smoking is associated with the induction of CYP1A2, by increasing the clearance of olanzapine, clozapine, chlorpromazine, and haloperidol. Smoking has a greater effect on clozapine metabolism than *CYP1A2**1C and *1F. Thus the influence of genetic and environmental factors on the CYP1A2-mediated drug metabolism makes it difficult to understand its effects.[206]

CYP3A enzymes are involved in the metabolism of many antipsychotics.[207] CYP3A5 is located on chromosome 7q22.1 and contains 13 coding exons, with more than 2000 genomic variants reported to date. Although the enzymatic activity of CYP3A4 and CYP3A5 is variable in the general population, most of the detected *CYP3A4* variant alleles are either very rare or have little effect on the CYP3A4 end-activity. The *CYP3A4* alleles tested include *13, *15A, and *22. The most common alleviation loss of function in Caucasian populations, *CYP3A5**3, was identified at frequencies ranging from 94% among Europeans to 18% among African-Americans. Other loss-of-function alleles include *CYP3A5**6 and *7. *CYP3A5**2 is considered a reduced-function allele.[208] Although there is some evidence that reduced-function alleles increase exposure to certain antipsychotics, other studies have not found such effects.

The *P*-glycoprotein (*P*-gp) drug transporter, encoded by the *ABCB1* gene, is a transmembrane transporter. *P*-gp may affect the absorption and distribution of substrates. Many antipsychotics, such as risperidone, olanzapine, aripiprazole, and paliperidone, are substrates of *P*-gp, and therefore the ABCB1 enzyme can play an essential role in their availability.[209] Three of the most commonly studied variants of the *ABCB1* gene are rs1128503, rs2032582, and rs1045642. Preliminary evidence suggests that patients who carry one of these rare alleles may have a higher risk of side effects and higher plasma levels in the case of olanzapine, risperidone, and clozapine. However, the clinical utility of these variants is still uncertain.

All approved antipsychotics are antagonists of dopamine D2 receptors. Hence, the genes of these receptors are essential candidates for pharmacogenomics studies. The rs1800497 and rs1799732 are the most well-known variants of the *DRD2* gene. In the case of rs17999732, patients carrying the deletion did not have a positive response to risperidone and olanzapine.[210–212] Studies also show a correlation of this genomic variant with side effects, such as hyperprolactinemia and weight gain.[213,214] Regarding rs1800497, Chinese and Korean patients that are treated with aripiprazole had better response rates and lower PANSS scores, whereas Chinese patients also have higher risk of TD.[211,212,214–216]

Of particular interest is the affinity of antipsychotics for D3 receptors. Pharmacogenomic research associated with antipsychotic drugs has focused on rs6280 (or p. S9G) of *DRD3* gene. Many studies have investigated the relationship between rs6280 and response to antipsychotic drugs, observing a nonsignificant trend between the p. 9S allele and reduced response to antipsychotics.[217–219] Apart from that, multiple studies have reported a significant correlation between the p. S9G variant and late dyskinesia, with three recent meta-analyses providing contradictory findings with moderate-to-no effect of p. 9G allele on TD appearance.[216]

COMT is a crucial enzyme for the clearance and metabolic termination of dopamine activity. The p. V158M variant is the most well-studied single-nucleotide variant of this gene. The p. 158M allele has been associated with an increased clinical response to antipsychotics, by improving negative symptoms and cognitive functions. Despite the controversial results, two latest studies indicate that carriers of the p. 158M allele and have a better response to atypical antipsychotics and improvement in positive symptoms.[220,221] However, the correlation between the p. 158M allele with protection against TD remains negligible.[222]

The serotonin A2 receptor is encoded by the *HTR2A* gene and is the most extensively studied serotonin receptor as the target of the second generation of antipsychotic drugs. The rs6313 (or c. C102T, where C is the ancestral allele) variant in the *HTR2A* gene shows little correlation with the response to atypical antipsychotics, with the variant allele (T) causing a moderate increase in response. Also, the relationship between the variant rs6313 allele[223–225] and a small increase in the risk of TD has been suggested. The wild-type G allele of rs6311 in the *HTR2A* gene has been found to be associated with reduced promoter activity and reciprocal response to atypical antipsychotics, but the results from various studies are ambiguous.[224,226,227] Other SNPs in the *HTR2A* gene, including p. H452T and the p. T25N, also have been associated with response to antipsychotics and their side effects, but with controversial results among different studies and uncertain clinical effects.[228]

The serotonin 2C receptor, encoded by *HTR2C*, has also been extensively studied. According to previous studies, genomic variants in the *HTR2C* gene are associated with the improvement of patients' negative symptoms, and possibly with amelioration of cognitive dysfunction,[225,229] as well as with antipsychotic-mediated weight gain.[230,231] Despite the various single-nucleotide variants of this gene associated with variable therapeutic effects, the most studied variant is rs3813929, due to its involvement in weight gain resulting from antipsychotics intake. The rs3813929C allele (wild-type allele) is associated with an increased chance of weight gain, whereas the rs3813929T allele (common allele) causes increased expression of the gene and thus less weight gain. Antipsychotics with high affinity to 5-HT2C, such as clozapine and olanzapine, lead to weight gain.

Overall, pharmacogenomic research of antipsychotic drugs is on the rise. Genetic variants of dopamine and serotonin receptors as well as genetic variants in drug metabolic pathways, including COMT and CYP2D6, may affect drug efficacy and adverse drug reactions. The clinical implementation of the previous research findings is at an early stage of development (see later), but at the same time promising to improve pharmacotherapy. Pharmacogenomics can offer personalized treatment based on patients' genetic background, by maximizing the therapeutic effects and minimizing drug-induced side effects. This type of personalized treatment can one day lead to lower medication costs by blocking the "trial-and-error" approach that usually occurs.

5.11.2 PHARMACOGENOMICS OF ANTIDEPRESSANT DRUGS AND MOOD STABILIZERS

Mental disorders are difficult to treat and very often require adjustments of medication and dose. Notably, only 50%–60% of patients receiving antidepressants and mood stabilizers will show complete or even satisfactory treatment response, even after long-term administration. Genetic factors play an essential role in both drug response and adverse drug reactions, with studies assessing that 42%–50% of the antidepressant response is due to the patients' genetic background. It is understood, therefore, that the variability in response to a drug remains a challenge in the treatment of mental disorders. Thus the field of pharmacogenomics finds good ground for application, with patients' genetic profile and their response to drugs to constitute the appropriate guideline for choosing the appropriate dosage and drug in order to maximize response and minimize the undesirable side effects.

As in the case of antipsychotic drugs, genes of the CYP450 family and their genomic variants are indicative biomarkers of antidepressant drugs. The most important of these, which are implicated in the pharmacokinetic pathway of antidepressants and mood regulators in patients of different nationalities, are CYP2D6, CYP2C19, CYP2C9, CYP1A2, and CYP3A4.[232]

*CYP2D6**4, *10, and *17 have been associated with poor metabolism of selective serotonin reuptake inhibitors (SSRIs),[233–235] and tricyclic antidepressants (TCAs),[233,235,236] and after reduction of the dose, the response to these drugs has shown to be increased. These variants seem also to be associated with serotonin syndrome, which is mainly characterized by euphoria, somnolence, dizziness, tremor, diarrhea, etc.[237] as well as cardiotoxicity, suicidal ideation, and liver dysfunctions.[237–240] In the case of serotonin–norepinephrine reuptake inhibitors (SNRIs), for example, venlafaxine, *CYP2D6**4 and *6 showed no statistically significant association in a cohort of both Caucasian and Asian populations.[241,242] However, these genetic variants were associated with serotonin syndrome, weight loss, decreased appetite, hyponatremia, seizures, cardiac abnormalities, anemia, increased risk of bleeding, and liver disease in cases where venlafaxine dose was not adjusted.[237,243]

*CYP2C19**2, *3, and *17 are also crucial for SSRIs-mediated metabolism,[233–235] and TCA.[233,235,236] Moreover, studies focusing on *CYP2C19**2, and *3, indicate that in case the dose is not adjusted, patients' response is affected, and side effects, such as somnolence, dizziness, and tremor, may occur.[237–240]

UDP-glucuronosyltransferases are involved in the transformation of the substrate into soluble and more polar metabolites for ABC transporters. The rs72551349, rs3821242, rs45625338, and rs6431625 variants in the *UGT1* gene, the rs2011425 variant in the *UGT1A4* gene, and the rs2942857 variant in the *UGT2B10* gene are associated with TCA metabolism.[232] The rs2032582 and rs2235015 variants in the *ABCB1* gene lead to increased drug levels and decreased the response, which is associated with the appearance of somnolence in patients under citalopram, venlafaxine, desipramine, paroxetine, and amitriptyline. SLC6A4 and 5-HTTLRP have been the subject of research in many candidate-gene studies. Patients have a poor response to SSRIs, SNRIs, and TCAs and are more sensitive in experiencing side effects.[232,244] Many studies have also focused on the rs7997012, rs7333412, rs6314, rs6311, and rs6313 variants in the *HTR2A* gene, and, even though there are inconsistent results among different studies, an association with increased response to SSRIs and SNRIs in Caucasian populations seems to exist.[232,244] Other genes that are associated with better response to antidepressants include rs1360780, rs5443, and rs6265 variants in the *FKBP5*, *GNB3*, and *BDNF* genes, respectively.[232]

According to GWASs, *UBE3C*, *BMP7*, *RORA*, *CDH17*, *AUTS2*, *HPRTP4*, *NRG1*, *STMN2*, *LOC644659*, *LOC100130766*, *MTMR12*, *CUX1*, and *BMP5* genes indicated high statistical significance and are associated with the efficacy of antidepressants.[245–250]

Regarding mood stabilizers, lithium is the one that has mostly received the attention of researchers. Candidate-gene studies have indicated that several genetic variations located in the *5-HTT*, *TPH*, *DRD1*, *SLC6A4*, *ASIC2* (or *ACCN1*), *INPP1*, *CREB1*, *ARNTL*, *TIM*, *DPB*, *NR3C1*, *BCR*, *XBP1*, *PLGC1*, and *CACNG2* genes are associated with lithium response.[235,251,252] Also, according to our own study, rs378448 located in *ASIC2* gene is also associated with kidney dysfunction in patients under long-term lithium treatment.[253]

Moreover, there are a lot of pharmacogenomic studies that focus on 5-HTT and indicate an association with patients' response to lithium.[254–256] The rs6265, rs3786282, rs334558, rs3730353, and rs4532 located in the *BDNF*, *IMPA2*, *GSK3β*, *FYN*, and *DRD1* genes, respectively, are associated with better response to lithium treatment.[254,257–260] The rs9663003, rs75222709, rs74795342, and rs78015114 variants in the *CYP2D6* gene are associated with lithium response and toxicity, such as tremor, confusion, cognitive dysfunction, kidney failure, gastrointestinal problems, nausea, and diarrhea.[235,261]

In 2008 the International Consortium on Lithium Genetics (ConLiGen consortium; http://www.conligen.org) was founded by eminent experts in the field of lithium genetics. The consortium aims to analyze data of patients who are treated with lithium, using high-throughput analysis approaches.[262] In particular, the consortium's first and most crucial goal is to define the phenotype of lithium response, which is a complex trait that requires researchers to make judgments about adequacy of treatment and tolerability as well as assess changes in episode frequency or symptom severity. With the largest combined sample to analyze lithium response to date, ConLiGen is poised to assess all aspects of the pharmacogenetics of lithium treatment in psychiatric disorders, including the study of genetic susceptibility to potential treatment-emergent adverse events (e.g., weight gain, hypothyroidism, and tremor).

Carbamazepine has also been associated with high blood levels and kidney dysfunction, as well as very severe dermatological syndromes, such as Stevens–Johnson, due to reduced metabolism by *CYP3A4**1B.[263] Furthermore, the *COMT* gene and rs4680 are associated with better response to carbamazepine.[264] Variants in the *DRD2*, *DBH*, *HRH1*, and *MCR2* genes have been associated with response to lamotrigine,[265] whereas patients' response to valproate has been correlated with variants in the *XBP1* gene.[266]

Given the vital role that drug metabolism in drug response, it is not surprising that variants in the *CYP2D6* and *CYP2C19* genes are the main subject of personalized therapy research. Also, due to significant differences in the prevalence of these genes variants among different populations, there is also the variability of antidepressant reactions between individuals that may even carry the same variant in some cases, indicating a possible polygenic effect. Therefore it is crucial to know the genetic variants related to absorption, distribution, metabolism, and excretion, which will reveal effects that may affect populations as a whole.[267] Future studies on the pharmacogenomics of antidepressants will benefit from including more individuals from subethnic groups.

Regarding the adverse antidepressant and mood stabilizers reactions, many of them can be potentially avoided, and a large number of patients can benefit from pharmacogenomics-guided dose adjustment, by receiving the appropriate medication and dosage based on their genetic profile. Finally, the pharmacogenomic test before treatment with antidepressants and mood regulators may soon be realistic, so that general guidelines linking genetic variants with treatment response can be issued. Taking these into account, the ideal planning of treatment is made possible so that patients can benefit as much as possible.

5.12 CLINICAL IMPLEMENTATION OF PHARMACOGENOMICS IN PSYCHIATRY

Despite the availability of a large number of pharmacogenomics studies, the implementation of pharmacogenomics into clinical practice remains still very limited, especially in the field of psychiatry.[268] The lack of guidelines for the clinical implementation of pharmacogenomic testing, appropriate bioinformatics translational tools, and proper knowledge of clinicians and health-care professionals are few of the main reasons.[269,270]

Since 2011 in the United States, a number of consortia that focus on the implementation of pharmacogenomics into clinical practice have been founded, such as IGNITE (Implementing Pharmacogenomics in Practice),[271] INGENIOUS (Indiana GENomics Implementation: an Opportunity for the UnderServed),[272] eMERGE-PGx (Electronic Medical records και genomics Network-Pharmacogenomics),[273] 1200 patients program, and[274] Cleveland Clinic's Personalized Medication Program.[275] In 2016 the Ubiquitous Pharmacogenomics project (U-PGx, www.upgx.eu) was launched. This was not only the first multicenter, open, randomized clustering, crossover study of pharmacogenomics application in clinical practice in Europe but also the very first study that focused on psychiatric drugs worldwide.

The study aims to guide the dose and the drug choice and to reduce the number of adverse drug reactions. Patients will avoid the administration of drugs that would not improve their symptoms and moreover cause side effects. Finally, and most importantly, patients' quality of life will be improved, and the overall cost will be reduced.

5.13 EDUCATIONAL, ETHICAL, AND LEGAL ISSUES

In view of the exponential increase in findings from genomic research, the issue of educating clinicians on how to interpret genomic information, how to implement in clinical practice, and how to communicate findings to the patients and the families has become more urgent than ever.[276] A large majority of psychiatrists have had minimal training on genetics, and their knowledge on the field is most probably outdated, at a level that for many it would be difficult of follow up with new genetic discoveries.[277] Although there is optimism among psychiatrists about the future role of genomics in aiding diagnosis and treatment, lack of awareness and experience with the genetic tests is also reported.[278] It is crucial that there will be initiatives for integration of a substantial component of genetics training in residency programs of psychiatry, so that future clinicians, active in the next 20 years, can fruitfully incorporate genomic information in their professional practice.[276]

Concerns about ethical issues are also expressed, such as fears of stigmatization, or psychological distress imposed on the individual undergoing genetic testing. Legal and insurance issues could arise, and employment discrimination could represent a future important risk.[279] The Genetic Information Nondiscrimination Act of 2008 is an example of a legal act aiming to protect the individual from discrimination due to genetic findings in the United States, but it seems that there still is a long way to go to ensure regulatory framework and individual safety.[280]

5.14 CONCLUSION AND FUTURE PERSPECTIVES

After struggling for decades in the quest of the elucidation of the genetic background underlying psychiatric diseases, it is now the time for the field to make a true and groundbreaking advancement, with large-scale GWASs as well as other high-throughput technologies, expected to yield increasingly more significant findings. This can set the basis for fulfilling the basic aims of genomic research, which expands from profound understanding of the nature of psychiatric illnesses and transformation of nosological classification, to inform clinical diagnosis and care and provide personalized treatment. Having in mind all the obstacles and caveats, it is reasonable to be optimistic that in the next few years personalized medicine, in the sense of personalized care based at large on genomic information, will be widely offered.

REFERENCES

1. Sullivan PF, Agrawal A, Bulik CM, et al. Psychiatric genomics: an update and an agenda. *Am J Psychiatry*. 2018;175:15–27. Available from: https://doi.org/10.1176/appi.ajp.2017.17030283.
2. Hoehe MR, Morris-Rosendahl DJ. The role of genetics and genomics in clinical psychiatry. *Dialogues Clin Neurosci*. 2018;20:169–177.
3. Gratten J, Wray NR, Keller MC, Visscher PM. Large-scale genomics unveils the genetic architecture of psychiatric disorders. *Nat Neurosci*. 2014;17:782–790. Available from: https://doi.org/10.1038/nn.3708.
4. Psychiatric GWAS Consortium Coordinating Committee, Cichon S, Craddock N, et al. Genomewide association studies: history, rationale, and prospects for psychiatric disorders. *Am J Psychiatry*. 2009;166:540–556. Available from: https://doi.org/10.1176/appi.ajp.2008.08091354.
5. Corvin A, Craddock N, Sullivan PF. Genome-wide association studies: a primer. *Psychol Med*. 2010;40:1063–1077. Available from: https://doi.org/10.1017/S0033291709991723.
6. Gratten. Biological insights from 108 schizophrenia-associated genetic loci. *Nature*. 2014;511:421–427. Available from: https://doi.org/10.1038/nature13595.
7. Cecchin E, Roncato R, Guchelaar HJ, Toffoli G, Ubiquitous Pharmacogenomics Consortium. Ubiquitous Pharmacogenomics (U-PGx): the time for implementation is now. An Horizon2020 program to drive pharmacogenomics into clinical practice. *Curr Pharm Biotechnol*. 2017;18:204–209. Available from: https://doi.org/10.2174/1389201018666170103103619.
8. Owen MJ, Sawa A, Mortensen PB. Schizophrenia. *Lancet (London, England)*. 2016;388:86–97. Available from: https://doi.org/10.1016/S0140-6736(15)01121-6.
9. Kahn RS, Sommer IE, Murray RM, et al. Schizophrenia. *Nat Rev Dis Prim*. 2015;1:15067. Available from: https://doi.org/10.1038/nrdp.2015.67.
10. Szkultecka-Dębek M, Walczak J, Augustyńska J, et al. Epidemiology and treatment guidelines of negative symptoms in schizo-phrenia in Central and Eastern Europe: a literature review. *Clin Pract Epidemiol Ment Health*. 2015;11:158–165. Available from: https://doi.org/10.2174/1745017901511010158.
11. Tsuang M. Schizophrenia: genes and environment. *Biol Psychiatry*. 2000;47:210–220.
12. Bush WS, Moore JH. Chapter 11: genome-wide association studies. *PLoS Comput Biol*. 2012;8:e1002822. Available from: https://doi.org/10.1371/journal.pcbi.1002822.
13. Bamshad MJ, Ng SB, Bigham AW, et al. Exome sequencing as a tool for Mendelian disease gene discovery. *Nat Rev Genet*. 2011;12:745–755. Available from: https://doi.org/10.1038/nrg3031.
14. Patnala R, Clements J, Batra J. Candidate gene association studies: a comprehensive guide to useful in silico tools. *BMC Genet*. 2013;14:39. Available from: https://doi.org/10.1186/1471-2156-14-39.

15. Zhang Z, Yu H, Jiang S, et al. Evidence for association of cell adhesion molecules pathway and NLGN1 polymorphisms with schizophrenia in Chinese Han population. *PLoS One*. 2015;10:e0144719. Available from: https://doi.org/10.1371/journal.pone.0144719.
16. Athanasiu L, Mattingsdal M, Kähler AK, et al. Gene variants associated with schizophrenia in a Norwegian genome-wide study are replicated in a large European cohort. *J Psychiatr Res*. 2010;44:748–753. Available from: https://doi.org/10.1016/j.jpsychires.2010.02.002.
17. Luan Z, Lu T, Ruan Y, Yue W, Zhang D. The human MSI2 gene is associated with schizophrenia in the Chinese Han population. *Neurosci Bull*. 2016;32:239–245. Available from: https://doi.org/10.1007/s12264-016-0026-9.
18. Li Z, Chen J, Xu Y, et al. Genome-wide analysis of the role of copy number variation in schizophrenia risk in Chinese. *Biol Psychiatry*. 2016;80:331–337. Available from: https://doi.org/10.1016/j.biopsych.2015.11.012.
19. Lee KY, Joo E-J, Ji YI, et al. Associations between DRDs and schizophrenia in a Korean population: multi-stage association analyses. *Exp Mol Med*. 2011;43:44–52. Available from: https://doi.org/10.3858/emm.2011.43.1.005.
20. Betcheva ET, Mushiroda T, Takahashi A, et al. Case–control association study of 59 candidate genes reveals the DRD2 SNPrs6277 (C957T) as the only susceptibility factor for schizophrenia in the Bulgarian population. *J Hum Genet*. 2009;54:98–107. Available from: https://doi.org/10.1038/jhg.2008.14.
21. Fan H, Zhang F, Xu Y, et al. An association study of DRD2 gene polymorphisms with schizophrenia in a Chinese Han population. *Neurosci Lett*. 2010;477:53–56. Available from: https://doi.org/10.1016/j.neulet.2009.11.017.
22. Gupta M, Chauhan C, Bhatnagar P, et al. Genetic susceptibility to schizophrenia: role of dopaminergic pathway gene polymorphisms. *Pharmacogenomics*. 2009;10:277–291. Available from: https://doi.org/10.2217/14622416.10.2.277.
23. Hoenicka J, Aragüés M, Rodríguez-Jiménez R, et al. C957T DRD2 polymorphism is associated with schizophrenia in Spanish patients. *Acta Psychiatr Scand*. 2006;114:435–438. Available from: https://doi.org/10.1111/j.1600-0447.2006.00874.x.
24. Lawford BR, Young RM, Swagell CD, et al. The C/C genotype of the C957T polymorphism of the dopamine D2 receptor is associated with schizophrenia. *Schizophr Res*. 2005;73:31–37. Available from: https://doi.org/10.1016/j.schres.2004.08.020.
25. Monakhov M, Golimbet V, Abramova L, Kaleda V, Karpov V. Association study of three polymorphisms in the dopamine D2 receptor gene and schizophrenia in the Russian population. *Schizophr Res*. 2008;100:302–307. Available from: https://doi.org/10.1016/j.schres.2008.01.007.
26. Zheng C, Shen Y, Xu Q. Rs1076560, a functional variant of the dopamine D2 receptor gene, confers risk of schizophrenia in Han Chinese. *Neurosci Lett*. 2012;518:41–44. Available from: https://doi.org/10.1016/j.neulet.2012.04.052.
27. Kukshal P, Kodavali VC, Srivastava V, et al. Dopaminergic gene polymorphisms and cognitive function in a north Indian schizophrenia cohort. *J Psychiatr Res*. 2013;47:1615–1622. Available from: https://doi.org/10.1016/j.jpsychires.2013.07.007.
28. Shi J, Gershon ES, Liu C. Genetic associations with schizophrenia: meta-analyses of 12 candidate genes. *Schizophr Res*. 2008;104:96–107. Available from: https://doi.org/10.1016/j.schres.2008.06.016.
29. Moens LN, De Rijk P, Reumers J, et al. Sequencing of DISC1 pathway genes reveals increased burden of rare missense variants in schizophrenia patients from a northern Swedish population. *PLoS One*. 2011;6:e23450. Available from: https://doi.org/10.1371/journal.pone.0023450.
30. Rastogi A, Zai C, Likhodi O, Kennedy JL, Wong AH. Genetic association and post-mortem brain mRNA analysis of DISC1 and related genes in schizophrenia. *Schizophr Res*. 2009;114:39–49. Available from: https://doi.org/10.1016/j.schres.2009.06.019.

31. He B-S, Zhang L-Y, Pan Y-Q, et al. Association of the DISC1 and NRG1 genetic polymorphisms with schizophrenia in a Chinese population. *Gene*. 2016;590:293–297. Available from: https://doi.org/10.1016/j.gene.2016.05.035.
32. Chen J, Wang M, Waheed Khan RA, et al. The GSK3B gene confers risk for both major depressive disorder and schizophrenia in the Han Chinese population. *J Affect Disord*. 2015;185:149–155. Available from: https://doi.org/10.1016/j.jad.2015.06.040.
33. Kawashima K, Ikeda M, Kishi T, et al. BDNF is not associated with schizophrenia: data from a Japanese population study and meta-analysis. *Schizophr Res*. 2009;112:72–79. Available from: https://doi.org/10.1016/j.schres.2009.03.040.
34. Kayahan B, Kaymaz BT, Altıntoprak AE, Aktan Ç, Veznedaroğlu B, Kosova B. The lack of association between catechol-*O*-methyltransferase (COMT) Val108/158Met and brain-derived neurotrophic factor (BDNF) Val66Met polymorphisms and schizophrenia in a group of Turkish population. *Neurol Psychiatry Brain Res*. 2013;19:102–108. Available from: https://doi.org/10.1016/J.NPBR.2013.05.004.
35. Naoe Y, Shinkai T, Hori H, et al. No association between the brain-derived neurotrophic factor (BDNF) Val66Met polymorphism and schizophrenia in Asian populations: evidence from a case–control study and meta-analysis. *Neurosci Lett*. 2007;415:108–112. Available from: https://doi.org/10.1016/j.neulet.2007.01.006.
36. Neves-Pereira M, Cheung JK, Pasdar A, et al. BDNF gene is a risk factor for schizophrenia in a Scottish population. *Mol Psychiatry*. 2005;10:208–212. Available from: https://doi.org/10.1038/sj.mp.4001575.
37. Pełka-Wysiecka J, Wroński M, Jasiewicz A, et al. BDNF rs 6265 polymorphism and COMT rs 4680 polymorphism in deficit schizophrenia in Polish sample. *Pharmacol Rep*. 2013;65:1185–1193.
38. Yi Z, Zhang C, Wu Z, et al. Lack of effect of brain derived neurotrophic factor (BDNF) Val66Met polymorphism on early onset schizophrenia in Chinese Han population. *Brain Res*. 2011;1417:146–150. Available from: https://doi.org/10.1016/j.brainres.2011.08.037.
39. Zhang XY, Chen D-C, Tan Y-L, et al. BDNF polymorphisms are associated with schizophrenia onset and positive symptoms. *Schizophr Res*. 2016;170:41–47. Available from: https://doi.org/10.1016/j.schres.2015.11.009.
40. Chen S-L, Lee S-Y, Chang Y-H, et al. The BDNF Val66Met polymorphism and plasma brain-derived neurotrophic factor levels in Han Chinese patients with bipolar disorder and schizophrenia. *Prog Neuropsychopharmacol Biol Psychiatry*. 2014;51:99–104. Available from: https://doi.org/10.1016/j.pnpbp.2014.01.012.
41. Dubertret C, Hanoun N, Adès J, Hamon M, Gorwood P. Family-based association studies between 5-HT5A receptor gene and schizophrenia. *J Psychiatr Res*. 2004;38:371–376. Available from: https://doi.org/10.1016/j.jpsychires.2004.01.002.
42. Ikeda M, Aleksic B, Kinoshita Y, et al. Genome-wide association study of schizophrenia in a Japanese population. *Biol Psychiatry*. 2011;69:472–478. Available from: https://doi.org/10.1016/j.biopsych.2010.07.010.
43. Higashiyama R, Ohnuma T, Takebayashi Y, et al. Association of copy number polymorphisms at the promoter and translated region of *COMT* with Japanese patients with schizophrenia. *Am J Med Genet, B: Neuropsychiatr Genet*. 2016;171:447–457. Available from: https://doi.org/10.1002/ajmg.b.32426.
44. Schwab SG, Kusumawardhani AAAA, Dai N, et al. Association of rs1344706 in the ZNF804A gene with schizophrenia in a case/control sample from Indonesia. *Schizophr Res*. 2013;147:46–52. Available from: https://doi.org/10.1016/j.schres.2013.03.022.
45. Zhang R, Yan J-D, Valenzuela RK, et al. Further evidence for the association of genetic variants of ZNF804A with schizophrenia and a meta-analysis for genome-wide significance variant rs1344706. *Schizophr Res*. 2012;141:40–47. Available from: https://doi.org/10.1016/j.schres.2012.07.013.
46. Li W, Ju K, Li Z, et al. Significant association of GRM7 and GRM8 genes with schizophrenia and major depressive disorder in the Han Chinese population. *Eur Neuropsychopharmacol*. 2016;26:136–146. Available from: https://doi.org/10.1016/j.euroneuro.2015.05.004.

47. Jajodia A, Kaur H, Kumari K, et al. Evidence for schizophrenia susceptibility alleles in the Indian population: an association of neurodevelopmental genes in case-control and familial samples. *Schizophr Res.* 2015;162:112–117. Available from: https://doi.org/10.1016/j.schres.2014.12.031.
48. Wang M, Chen J, He K, et al. The NVL gene confers risk for both major depressive disorder and schizophrenia in the Han Chinese population. *Prog Neuropsychopharmacol Biol Psychiatry.* 2015;62:7–13. Available from: https://doi.org/10.1016/j.pnpbp.2015.04.001.
49. Wang X, He G, Gu N, et al. Association of G72/G30 with schizophrenia in the Chinese population. *Biochem Biophys Res Commun.* 2004;319:1281–1286. Available from: https://doi.org/10.1016/j.bbrc.2004.05.119.
50. Tang R, Zhao X, Fang C, et al. Investigation of variants in the promoter region of PIK3C3 in schizophrenia. *Neurosci Lett.* 2008;437:42–44. Available from: https://doi.org/10.1016/j.neulet.2008.03.043.
51. Lim CH, Zain SM, Reynolds GP, et al. Genetic association of LMAN2L gene in schizophrenia and bipolar disorder and its interaction with ANK3 gene polymorphism. *Prog Neuropsychopharmacol Biol Psychiatry.* 2014;54:157–162. Available from: https://doi.org/10.1016/j.pnpbp.2014.05.017.
52. He K, Wang Q, Chen J, et al. ITIH family genes confer risk to schizophrenia and major depressive disorder in the Han Chinese population. *Prog Neuropsychopharmacol Biol Psychiatry.* 2014;51:34–38. Available from: https://doi.org/10.1016/j.pnpbp.2013.12.004.
53. Bangel FN, Yamada K, Arai M, et al. Genetic analysis of the glyoxalase system in schizophrenia. *Prog Neuropsychopharmacol Biol Psychiatry.* 2015;59:105–110. Available from: https://doi.org/10.1016/j.pnpbp.2015.01.014.
54. Yamada K, Hattori E, Iwayama Y, et al. Population-dependent contribution of the major histocompatibility complex region to schizophrenia susceptibility. *Schizophr Res.* 2015;168:444–449. Available from: https://doi.org/10.1016/j.schres.2015.08.018.
55. Su L, Shen T, Huang G, et al. Genetic association of GWAS-supported MAD1L1 gene polymorphism rs12666575 with schizophrenia susceptibility in a Chinese population. *Neurosci Lett.* 2016;610:98–103. Available from: https://doi.org/10.1016/j.neulet.2015.10.061.
56. Zhang F, Wang G, Shugart YY, et al. Association analysis of a functional variant in ATXN2 with schizophrenia. *Neurosci Lett.* 2014;562:24–27. Available from: https://doi.org/10.1016/j.neulet.2013.12.001.
57. Khan RAW, Chen J, Wang M, et al. A new risk locus in the ZEB2 gene for schizophrenia in the Han Chinese population. *Prog Neuropsychopharmacol Biol Psychiatry.* 2016;66:97–103. Available from: https://doi.org/10.1016/j.pnpbp.2015.12.001.
58. Luo X-J, Huang L, Oord EJ, et al. Common variants in the MKL1 gene confer risk of schizophrenia. *Schizophr Bull.* 2015;41:715–727. Available from: https://doi.org/10.1093/schbul/sbu156.
59. Sun S, Wei J, Li H, Jin S, Li P, Ju G, et al. A family-based study of the IL3RA gene on susceptibility to schizophrenia in a Chinese Han population. *Brain Res.* 2009;1268:13–16.
60. Zhang JP, Gallego JA, Robinson DG, Malhotra AK, Kane JM, Correll CU. Efficacy and safety of individual second-generation vs. first-generation antipsychotics in first-episode psychosis: a systematic review and meta-analysis. *Int J Neuropsychopharmacol.* 2013;16(6):1205–1218.
61. Khan RA, Chen J, Wang M, Li Z, Shen J, Wen Z, et al. A new risk locus in the ZEB2 gene for schizophrenia in the Han Chinese population. *Prog Neuropsychopharmacol Biol Psychiatry.* 2016;66:97–103.
62. Niu W, Huang X, Yu T, Chen S, Li X, Wu X, et al. Association study of GRM7 polymorphisms and schizophrenia in the Chinese Han population. *Neurosci Lett.* 2015;604:109–112.
63. Gilks WP, Allott EH, Donohoe G, Cummings E. International Schizophrenia Consortium, Gill M, et al. Replicated genetic evidence supports a role for HOMER2 in schizophrenia. *Neurosci Lett.* 2010;468(3):229–233.
64. Dubertret C, Bardel C, Ramoz N, Martin PM, Deybach JC, Adès J, et al. A genetic schizophrenia-susceptibility region located between the ANKK1 and DRD2 genes. *Prog Neuropsychopharmacol Biol Psychiatry.* 2010;34(3):492–499.

65. Xu St M, Clair D, He L. Testing for genetic association between the ZDHHC8 gene locus and susceptibility to schizophrenia: An integrated analysis of multiple datasets. *Am J Med Genet B Neuropsychiatr Genet*. 2010;153B(7):1266–1275.
66. Liu JB, Li YM, Lü J, Wei YS, Yang M, Yu DW. Performance and Factors Analysis of Sludge Dewatering in Different Wastewater Treatment Processes. *Huan Jing Ke Xue*. 2015;36(10):3794–3800 [Article in Chinese].
67. Autism Spectrum Disorders Working Group of The Psychiatric Genomics Consortium. Meta-analysis of GWAS of over 16,000 individuals with autism spectrum disorder highlights a novel locus at 10q24.32 and a significant overlap with schizophrenia. *Mol Autism*. 2017;8:21.
68. Lochman J, Balcar VJ, St'astný F, Serý O. Preliminary evidence for association between schizophrenia and polymorphisms in the regulatory Regions of the ADRA2A, DRD3 and SNAP-25 Genes. *Psychiatry Res*. 2013;205(1–2):7–12.
69. Forero DA, Herteleer L, De Zutter S, Norrback KF, Nilsson LG, Adolfsson R, et al. A network of synaptic genes associated with schizophrenia and bipolar disorder. *Schizophr Res*. 2016;172(1–3):68–74.
70. Maaser A, Forstner AJ, Strohmaier J, et al. Exome sequencing in large, multiplex bipolar disorder families from Cuba. *PLoS One*. 2018;13:e0205895. Available from: https://doi.org/10.1371/journal.pone.0205895.
71. Müller JK, Leweke FM. Bipolar disorder: clinical overview. *Med Monatsschr Pharm*. 2016;39:363–369.
72. Szczepankiewicz A. Evidence for single nucleotide polymorphisms and their association with bipolar disorder. *Neuropsychiatr Dis Treat*. 2013;9:1573. Available from: https://doi.org/10.2147/NDT.S28117.
73. Harrison PJ. The neuropathology of primary mood disorder. *Brain*. 2002;125:1428–1449.
74. Lewis CM, Levinson DF, Wise LH, et al. Genome scan meta-analysis of schizophrenia and bipolar disorder, Part II: Schizophrenia. *Am J Hum Genet*. 2003;73:34–48. Available from: https://doi.org/10.1086/376549.
75. Khalid M, Driessen TM, Lee JS, et al. Association of CACNA1C with bipolar disorder among the Pakistani population. *Gene*. 2018;664:119–126. Available from: https://doi.org/10.1016/j.gene.2018.04.061.
76. Mühleisen TW, Leber M, Schulze TG, et al. Genome-wide association study reveals two new risk loci for bipolar disorder. *Nat Commun*. 2014;5:3339. Available from: https://doi.org/10.1038/ncomms4339.
77. Dima D, Jogia J, Collier D, Vassos E, Burdick KE, Frangou S. Independent modulation of engagement and connectivity of the facial network during affect processing by CACNA1C and ANK3 risk genes for bipolar disorder. *JAMA Psychiatry*. 2013;70:1303–1311. Available from: https://doi.org/10.1001/jamapsychiatry.2013.2099.
78. Ruberto G, Vassos E, Lewis CM, et al. The cognitive impact of the ANK3 risk variant for bipolar disorder: initial evidence of selectivity to signal detection during sustained attention. *PLoS One*. 2011;6:e16671. Available from: https://doi.org/10.1371/journal.pone.0016671.
79. Szczepankiewicz A. Evidence for single nucleotide polymorphisms and their association with bipolar disorder. *Neuropsychiatr Dis Treat*. 2013;9:1573–1582.
80. Kim S, Cho CH, Geum D, Lee HJ. Association of CACNA1C Variants with Bipolar Disorder in the Korean Population. *Psychiatry Investig*. 2016 Jul;13(4):453–457.
81. Lohoff FW, Sander T, Ferraro TN, Dahl JP, Gallinat J, Berrettini WH. Confirmation of association between the Val66Met polymorphism in the brain-derived neurotrophic factor (BDNF) gene and bipolar I disorder. *Am J Med Genet B Neuropsychiatr Genet*. 2005;139B(1):51–53.
82. Mandolini GM, Lazzaretti M, Pigoni A, Delvecchio G, Soares JC, Brambilla P. The impact of BDNF Val66Met polymorphism on cognition in Bipolar Disorder: A review: Special Section on "Translational and Neuroscience Studies in Affective Disorders" Section Editor, Maria Nobile MD, PhD. This Section of JAD focuses on the relevance of translational and neuroscience studies in providing a better

understanding of the neural basis of affective disorders. The main aim is to briefly summaries relevant research findings in clinical neuroscience with particular regards to specific innovative topics in mood and anxiety disorders. *J Affect Disord*. 2019;243:552–558.

83. Taylor S. Association between COMT Val158Met and psychiatric disorders: A comprehensive meta-analysis. *Am J Med Genet B Neuropsychiatr Genet*. 2018;177(2):199–210.
84. Chen DT, Jiang X, Akula N, Shugart YY, Wendland JR, Steele CJ, et al. Genome-wide association study meta-analysis of European and Asian-ancestry samples identifies three novel loci associated with bipolar disorder. *Mol Psychiatry*. 2013;18(2):195–205.
85. Zhao L, Chang H, Zhou DS, Cai J, Fan W, Tang W, et al. Replicated associations of FADS1, MAD1L1, and a rare variant at 10q26.13 with bipolar disorder in Chinese population. *Transl Psychiatry*. 2018;8(1):270.
86. Soronen P, Mantere O, Melartin T, Suominen K, Vuorilehto M, Rytsälä H, et al. P2RX7 gene is associated consistently with mood disorders and predicts clinical outcome in three clinical cohorts. *Am J Med Genet B Neuropsychiatr Genet*. 2011;156B(4):435–447.
87. Roussos P, Giakoumaki SG, Georgakopoulos A, Robakis NK, Bitsios P. The CACNA1C and ANK3 risk alleles impact on affective personality traits and startle reactivity but not on cognition or gating in healthy males. *Bipolar Disord*. 2011;13:250–259. Available from: https://doi.org/10.1111/j.1399-5618.2011.00924.x.
88. Erk S, Meyer-Lindenberg A, Schnell K, et al. Brain function in carriers of a genome-wide supported bipolar disorder variant. *Arch Gen Psychiatry*. 2010;67:803–811. Available from: https://doi.org/10.1001/archgenpsychiatry.2010.94.
89. Fan M, Liu B, Jiang T, Jiang X, Zhao H, Zhang J. Meta-analysis of the association between the monoamine oxidase-A gene and mood disorders. *Psychiatr Genet*. 2010;20:1–7. Available from: https://doi.org/10.1097/YPG.0b013e3283351112.
90. Eslami Amirabadi MR, Rajezi Esfahani S, Davari-Ashtiani R, et al. Monoamine oxidase a gene polymorphisms and bipolar disorder in Iranian population. *Iran Red Crescent Med J*. 2015;17:e23095. Available from: https://doi.org/10.5812/ircmj.23095.
91. Neves-Pereira M, Mundo E, Muglia P, King N, Macciardi F, Kennedy JL. The brain-derived neurotrophic factor gene confers susceptibility to bipolar disorder: evidence from a family-based association study. *Am J Hum Genet*. 2002;71:651–655. Available from: https://doi.org/10.1086/342288.
92. Lelli-Chiesa G, Kempton MJ, Jogia J, et al. The impact of the Val158Met catechol-*O*-methyltransferase genotype on neural correlates of sad facial affect processing in patients with bipolar disorder and their relatives. *Psychol Med*. 2011;41:779–788. Available from: https://doi.org/10.1017/S0033291710001431.
93. Palo OM, et al. Association of distinct allelic haplotypes of DISC1 with psychotic and bipolar spectrum disorders and with underlying cognitive impairments. *Hum Mol Genet*. 2007;16(20):2517–2528. Available from: https://doi.org/10.1093/hmg/ddm207.
94. Schosser A, Gaysina D, Cohen-Woods S, et al. Association of DISC1 and TSNAX genes and affective disorders in the depression case-control (DeCC) and bipolar affective case-control (BACCS) studies. *Mol Psychiatry*. 2010;15:844–849. Available from: https://doi.org/10.1038/mp.2009.21.
95. Joo EJ, Lee KY, Jeong SH, et al. Dysbindin gene variants are associated with bipolar I disorder in a Korean population. *Neurosci Lett*. 2007;418:272–275. Available from: https://doi.org/10.1016/j.neulet.2007.03.037.
96. Cichon S, Mühleisen TW, Degenhardt FA, et al. Genome-wide association study identifies genetic variation in neurocan as a susceptibility factor for bipolar disorder. *Am J Hum Genet*. 2011;88:372–381. Available from: https://doi.org/10.1016/j.ajhg.2011.01.017.
97. Brasil Rocha PM, Campos SB, Neves FS, da Silva Filho HC. Genetic association of the PERIOD3 (Per3) clock gene with bipolar disorder. *Psychiatry Investig*. 2017;14:674–680. Available from: https://doi.org/10.4306/pi.2017.14.5.674.

98. Shinozaki G, Potash JB. New developments in the genetics of bipolar disorder. *Curr Psychiatry Rep*. 2014;16:493. Available from: https://doi.org/10.1007/s11920-014-0493-5.
99. Tecelão D, Mendes A, Martins D, et al. The effect of psychosis associated CACNA1C, and its epistasis with ZNF804A, on brain function. *Genes Brain Behav*. 2019;18:e12510. Available from: https://doi.org/10.1111/gbb.12510.
100. Mandolini GM, Lazzaretti M, Pigoni A, Delvecchio G, Soares JC, Brambilla P. The impact of BDNF Val66Met polymorphism on cognition in bipolar disorder: a review. *J Affect Disord*. 2019;243:552–558. Available from: https://doi.org/10.1016/j.jad.2018.07.054.
101. Taylor S. Association between COMT Val158Met and psychiatric disorders: a comprehensive meta-analysis. *Am J Med Genet, B: Neuropsychiatry Genet*. 2018;177:199–210. Available from: https://doi.org/10.1002/ajmg.b.32556.
102. Serretti A, Mandelli L. The genetics of bipolar disorder: genome "hot regions," genes, new potential candidates and future directions. *Mol Psychiatry*. 2008;13:742–771. Available from: https://doi.org/10.1038/mp.2008.29.
103. Palo OM, Antila M, Silander K, et al. Association of distinct allelic haplotypes of DISC1 with psychotic and bipolar spectrum disorders and with underlying cognitive impairments. *Hum Mol Genet*. 2007;16:2517–2528. Available from: https://doi.org/10.1093/hmg/ddm207.
104. Orrù G, Carta MG. Genetic variants involved in bipolar disorder, a rough road ahead. *Clin Pract Epidemiol Ment Health*. 2018;14:37–45. Available from: https://doi.org/10.2174/1745017901814010037.
105. Chen DT, Jiang X, Akula N, et al. Genome-wide association study meta-analysis of European and Asian-ancestry samples identifies three novel loci associated with bipolar disorder. *Mol Psychiatry*. 2013;18:195–205. Available from: https://doi.org/10.1038/mp.2011.157.
106. Bigdeli TB, Ripke S, Peterson RE, et al. Genetic effects influencing risk for major depressive disorder in China and Europe. *Transl Psychiatry*. 2017;7:e1074. Available from: https://doi.org/10.1038/tp.2016.292.
107. López-León S, Janssens ACJW, González-Zuloeta Ladd AM, et al. Meta-analyses of genetic studies on major depressive disorder. *Mol Psychiatry*. 2008;13:772–785. Available from: https://doi.org/10.1038/sj.mp.4002088.
108. Hettema JM. Genetics of depression. *Focus (Madison)*. 2010;8:316–322. Available from: https://doi.org/10.1176/foc.8.3.foc316.
109. Anguelova M, Benkelfat C, Turecki G. A systematic review of association studies investigating genes coding for serotonin receptors and the serotonin transporter: I. Affective disorders. *Mol Psychiatry*. 2003;8(6):574–591.
110. Peerbooms OL, van Os J, Drukker M, Kenis G, Hoogveld L. MTHFR in Psychiatry Group, et al. Meta-analysis of MTHFR gene variants in schizophrenia, bipolar disorder and unipolar depressive disorder: evidence for a common genetic vulnerability? *Brain Behav Immun*. 2011;25(8):1530–1543.
111. Shi M, Hu J, Dong X, Gao Y, An G, Liu W. Association of unipolar depression with gene polymorphisms in the serotonergic pathways in Han Chinese. *Acta Neuropsychiatr*. 2008;20(3):139–144.
112. Clarke H, Flint J, Attwood AS, Munafò MR. Association of the 5- HTTLPR genotype and unipolar depression: a meta-analysis. *Psychol Med*. 2010;40(11):1767–1778.
113. Gyekis JP, Yu W, Dong S, Wang H, Qian J, Kota P, Yang J. No association of genetic variants in BDNF with major depression: a meta- and gene-based analysis. *Am J Med Genet B Neuropsychiatr Genet*. 2013;162B(1):61–70.
114. Kishi T, Tsunoka T, Ikeda M, Kawashima K, Okochi T, Kitajima T. Serotonin 1A receptor gene and major depressive disorder: an association study and meta-analysis. *J Hum Genet*. 2009;54(11):629–633.
115. Zhao X, Huang Y, Ma H, Jin Q, Wang Y, Zhu G. Association between major depressive disorder and the norepinephrine transporter polymorphisms T-182C and G1287A: a meta-analysis. *J Affect Disord*. 2013;150(1):23–28.

116. Levinson DF. The genetics of depression: a review. *Biol Psychiatry*. 2006;60:84–92. Available from: https://doi.org/10.1016/j.biopsych.2005.08.024.
117. Wang X, Lin Y, Song C, Sibille E, Tseng GC. Detecting disease-associated genes with confounding variable adjustment and the impact on genomic meta-analysis: with application to major depressive disorder. *BMC Bioinf*. 2012;13:52. Available from: https://doi.org/10.1186/1471-2105-13-52.
118. Mullins N, Lewis CM. Genetics of depression: progress at last. *Curr Psychiatry Rep*. 2017;19:43. Available from: https://doi.org/10.1007/s11920-017-0803-9.
119. American Psychiatric Association. *Diagnostic and Statistical Manual of Mental Disorders (DSM-5 (R))*. 2013.
120. Scaini S, Belotti R, Ogliari A. Genetic and environmental contributions to social anxiety across different ages: a meta-analytic approach to twin data. *J Anxiety Disord*. 2014;28:650–656. Available from: https://doi.org/10.1016/j.janxdis.2014.07.002.
121. Smoller JW. The genetics of stress-related disorders: PTSD, depression, and anxiety disorders. *Neuropsychopharmacology*. 2016;41:297–319. Available from: https://doi.org/10.1038/npp.2015.266.
122. Tambs K, Czajkowsky N, Røysamb E, et al. Structure of genetic and environmental risk factors for dimensional representations of DSM-IV anxiety disorders. *Br J Psychiatry*. 2009;195:301–307. Available from: https://doi.org/10.1192/bjp.bp.108.059485.
123. Shimada-Sugimoto M, Otowa T, Hettema JM. Genetics of anxiety disorders: genetic epidemiological and molecular studies in humans. *Psychiatry Clin Neurosci*. 2015;69:388–401. Available from: https://doi.org/10.1111/pcn.12291.
124. Bastiaansen JA, Servaas MN, Marsman JBC, et al. Filling the gap. *Psychol Sci*. 2014;25:2058–2066. Available from: https://doi.org/10.1177/0956797614548877.
125. Howe AS, Buttenschøn HN, Bani-Fatemi A, et al. Candidate genes in panic disorder: meta-analyses of 23 common variants in major anxiogenic pathways. *Mol Psychiatry*. 2016;21:665–679. Available from: https://doi.org/10.1038/mp.2015.138.
126. Haaker J, Lonsdorf TB, Raczka KA, Mechias M-L, Gartmann N, Kalisch R. Higher anxiety and larger amygdala volumes in carriers of a TMEM132D risk variant for panic disorder. *Transl Psychiatry*. 2014;4:e357. Available from: https://doi.org/10.1038/TP.2014.1.
127. Erhardt A, Akula N, Schumacher J, et al. Replication and meta-analysis of TMEM132D gene variants in panic disorder. *Transl Psychiatry*. 2012;2:e156. Available from: https://doi.org/10.1038/tp.2012.85.
128. Walter S, Glymour MM, Koenen K, et al. Performance of polygenic scores for predicting phobic anxiety. *PLoS One*. 2013;8:e80326. Available from: https://doi.org/10.1371/journal.pone.0080326.
129. de Moor MHM, Costa PT, Terracciano A, et al. Meta-analysis of genome-wide association studies for personality. *Mol Psychiatry*. 2012;17:337–349. Available from: https://doi.org/10.1038/mp.2010.128.
130. de Moor MHM, van den Berg SM, Verweij KJH, et al. Meta-analysis of genome-wide association studies for neuroticism, and the polygenic association with major depressive disorder. *JAMA Psychiatry*. 2015;72:642. Available from: https://doi.org/10.1001/jamapsychiatry.2015.0554.
131. Breslau N. Epidemiologic studies of trauma, posttraumatic stress disorder, and other psychiatric disorders. *Can J Psychiatry*. 2002;47:923–929. Available from: https://doi.org/10.1177/070674370204701003.
132. Stein MB, Jang KL, Taylor S, Vernon PA, Livesley WJ. Genetic and environmental influences on trauma exposure and posttraumatic stress disorder symptoms: a twin study. *Am J Psychiatry*. 2002;159:1675–1681. Available from: https://doi.org/10.1176/appi.ajp.159.10.1675.
133. Klengel T, Mehta D, Anacker C, et al. Allele-specific FKBP5 DNA demethylation mediates gene–childhood trauma interactions. *Nat Neurosci*. 2013;16:33–41. Available from: https://doi.org/10.1038/nn.3275.
134. Mehta D, Gonik M, Klengel T, et al. Using polymorphisms in FKBP5 to define biologically distinct subtypes of posttraumatic stress disorder. *Arch Gen Psychiatry*. 2011;68:901. Available from: https://doi.org/10.1001/archgenpsychiatry.2011.50.

135. Binder EB, Bradley RG, Liu W, et al. Association of FKBP5 polymorphisms and childhood abuse with risk of posttraumatic stress disorder symptoms in adults. *JAMA*. 2008;299:1291. Available from: https://doi.org/10.1001/jama.299.11.1291.
136. Ressler KJ, Mercer KB, Bradley B, et al. Post-traumatic stress disorder is associated with PACAP and the PAC1 receptor. *Nature*. 2011;470:492–497. Available from: https://doi.org/10.1038/nature09856.
137. Uddin M, Chang S-C, Zhang C, et al. Adcyap1r1 genotype, posttraumatic stress disorder, and depression among women exposed to childhood maltreatment. *Depress Anxiety*. 2013;30:251–258. Available from: https://doi.org/10.1002/da.22037.
138. Liberzon I, King AP, Ressler KJ, et al. Interaction of the iADRB2/i gene polymorphism with childhood trauma in predicting adult symptoms of posttraumatic stress disorder. *JAMA Psychiatry*. 2014;71:1174. Available from: https://doi.org/10.1001/jamapsychiatry.2014.999.
139. Koenen KC, Aiello AE, Bakshis E, et al. Modification of the association between serotonin transporter genotype and risk of posttraumatic stress disorder in adults by county-level social environment. *Am J Epidemiol*. 2009;169:704–711. Available from: https://doi.org/10.1093/aje/kwn397.
140. Mercer KB, Orcutt HK, Quinn JF, et al. Acute and posttraumatic stress symptoms in a prospective gene × environment study of a university campus shooting. *Arch Gen Psychiatry*. 2012;69:89. Available from: https://doi.org/10.1001/archgenpsychiatry.2011.109.
141. Wang Z, Baker DG, Harrer J, Hamner M, Price M, Amstadter A. The relationship between combat-related posttraumatic stress disorder and the 5-HTTLPR/rs25531 polymorphism. *Depress Anxiety*. 2011;28:1067–1073. Available from: https://doi.org/10.1002/da.20872.
142. Segman RH, Cooper-Kazaz R, Macciardi F, et al. Association between the dopamine transporter gene and posttraumatic stress disorder. *Mol Psychiatry*. 2002;7:903–907. Available from: https://doi.org/10.1038/sj.mp.4001085.
143. Valente NLM, Vallada H, Cordeiro Q, et al. Candidate-gene approach in posttraumatic stress disorder after urban violence: association analysis of the genes encoding serotonin transporter, dopamine transporter, and BDNF. *J Mol Neurosci*. 2011;44:59–67. Available from: https://doi.org/10.1007/s12031-011-9513-7.
144. Drury SS, Theall KP, Keats BJB, Scheeringa M. The role of the dopamine transporter (DAT) in the development of PTSD in preschool children. *J Trauma Stress*. 2009;22. Available from: https://doi.org/10.1002/jts.20475. n/a–n/a.
145. Young RM, Lawford BR, Noble EP, et al. Harmful drinking in military veterans with post-traumatic stress disorder: association with the D2 dopamine receptor A1 allele. *Alcohol Alcohol*. 2002;37:451–456. Available from: https://doi.org/10.1093/alcalc/37.5.451.
146. Comings DE, Muhleman D, Gysin R. Dopamine D2 receptor (DRD2) gene and susceptibility to post-traumatic stress disorder: a study and replication. *Biol Psychiatry*. 1996;40:368–372. Available from: https://doi.org/10.1016/0006-3223(95)00519-6.
147. Gelernter J, Southwick S, Goodson S, Morgan A, Nagy L, Charney DS. No association between D2 dopamine receptor (DRD2) "A" system alleles, or DRD2 haplotypes, and posttraumatic stress disorder. *Biol Psychiatry*. 1999;45:620–625. Available from: https://doi.org/10.1016/S0006-3223(98)00087-0.
148. Bailey JN, Goenjian AK, Noble EP, Walling DP, Ritchie T, Goenjian HA. PTSD and dopaminergic genes, DRD2 and DAT, in multigenerational families exposed to the Spitak earthquake. *Psychiatry Res*. 2010;178:507–510. Available from: https://doi.org/10.1016/J.PSYCHRES.2010.04.043.
149. Valente NLM, Vallada H, Cordeiro Q, et al. Catechol-*O*-methyltransferase (COMT) val158met polymorphism as a risk factor for PTSD after urban violence. *J Mol Neurosci*. 2011;43:516–523. Available from: https://doi.org/10.1007/s12031-010-9474-2.
150. Kolassa I-T, Kolassa S, Ertl V, Papassotiropoulos A, De Quervain DJ-F. The risk of posttraumatic stress disorder after trauma depends on traumatic load and the catechol-*O*-methyltransferase Val158Met polymorphism. *Biol Psychiatry*. 2010;67:304–308. Available from: https://doi.org/10.1016/J.BIOPSYCH.2009.10.009.

151. Lonsdorf TB, Weike AI, Nikamo P, Schalling M, Hamm AO, Öhman A. Genetic gating of human fear learning and extinction. *Psychol Sci*. 2009;20:198–206. Available from: https://doi.org/10.1111/j.1467-9280.2009.02280.x.
152. Logue MW, Baldwin C, Guffanti G, et al. A genome-wide association study of post-traumatic stress disorder identifies the retinoid-related orphan receptor alpha (RORA) gene as a significant risk locus. *Mol Psychiatry*. 2013;18:937–942. Available from: https://doi.org/10.1038/mp.2012.113.
153. Almli LM, Srivastava A, Fani N, et al. Follow-up and extension of a prior genome-wide association study of posttraumatic stress disorder: gene × environment associations and structural magnetic resonance imaging in a highly traumatized African-American civilian population. *Biol Psychiatry*. 2014;76: e3–e4. Available from: https://doi.org/10.1016/j.biopsych.2014.01.017.
154. Guffanti G, Galea S, Yan L, et al. Genome-wide association study implicates a novel RNA gene, the lincRNA AC068718.1, as a risk factor for post-traumatic stress disorder in women. *Psychoneuroendocrinology*. 2013;38:3029–3038. Available from: https://doi.org/10.1016/J.PSYNEUEN.2013.08.014.
155. Nievergelt CM, Maihofer AX, Mustapic M, et al. Genomic predictors of combat stress vulnerability and resilience in U.S. Marines: a genome-wide association study across multiple ancestries implicates PRTFDC1 as a potential PTSD gene. *Psychoneuroendocrinology*. 2015;51:459–471. Available from: https://doi.org/10.1016/J.PSYNEUEN.2014.10.017.
156. Logue MW, Amstadter AB, Baker DG, et al. The psychiatric genomics consortium posttraumatic stress disorder workgroup: posttraumatic stress disorder enters the age of large-scale genomic collaboration. *Neuropsychopharmacology*. 2015;40:2287–2297. Available from: https://doi.org/10.1038/npp.2015.118.
157. Duncan LE, Ratanatharathorn A, Aiello AE, et al. Largest GWAS of PTSD ($N = 20\,070$) yields genetic overlap with schizophrenia and sex differences in heritability. *Mol Psychiatry*. 2017. Available from: https://doi.org/10.1038/mp.2017.77.
158. Uhl GR, Drgon T, Johnson C, et al. Molecular genetics of addiction and related heritable phenotypes. *Ann NY Acad Sci*. 2008;1141:318–381. Available from: https://doi.org/10.1196/annals.1441.018.
159. Thorgeirsson TE, Gudbjartsson DF, Surakka I, et al. Sequence variants at CHRNB3–CHRNA6 and CYP2A6 affect smoking behavior. *Nat Genet*. 2010;42:448–453. Available from: https://doi.org/10.1038/ng.573.
160. Weiss RB, Baker TB, Cannon DS, et al. A candidate gene approach identifies the CHRNA5-A3-B4 region as a risk factor for age-dependent nicotine addiction. *PLoS Genet*. 2008;4:e1000125. Available from: https://doi.org/10.1371/journal.pgen.1000125.
161. Yilmaz Z, Hardaway JA, Bulik CM. Genetics and epigenetics of eating disorders. *Adv Genomics Genet*. 2015;5:131–150. Available from: https://doi.org/10.2147/AGG.S55776.
162. Bulik CM, Kleiman SC, Yilmaz Z. Genetic epidemiology of eating disorders. *Curr Opin Psychiatry*. 2016;29:383–388. Available from: https://doi.org/10.1097/YCO.0000000000000275.
163. Gelernter J. Genetics of complex traits in psychiatry. *Biol Psychiatry*. 2015;77:36–42. Available from: https://doi.org/10.1016/J.BIOPSYCH.2014.08.005.
164. Boraska V, Franklin CS, Floyd JAB, et al. A genome-wide association study of anorexia nervosa. *Mol Psychiatry*. 2014;19:1085–1094. Available from: https://doi.org/10.1038/mp.2013.187.
165. Wang K, Zhang H, Bloss CS, et al. A genome-wide association study on common SNPs and rare CNVs in anorexia nervosa. *Mol Psychiatry*. 2011;16:949–959. Available from: https://doi.org/10.1038/mp.2010.107.
166. Duncan L, Yilmaz Z, Gaspar H, et al. Significant locus and metabolic genetic correlations revealed in genome-wide association study of anorexia nervosa. *Am J Psychiatry*. 2017;174:850–858. Available from: https://doi.org/10.1176/appi.ajp.2017.16121402.
167. Breithaupt L, Hubel C, Bulik CM. Updates on genome-wide association findings in eating disorders and future application to precision medicine. *Curr Neuropharmacol*. 2018;16:1102–1110. Available from: https://doi.org/10.2174/1570159X16666180222163450.

168. Elsabbagh M, Divan G, Koh Y-J, et al. Global prevalence of autism and other pervasive developmental disorders. *Autism Res*. 2012;5:160. Available from: https://doi.org/10.1002/AUR.239.
169. Howes OD, Rogdaki M, Findon JL, et al. Autism spectrum disorder: consensus guidelines on assessment, treatment and research from the British Association for Psychopharmacology. *J Psychopharmacol*. 2018;32:3–29. Available from: https://doi.org/10.1177/0269881117741766.
170. Ozonoff S, Young GS, Carter A, et al. Recurrence risk for autism spectrum disorders: a Baby Siblings Research Consortium study. *Pediatrics*. 2011;128:e488–e495. Available from: https://doi.org/10.1542/peds.2010-2825.
171. Sandin S, Lichtenstein P, Kuja-Halkola R, Larsson H, Hultman CM, Reichenberg A. The familial risk of autism. *JAMA*. 2014;311:1770–1777. Available from: https://doi.org/10.1001/jama.2014.4144.
172. Colvert E, Tick B, McEwen F, et al. Heritability of autism spectrum disorder in a UK population-based twin sample. *JAMA Psychiatry*. 2015;72:415–423. Available from: https://doi.org/10.1001/jamapsychiatry.2014.3028.
173. Hallmayer J, Cleveland S, Torres A, et al. Genetic heritability and shared environmental factors among twin pairs with autism. *Arch Gen Psychiatry*. 2011;68:1095–1102. Available from: https://doi.org/10.1001/archgenpsychiatry.2011.76.
174. Anney R, Klei L, Pinto D, et al. Individual common variants exert weak effects on the risk for autism spectrum disorders. *Hum Mol Genet*. 2012;21:4781–4792. Available from: https://doi.org/10.1093/hmg/dds301.
175. Anney R, Klei L, Pinto D, et al. A genome-wide scan for common alleles affecting risk for autism. *Hum Mol Genet*. 2010;19:4072–4082. Available from: https://doi.org/10.1093/hmg/ddq307.
176. De Rubeis S, Buxbaum JD. Genetics and genomics of autism spectrum disorder: embracing complexity. *Hum Mol Genet*. 2015;24:R24–R31. Available from: https://doi.org/10.1093/hmg/ddv273.
177. Pinto D, Delaby E, Merico D, et al. Convergence of genes and cellular pathways dysregulated in autism spectrum disorders. *Am J Hum Genet*. 2014;94:677–694. Available from: https://doi.org/10.1016/j.ajhg.2014.03.018.
178. Pinto D, Pagnamenta AT, Klei L, et al. Functional impact of global rare copy number variation in autism spectrum disorder. *Nature*. 2010;466:368. Available from: https://doi.org/10.1038/NATURE09146.
179. Sanders SJ, He X, Willsey AJ, et al. Insights into autism spectrum disorder genomic architecture and biology from 71 risk loci. *Neuron*. 2015;87:1215–1233. Available from: https://doi.org/10.1016/j.neuron.2015.09.016.
180. Thapar A. Discoveries on the genetics of ADHD in the 21st century: new findings and their implications. *Am J Psychiatry*. 2018;175:943–950. Available from: https://doi.org/10.1176/appi.ajp.2018.18040383.
181. Faraone SV, Ghirardi L, Kuja-Halkola R, Lichtenstein P, Larsson H. The familial co-aggregation of attention-deficit/hyperactivity disorder and intellectual disability: a register-based family study. *J Am Acad Child Adolesc Psychiatry*. 2017;56:167–174.e1. Available from: https://doi.org/10.1016/j.jaac.2016.11.011.
182. Ghirardi L, Brikell I, Kuja-Halkola R, et al. The familial co-aggregation of ASD and ADHD: a register-based cohort study. *Mol Psychiatry*. 2018;23:257–262. Available from: https://doi.org/10.1038/mp.2017.17.
183. Demontis D, Walters RK, Martin J, et al. Discovery of the first genome-wide significant risk loci for attention deficit/hyperactivity disorder. *Nat Genet*. 2019;51:63–75. Available from: https://doi.org/10.1038/s41588-018-0269-7.
184. Breiderhoff T, Christiansen GB, Pallesen LT, et al. Sortilin-related receptor SORCS3 is a postsynaptic modulator of synaptic depression and fear extinction. *PLoS One*. 2013;8:e75006. Available from: https://doi.org/10.1371/journal.pone.0075006.

185. Tsui D, Vessey JP, Tomita H, Kaplan DR, Miller FD. FoxP2 regulates neurogenesis during embryonic cortical development. *J Neurosci.* 2013;33:244–258. Available from: https://doi.org/10.1523/JNEUROSCI.1665-12.2013.
186. Lionel AC, Crosbie J, Barbosa N, et al. Rare copy number variation discovery and cross-disorder comparisons identify risk genes for ADHD. *Sci Transl Med.* 2011;3. Available from: https://doi.org/10.1126/scitranslmed.3002464. 95ra75–95ra75.
187. Williams NM, Franke B, Mick E, et al. Genome-wide analysis of copy number variants in attention deficit hyperactivity disorder: the role of rare variants and duplications at 15q13.3. *Am J Psychiatry.* 2012;169:195–204. Available from: https://doi.org/10.1176/appi.ajp.2011.11060822.
188. Stroup TS, Gerhard T, Crystal S, Huang C, Olfson M. Comparative effectiveness of clozapine and standard antipsychotic treatment in adults with schizophrenia. *Am J Psychiatry.* 2016;173:166–173. Available from: https://doi.org/10.1176/appi.ajp.2015.15030332.
189. Pratt J, Winchester C, Dawson N, Morris B. Advancing schizophrenia drug discovery: optimizing rodent models to bridge the translational gap. *Nat Rev Drug Discovery.* 2012;11:560–579. Available from: https://doi.org/10.1038/nrd3649.
190. Dunlop J, Brandon NJ. Schizophrenia drug discovery and development in an evolving era: are new drug targets fulfilling expectations? *J Psychopharmacol.* 2015;29:230–238. Available from: https://doi.org/10.1177/0269881114565806.
191. Karam CS, Ballon JS, Bivens NM, et al. Signaling pathways in schizophrenia: emerging targets and therapeutic strategies. *Trends Pharmacol Sci.* 2010;31:381–390. Available from: https://doi.org/10.1016/j.tips.2010.05.004.
192. Ravyn D, Ravyn V, Lowney R, Nasrallah HA. CYP450 pharmacogenetic treatment strategies for antipsychotics: a review of the evidence. *Schizophr Res.* 2013;149:1–14. Available from: https://doi.org/10.1016/j.schres.2013.06.035.
193. Sistonen J, Sajantila A, Lao O, Corander J, Barbujani G, Fuselli S. CYP2D6 worldwide genetic variation shows high frequency of altered activity variants and no continental structure. *Pharmacogenet Genomics.* 2007;17:93–101. Available from: https://doi.org/10.1097/01.fpc.0000239974.69464.f2.
194. Llerena A, Berecz R, Dorado P, de la Rubia A. QTc interval, CYP2D6 and CYP2C9 genotypes and risperidone plasma concentrations. *J Psychopharmacol.* 2004;18:189–193. Available from: https://doi.org/10.1177/0269881104042618.
195. Mas S, Gassò P, Alvarez S, Parellada E, Bernardo M, Lafuente A. Intuitive pharmacogenetics: spontaneous risperidone dosage is related to CYP2D6, CYP3A5 and ABCB1 genotypes. *Pharmacogenomics J.* 2012;12:255–259. Available from: https://doi.org/10.1038/tpj.2010.91.
196. Mihara K, Kondo T, Yasui-Furukori N, et al. Effects of various CYP2D6 genotypes on the steady-state plasma concentrations of risperidone and its active metabolite, 9-hydroxyrisperidone, in Japanese patients with schizophrenia. *Ther Drug Monit.* 2003;25:287–293.
197. Novalbos J, López-Rodríguez R, Román M, Gallego-Sandín S, Ochoa D, Abad-Santos F. Effects of CYP2D6 genotype on the pharmacokinetics, pharmacodynamics, and safety of risperidone in healthy volunteers. *J Clin Psychopharmacol.* 2010;30:504–511. Available from: https://doi.org/10.1097/JCP.0b013e3181ee84c7.
198. Suzuki Y, Fukui N, Tsuneyama N, et al. Effect of the cytochrome P450 2D6*10 allele on risperidone metabolism in Japanese psychiatric patients. *Hum Psychopharmacol.* 2012;27:43–46.
199. Xiang Q, Zhao X, Zhou Y, Duan JL, Cui YM. Effect of CYP2D6, CYP3A5, and MDR1 genetic polymorphisms on the pharmacokinetics of risperidone and its active moiety. *J Clin Pharmacol.* 2010;50:659–666. Available from: https://doi.org/10.1177/0091270009347867.
200. Yoo H-D, Cho H-Y, Lee S-N, Yoon H, Lee Y-B. Population pharmacokinetic analysis of risperidone and 9-hydroxyrisperidone with genetic polymorphisms of CYP2D6 and ABCB1. *J Pharmacokinet Pharmacodyn.* 2012;39:329–341. Available from: https://doi.org/10.1007/s10928-012-9253-5.

201. de Leon J, Susce MT, Pan R-M, Fairchild M, Koch WH, Wedlund PJ. The CYP2D6 poor metabolizer phenotype may be associated with risperidone adverse drug reactions and discontinuation. *J Clin Psychiatry*. 2005;66:15–27.
202. Kobylecki CJ, Jakobsen KD, Hansen T, Jakobsen IV, Rasmussen HB, Werge T. CYP2D6 genotype predicts antipsychotic side effects in schizophrenia inpatients: a retrospective matched case-control study. *Neuropsychobiology*. 2009;59:222–226. Available from: https://doi.org/10.1159/000223734.
203. Rasmussen BB, Brix TH, Kyvik KO, Brøsen K. The interindividual differences in the 3-demthylation of caffeine alias CYP1A2 is determined by both genetic and environmental factors. *Pharmacogenetics*. 2002;12:473–478.
204. Schulze TG, Schumacher J, Müller DJ, et al. Lack of association between a functional polymorphism of the cytochrome P450 1A2 (CYP1A2) gene and tardive dyskinesia in schizophrenia. *Am J Med Genet*. 2001;105:498–501.
205. Fu Y, Fan C, Deng H, et al. Association of CYP2D6 and CYP1A2 gene polymorphism with tardive dyskinesia in Chinese schizophrenic patients. *Acta Pharmacol Sin*. 2006;27:328–332. Available from: https://doi.org/10.1111/j.1745-7254.2006.00279.x.
206. Hartz SM, Pato CN, Medeiros H, et al. Comorbidity of severe psychotic disorders with measures of substance use. *JAMA Psychiatry*. 2014;71:248. Available from: https://doi.org/10.1001/jamapsychiatry.2013.3726.
207. Zanger UM, Turpeinen M, Klein K, Schwab M. Functional pharmacogenetics/genomics of human cytochromes P450 involved in drug biotransformation. *Anal Bioanal Chem*. 2008;392:1093–1108. Available from: https://doi.org/10.1007/s00216-008-2291-6.
208. Lamba J, Hebert JM, Schuetz EG, Klein TE, Altman RB. PharmGKB summary: very important pharmacogene information for CYP3A5. *Pharmacogenet Genomics*. 2012;22:555–558. Available from: https://doi.org/10.1097/FPC.0b013e328351d47f.
209. Moons T, de Roo M, Claes S, Dom G. Relationship between *P*-glycoprotein and second-generation antipsychotics. *Pharmacogenomics*. 2011;12:1193–1211. Available from: https://doi.org/10.2217/pgs.11.55.
210. Lencz T, Robinson DG, Xu K, et al. DRD2 promoter region variation as a predictor of sustained response to antipsychotic medication in first-episode schizophrenia patients. *Am J Psychiatry*. 2006;163:529–531. Available from: https://doi.org/10.1176/appi.ajp.163.3.529.
211. Shen Y-C, Chen S-F, Chen C-H, et al. Effects of DRD2/ANKK1 gene variations and clinical factors on aripiprazole efficacy in schizophrenic patients. *J Psychiatr Res*. 2009;43:600–606. Available from: https://doi.org/10.1016/j.jpsychires.2008.09.005.
212. Xing Q, Qian X, Li H, et al. The relationship between the therapeutic response to risperidone and the dopamine D2 receptor polymorphism in Chinese schizophrenia patients. *Int J Neuropsychopharmacol*. 2007;10:631–637. Available from: https://doi.org/10.1017/S146114570600719X.
213. Kurylev AA, Brodyansky VM, Andreev VB, Kibitov OA, Limankin VO, Mosolov NS. The combined effect of CYP2D6 and DRD2 Taq1A polymorphisms on the antipsychotics daily doses and hospital stay duration in schizophrenia inpatients (observational naturalistic study). *Psychiatr Danub*. 2018;30:157–163. Available from: https://doi.org/10.24869/psyd.2018.157.
214. Liou Y-J, Lai I-C, Liao D-L, et al. The human dopamine receptor D2 (DRD2) gene is associated with tardive dyskinesia in patients with schizophrenia. *Schizophr Res*. 2006;86:323–325. Available from: https://doi.org/10.1016/j.schres.2006.04.008.
215. Kwon JS, Kim E, Kang D-H, et al. Taq1A polymorphism in the dopamine D2 receptor gene as a predictor of clinical response to aripiprazole. *Eur Neuropsychopharmacol*. 2008;18:897–907. Available from: https://doi.org/10.1016/j.euroneuro.2008.07.010.
216. Hong C-J, Liou Y-J, Bai YM, Chen T-T, Wang Y-C, Tsai S-J. Dopamine receptor D2 gene is associated with weight gain in schizophrenic patients under long-term atypical antipsychotic treatment. *Pharmacogenet Genomics*. 2010;20:359–366. Available from: https://doi.org/10.1097/FPC.0b013e3283397d06.

217. Lane H-Y, Hsu S-K, Liu Y-C, Chang Y-C, Huang C-H, Chang W-H. Dopamine D3 receptor Ser9Gly polymorphism and risperidone response. *J Clin Psychopharmacol*. 2005;25:6–11.
218. Reynolds GP, Yao Z, Zhang X, Sun J, Zhang Z. Pharmacogenetics of treatment in first-episode schizophrenia: D3 and 5-HT2C receptor polymorphisms separately associate with positive and negative symptom response. *Eur Neuropsychopharmacol*. 2005;15:143–151. Available from: https://doi.org/10.1016/j.euroneuro.2004.07.001.
219. Hwang R, Zai C, Tiwari A, et al. Effect of dopamine D3 receptor gene polymorphisms and clozapine treatment response: exploratory analysis of nine polymorphisms and meta-analysis of the Ser9Gly variant. *Pharmacogenomics J*. 2010;10:200–218. Available from: https://doi.org/10.1038/tpj.2009.65.
220. Bertolino A, Caforio G, Blasi G, et al. COMT Val158Met polymorphism predicts negative symptoms response to treatment with olanzapine in schizophrenia. *Schizophr Res*. 2007;95:253–255. Available from: https://doi.org/10.1016/j.schres.2007.06.014.
221. Woodward ND, Jayathilake K, Meltzer HY. COMT val108/158met genotype, cognitive function, and cognitive improvement with clozapine in schizophrenia. *Schizophr Res*. 2007;90:86–96. Available from: https://doi.org/10.1016/j.schres.2006.10.002.
222. Srivastava V, Varma PG, Prasad S, et al. Genetic susceptibility to tardive dyskinesia among schizophrenia subjects: IV. Role of dopaminergic pathway gene polymorphisms. *Pharmacogenet Genomics*. 2006;16:111–117.
223. Kim B, Choi EY, Kim CY, Song K, Joo YH. Could HTR2A T102C and DRD3 Ser9Gly predict clinical improvement in patients with acutely exacerbated schizophrenia? Results from treatment responses to risperidone in a naturalistic setting. *Hum Psychopharmacol Clin Exp*. 2008;23:61–67. Available from: https://doi.org/10.1002/hup.897.
224. Chen S-F, Shen Y-C, Chen C-H. HTR2A A-1438G/T102C polymorphisms predict negative symptoms performance upon aripiprazole treatment in schizophrenic patients. *Psychopharmacology (Berl)*. 2009;205:285–292. Available from: https://doi.org/10.1007/s00213-009-1538-z.
225. Ikeda M, Yamanouchi Y, Kinoshita Y, et al. Variants of dopamine and serotonin candidate genes as predictors of response to risperidone treatment in first-episode schizophrenia. *Pharmacogenomics*. 2008;9:1437–1443. Available from: https://doi.org/10.2217/14622416.9.10.1437.
226. Benmessaoud D, Hamdani N, Boni C, et al. Excess of transmission of the G allele of the -1438A/G polymorphism of the 5-HT2A receptor gene in patients with schizophrenia responsive to antipsychotics. *BMC Psychiatry*. 2008;8:40. Available from: https://doi.org/10.1186/1471-244X-8-40.
227. Al Hadithy AFY, Ivanova SA, Pechlivanoglou P, et al. Tardive dyskinesia and DRD3, HTR2A and HTR2C gene polymorphisms in Russian psychiatric inpatients from Siberia. *Prog Neuropsychopharmacol Biol Psychiatry*. 2009;33:475–481. Available from: https://doi.org/10.1016/j.pnpbp.2009.01.010.
228. Ellingrod VL, Perry PJ, Lund BC, et al. 5HT2A and 5HT2C receptor polymorphisms and predicting clinical response to olanzapine in schizophrenia. *J Clin Psychopharmacol*. 2002;22:622–624.
229. Reynolds GP, Templeman LA, Zhang ZJ. The role of 5-HT2C receptor polymorphisms in the pharmacogenetics of antipsychotic drug treatment. *Prog Neuropsychopharmacol Biol Psychiatry*. 2005;29:1021–1028. Available from: https://doi.org/10.1016/j.pnpbp.2005.03.019.
230. Godlewska BR, Olajossy-Hilkesberger L, Ciwoniuk M, et al. Olanzapine-induced weight gain is associated with the -759C/T and -697G/C polymorphisms of the HTR2C gene. *Pharmacogenomics J*. 2009;9:234–241. Available from: https://doi.org/10.1038/tpj.2009.18.
231. Gunes A, Melkersson KI, Scordo MG, Dahl M-L. Association between HTR2C and HTR2A polymorphisms and metabolic abnormalities in patients treated with olanzapine or clozapine. *J Clin Psychopharmacol*. 2009;29:65–68. Available from: https://doi.org/10.1097/JCP.0b013e31819302c3.
232. Fabbri C, Crisafulli C, Calabrò M, Spina E, Serretti A. Progress and prospects in pharmacogenetics of antidepressant drugs. *Expert Opin Drug Metab Toxicol*. 2016;12:1157–1168. Available from: https://doi.org/10.1080/17425255.2016.1202237.

233. Pérez V, Salavert A, Espadaler J, et al. Efficacy of prospective pharmacogenetic testing in the treatment of major depressive disorder: results of a randomized, double-blind clinical trial. *BMC Psychiatry*. 2017;17:250. Available from: https://doi.org/10.1186/s12888-017-1412-1.
234. Hicks JK, Sangkuhl K, Swen JJ, et al. Clinical pharmacogenetics implementation consortium guideline (CPIC) for CYP2D6 and CYP2C19 genotypes and dosing of tricyclic antidepressants: 2016 update. *Clin Pharmacol Ther*. 2017;102:37–44. Available from: https://doi.org/10.1002/cpt.597.
235. Hou L, Heilbronner U, Degenhardt F, et al. Genetic variants associated with response to lithium treatment in bipolar disorder: a genome-wide association study. *Lancet (London, England)*. 2016. Available from: https://doi.org/10.1016/S0140-6736(16)00143-4.
236. Hicks JK, Bishop JR, Sangkuhl K, et al. Clinical pharmacogenetics implementation consortium (CPIC) guideline for CYP2D6 and CYP2C19 genotypes and dosing of selective serotonin reuptake inhibitors. *Clin Pharmacol Ther*. 2015;98:127–134. Available from: https://doi.org/10.1002/cpt.147.
237. Gressier F, Ellul P, Dutech C, et al. Serotonin toxicity in a CYP2D6 poor metabolizer, initially diagnosed as a drug-resistant major depression. *Am J Psychiatry*. 2014;171. Available from: https://doi.org/10.1176/appi.ajp.2014.13101377. 890–890.
238. Dean L. *Imipramine Therapy and CYP2D6 and CYP2C19 Genotype, Medical Genetics Summaries*. 2012a.
239. Dean L. *Amitriptyline Therapy and CYP2D6 and CYP2C19 Genotype, Medical Genetics Summaries*. 2012b.
240. Ingelman-Sundberg M, Oscarson M, McLellan RA. Polymorphic human cytochrome P450 enzymes: an opportunity for individualized drug treatment. *Trends Pharmacol Sci*. 1999;20:342–349.
241. Ng C, Sarris J, Singh A, et al. Pharmacogenetic polymorphisms and response to escitalopram and venlafaxine over 8 weeks in major depression. *Hum Psychopharmacol*. 2013;28:516–522. Available from: https://doi.org/10.1002/hup.2340.
242. Shams MEE, Arneth B, Hiemke C, et al. CYP2D6 polymorphism and clinical effect of the antidepressant venlafaxine. *J Clin Pharm Ther*. 2006;31:493–502. Available from: https://doi.org/10.1111/j.1365-2710.2006.00763.x.
243. Dean L. *Venlafaxine Therapy and CYP2D6 Genotype, Medical Genetics Summaries*. 2012.
244. Zhou S-F, Di YM, Chan E, et al. Clinical pharmacogenetics and potential application in personalized medicine. *Curr Drug Metab*. 2008;9:738–784.
245. Myung W, Kim J, Lim S-W, et al. A genome-wide association study of antidepressant response in Koreans. *Transl Psychiatry*. 2015;5: e633–e633. Available from: https://doi.org/10.1038/tp.2015.127.
246. Biernacka JM, Sangkuhl K, Jenkins G, et al. The International SSRI Pharmacogenomics Consortium (ISPC): a genome-wide association study of antidepressant treatment response. *Transl Psychiatry*. 2015;5. Available from: https://doi.org/10.1038/tp.2015.47. e553–e553.
247. Sasayama D, Hiraishi A, Tatsumi M, et al. Possible association of CUX1 gene polymorphisms with antidepressant response in major depressive disorder. *Pharmacogenomics J*. 2013;13:354–358. Available from: https://doi.org/10.1038/tpj.2012.18.
248. Tammiste A, Jiang T, Fischer K, et al. Whole-exome sequencing identifies a polymorphism in the BMP5 gene associated with SSRI treatment response in major depression. *J Psychopharmacol*. 2013;27:915–920. Available from: https://doi.org/10.1177/0269881113499829.
249. Garriock HA, Kraft JB, Shyn SI, et al. A genomewide association study of citalopram response in major depressive disorder. *Biol Psychiatry*. 2010;67:133–138. Available from: https://doi.org/10.1016/j.biopsych.2009.08.029.
250. Ising M, Lucae S, Binder EB, et al. A genomewide association study points to multiple loci that predict antidepressant drug treatment outcome in depression. *Arch Gen Psychiatry*. 2009;66:966. Available from: https://doi.org/10.1001/archgenpsychiatry.2009.95.

251. Pisanu C, Tsermpini E-E, Mavroidi E, Katsila T, Patrinos GP, Squassina A. Assessment of the pharmacogenomics educational environment in southeast Europe. *Public Health Genomics*. 2014;17:272–279. Available from: https://doi.org/10.1159/000366461.
252. Squassina A, Manchia M, Borg J, et al. Evidence for association of an ACCN1 gene variant with response to lithium treatment in Sardinian patients with bipolar disorder. *Pharmacogenomics*. 2011;12:1559–1569. Available from: https://doi.org/10.2217/pgs.11.102.
253. Tsermpini EE, Zhang Y, Niola P, et al. Pharmacogenetics of lithium effects on glomerular function in bipolar disorder patients under chronic lithium treatment: a pilot study. *Neurosci Lett*. 2017;638:1–4. Available from: https://doi.org/10.1016/j.neulet.2016.12.001.
254. Rybakowski JK, Suwalska A, Skibinska M, et al. Prophylactic lithium response and polymorphism of the brain-derived neurotrophic factor gene. *Pharmacopsychiatry*. 2005;38:166–170. Available from: https://doi.org/10.1055/s-2005-871239.
255. Serretti A, Lilli R, Mandelli L, Lorenzi C, Smeraldi E. Serotonin transporter gene associated with lithium prophylaxis in mood disorders. *Pharmacogenomics J*. 2001;1:71–77.
256. Serretti A, Malitas PN, Mandelli L, et al. Further evidence for a possible association between serotonin transporter gene and lithium prophylaxis in mood disorders. *Pharmacogenomics J*. 2004;4:267–273. Available from: https://doi.org/10.1038/sj.tpj.6500252.
257. Benedetti F, Serretti A, Pontiggia A, et al. Long-term response to lithium salts in bipolar illness is influenced by the glycogen synthase kinase 3-beta -50 T/C SNP. *Neurosci Lett*. 2005;376:51–55. Available from: https://doi.org/10.1016/j.neulet.2004.11.022.
258. Rybakowski JK, Dmitrzak-Weglarz M, Suwalska A, Leszczynska-Rodziewicz A, Hauser J. Dopamine D1 receptor gene polymorphism is associated with prophylactic lithium response in bipolar disorder. *Pharmacopsychiatry*. 2009;42:20–22. Available from: https://doi.org/10.1055/s-0028-1085441.
259. Szczepankiewicz A, Skibinska M, Suwalska A, Hauser J, Rybakowski JK. The association study of three FYN polymorphisms with prophylactic lithium response in bipolar patients. *Hum Psychopharmacol*. 2009;24:287–291. Available from: https://doi.org/10.1002/hup.1018.
260. Dimitrova A, Milanova V, Krastev S, et al. Association study of myo-inositol monophosphatase 2 (IMPA2) polymorphisms with bipolar affective disorder and response to lithium treatment. *Pharmacogenomics J*. 2005;5:35–41. Available from: https://doi.org/10.1038/sj.tpj.6500273.
261. Kawanishi C, Lundgren S, Agren H, Bertilsson L. Increased incidence of CYP2D6 gene duplication in patients with persistent mood disorders: ultrarapid metabolism of antidepressants as a cause of nonresponse. A pilot study. *Eur J Clin Pharmacol*. 2004;59:803–807. Available from: https://doi.org/10.1007/s00228-003-0701-4.
262. Schulze TG, Alda M, Adli M, et al. The International Consortium on Lithium Genetics (ConLiGen): an initiative by the NIMH and IGSLI to study the genetic basis of response to lithium treatment. *Neuropsychobiology*. 2010;62:72–78. Available from: https://doi.org/10.1159/000314708.
263. Saruwatari J, Ishitsu T, Nakagawa K. Update on the genetic polymorphisms of drug-metabolizing enzymes in antiepileptic drug therapy. *Pharmaceuticals (Basel)*. 2010;3:2709–2732. Available from: https://doi.org/10.3390/ph3082709.
264. Lee H-Y, Kim Y-K. Catechol-*O*-methyltransferase Val158Met polymorphism affects therapeutic response to mood stabilizer in symptomatic manic patients. *Psychiatry Res*. 2010;175:63–66. Available from: https://doi.org/10.1016/j.psychres.2008.09.011.
265. Perlis RH, Adams DH, Fijal B, et al. Genetic association study of treatment response with olanzapine/fluoxetine combination or lamotrigine in bipolar I depression. *J Clin Psychiatry*. 2010;71:599–605. Available from: https://doi.org/10.4088/JCP.08m04632gre.
266. Kim B, Kim CY, Lee MJ, Joo YH. Preliminary evidence on the association between XBP1-116C/G polymorphism and response to prophylactic treatment with valproate in bipolar disorders. *Psychiatry Res*. 2009;168:209–212. Available from: https://doi.org/10.1016/j.psychres.2008.05.010.

267. Mizzi C, Dalabira E, Kumuthini J, et al. A European spectrum of pharmacogenomic biomarkers: implications for clinical pharmacogenomics. *PLoS One*. 2016;11:e0162866. Available from: https://doi.org/10.1371/journal.pone.0162866.
268. Relling MV, Evans WE. Pharmacogenomics in the clinic. *Nature*. 2015;526:343–350. Available from: https://doi.org/10.1038/nature15817.
269. Mai Y, Mitropoulou C, Papadopoulou XE, et al. Critical appraisal of the views of healthcare professionals with respect to pharmacogenomics and personalized medicine in Greece. *Per Med*. 2014;11:15–26. Available from: https://doi.org/10.2217/pme.13.92.
270. Kampourakis K, Vayena E, Mitropoulou C, et al. Key challenges for next-generation pharmacogenomics: Science & Society series on Science and Drugs. *EMBO Rep*. 2014;15:472–476. Available from: https://doi.org/10.1002/embr.201438641.
271. Weitzel KW, Alexander M, Bernhardt BA, et al. The IGNITE network: a model for genomic medicine implementation and research. *BMC Med Genomics*. 2016;9:1. Available from: https://doi.org/10.1186/s12920-015-0162-5.
272. Eadon MT, Desta Z, Levy KD, et al. Implementation of a pharmacogenomics consult service to support the INGENIOUS trial. *Clin Pharmacol Ther*. 2016;100:63–66. Available from: https://doi.org/10.1002/cpt.347.
273. Gottesman O, Kuivaniemi H, Tromp G, et al. The Electronic Medical Records and Genomics (eMERGE) Network: past, present, and future. *Genet Med*. 2013;15:761–771. Available from: https://doi.org/10.1038/gim.2013.72.
274. O'Donnell PH, Bush A, Spitz J, et al. The 1200 patients project: creating a new medical model system for clinical implementation of pharmacogenomics. *Clin Pharmacol Ther*. 2012;92:446–449. Available from: https://doi.org/10.1038/clpt.2012.117.
275. Teng K, DiPiero J, Meese T, et al. Cleveland Clinic's Center for personalized healthcare: setting the stage for value-based care. *Pharmacogenomics*. 2014;15:587–591. Available from: https://doi.org/10.2217/pgs.14.31.
276. Nurnberger JI, Austin J, Berrettini WH, et al. What should a psychiatrist know about genetics? *J Clin Psychiatry*. 2018;80. Available from: https://doi.org/10.4088/JCP.17nr12046.
277. Winner JG, Goebert D, Matsu C, Mrazek DA. Training in psychiatric genomics during residency: a new challenge. *Acad Psychiatry*. 2010;34:115–118. Available from: https://doi.org/10.1176/appi.ap.34.2.115.
278. Finn CT, Wilcox MA, Korf BR, et al. Psychiatric genetics: a survey of psychiatrists' knowledge, opinions, and practice patterns. *J Clin Psychiatry*. 2005;66:821–830.
279. Ward ET, Kostick KM, Lázaro-Muñoz G. Integrating genomics into psychiatric practice. *Harv Rev Psychiatry*. 2019;27:53–64. Available from: https://doi.org/10.1097/HRP.0000000000000203.
280. Laedtke AL, O'Neill SM, Rubinstein WS, Vogel KJ. Family physicians' awareness and knowledge of the Genetic Information Non-Discrimination Act (GINA). *J Genet Couns*. 2012;21:345–352. Available from: https://doi.org/10.1007/s10897-011-9405-6.

CHAPTER

PHARMACOGENOMICS IN CLINICAL CARE: IMPLICATIONS FOR PUBLIC HEALTH

6

George P. Patrinos[1,2,3], Asimina Andritsou[1], Konstantina Chalikiopoulou[1], Effrosyni Mendrinou[1] and Evangelia-Eirini Tsermpini[1]

[1]Department of Pharmacy, School of Health Sciences, University of Patras, Patras, Greece [2]Department of Pathology, College of Medicine and Health Sciences, United Arab Emirates University, Al Ain, United Arab Emirates [3]Zayed Center of Health Sciences, United Arab Emirates University, Al Ain, United Arab Emirates

6.1 INTRODUCTION

There has been a significant progress, especially during the last decades, in scientific fields, such as molecular biology, genetics and genomics, accompanied with great advances in related methods and technologies that contribute to a better knowledge of the mechanism of actions of biological molecules, including xenobiotics and drugs. Today, we are fully aware of genetic variability among individuals and its substantial role in drug response and efficacy.[1,2] In particular, as far as genetic variability in humans is concerned, there are estimated to be 3.5–5 million genomic variants, naturally occurring in each genome, a significant number of which is rare and potentially resulting in observable phenotypes. Variants have, also, been observed in genes encoding various enzymes that have significant role in metabolic pathways related to the absorption, distribution, metabolism, excretion, and toxicity of medicines. These genes are also referred to as ADMET genes due to their association with the earlier procedures. It is remarkable that, on average, half of the patients are properly responding to their medications, varying between 25% and 60% among individuals, underlining that a significant proportion of patients are not given a suitable medication or are probably experiencing adverse drug reactions by changing medication till a clinical benefit is, actually, observed.[3]

Pharmacogenomics refers to "... the delivery of the right drug to the right patient at the right dose" (Fig. 6.1). However, the complexity owing to both disease heterogeneity (several phenotypes observed in a single disease) and genetic complexity (several genes contributing to a single phenotype), the difficulty in the establishment of clinically actionable pharmacogenomic biomarkers, and adjacent public health genomics–related issues, are some of the obstacles that hamper translation of pharmacogenomics into clinical practice.[4] Also, pharmacogenomic tests are being developed in parallel with new drug candidates, also known as companion diagnostics.[5] Several randomized controlled trials have provided evidence for the clinical utility of single drug–gene pharmacogenomic testing (Table 6.1).

Applied Genomics and Public Health. DOI: https://doi.org/10.1016/B978-0-12-813695-9.00006-6

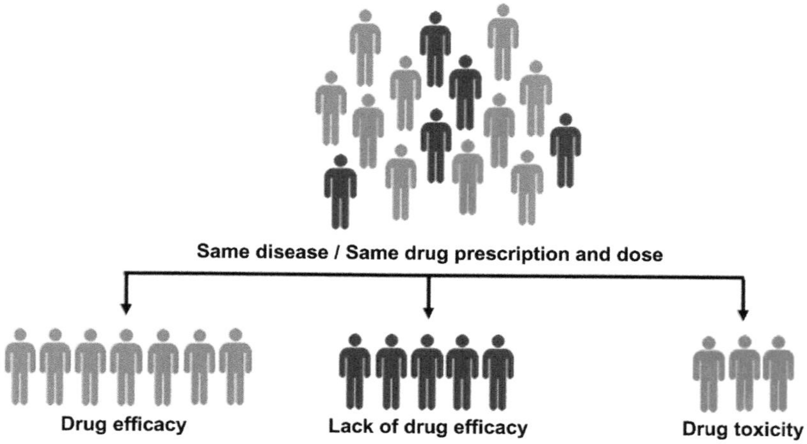

FIGURE 6.1

Outline of the pharmacogenomics principle. In a group of patients suffering from the same disease and being prescribed the same drug prescription (same regimen, same dose), a group of patients is expected to respond well to the drug prescription (depicted in green), others will partly or fully fail to respond to drug treatment (depicted in blue), while others will develop mild or even lethal adverse drug reactions (depicted in red).

Table 6.1 Randomized Controlled Trials Providing Evidence for the Clinical Utility of Single Drug–Gene Pharmacogenomic Testing

Medication	Guidance	References
Warfarin	Drug dosing	[6] [7]
Acenocoumarol and phenprocoumon	Drug dosing	[8]
Carbamazepine	Drug dosing	[9]
Thiopurines	Drug dosing	[10]
Abacavir	Drug selection	[11]

This chapter aims to highlight the importance of introducing pharmacogenomics into clinical care. In particular, the chapter provides key examples of the application of pharmacogenomics in the clinic pertaining to various medical specialties, such as oncology, cardiology, psychiatry, and also touches upon the adjacent, to pharmacogenomics, disciplines that if properly addressed will catalyze smooth integration of pharmacogenomics in the clinic.

6.2 APPLICATIONS OF PHARMACOGENOMICS IN CLINICAL CARE

Clinical pharmacology aims at personalizing the dosage of many drugs with a low therapeutic index in different conditions, such as cancer, cardiovascular, infectious, and neuropsychiatric disorders. Pharmacogenomics has been currently applied in several medical specialities, and this is the reason why regulatory bodies, such as the US Food and Drug Administration (FDA; www.fda.gov) and the European Medicines Agency (EMA; www.ema.europa.eu), have approved the use of pharmacogenomic tests. Today, there is, in many drug labels, a recommendation for patients and clinicians to conduct pharmacogenomic tests, which precede prescribing a particular drug, while experimental evidence indicate that genomic variants may affect the efficacy and/or toxicity of many more drugs than we already know (Ref. [2]; see also Chapter 20).

In the following paragraphs, we will be highlighting the most promising pharmacogenomic applications, accompanied with some examples of currently used tests in the clinic to define drug doses and to individualize treatment, namely, ensuring drug efficacy and/or avoiding drug toxicity.

6.2.1 PHARMACOGENOMICS FOR CANCER THERAPEUTICS

6.2.1.1 Tamoxifen

Tamoxifen is classified as a selective estrogen receptor (ER) modulator and is the standard endocrine therapy for ER-positive (ER +) breast cancer in premenopausal women. In clinical practice, tamoxifen is used in consecutive treatment with an aromatase inhibitor in postmenopausal women[12]. Through the metabolic conversion of tamoxifen, its metabolites are formed mainly, due to the hepatic enzyme CYP2D6. Both the prodrug and its active metabolite [4-hydroxy-*N*-desmethyl-tamoxifen (endoxifen) and 4-hydroxytamoxifen] have pharmacologic activity, with the metabolites of tamoxifen presenting 30- to 100-fold greater affinity to the ER. In the case of tamoxifen and CYP2D6, there are plenty of research data that suggest a benefit of pharmacogenomic testing in drug dosing and/or treatment efficacy.

The most common allele found in the Caucasian population is *CYP2D6*4*, while *CYP2D6*10* polymorphism is mainly found in Asians.[13–15] As far as tamoxifen dosing is concerned, there were a series of research efforts where the association of these alleles with endoxifen concentration was examined. One of these was the study of Borges et al.,[16] who revealed that individuals carry at least one null (*4) of reduced function allele (*10), which indicated similar endoxifen concentrations, while another group of people with multiple copies of the gene presented significantly higher endoxifen plasma concentration. The differences in active metabolite concentration levels displayed a variation in drug efficacy. Goetz et al.[17] presented results of women bearing the *CYP2D6*4/*4* genotype. Specifically, this group of women had showed reduced relapse-free time and worse disease-free survival compared to women who were not carriers of this allele. These findings were in agreement with the findings of Schroth et al.,[18] who showed that women characterized by a null or reduced function allele had significantly more recurrence, shorter relapse-free survival, and worse event-free survival unlike the functional allele carriers. Taking into consideration the findings with respect to tamoxifen dosing, the FDA advisory committee updated tamoxifen's drug label with recommendations for drug use.[19]

6.2.1.2 Irinotecan

Irinotecan is a camptothecin analog, which is approved worldwide for the treatment of metastatic colorectal cancer. It is often used in combination with other drugs such as 5-fluorouracil (5-FU), leucovorin, bevacizumab, and/or cetuximab. Irinotecan is delivered as prodrug, with therapeutic activity found in its active metabolite, 7-ethyl-10-hydroxycamptothecin (SN-38), which is a potent topoisomerase I inhibitor[20,21] being 100–1000 times more cytotoxic than the parent drug. The SN-38 concentration-time curve (AUC) has been correlated to neutropenia.[22,23]

The phase II enzymes, uridine diphosphate glucuronosyltransferases (UGTs), are responsible for the clearance of SN-38 via glucuronidation (SN-38 glucuronide, SN-38G) and specifically the hepatic UGT1A1, although it is possible that UGT1A9 and UGT1A7 (extrahepatic) may also contribute to the detoxification of SN-38.[24] The first clinical evidence about irinotecan toxicity and SN-38 glucuronidation association was described by Gupta et al.[20] Later evidence came from two case reports in patients with Gilbert's syndrome[25] where variability in UGT1A1 was noticed, as well as results indicating that SN-38 is a substrate for UGT1A1.[26]

In the case of the *UGT1A1* gene, more than 100 genomic variants have been identified.[27,28] UGT1A1 is known for its role in bilirubin's drug clearance. Several investigators identified a number of *UGT1A1* genomic variants related to a benign familial condition of reduced bilirubin glucuronidation (Gilbert's syndrome).[29]

The *UGT1A1*28* variant is characterized by an extra thymine–adenine (TA) repeat in the *UGT1A1* gene promoter. The length of TA repeat is inversely correlated with the UGT1A1 enzyme's expression. This insertion polymorphism modifies the TATA box close to *UGT1A1* transcription initiation site where the general transcription factor IID is binding and significantly affects the initiation of transcription.[29] As a result, individuals that are homozygous for seven TA repeats in the *UGT1A1* gene promoter [i.e., $(TA)_7$] are characterized by a 70% reduction in *UGT1A1* expression in contrast with those who bear the $(TA)_6$ allele.[30] It seems that up to 40% of the variability in the in vitro UGT1A1 enzyme activity can be explained by the *UGT1A1*28* polymorphism[31]. Also, two, not so common, variant alleles at the same location have been initially found in individuals of African ancestry, namely, $(TA)_5$ or *UGT1A1*36* and $(TA)_8$ or *UGT1A1*37*.[32]

Interestingly, the level of irinotecan metabolism was inversely correlated to the number of TA repeats in the UGT1A1 promoter region.[33] The correlation between the *UGT1A1*28* allele and the risk of severe adverse drug reactions, including diarrhea and/or neutropenia in patients under irinotecan treatment, was estimated in a retrospective study.[34] Iyer et al.[35] were the first to conduct a prospective trial of irinotecan pharmacogenomics, demonstrating an association between *UGT1A1*28* and reduced SN-38 glucuronidation rate, while Innocenti et al.[36] found a significant correlation between *UGT1A1*28* and grade 4 neutropenia (see also Ref. [37]). This led the FDA to amend the irinotecan label to include *UGT1A1*28* as a risk factor for cancer patients to develop neutropenia upon irinotecan treatment.

6.2.1.3 5-Fluorouracil

5-FU is known for its antineoplastic activity and is approved as a chemotherapeutic agent. Currently, it is widely used by clinicians as treatment for malignancies, such as breast, prostate, skin, and colorectal cancer, both in the adjuvant and metastatic setting. However, patients who receive 5-FU treatment tend to suffer from harmful side effects, such as cardiac side effects.[38] Dihydropyrimidine dehydrogenase (DPD) enzyme is responsible for the catabolism of 5-FU in 80% of the administered dose.[39]

DPD is encoded by the *DPYD* gene, composed of 23 exons, and located in chromosome 1p22 region.[40] It is found that genetic variations in the *DPYD* gene, could potentially, result in DPD enzyme deficiency in approximately 4%–5% of the population. Under those circumstances, there is an increase in the half-life of the drug and ultimately, the bioaccumulation of the drug and severe fluoropyrimidine-related toxicity in patients.

Three independent studies and metaanalysis[41–43] indicated notable association between *DPYD* variants and 5-FU-induced toxicity. Absent or reduced enzyme activity has been associated with several SNPs in the *DPYD* gene, such as *DPYD*2A*, *DPYD*13*, c.2846A > T, and c.1236G > A/ haplotype B3.[44] *DPYD*2A* is the most well-known variant for fluoropyrimidine-associated toxicity[45].

Although single-nucleotide polymorphisms (SNPs) of the *DPYD* gene have been proposed as pharmacogenomic biomarkers, prospective genotyping has not yet been implemented in the daily clinic. Deenen et al.[46] performed a prospective study with respect to dose adjustment of 5-FU, in which patients were preemptively genotyped for the *DPYD*2A* biomarker, based on which the dose was adjusted accordingly. This study consequently suggested that genotyping prior to 5-FU treatment increases patient safety. While the use of 5-FU is contraindicated in patients with a known DPD deficiency, by both the FDA and EMA, no recommendation has been proposed for preemptive genotyping in the *DPYD* gene in cancer patients before receiving fluoropyrimidine treatment.

6.2.2 PHARMACOGENOMICS FOR DRUG TREATMENT OF CARDIOVASCULAR DISEASES

It is estimated that cardiovascular diseases are the leading cause of death worldwide. Taking this into account, there is clearly a great need for patients to be administered the most effective drug according to their genetic profile, and hence, precision medicine to be implemented in individuals suffering from cardiovascular diseases. There is ample scientific evidence suggesting that genomic variants can affect the response to certain drugs used in treating cardiovascular disorders,[47] which are outlined below.

6.2.2.1 Clopidogrel

Clopidogrel, when combined with aspirin, is used to prevent atherothrombotic events and cardiac stent thrombosis.[48] Despite the fact that clopidogrel was approved as an antiplatelet drug, a subtherapeutic response is observed in almost one fourth of individuals treated with clopidogrel. Enzymes participating in the oxidation, reduction, or hydrolysis of substrates or enzymes, which have the capability to acetylate, glucuronidate, sulfate, or methylate their substrates, have been revealed in the past 20 years through pharmacogenomics studies. Human cytochrome P450 enzymes (CYP450) are a quite common paradigm.[49]

CYP2C19 plays a significant role in the bioactivation of clopidogrel. Clopidogrel is a prodrug, and *CYP2C19* is responsible for its metabolic activation.[50] A common loss-of-function *CYP2C19* genomic variant is *CYP2C19*2* (rs4244285 or c.681G > A[50]). This variant is shown to generate a cryptic splice site and a premature stop codon,[51] which causes loss of function. It seems that there is an association between low concentration levels of the clopidogrel's active metabolite and the *CYP2C19*2* allele, while cohort studies have shown that not only *CYP2C19*2* but also other *CYP2C19* loss-of-function alleles contribute to nonresponsiveness and adverse clinical outcomes of clopidogrel.[52] *CYP2C19*3* (rs4986893, c.636G > A) has also been reported to be a loss-of-function variant, since it creates a premature stop codon[51] Other loss-of-function alleles observed in much lower prevalence are *CYP2C19* **4*, **5*, **6*, **7*, and **8*.[53] Another variant, namely, the *CYP2C19***17* allele,[53] alters the interaction between different transcription factors and the *CYP2C19* promoter region, reflected on the extent of gene transcription.[49]

Taking into consideration the previously mentioned *CYP2C19***2*, **3*, **4*, **5*, **8*, and **17* alleles that have been referred earlier, individuals can be categorized, with respect to their metabolic status, into the following groups: potentially ultrarapid (**17*/**17* and **1*/**17*), extensive (good) (**1*/**1*), intermediate (**1*/**2*, **1*/**3*, **1*/**4*, and **1*/**8*), and poor metabolizers (**2*/**2*, **2*/**3*, **2*/**4*, **2*/**5*, **2*/**8*, and **3*/**3*).

Also, Bouman et al.[54] revealed that the enzyme paraoxonase-1 (*PON1*) has a crucial role in clopidogrel's response. p. Q192R is a variant *PON1* allele causing nonfunctional alteration but affecting the platelet response, clopidogrel pharmacokinetics, and the risk for thrombosis.[54] Genetic variants of *CYP3A4*[55–57] and *CYP3A5*[56] have also been held responsible for affecting response to clopidogrel treatment. These variants encode for CYP enzymes, which are involved in the formation of the active metabolite of clopidogrel. Additional genetic factors include *P2RY12* variants encoding the platelet ADP receptor P2Y12 and *ABCB1* variants,[58] encoding for P-glycoprotein. Notably, environmental and pathophysiological clinical factors (age, smoking, diet, and drug–drug interactions) could possibly influence the way patients respond to the administration of clopidogrel.

From the earlier fact, it can be deduced that determining a patient's response to clopidogrel treatment is a multifaceted issue, which involves many genetic variants and nongenetic factors. Also, truly personalized care in precision cardiovascular medicine will necessitate integration of genomics, transcriptomics, epigenomics, microRNA (miRNA) regulomics, proteomics, metabolomics, microbiomics, and mathematical and computational modeling to prevent, diagnose, prognose, and manage disease.[49]

6.2.2.2 Coumarinic oral anticoagulants

Warfarin, acenocoumarol, and phenprocoumon are known as coumarinic oral anticoagulants. These drugs are considered to be life-saving drugs, but a great interindividual variability is shown, concerning coumarinic drug response, mainly owing to several factors, including genomic variants in the *CYP2C9* and *VKORC1* genes. Specifically, *CYP2C9***2* and **3* influence coumarin pharmacokinetics by reducing enzymatic activity of CYP2C9, while the *VKORC1* variant modifies the pharmacodynamic response to coumarins.[59] Although warfarin is administered as a long-term treatment to prevent thromboembolic events, it is characterized by a narrow therapeutic window. There are plenty of studies that emphasize on *CYP2C9* variants (*CYP2C9***2*, *CYP2C9***3*).[60,61] Extensive warfarin metabolizers are those that are able to metabolize warfarin well and bear two copies of *CYP2C9***1*. Heterozygotes for the *CYP2C9***2* and *CYP2C9***3* variant alleles (these variants include point mutations in exons 3 and 7 of *CYP2C9*, respectively) show minor metabolic capacity and are classified as intermediate warfarin metabolizers. Patients who are *CYP2C9***3*/**3* homozygotes show S-warfarin clearance at a rate that is 90% lower, while individuals with the following genotypes, *CYP2C9***1*/**3*, *CY-P2C9***1*/**2*, *CYP2C9***2*/**2*, or *CYP2C9***2*/**3*, exhibit drug clearance ranging from 50% to 75% lower[48,66].

VKORC1 encodes the target receptor of warfarin. The c.1173C > T variant has been associated with variability in warfarin dose[63] as well as with lower maintenance doses of acenocoumarol and phenprocoumon.[64] Another *VKORC1* variant that should be taken into account to amend warfarin dose is c.-1639G > A. In particular, patients bearing the *VKORC1* c.-1639AA, AG, and GG genotypes are associated with high, intermediate, and low sensitivity to warfarin, respectively.[65]

There are several other genetic variants that could probably contribute to warfarin dosage adjustment. For instance, recent studies suggest the implication of *CYP4F2* in order to implement warfarin dosage regulation.[66] Also, there have been other studies showing that the *CYP4F2* variant (rs2108622, c.1297G > A) is responsible for reduced vitamin K metabolism indirectly leading to greater vitamin K availability.[67] Studies, in which both Caucasians and Asians participated, confirmed that the c.1297G > A carriers require higher warfarin dose.[66] Later, Gschwind et al.[68] noticed that a group of individuals who carried *ABCB1* c.2677TT were in greater need of a daily dose of acenocoumarol compared to those with the wild-type genotype.

Algorithms have emerged aiming to estimate the best dosage of coumarinic anticoagulants by considering clinical features, drug–drug interactions, and genetic variants. A genome-wide association study (GWAS) highlighted the significance of *VKORC1*, *CYP2C9*, and *CYP4F2* for interindividual drug variability,[69] while another study suggested preemptive genotyping for *VKORC1*, *CYP4F2*, *CYP2C9*2*, and *CYP2C9*3* in patients on demand to extreme doses of acenocoumarol.[70]

6.2.2.3 Statins

Statins, which are HMG-CoA reductase inhibitors, are one of the most prescribed cardiovascular drugs worldwide. Their wide administration in patients suffering from cardiovascular disorders is interwoven with their efficacy at lowering the concentration of low-density lipoprotein (LDL) up to 55% and minimizing the risk of cardiovascular events by 20%–30%. Pharmacogenomic studies of statins are mostly focused on investigating the efficacy of lowering LDL levels and the extent to which they are able to prevent patients from myocardial infarction and other important vascular events as well as myotoxicity.[71,72]

There is ample scientific evidence that individuals bearing common variants in the *SLCO1B1* gene show intolerance when receiving the first statin prescribed to them, which is misleading and may result in trial-and-error prescribing.[73,74] *SLCO1B1* encodes the hepatic uptake transporter OAT1B1, which has a key role in mediated hepatocyte uptake of multiple statins. Genomic variants in *SLCO1B1* affect statin delivery into the liver cells by OATP1B1, resulting in the elevation of statin's concentration in the bloodstream. Increased plasma concentrations of statins enhance the possibility of adverse drug reactions, giving as an example the statin-related myopathy. Voora et al.,[75] while investigating *SLCO1B1*5* (p.V174A, c.521T > C), found that this variant allele is correlated to a higher risk for adverse statin-induced effects, such as myopathy, in patients who were under atorvastatin, simvastatin, and pravastatin treatment. As a result, it has been proposed that *SLCO1B1*5* allele has a gene-dosage effect causing adverse drug reactions upon statin treatment.

While *SLCO1B1* is known for statin-related toxicity, *KIF6* is related to statin efficacy. Kinesin-like protein 6 is the gene product of KIF6 and is known for its role in intracellular transport. Numerous publications mention that patients bearing the rs20455 allele could potentially benefit from statin treatment in comparison with those patients that do not bear this allele.[76] Notwithstanding the earlier findings, there are some results from a number of clinical trials with atorvastatin, rosuvastatin, and simvastatin, which exhibit nonsignificant correlation between *KIF6* variants and the potential of developing coronary events or the likelihood of statin response.[77] CETP is also related to cholesterol metabolism and is considered an agent of delivering cholesteryl esters into the liver. It is also responsible for transporting triglycerides from LDL to high density lipoprotein (HDL) cholesterol. Alterations of cholesterol levels, cardiovascular events, and aberrant response to statins have been observed due to genetic variants in *CETP*. One of the most notable examples is *Taq 1B* carriers. Carriers of this variant allele tend to express less CETP, which leads to higher HDL levels and minimizes the risk of coronary artery disease progression when compared with noncarriers.[78] Papp et al.[79] have managed to identify *CETP* variants showing strong association with cholesterol levels and sex-dependent cardiovascular risk, as well.

Last but not the least, CYP3A enzymes are also involved in statins' metabolism, such as in atorvastatin, lovastatin, and simvastatin. Indicatively, concentration levels of the simvastatin and its active metabolite found about 50% higher in *CYP3A4*22* allele carriers in contrast with the group of wild-type homozygous patients.[80]

6.2.3 PHARMACOGENOMICS FOR PSYCHIATRIC DISEASES

The pharmacological treatment of patients suffering from psychiatric disorders has been severely hampered by the large interindividual variation in drug response and/or severe side effects, similar to the case of patients suffering from cardiovascular diseases and cancer. For the earlier reason, a plethora of studies attempted to correlate genomic variants with psychiatric drug response and toxicity (see also Chapter 5). Some key examples of genomic variants affecting psychiatric drug treatment modalities are summarized later.

6.2.3.1 Lithium

Lithium chloride, one of the most well-known mood stabilizers with antisuicidal effects, is currently being utilized as an agent for acute mania and as maintenance treatment in bipolar disorder (BD).[81,82] Unfortunately, there are few pharmacogenomic studies that address the issue of response to lithium treatment (reviewed in Ref. [83]), while previous GWASs, also dealing with lithium treatment response, have added a few genetic factors affecting lithium response including only limited criteria for the phenotypic characterization of treatment response (Ref. [2] and references therein).

According to published candidate-gene studies, several genomic variants in different genes, such as *5-HTT*, *TPH*, *DRD1*, *FYN*, *INPP1*, *CREB1*, *BDNF*, *GSK3β*, *ARNTL*, *TIM*, *DPB*, *NR3C1*, *BCR*, *XBP1*, and *CACNG2* genes, have been shown to be associated with lithium treatment response. Moreover, *SLC6A4* and *ACCN1* gene variants have also been associated with lithium treatment outcomes in patients with BD.

In recent studies, it has been shown that the rs1800532 variant in the *TPH1* gene was associated with poor lithium response in patients with the rs1800532A/A genotype. Furthermore, the rs4532 variant in the *DRD1* gene is associated with poor response to lithium.[84] The rs3730353 variant in the *FYN* gene showed prophylactic response to lithium in patients diagnosed with BD, while the c. C973A variant in the *INPP1* gene affected lithium treatment efficacy. Notably, an association was also observed in the lithium response and the rs206472 variant[85] As for *CREB1* gene, two variants (rs6740584/rs2551710) have been correlated with response to lithium treatment, but further investigations are needed to confirm these findings.[84] Furthermore, in a large-scale study, including 3874 psychiatric patients from Sweden and the United Kingdom, it was shown that variants in the *PLET1* gene are significantly associated with response to lithium treatment.[85]

Notably, some of the genes, already mentioned to be associated with lithium response, have overlapping effects in response to antidepressants in major depressive disorder and lithium treatment response in BD, such as *SLC6A4* genomic variants.[85]

Unfortunately, despite the large amount of genetic data on lithium response published so far, we still miss conclusive and robust evidence suggesting that certain genomic variants could be reliably used to predict the probability of responding to lithium,[84] and currently, there is no single pharmacogenomic biomarker that has been approved by the FDA or the EMA yet.

6.2.3.2 Antipsychotics

Antipsychotics are prescribed to treat psychosis and related symptoms, which also vary in terms of efficacy and extent of adverse drug reactions. There are several genomic variants that can potentially serve as pharmacogenomic biomarkers for antipsychotic drug treatment. Indicatively, genomic biomarkers include polymorphisms in genes encoding the CYP450 metabolizing enzymes, in the *ABCB1* transporter gene, in genes of dopaminergic and serotonergic receptors, which are targets for several antipsychotic drugs, such as *DRD2, DRD3, DRD4, 5-HT1, 5HT-2A, 5HT-2C,* and *5HT6,* including the ones related to the metabolism of neurotransmitters and G-signaling pathways, such as *5-HTTLPR, BDNF, COMT, RGS4.*

There have been numerous studies envisioning to find reliable pharmacogenomic markers for antipsychotic drug treatment. A large study including Chinese patients, a total of 77 SNPs of 25 genes, have been examined for four commonly prescribed antipsychotic drugs, namely, risperidone, clozapine, quetiapine, and chlorpromazine.[86] Genomic variants in the *CYP2D6, CYP2C19, COMT, ABCB1, DRD3,* and *HTR2C* genes have been significantly associated with antipsychotic drug treatment response. In addition, a number of candidate genes were also explored, namely, *TNIK, RELN, NOTCH4,* and *SLC6A2,* along with haplotypes (e.g., the rs1544325/rs5993883/rs6269/rs4818 block in the *COMT* gene), which are related with treatment response of the aforementioned drugs. Furthermore, evidence from multivariate analysis suggested the combination of rs6269 in the *COMT* gene and rs3813929 in the *HTR2C* gene as predictors of clinical antipsychotic treatment and response.[86]

As far as the adverse drug reactions that result from antipsychotic drug treatment, the most common ones are tardive dyskinesia, antipsychotic-induced weight gain (AIWG), and clozapine-induced agranulocytosis.[87] Several genomic variants in the *HSPG2, CNR1,* and *DPP6* genes were associated with tardive dyskinesia, including variants in the *SLC18A2* gene, which constitutes a target of valbenazine, a new agent for tardive dyskinesia treatment. Also, variants in the HLA and *MC4R* loci have been associated with atypical weight gain resulting from antipsychotic treatment.[88] Interestingly, the rs6971 TSPO genomic variant was found to be independently associated with AIWG, suggesting that *TSPO* could be utilized as a reliable biomarker for AIWG.[89]

Several other genes, including some already mentioned ones, such as, *HTR2C, DRD2, ADRA2A, MC4R,* and *GNB3,* showed significant association with antipsychotic-related weight gain, indicating that antipsychotic-related weight gain is polygenic and also influenced by environmental factors.[90]

Notably, a large number of these associations failed to be replicated in independent studies and as such, only a small number of pharmacogenomic biomarkers are currently of clinical utility, residing mostly in the *CYP2D6* and *CYP2C19* genes (see also Chapter 5: Genomic Basis of Psychiatric Illnesses and Response to Psychiatric Drug-Treatment Modalities).

6.2.4 PHARMACOGENOMICS AND TRANSPLANTATIONS

Organ transplantation is the optimum choice for patients dealing with terminal and irreversible organ failure. The impact of solid organ transplantations in clinical care could be carefully estimated, by accounting the rate on patient survival, the decrease of comorbidities, as well as the general quality of life of patients, which have been submitted to transplantations.[91]

Currently, the most widely used immunosuppressant medications for solid organ transplantation are a combination of a calcineurin inhibitor (cyclosporine or tacrolimus) coadministered with an antiproliferative agent (e.g., mycophenolic acid). The immunosuppressants are an indicative category of drugs with narrow therapeutic index, and they present with remarkable interpatient variability; hence, in order to reach favorable clinical outcomes, it is important to reach an equilibrium between efficacy and toxicity.

Therapeutic drug monitoring (TDM) has been involved in routine clinical practice for the last 30 years. Currently, in TDM practice, blood concentrations of several drugs are being monitored before the next dosing.[92] TDM limits the pharmacokinetic component of variability since concentrations of drug in blood can be estimated, but only after the drug has been received from the patient. Thus complementary strategies are required.

One such approach would be to combine information from drug dosing and genotyping data obtained from transplant recipients. However, so far, the only clinical recommendation is between *CYP3A5* polymorphisms and tacrolimus dose. Tacrolimus is a calcineurin inhibitor, which is considered to be in the first line of treatment for the most kidney transplantations carried out in centers, both in Europe and the United States.[93,94] Tacrolimus' pharmacokinetics have been characterized by narrow therapeutic range and extensive interindividual variability, resulting in suboptimal immunosuppression or toxicity. Due to its narrow therapeutic index, it is quite hard to reach the desirable tacrolimus concentration levels in blood with TDM.[95] The main cause of tacrolimus' interindividual variability in drug response can be attributed to genomic variants in the *CYP3A5* gene.[93] More specifically, *CYP3A5*3* (rs776746A/G), a genomic variant residing in intron 3 of the *CYP3A5* gene,[92] creates a cryptic splice site, resulting in abnormal splicing, and hence, low CYP3A5 protein levels.[96] Individuals bearing at least one copy of the rs776746A allele (*CYP3A5*1*) are defined as CYP3A5 expressers, while those bearing the rs776746G allele (*CYP3A5*3*) in homozygosity are defined as CYP3A5 nonexpressors (Table 6.2).[97]

From Table 6.2, it can be deduced that transplant recipients, characterized as extensive or intermediate metabolizers, require a higher dose of tacrolimus in order to reach the required therapeutic drug concentrations in blood, estimated to be 1.5–2 times higher than the usual dose, with the restriction not exceeding 0.3 mg/kg/day and always accompanied by TDM.

6.3 LARGE-SCALE PROGRAMS ON THE CLINICAL APPLICATION OF PHARMACOGENOMICS

During the last decade, there have been several pharmacogenomic implementation programs initiated worldwide, aiming to demonstrate the feasibility and benefits from incorporating pharmacogenomics into routine clinical practice.

Table 6.2 Assignment of the Likely Metabolic Status of Patients Depending on the *CYP3A5*3* Genotype and the Corresponding Implications for Tacrolimus Levels and Dosing Recommendations

Likely Phenotype	Genotypes	Implications for Tacrolimus' Pharmacologic Measures	Therapeutic Recommendations
Extensive metabolizer (CYP3A5 expressor)	*CYP3A5*1/*1*	Lower dose and decreased chance of achieving target tacrolimus concentrations	Increase starting dose 1.5 to 2-fold the recommended starting dose. Use TDM to guide dose adjustment
Intermediate metabolizer (CYP3A5 expressor)	*CYP3A5*1/*3*		
Poor metabolizer (CYP3A5 nonexpressor)	*CYP3A5*3/*3*	Higher (normal) dose and increased chance of achieving target tacrolimus concentrations	Initiate therapy with the standard recommended dose. Use TDM to guide dose adjustment

TDM, *Therapeutic drug monitoring.*

In the United States the Electronic Medical Records and Genomics Network (eMERGE)-PGx (https://emerge.mc.vanderbilt.edu/emerge-pgx) is a partnership of eMERGE and the Pharmacogenomics Research Network. eMERGE-PGx is a multicenter project, which aims to implement targeted sequencing of 84 pharmacogenes and assess process and clinical outcomes of this implementation at 10 academic medical centers in the United States.[98] Also, as part of the eMERGE-PGx project, the Icahn School of Medicine at Mount Sinai has initiated the CLIPMERGE PGx project for implementing pharmacogenomic testing into the electronic health-care record and clinical decision support by using a *BioMe* biobank–derived patient cohort of 1500 patients. The main goal of CLIPMERGE PGx is to provide valuable insight into the mechanisms, tools, and processes that will best support the use of genomic information in clinical care and in optimizing drug safety and efficacy. Also, the Pharmacogenomics Resource for Enhanced Decisions in Care and Treatment study is part of the eMERGE-PGx project and was initiated by the Vanderbilt University. The aim of this study is to develop the infrastructure and framework for incorporating pharmacogenomic results into the electronic medical record and making these available to health-care professionals at the time of prescribing. The initial focus of the project was *CYP2C19* genotyping for patients receiving antiplatelet therapy (20), and, as of November 2013, 10,000 patients have been genotyped, and several other drug–gene pairs have been implemented.[99]

Furthermore, INGENIOUS is an National Institutes of Health (NIH)-funded randomized trial conducted by the Indiana University School of Medicine and the Indiana Institute for Personalized Medicine in collaboration with the Eskenazi Health system. The INGENIOUS trial aims to implement preemptive PGx genotyping of a panel of pharmacogenes in a total of 6000 patients, with 2000 patients assigned to a pharmacogenetic testing arm and 4000 to a control arm. Both arms will be followed for a year after being prescribed a targeted medication.[100]

The University of Chicago has initiated the 1200 Patients Project and aims to determine the feasibility and utility of incorporating preemptive pharmacogenomic testing in clinical care. This observational study implements a novel genomic prescribing system, also known as Genomic Prescribing System, to deliver patient-specific interpretation of complex genomic data for a particular drug.[101]

The University of Florida and Shands Hospital launched the Personalized Medicine Program in 2011 to ensure the clinical implementation of pharmacogenomics-based prescribing. The pilot implementation project focused on implementation of *CYP2C19*—clopidogrel gene—drug pair, and future plans include expansion to additional gene—drug pairs.[102]

The PG4KDS of St. Jude Children's Research Hospital aims to selectively use Drug Metabolism Enzymes and Transporters (DMET) + microarray-based pharmacogenomic test in routine patient care and to migrate all Clinical Pharmacogenomics Implementation Consortium (CPIC; www.cpicpgx.org) gene—drug pairs into the electronic health-care records as an aim toward making genome-informed drug prescribing become routine care.[103]

Contrary to the United States, pharmacogenomic implementation projects within the European Union may encounter even more challenges as a result of the multilinguistic setting, different legal environments, and highly different health-care systems across the different European countries. In Europe the PREemptive Pharmacogenomic testing for Prevention of Adverse drug Reactions (PREPARE) study is run by the Ubiquitous Pharmacogenomics (U-PGx; www.upgx.eu) consortium, an established network of European experts aiming to address the challenges and obstacles for the clinical implementation of pharmacogenomics in patient care.[104] This European Commission—funded project is perhaps the only one to investigate the impact of population-wide implementation of preemptive PGx testing, of a panel of clinically relevant pharmacogenomic biomarkers, on both the incidence of adverse drug reactions and health-care costs. Compared to many other implementation initiatives, U-PGx aims to implement pharmacogenomics through the preemptive panel strategy as opposed to an individual drug—gene pair strategy. For the reasons stated earlier, this approach provides relevant evidence supporting the implementation of pharmacogenomics in routine care, in the context of a large prospective, international, randomized, controlled study including more than 8000 patients.

6.4 PUBLIC HEALTH PHARMACOGENOMICS

As previously mentioned, in recent years, we have witnessed significant progress in pharmacogenomic research, facilitated by the advent of genomic technologies. However, these discoveries proceed with an asynchronous pace, as far as translation of research findings into the clinic is concerned, hampering the smooth incorporation of pharmacogenomic research findings into daily medical practice (see also Chapter 1: Applied Genomics and Public Health). Public health pharmacogenomics touches upon disciplines, such as ethics in pharmacogenomics, economic evaluation in genome-guided treatment modalities, translational tools in pharmacogenomics and knowledge bases, and the advancement of pharmacogenomic education of clinicians and other stakeholders with interest in this field, which would eventually facilitate pharmacogenomics research to be incorporated in the clinic.[105]

Individual genomic profiling is anticipated to provide a highly efficacious therapeutic strategy. However, a gap still exists between the pharmacogenomic testing per se and the interpretation and utilization of its results in a clinical setting (also known as translation of genomic results into patient care[106]). In this context, few studies have demonstrated a significant lack of harmonization of pharmacogenomics education,[107] while several international organizations have called for the integration of pharmacogenomics and personalized medicine education into core medical curricula. Today, several studies indicate that both the patients and the general public desire to receive pharmacogenomic services from health-care professionals who can confidently explain the test and interpret its implications for prescriptions. However, a gap still exists between patients' high expectations and health-care professionals' knowledge, while the stakeholders in question appear to have a generally positive attitude, despite concerns regarding privacy issues.[108] Proper interventions at the educational level will definitely increase understanding, facilitating the incorporation of pharmacogenomics into patient care.

Cost-effectiveness of pharmacogenomic tests is another crucial factor, if pharmacogenomic testing is to be adopted in a clinical setting. Recent reports on drug treatments for cardiology[109,110] and cancer[111] have demonstrated cost-effectiveness of these interventions (see also Ref. [112]). Based on these findings, pricing and reimbursement policies of health-care system payers can be established for the rapid dissemination of pharmacogenomics (see also Chapter 17) with the potential of yielding improved quality of clinical care, along with increased economic benefits, both for the pharmaceutical industry and public health. It should be noted, however, that with few exceptions, pharmacogenomic tests will probably prove to be more cost-effective than cost-saving or eventually cost-effective only for certain combinations of disease, treatment, and test and gene characteristics (see also Chapter 17).

Similarly, pharmacogenomics has to deal with several ethical issues that relate to genetic discrimination, privacy, possible implications for access to life and health insurance as well as genetic discrimination.[113] In particular, when an individual is defined as a "responder" or a "nonresponder" to therapeutics, this categorization serves as the new disease labels with social consequences, which involve interpersonal discrimination or identity issues. For economic reasons, pharmaceutical companies could voluntarily ignore patients with rare or complex genetic conditions or those who are not responding to any known treatment, resulting in consequent deprivation of effective treatments.[114]

Another crucial element of concern is the storage of genomic information in databases, considering the potential loss of confidentiality or privacy, since databases link an enormous quantity of genotypic, phenotypic, and demographic data regarding individuals.[115] In this regard, protection for privacy and confidentiality has to be ensured, particularly in the whole-genome sequencing era, as pharmacogenomic tests may carry several types of secondary information that represent a risk of psychosocial harm or adverse insurance and/or employment implications. Moreover, particular genetic subgroups may face discrimination in accessing health care or health insurance.[116]

Lastly, harmonization of the guidelines that have been issued for genome-informed drug prescription is of paramount importance for the accurate interpretation of pharmacogenomic testing results. So far, there are differences that have been identified in recommendations and guidelines, issued by major research consortia, such as the CPIC and the Dutch Pharmacogenetics Working Group,[117] while the same is true for major regulatory bodies such as the US FDA and the EMA. Depicting these discrepancies can assist toward harmonization of these guidelines, especially since they derive from a single evidence base.

The aforementioned are some of the key factors that relate to public-health pharmacogenomics, which need to be adequately addressed so that implementation of pharmacogenomics in the clinical setting is expedited.

6.5 CONCLUSIONS AND FUTURE PERSPECTIVES

Currently, research findings have solidified genotype associations with drug efficacy and/or toxicity in a few cases, which have been further supported by the updates in drug labeling instituted by the regulatory agencies. Notwithstanding the clinical utility of pharmacogenomic testing is far from optimal. Further prospective clinical studies as well as discovery research and economic evaluation data are needed to establish utility to genetic information in the clinic and drug research and development setting. The clinical implications of pharmacogenomics also rely heavily on "genethics," patient awareness as well as the education of health-care professionals. We stand on the cusp of personalized medicine, and the success of this endeavor will largely depend on more than knowing the genetic code. Indeed, a deeper understanding of the intricacies that regulate and underlie the code, especially as far as drug treatment interventions are concerned, becomes more than even fundamental.

ACKNOWLEDGMENTS

Part of our own work has been funded by a European Commission grant (H2020-668353; U-PGx) to GPP.

REFERENCES

1. Vesell ES, Page JG. Genetic control of drug levels in man: phenylbutazone. *Science.* 1968;159:1479–1480.
2. Squassina A, Manchia M, Manolopoulos VG, et al. Realities and expectations of pharmacogenomics and personalized medicine: impact of translating genetic knowledge into clinical practice. *Pharmacogenomics.* 2010;11:1149–1167.
3. Spear BB, Heath-Chiozzi M, Huff J. Clinical application of pharmacogenomics. *Trends Mol Med.* 2001;7:201–204.
4. Piquette-Miller M, Grant DM. The art and science of personalized medicine. *Clin Pharmacol Ther.* 2007;81:311–315.
5. Giacomini KM, Brett CM, Altman RB, et al. The pharmacogenetics research network: from SNP discovery to clinical drug response. *Clin Pharmacol Ther.* 2007;81:328–345.
6. Pirmohamed M, Burnside G, Eriksson N, et al. A randomized trial of genotype-guided dosing of warfarin. *N Engl J Med.* 2013;369(24):2294–2303.
7. Wu AH. Pharmacogenomic testing and response to warfarin. *Lancet.* 2015;385(9984):2231–2232.
8. Verhoef TI, Ragia G, de Boer A, et al. A randomized trial of genotype-guided dosing of acenocoumarol and phenprocoumon. *N Engl J Med.* 2013;369(24):2304–2312.
9. Chen P, Lin JJ, Lu CS, et al. Carbamazepine-induced toxic effects and HLA-B*1502 screening in Taiwan. *N Engl J Med.* 2011;364(12):1126–1133.

10. Coenen MJ, de Jong DJ, van Marrewijk CJ, et al. Identification of patients with variants in TPMT and dose reduction reduces hematologic events during thiopurine treatment of Inflammatory Bowel Disease. *Gastroenterology*. 2015;149(4):907–917.e7.
11. Mallal S, Phillips E, Carosi G, et al. HLA-B*5701 screening for hypersensitivity to abacavir. *N Engl J Med*. 2008;358(6):568–579.
12. Burstein HJ, Griggs JJ, Prestrud AA, Temin S. American society of clinical oncology clinical practice guideline update on adjuvant endocrine therapy for women with hormone receptor-positive breast cancer. *J Oncol Pract*. 2010;6:243–246.
13. Chamnanphon M, Pechatanan K, Sirachainan E, et al. Association of CYP2D6 and CYP2C19 polymorphisms and disease-free survival of Thai post-menopausal breast cancer patients who received adjuvant tamoxifen. *Pharmacogenomics Pers Med*. 2013;6:37–48.
14. Fernandez-Santander A, Gaibar M, Novillo A, et al. Relationship between genotypes SULT1A2 and CYP2D6 and tamoxifen metabolism in breast cancer patients. *PLoS One*. 2013;8:e70183.
15. Lim JS, Chen XA, Singh O, et al. Impact of CYP2D6, CYP3A5, CYP2C9 and CYP2C19 polymorphisms on tamoxifen pharmacokinetics in Asian breast cancer patients. *Br J Clin Pharmacol*. 2011;71(5):737–750.
16. Borges S, Desta Z, Li L, et al. Quantitative effect of CYP2D6 genotype and inhibitors on tamoxifen metabolism: implication for optimization of breast cancer treatment. *Clin Pharmacol Ther*. 2006;80 (1):61–74.
17. Goetz MP, Rae JM, Suman VJ, et al. Pharmacogenetics of tamoxifen biotransformation is associated with clinical outcomes of efficacy and hot flashes. *J Clin Oncol*. 2005;23(36):9312–9318.
18. Schroth W, Antoniadou L, Fritz P, et al. Breast cancer treatment outcome with adjuvant tamoxifen relative to patient CYP2D6 and CYP2C19 genotypes. *J Clin Oncol*. 2007;25(33):5187–5193.
19. Goetz MP, Kamal A, Ames MM. Tamoxifen pharmacogenomics: the role of CYP2D6 as a predictor of drug response. *Clin Pharmacol Ther*. 2008;83(1):160–166.
20. Fukui T, Mitsufuji H, Kubota M, et al. Prevalence of topoisomerase I genetic mutations and UGT1A1 polymorphisms associated with irinotecan in individuals of Asian descent. *Oncol Lett*. 2011;2(5):923–928.
21. Gupta E, Lestingi TM, Mick R, Ramirez J, Vokes EE, Ratain MJ. Metabolic fate of irinotecan in humans: correlation of glucuronidation with diarrhea. *Cancer Res*. 1994;54(14):3723–3725.
22. Hirose K, Kozu C, Yamashita K, et al. Correlation between plasma concentration ratios of SN-38 glucuronide and SN-38 and neutropenia induction in patients with colorectal cancer and wild-type UGT1A1 gene. *Oncol Lett*. 2012;3(3):694–698.
23. Pitot HC, Goldberg RM, Reid JM, et al. Phase I dose-finding and pharmacokinetic trial of irinotecan hydrochloride (CPT-11) using a once-every-three-week dosing schedule for patients with advanced solid tumor malignancy. *Clin Cancer Res*. 2000;6(6):2236–2244.
24. Hahn KK, Wolff JJ, Kolesar JM. Pharmacogenetics and irinotecan therapy. *Am J Health Syst Pharm*. 2006;63(22):2211–2217.
25. Wasserman E, Myara A, Lokiec F, et al. Severe CPT-11 toxicity in patients with Gilbert's syndrome: two case reports. *Ann Oncol*. 1997;8(10):1049–1051.
26. Iyer L, King CD, Whitington PF, et al. Genetic predisposition to the metabolism of irinotecan (CPT-11). Role of uridine diphosphate glucuronosyltransferase isoform 1A1 in the glucuronidation of its active metabolite (SN-38) in human liver microsomes. *J Clin Invest*. 1998;101(4):847–854.
27. Hasegawa Y, Ando Y, Shimokata K. Screening for adverse reactions to irinotecan treatment using the Invader UGT1A1 Molecular Assay. *Expert Rev Mol Diagn*. 2006;6(4):527–533.
28. Takano M, Suriyama T. UGT1A1 polymorphisms in cancer: impact on irinotecan treatment. *Pharmgenomics Pers Med*. 2017;10:61–68.
29. Bosma PJ, Chowdhury JR, Bakker C, et al. The genetic basis of the reduced expression of bilirubin UDP-glucuronosyltransferase 1 in Gilbert's syndrome. *N Engl J Med*. 1995;333(18):1171–1175.

30. Beutler E, Gelbart T, Demina A. Racial variability in the UDP-glucuronosyltransferase 1 (UGT1A1) promoter: a balanced polymorphism for regulation of bilirubin metabolism? *Proc Natl Acad Sci USA*. 1998;95(14):8170–8174.
31. Peterkin VC, Bauman JN, Goosen TC, Menning L, Man MZ, Paulauskis JD, et al. Limited influence of UGT1A1*28 and no effect of UGT2B7*2 polymorphisms on UGT1A1 or UGT2B7 activities and protein expression in human liver microsomes. *Br J Clin Pharmacol*. 2007;64(4):458–468.
32. Horsfall LJ, Zeitlyn D, Tarekegn A, et al. Prevalence of clinically relevant UGT1A alleles and haplotypes in African populations. *Ann Hum Genet*. 2011;75(2):236–246.
33. Iyer L, Hall D, Das S, et al. Phenotype-genotype correlation of in vitro SN-38 (active metabolite of irinotecan) and bilirubin glucuronidation in human liver tissue with UGT1A1 promoter polymorphism. *Clin Pharmacol Ther*. 1999;65(5):576–582.
34. Ando Y, Saka H, Ando M, et al. Polymorphisms of UDP-glucuronosyltransferase gene and irinotecan toxicity: a pharmacogenetic analysis. *Cancer Res*. 2000;60(24):6921–6926.
35. Iyer L, Das S, Janisch L, Wen M, Ramírez J, Karrison T, et al. UGT1A1*28 polymorphism as a determinant of irinotecan disposition and toxicity. *Pharmacogenomics J*. 2002;2(1):43–47.
36. Innocenti F, Undevia SD, Iyer L, et al. Genetic variants in the UDP-glucuronosyltransferase 1A1 gene predict the risk of severe neutropenia of irinotecan. *J Clin Oncol*. 2004;22(8):1382–1388.
37. Kim TW, Innocenti F. Insights, challenges, and future directions in irinogenetics. *Ther Drug Monit*. 2007;29(3):265–270.
38. Alter P, Herzum M, Soufi M, Schaefer JR, Maisch B. Cardiotoxicity of 5-fluorouracil. *Cardiovasc Hematol Agents Med Chem*. 2006;4(1):1–5.
39. Pandey K, Dubey RS, Prasad BB. A critical review on clinical application of separation techniques for selective recognition of uracil and 5-fluorouracil. *Indian J Clin Biochem*. 2016;31(1):3–12.
40. Lunenburg CA, Henricks LM, Guchelaar HJ, et al. Prospective DPYD genotyping to reduce the risk of fluoropyrimidine-induced severe toxicity: ready for prime time. *Eur J Cancer*. 2016;54(2):40–48.
41. Terrazzino S, Cargnin S, Del Re M, Danesi R, Canonico PL, Genazzani AA. DPYD IVS14 + 1G > A and 2846A > T genotyping for the prediction of severe fluoropyrimidine-related toxicity: a meta-analysis. *Pharmacogenomics*. 2013;14(11):1255–1272.
42. Rosmarin D, Palles C, Pagnamenta A, et al. A candidate gene study of capecitabine-related toxicity in colorectal cancer identifies new toxicity variants at DPYD and a putative role for ENOSF1 rather than TYMS. *Gut*. 2015;64(1):111–1120.
43. Meulendijks D, Henricks LM, Sonke GS, et al. Clinical relevance of DPYD variants c.1679T > G, c.1236G > A/HapB3, and c.1601G > A as predictors of severe fluoropyrimidine-associated toxicity: a systematic review and meta-analysis of individual patient data. *Lancet Oncol*. 2015;16(16):1639–1650.
44. Henricks LM, Lunenburg CA, Meulendijks D, et al. Translating DPYD genotype into DPD phenotype: using the DPYD gene activity score. *Pharmacogenomics*. 2015;16(11):1277–1286.
45. van Kuilenburg AB, Dobritzsch D, Meinsma R, et al. Novel disease-causing mutations in the dihydropyrimidine dehydrogenase gene interpreted by analysis of the three-dimensional protein structure. *Biochem J*. 2002;364(Pt1):157–163.
46. Deenen MJ, Cats A, Mandigers CM, et al. Prevention of severe toxicity from capecitabine, 5-fluorouracil and tegafur by screening for DPD-deficiency. *Ned Tijdschr Geneeskd*. 2012;156(48):A4934.
47. Shukla H, Mason JL, Sabyah A. A Literature review of genetic markers conferring impaired response to cardiovascular drugs. *Am J Cardiovasc Drugs*. 2018;18(4):259–269.
48. Kitzmiller JP, Groen DK, Phelps MA, Sadee W. Pharmacogenomic testing: relevance in medical practice: why drugs work in some patients but not in others. *Cleve Clin J Med*. 2011;78(4):243–257.
49. Brown SA, Pereira N. Pharmacogenomic impact of CYP2C19 variation on clopidogrel therapy in precision cardiovascular medicine. *J Pers Med*. 2018;8(1):pii:E8.

50. Hulot JS, Bura A, Villard E, et al. Cytochrome P450 2C19 loss-of-function polymorphism is a major determinant of clopidogrel responsiveness in healthy subjects. *Blood.* 2006;108:2244–2247.
51. De Morais SM, Wilkinson GR, Blaisdell J, Meyer UA, Nakamura K, Goldstein JA. Identification of a new genetic defect responsible for the polymorphism of (*S*)-mephenytoin metabolism in Japanese. *Mol Pharmacol.* 1994;46:594–598.
52. Yin T, Miyata T. Pharmacogenomics of clopidogrel: evidence and perspectives. *Thrombosis Res.* 2011;128:307–316.
53. Scott SA, Sangkuhl K, Shuldiner AR, et al. PharmGKB summary: very important pharmacogene information for cytochrome P450, family 2, subfamily C, polypeptide 19. *Pharmacogenet Genomics.* 2012;22:159–165.
54. Bouman HJ, Schömig E, van Werkum JW, et al. Paraoxonase-1 is a major determinant of clopidogrel efficacy. *Nat Med.* 2011;17(1):110–116.
55. Fontana P, Hulot JS, De Moerloose P, Gaussem P. Influence of CYP2C19 and CYP3A4 gene polymorphisms on clopidogrel responsiveness in healthy subjects. *J Thromb Haemost.* 2007;5:2153–2155.
56. Frere C, Cuisset T, Morange PE, et al. Effect of cytochrome p450 polymorphisms on platelet reactivity after treatment with clopidogrel in acute coronary syndrome. *Am J Cardiol.* 2008;101:1088–1093.
57. Gladding P, Panattoni L, Webster M, Cho L, Ellis S. Clopidogrel pharmacogenomics: next steps. *J Am Coll Cardiol Interv.* 2010;3:995–1000.
58. Simon T, Verstuyft C, Mary-Krause M, et al. Genetic determinants of response to clopidogrel and cardiovascular events. *N Engl J Med.* 2009;360:363–375.
59. Manolopoulos VG, Ragia G, Tavridou A. Pharmacogenetics of coumarinic oral anticoagulants. *Pharmacogenomics.* 2010;11(4):493–496.
60. Carlquist JF, Anderson JL. Using pharmacogenetics in real time to guide warfarin initiation: a clinician update. *Circulation.* 2011;124(23):2554–2559.
61. Crespi CL, Miller VP. The R144C change in the CYP2C9*2 allele alters interaction of the cytochrome P450 with NADPH: cytochrome P450 oxidoreductase. *Pharmacogenetics.* 1997;7:203–210.
62. Rettie AE, Haining RL, Bajpai M, Levy RH. A common genetic basis for idiosyncratic toxicity of warfarin and phenytoin. *Epilepsy Res.* 1999;35(3):253–255.
63. Rieder MJ, Reiner AP, Gage BF, et al. Effect of VKORC1 haplotypes on transcriptional regulation and warfarin dose. *N Engl J Med.* 2005;352(22):2285–2293.
64. Au N, Rettie AE. Pharmacogenomics of 4-hydroxycoumarin anticoagulants. *Drug Metab Rev.* 2008;40(2):355–375.
65. Limdi NA, Wadelius M, Cavallari L, et al. Warfarin pharmacogenetics: a single VKORC1 polymorphism is predictive of dose across 3 racial groups. *Blood.* 2010;115:3827–3834.
66. Caldwell MD, Awad T, Johnson JA, et al. CYP4F2 genetic variant alters required warfarin dose. *Blood.* 2008;111(8):4106–4112.
67. McDonald MG, Rieder MJ, Nakano M, Hsia CK, Rettie AE. CYP4F2 is a vitamin K1 oxidase: An explanation for altered warfarin dose in carriers of the V433M variant. *Mol Pharmacol.* 2009;75:1337–1346.
68. Gschwind L, Rollason V, Boehlen F, et al. *P*-Glycoprotein: a clue to vitamin K antagonist stabilization. *Pharmacogenomics.* 2015;16(2):129–136.
69. Takeuchi F, McGinnis R, Bourgeois S, et al. A genome-wide association study confirms VKORC1, CYP2C9, and CYP4F2 as principal genetic determinants of warfarin dose. *PLoS Genet.* 2009;5:e1000433.
70. Perez-Andreu V, Roldan V, Lopez-Fernandez MF, et al. Pharmacogenetics of acenocoumarol in patients with extreme dose requirements. *J Thromb Haemost.* 2010;8:1012–1017.
71. Postmus I, Verschuren JJ, de Craen AJM, et al. Pharmacogenetics of statins: achievements, whole-genome analyses and future perspectives. *Pharmacogenomics.* 2012;13(7):831–840.

72. Roden DM. Current status and future directions of cardiovascular pharmacogenomics. *J Hum Genet.* 2016;61:79–85.
73. Donnelly LA, Doney AS, Tavendale R, et al. Common nonsynonymous substitutions in slco1b1 predispose to statin intolerance in routinely treated individuals with type 2 diabetes: a go-darts study. *Clin Pharmacol Ther.* 2011;89:210–216.
74. Legault MA, Tardif JC, Dubé MP. Pharmacogenomics of blood lipid regulation. *Pharmacogenomics.* 2018;19(7):651–665.
75. Voora D, Shah SH, Spasojevic I, et al. The SLCO1B1*5 genetic variant is associated with statin-induced side effects. *J Am Coll Cardiol.* 2009;54:1609–1616.
76. Li Y, Iakoubova OA, Shiffman D, Devlin JJ, Forrester JS, Superko HR. KIF6 polymorphism as a predictor of risk of coronary events and of clinical event reduction by statin therapy. *Am J Cardiol.* 2010;106 (7):994–998.
77. Hopewell JC, Parish S, Clarke R, et al. No impact of KIF6 genotype on vascular risk and statin response among 18,348 randomized patients in the heart protection study. *J Am Coll Cardiol.* 2011;57:2000–2007.
78. Willer CJ, Sanna S, Jackson AU, et al. Newly identified loci that influence lipid concentrations and risk of coronary artery disease. *Nat Genet.* 2008;40(2):161–169.
79. Papp AC, Pinsonneault JK, Wang D, et al. Cholesteryl ester transfer protein (CETP) polymorphisms affect mRNA splicing, HDL levels, and sex-dependent cardiovascular risk. *PLoS One.* 2012;7(3):e31930.
80. Kitzmiller JP, Binkley PF, Pandey SR, Suhy AM, Baldassarre D, Hartmann K. Statin pharmacogenomics: pursuing biomarkers for predicting clinical outcomes. *Discov Med.* 2013;16(86):45–51.
81. Yatham LN, Kennedy SH, Parikh SV, et al. Canadian network for mood and anxiety treatments (CANMAT) and international society for bipolar disorders (ISBD) collaborative update of CANMAT guidelines for the management of patients with bipolar disorder: update 2013. *Bipolar Disord.* 2013;15(1):1–44.
82. Aral H, Vecchio-Sadus A. Toxicity of lithium to humans and the environment—a literature review. *Ecotoxicol Environ Saf.* 2008;70(3):349–356.
83. Squassina A, Costa M, Congiu D, et al. Insulin-like growth factor 1 (IGF-1) expression is up-regulated in lymphoblastoid cell lines of lithium responsive bipolar disorder patients. *Pharmacol Res.* 2013;73:1–7.
84. Pisanu C, Heilbronner U, Squassina A. The role of pharmacogenomics in bipolar disorder: moving towards precision medicine. *Mol Diagn Ther.* 2018;22(4):409–420.
85. Amare AT, Schubert KO, Klingler-Hoffmann M, Cohen-Woods S, Baune BT. The genetic overlap between mood disorders and cardiometabolic diseases: a systematic review of genome wide and candidate gene studies. *Transl Psychiatry.* 2017;7(1):e1007.
86. Xu Q, Wu X, Li M, et al. Association studies of genomic variants with treatment response to risperidone, clozapine, quetiapine and chlorpromazine in the Chinese Han population. *Pharmacogenomics J.* 2016;16 (4):357–365.
87. Zai CC, Maes MS, Tiwari AK, Zai GC, Remington G, Kennedy JL. Genetics of tardive dyskinesia: promising leads and ways forward. *J Neurol Sci.* 2018;389:28–34.
88. Hamilton SP. The promise of psychiatric pharmacogenomics. *Biol Psychiatry.* 2015;77(1):29–35.
89. Pouget JG, Gonçalves VF, Nurmi EL, et al. Investigation of TSPO variants in schizophrenia and antipsychotic treatment outcomes. *Pharmacogenomics.* 2015;16(1):5–22.
90. Zhang JP, Lencz T, Zhang RX, et al. Pharmacogenetic associations of antipsychotic drug-related weight gain: a systematic review and meta-analysis. *Schizophr Bull.* 2016;42(6):1418–1437.
91. Grinyó JM. Why is organ transplantation clinically important? *Cold Spring Harb Perspect Med.* 2013;3(6):a014985.
92. Picard N, Bergan S, Marquet P, et al. Pharmacogenetic biomarkers predictive of the pharmacokinetics and pharmacodynamics of immunosuppressive drugs. *Ther Drug Monit.* 2016;38(suppl 1):S57–S69.

93. van Gelder T, van Schaik RH, Hesselink DA. Pharmacogenetics and immunosuppressive drugs in solid organ transplantation. *Nat Rev Nephrol*. 2014;10(12):725–731.
94. Chen L, Ramesh Prasad GV. CYP3A5 polymorphisms in renal transplant recipients: influence on tacrolimus treatment. *Pharmgenomics Pers Med*. 2018;11:23–33.
95. Hendijani F, Azarpira N, Kaviani M. Effect of CYP3A5*1 expression on tacrolimus required dose for transplant pediatrics: a systematic review and meta-analysis. *Pediatr Transplant*. 2018;19:e13248.
96. Burckart GJ. Pharmacogenomics: the key to improved drug therapy in transplant patients. *Clin Lab Med*. 2008;28(3):411–422.
97. Birdwell KA, Decker B, Barbarino JM, et al. Clinical pharmacogenetics implementation consortium (CPIC) guidelines for CYP3A5 genotype and tacrolimus dosing. *Clin Pharmacol Ther*. 2015;98 (1):19–24.
98. Rasmussen-Torvik LJ, Stallings SC, Gordon AS, et al. Design and anticipated outcomes of the eMERGE-PGx project: a multicenter pilot for preemptive pharmacogenomics in electronic health record systems. *Clin Pharmacol Ther*. 2014;96(4):482–489.
99. Pulley JM, Denny JC, Peterson JF, et al. *Clin Pharmacol Ther*. 2012;92(1):87–95.
100. Eadon MT, Desta Z, Levy KD, et al. Implementation of a pharmacogenomics consult service to support the INGENIOUS trial. *Clin Pharmacol Ther*. 2016;100(1):63–66.
101. Ratain MJ. Personalized medicine: building the GPS to take us there. *Clin Pharmacol Ther*. 2007;81:321–322.
102. Johnson JA, Burkley BM, Langaee TY, Clare-Salzler MJ, Klein TE, Altman RB. Implementing personalized medicine: development of a cost-effective customized pharmacogenetics genotyping array. *Clin Pharmacol Ther*. 2012;92(4):437–439.
103. Hoffman JM, Haidar CE, Wilkinson MR, et al. PG4KDS: a model for the clinical implementation of pre-emptive pharmacogenetics. *Am J Med Genet, C*. 2014;166C(1):45–55.
104. van der Wouden CH, Cambon-Thomsen A, Cecchin E, et al. Implementing pharmacogenomics in Europe: design and implementation strategy of the ubiquitous pharmacogenomics consortium. *Clin Pharmacol Ther*. 2017;101(3):341–358.
105. Patrinos GP. Public health pharmacogenomics. *Public Health Genomics*. 2014;17(5–6):245–247.
106. Reydon TA, Kampourakis K, Patrinos GP. Genetics, genomics and society: the responsibilities of scientists for science communication and education. *Per Med*. 2012;9(6):633–643.
107. Pisanu C, Tsermpini EE, Mavroidi E, Katsila T, Patrinos GP, Squassina A. Assessment of the pharmacogenomics educational environment in Southeast Europe. *Public Health Genomics*. 2014;17 (5–6):272–279.
108. Mai Y, Mitropoulou C, Papadopoulou XE, et al. Critical appraisal of the views of healthcare professionals with respect to pharmacogenomics and personalized medicine in Greece. *Per Med*. 2014;11(1):15–26.
109. Fragoulakis V, Bartsakoulia M, Díaz-Villamarín X, et al. Cost-effectiveness analysis of pharmacogenomics-guided clopidogrel treatment in Spanish patients undergoing percutaneous coronary intervention. *Pharmacogenomics J*. 2019;. in press.
110. Mitropoulou C, Fragoulakis V, Bozina N, et al. Economic evaluation for pharmacogenomic-guided warfarin treatment for elderly Croatian patients with atrial fibrillation. *Pharmacogenomics*. 2015;16(2):137–148.
111. Fragoulakis V, Roncato R, Fratte CD, et al. Estimating the effectiveness of DPYD genotyping in Italian individuals suffering from cancer based on the cost of chemotherapy-induced toxicity. *Am J Hum Genet*. 2019;104(6):1158–1168.
112. Symeonidis S, Koutsilieri S, Vozikis A, Cooper DN, Mitropoulou C, Patrinos GP. Application of economic evaluation to assess feasibility for reimbursement of genomic testing as part of personalized medicine interventions. *Front Pharmacol*. 10, 2019;830.

113. Issa AM. Ethical perspectives on pharmacogenomic profiling in the drug development process. *Nat Rev Drug Discov*. 2002;1:300–308.
114. Rothstein MA, Epps PG. Ethical and legal implications of pharmacogenomics. *Nat Rev Genet*. 2001;2:228–231.
115. Joly Y, Saulnier KM, Osien G, Knoppers BM. The ethical framing of personalized medicine. *Curr Opin Allergy Clin Immunol*. 2014;14(5):404–408.
116. Smart A, Martin P, Parker M. Tailored medicine: whom will it fit? The ethics of patient and disease stratification. *Bioethics*. 2004;18:322–342.
117. Bank PC, Caudle KE, Swen JJ, et al. Comparison of the Guidelines of the Clinical Pharmacogenetics Implementation Consortium and the Dutch Pharmacogenetics Working Group. *Clin Pharmacol Ther*. 2018;103(4):599–618.

CHAPTER

7 MICROBIAL GENOMICS IN PUBLIC HEALTH: A TRANSLATIONAL RISK-RESPONSE ASPECT

Manousos E. Kambouris[1], Yiannis Manoussopoulos[2], George P. Patrinos[1,3,4] and Aristea Velegraki[5]

[1]Department of Pharmacy, School of Health Sciences, University of Patras, Patras, Greece [2]Plant Protection Division of Patras, Institute of Industrial and Forage Plants, Patras, Greece [3]Department of Pathology, College of Medicine and Health Sciences, United Arab Emirates University, Al-Ain, United Arab Emirates [4]Zayed Center of Health Sciences, United Arab Emirates University, Al-Ain, United Arab Emirates [5]Department of Microbiology, School of Medicine, National and Kapodistrian University of Athens, Athens, Greece

ABBREVIATIONS

CIDTs	culture-independent diagnostic tests
IA	immunoassay
MALDI-TOF	matrix-assisted laser desorption/ionization-time of flight
NAATs	nucleic-acid amplification tests
NGS	next-generation sequencing
PCR	polymerase chain reaction
RT-PCR	real-time polymerase chain reaction
TAT	turn-around time
WGS	whole-genome sequencing

7.1 INTRODUCTION

Infectious diseases have been plighting human communities, less or more evolved, for millennia, being the result of spontaneous cumulative events potentially exacerbated by natural or social factors, such as floods and dense population growth, respectively. Although the Western World experienced a temporary remittance due to the development of antibiotics, this hope was short-lived as misdiagnosis and, most of all, pathogen resistance developed in a few short years. High population turnovers, decreased health standard due to increased percentages of immunocompromised individuals, densely populated areas by poor and, in times, illiterate populations, and old-aged physically and immunologically weakened societies in the affluent zones, combine with a new era in microbiomic applications,[1,2] to turn the prospects more bleak (Fig. 7.1). The last factor is exacerbated by the emergence of new and rare pathogens by massive and rapid transportation of goods and travelers and migration,[3,4] which transcend state borders and natural geographical barriers and alter

Applied Genomics and Public Health. DOI: https://doi.org/10.1016/B978-0-12-813695-9.00007-8

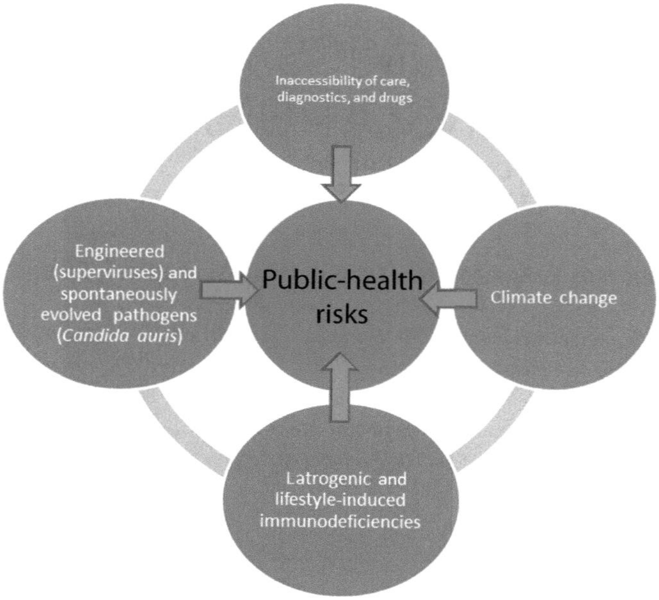

FIGURE 7.1

Factors increasing the microbiomatic dimension of public health risk.

regional microbiomes, along with an explosive development projected for synthetic microbiology and microbiomic engineering.[5,6] These pathogens, collectively called *neopathogens*, differ in physiology, phylogeny, and ecology. Some of them become opportunistic pathogens in case of deficiencies in the host's immunity and others, by rehosting onto novel host species, previously inaccessible,[7] and by spontaneous or perpetrated evolution,[8] that is, by acquisition of virulence factors. Facultative and opportunistic pathogens causing spontaneous outbreaks and metapathogens (agents custom-engineered for pathogenicity) compose the prospective threat list upon access or release on susceptible populations, be they human, livestock, or crops, either exclusively or inclusively, depending on the pathogen, its virulence, hosting patterns, and dispersion dynamics.[3,9–12]

7.2 NEW RULES FOR AN OLD GAME

The conventional approach toward the management of infectious agents is deterministic: sensor systems acquire a signature characteristic for an agent, which, though, may be degraded or of compromised specificity and thus requiring interpretation. The latter introduces additional uncertainties and unknowns. More robust approaches were called upon to tackle the issue, such as the combination of multiple signatures, either of the same type, as in *multilocus sequencing*[13,14] or of different types, as in *polyphasic identification*.[15] Spatial and temporal patterns (geography, season) may help resolve or complicate uncertainties.

Engineered or fully artificial agents are not matched to existing biosignatures or event patterns as a general rule, and their own respective ones are unrecorded and/or intentionally subversive. Thus *agnostic* approaches become essential, based on ultrahigh throughput methods and/or collective biosignatures made up from multiple similar signals.

Situational awareness is the Holy Grail of public health, both in the original, human-centric concept and in the novel, multispectral (bio-/radio/chemical agents), multisectorial (different stakeholders of the public and the private sector), and transdisciplinary (human/animal/plant) one.[16,17] It is usually implemented through surveillance, which entails a notion of temporal and/or spatial continuum. As an operational concept, (bio)surveillance may be described as *the constant or repetitive inspection of an entity of interest so as to detect and then process drifts from a factual baseline*. The baseline is arbitrary and depends on the normal horizon of events corrected for applicable noise levels.[18,19]

Surveillance basically provides early warning[20] so as to formulate and enact response,[4] both proactive (protection and/or prophylaxis) and reactive (containment, decontamination/disinfection, and therapeutic treatment).

Detectors can provide constant surveillance by fixed deployment[21] or intermittently by being permanently mounted or occasionally loaded on different mobile platforms so as to maximize the surveillance footprint.[22,23] Still, detectors are usually unable to identify pathogens, at least to any usable detail.[24–26]

To establish a divergence from normality, which is focal in the surveillance context, normality must be established, even arbitrarily. A background image, condition, or value/figure may do the trick, but it is imperative that each of these must have been collected or compiled well in advance, although background readings change over time, sometimes unexpectedly or irregularly, calling for updates and projections/predictions by dedicated algorithms between successive updates.

Thus a suspect signal must be detected and compared against the *current* background, possibly through fusing data from diverse sources, the most important but not the only one being syndromic surveillance.[27] Subsequent identification is essential for projecting number and severity of health casualties and for initiating proper countermeasures including, but not restricted to, treatment, as mentioned earlier; quarantines and directions to the public are the most efficient protective approaches if applied in a targeted and specific manner.[4]

Identification is accomplished by dedicated instrumentation, the identifiers,[23,28] which exploit diverse technological and scientific principles to process live or ex vivo samples collected by sampling devices, either in stand-alone format or attached to the identifier[23,28] (Table. 7.1).

Table 7.1 Comparative Merits of Indicative Outbreak Resolution Technologies

Method	Fast	Specific	Toxin-Compatible	Self-Contained	Massive	Discrete	Agnostic	PoC/Field Use
IA	+	+	+	+	+	+		+
M/array	+	+		+	+	+		+
NAAT	+	+			+	+		?
NGS		+					+	
LRS		+		+			+	+

IA, Immunoassay; *LRS*, long-range sequencing; *NAAT*, nucleic-acid amplification test; *NGS*, next-generation *sequencing*; PoC, *point-of-care*.

Diagnosis is a retroactive cognitive process, which determines the cause of health casualties. In the context of infections, it presupposes a fair level of direct or indirect identification so as to allow initiation of the optimum response in protective, prophylactic, and therapeutic terms.

The background in such cases consists mainly of the biological signatures of the host and of occurring (appendage or not) microbiomes, which produce noise or directly degrade the specific signal/biosignature. To alleviate its effect, signal enhancement and background depletion are the prescribed approaches, taken selectively or in combination.[20]

7.2.1 COUNTERING INTELLIGENCE OBSOLESCENCE: THE TEMPORAL WINDOW OF OPPORTUNITY IN INFECTIOUS DISEASE OUTBREAKS

Longer detection ranges for standoff detectors increase the margin for timely response, although, strictly speaking, the issue may not be spatial but temporal. It is the time that is of essence, not any amount of physical distance between the threat pathogens and the target subjects per se. Expansive formats of persistent surveillance, including nonphysical inputs by estimations, information, reporting and database-browsing networks, and any other conceivable means and on-the-spot, quick turnover bioassays, partially contribute toward resolving the problem of temporal management of public-health threats and crises. The Internet of Things (see also Chapter 15), which can make instantly available database assets and disseminate raw data among sensors and processors, has the potential to enhance, optimize, and overstep previous concepts such as online surveillance formats based on shared and dedicated nodes, servers, and pipelines.[20]

7.3 REQUIREMENTS

Candidate solutions are constantly improving in terms of availability/accessibility, user-friendliness,[20] reliability, discriminatory power,[22] throughput,[29] dependency on infrastructure and supplies,[22,30–33] and turn-around time (TAT).[20] Culture-dependent assays, such as culturomics[34] and matrix-assisted laser desorption/ionization-time of flight (MALDI-TOF),[35] may seem inapplicable in principle; others, focusing on direct analysis of the core biosignatures, are conditionally considered.[18,22]

On the contrary, the capability for agnostic identification is much sought after in public health, although not so in conventional evidence-based medicine. It allows detection and classification of spontaneously evolved, engineered, or artificial microbiota.[36]

Just as important is the capability for simultaneous multiagent resolution, especially in a context of perpetrated outbreaks, most probably, but not exclusively, as a means of spiking supplies and commodities from food, feed, and water to ambient and supplied air, tools, benches, and handles.[26,30,36–39]

Genomics appears ideal due to an inherent flexibility in specifications and capabilities; an ever-expanding range of technologies, methods, and brands offers different output, performance, and logistics particulars so as to satisfy any set of requirements. On the contrary, the quintessential advantage is that genomics targets probably the most robust biosignature, the DNA, a chemically resilient molecule coding both phylogenetic and functional information about its carrier, including, but not restricted to, virulence factors to the like of resistance determinants, biotoxin production, fast propagation.

All molecular methods, interpreted sensu stricto as methods involved with nucleic acids study, research, and manipulation, including, but not restricted to, nucleic-acid amplification tests (NAATs), are plagued by an inherent limitation: they are utterly inapplicable in the case of free (bio)toxins; "free" being defined as functional even when in cell-free contexts. Toxins do not propagate by themselves but may be transmitted recessively. Biotoxins remain feared agents because (1) of the very high specific activity and potency, which lead to low drastic doses, (2) of their extreme variability and distribution in any environment. Immunodiagnostics,[28,33,40] immunoglobulins, basically antibodies, and aptamers,[26] are structure-sensitive molecules which sense a combination of static electric charges and a three-dimensional molecular conformation to achieve specific binding and thus seem best fit for the task. Mass spectrometrical analyses, which are nowadays able to tackle large moieties (up to the size of a cell) in atomic and molecular levels by fracturing molecules and finely discriminating mass to charge ratios,[22,38,40,41] are also suitable.

7.4 CURRENT AND PROJECTED OUTBREAK RESOLUTION APPROACHES

7.4.1 MASS SPECTROMETRY

Mass spectrometry is plagued with even more challenging agnostic applications; cutting edge methods, such as MALDI-TOF and electrospray ionization (ESI), are able to accomplish an objective molecular scan from an environmental or multispecies sample.[35] This scan records the molecular context of the sample and does so with adequate resolution. However, the said molecules are grouped—not always exclusively—in multiple patterns, and thus both form and define different cells and organisms; such datasets are limited in number and variety and are occasionally proprietary. Thus by definition, they cannot tackle an unknown agent since phylogenetic diversion is not always proportional to molecular variability, both qualitatively and quantitatively.[42,43]

7.4.2 IMMUNOASSAYS

Cross-reaction is a phenomenon causally linked with immunoassays (IAs),[44–46] and its negative impact on detection and diagnosis will only exacerbate as more possible target microbiota emerge, by the sum of perpetrated or spontaneous processes. This fact mars their prospects for point-of-care assays meant for the detection and identification of known analytes in unknown, complicated, and/or degraded samples with high scores of specificity and discrimination,[33] although most IA tests show impressive specificity if panels of cross-reactions have been compiled, and sensitivity in many cases, such as, but not restricted to, sandwich enzyme-linked immunosorbent assay is satisfactory at the very least.

Although the use of conventional antibody-based methods cannot tackle novel agents, artificial or natural, should pure samples of novel, artificial agents become available, an epitopic profile may be compiled by antibody microarrays containing thousands of paratopic loci. In such a case the collective signal may compensate for the perceived lack of specificity of the individual specific ones, providing a better alternative to differential detection of known pathogens than current formats with mutually exclusive and highly specific loci. By fusing the binary signals of the reacting array loci, collective epitopic imaging may be achievable. Thus typing becomes possible even for novel agents. The relation between phylogenetic distance and immunoprofile is not proportional, nor causal, and so in silico methods cannot correct any raw data to mine relationships in a reproducible way to extend applicability from typing to identification/classification.

On the other hand, IAs are uniquely suited to field, point-of-care applications. Many IA-based tests require neither instrumentation nor electricity and other infrastructure to work and no refrigeration or other active conservation method (Table 7.1). Furthermore, IAs are adequately specific and can detect target epitopes in the high clutter of degraded, mixed, and impure samples[33,47] and are thus able to positively identify known pathogens in such conditions while processing high volumes of clinical samples to determine syndromes.[32] Fast-track immunology is prone to tackle with emerging pathogens, provided that they expose epitopes stable over time. This format is a medium-term solution as it takes a minimum of 1−2 months to develop monoclonal antibodies de novo,[48]resulting in perhaps double that time for assay deployment in field conditions, as compilation, testing, amending, and mass production have to follow. As mentioned, IAs offer a cardinal advantage: they are likely to detect biotoxins, although not all of them,[49] and presumably other acellular agents not detectable through NAAT, such as prions and, possibly, viroids.[28,50]

7.4.3 GENOMICS

The term *genomics* has been introduced since the turn of the century; still, its predecessor, the "molecular" detection/identification of pathogens, had already been established as a valid, if not an advantageous approach, once molecular biology came of age, especially with the advent of amplification formats, that is, NAATs. It allowed the detection and identification of spawning bioagents with high sensitivity and specificity, relatively fast (compared to culture-based methods), and more objectively, being unaffected by phenotypic shifts and antigenic variation[51,52] while also tackling noncultivable/fastidious pathogens.

Generally, detection by nucleic acid signature is easy, but simple detection does not reveal or reject the existence of a pathogen, as at least two and usually multiple genomes coexist in clinical and environmental samples. Thus analyzing a nucleic acid sequence to achieve classification or identification presupposes the recording of a number of known/existing similar/analogous sequences assigned to different organisms so as to allow comparisons.[20] Consequently, molecular identification of pathogens and diagnosis of infections required microbiome sequences extracted from pure cultures to be used as targets for oligonucleotide polymerase chain reaction (PCR) primers and probes designed accordingly and cross-tested with abundant genomes (i.e., of hosts) so as to establish specificity. Further noncompetitive or competitive testing along with clinically or phylogenetically relevant taxa determined specificity and discriminative power.[53] Once new modalities started producing massively sequencing data from diverse organisms, which were accessible in public depositories, the previous process was simplified and implemented in silico instead of massive testing of diverse isolates.[29,54]

Simultaneously, the informational context of the NAAT assays has increased manifold and multidimensionally. The first dimension has been the quantification of the examined sample, in relative—but also, if proper standards are provided, in absolute—terms, mainly but not exclusively through real-time PCR (RT-PCR).[20,40,55] Another dimension has been high throughput, a result not so much of decreasing the implementation and TATs of usual tests, as of massively parallel formats, which analyze large numbers of samples simultaneously. This was brought about initially by multiplexing, which is the similar but not identical coprocessing of multiple analytes present within the same sample, either in mixed or in pooled samples.[27] Initially, multiplexing was a prerogative of PCR[56] followed by microarrays,[57] to be promptly superseded by a combination of the two,[29,39] a development that actually spawned the term "genomics" in wet lab context.

Unfortunately, these two dimensions were never convincingly combined to an integrated approach: as the microarrays (see Section 7.4.4.) proved too cumbersome and went out of fashion, despite a cardinal advantage in social and legal terms—genetic discretion, where any number of loci can be interrogated, but the interrogation is limited to these, thus not exposing the sum of the genomic background of the individual, the privacy of which is otherwise compromised and raises multiple legal and ethical issues.[58]

NAATs are by definition characterized by the capability to detect and possibly classify, if not identify, dormant and fastidious agents, exemplified by viruses and spores, conidia and cysts, and thus have become the preferred culture-independent diagnostic tests (CIDTs).[20] The high representation of viral agents in successive iterations of the select agents and toxins list[59] furthered this attractiveness to biosecurity and even to biosafety contexts and applications. The first steps in the spiral evolution of NAAT have been the increase in throughput and the adoption of multiplexing to achieve parallel processing, up to high-order multiplexing, or "hi-plex," tackling simultaneously four-digit singular analyses.[20,29] Hi-plexing was indispensable for tackling unknown agents and multiagent samples[60] and reached its apogee with the development and testing in real outbreaks of *panmicrobial microarrays*[20,61,62]—which were rather *multimicrobial* in real terms.

A third dimension of evolution referred to improvements in sensitivity, which achieved reliable detection of single-digit genomes/cells, and in TAT that was brought down to 30 minutes or less, along with a reliable quantification process. The invention of RT-PCR[20] brought this combination of evolutionary improvements into being, with the concomitant handicap of low-plexing (low double-digit number of coprocessed interrogations per sample) and at a cost of nearly $50,000 for cutting-edge benchtop cyclers. The investment would be double for automated solutions, which embedded sample preparation and nucleic acid–extraction steps.[20,27] Further improvements in the field of deployability, portability, and ruggedization allowed the expeditionary use of RT-PCR, with portable, battery-operated instruments becoming gradually marketed.[63] Subsequently, fully automated solutions were developed, able to integrate nucleic acid–extraction, amplification, and reading by prepackaged reaction-mixture cartridges,[20] which brought NAAT close to eligibility for point-of-care solution, thus competing with IAs.[20,27,28]

On the contrary, NAATs are considered inappropriate for massive use to screen relatively large population fractions, as in the event of an outbreak. High assay costs are currently manageable in individual diagnostic formats for—mostly insured—subjects. But in massive applications the price tag is always an issue for a public-health budget, making high resolution and throughput a necessity, primarily in terms of operational efficiency, but of logistics/sustainability as well. As a result, approaches, such as multiplex PCR and nucleic-acid microarrays, should be included in the operational planning of expanded public-health formats for (proactive) surveillance and (reactive) examination plans.[19,27]

A spin-off of NAATs, the isothermal assay chemistries are easier to miniaturize and to dissociate from energy supply and other infrastructure, thus being more suitable for portable, and/or instrumentation-free assays. This implies a much easier pitch for point-of-care tests for affordable in situ processing of samples and subjects.[20,30,31,64]

7.4.4 ARRAYS: THE GENERATION - X OF MICROBIAL GENOMICS

The microarrays were a concerted effort to allow massive use of NAATs by allowing massively parallel formats, which would explore the potential of multiplexing. The panmicrobial microarrays was a concept pitched in terms of a promise addressing a need; more specifically, the promise of early OMICS, headed by genomics, to tackle the issue of infectious diseases, both spontaneous and perpetrated.[62,65–68] The concept evolved gradually. Initially, the prefix "pan-" (meaning "every" in Greek) implied, rather, "all sequenced," as the population of microbial sequences had been the restrictive factor.[69]

The concept evolved as follows: when sequences became more abundant, the implication changed to "all relevant"; the "relevant" clause, though, was the subject of some fuss. One interpretation lies with taxonomy and results in homogenous target groups (Fig. 7.2); the other is rather functional, pertaining to physiology, pathology, epidemiology, and microecology and includes by definition, heterogeneous target groups. The former was easier to approach, with microarrays designed specifically for viruses being the most common.[62,70–72] Still, examples meant for prokaryotes (bacteria) and eukaryotes (fungi and protozoa) became available eventually.[73–75] The latter

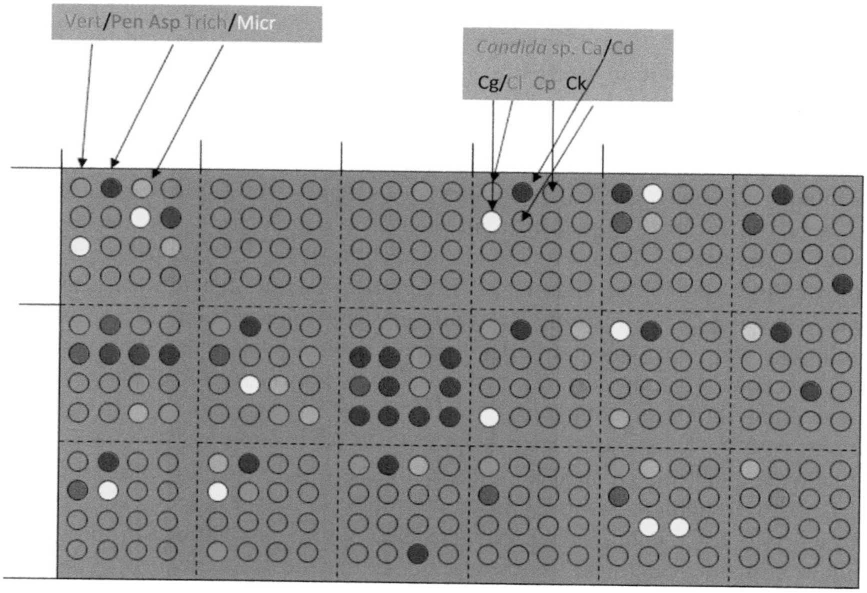

FIGURE 7.2

One purpose for using DNA microarrays formed by homogenous probes (grouped in subarrays by phylogeny) is to establish the presence of a microbiome by detection or discovery. In here a hypothetical fungal application with positional and differential resolution of analytes.

interpretation was more applicable in real-world cases as it allowed assortment, arraying, and comparison of diverse agents, which were possibly involved in a specific kind of outbreak, as, for example, in respiratory infections.[66,68,71,72]

Improved array formats and probe designs brought about a more literary use of the adjective "panmicrobial."[62,69] Long probes were designed, meant to attach to generic, conserved barcoding sequences even when a fraction of mismatches occurred.[66] This allowed the attachment of—more or less distant—homologues, thus detecting strains of rare or novel (unknown or simply unsequenced) lower taxa.[74]

In the context of microbiome, DNA microarrays containing probes for detection, discovery, and functionality assessment had been developed. Differentiations occur based on the range of the targeted pathogens, the role, production, and metrics of the probes, the purpose and specifications of use of the selected platform.[69] The usual purpose for using DNA microarrays is, most of the times, either establishing presence (by detection or by discovery) of a microbiome (Fig. 7.2), or determining the functionality, not necessarily of one microbiome, but possibly within the collective microbial genome.

For establishing presence, target selection and probe design must balance between individual and collective characteristics. Within these premises, detection-oriented designs are able to identify represented organisms with operationally usable resolution (lower taxon, usually species, level). The discovery-purposed arrays, on the other hand, determine if there are any and use purposely low resolution (higher taxon) in order to be collective rather than selective. The functionality assessment arrays scrutinize the sample for indications of qualities, capabilities, and traits these organisms might possess, such as virulence factors, resistance genes,[69,76] adaptability features, and in general genomic roots of translational (metabolomics, proteomic), interactomic, and infectiomic features, which may foretell anticipated niche(s) within a habitat.

Combined D2 arrays (detection and discovery) consist of a combination of probes, some of them targeting species-specific regions and able to detect and identify known pathogens (Table 7.1 and Fig. 7.2); other probes target conserved/generic regions to enable binding of novel organisms with a degree of homology to known ones.[69,70,74] On the contrary, truly novel, naturally occurring or engineered microbiomes are unlikely to bind to probes designed to target known genomes[69] and this is a strong argument in favor of next-generation sequencing (NGS) and other agnostic-oriented methods[18,77] (Table 7.1), further extended by the need to regularly recapitalize array design to include new sequence data as it becomes available.[69]

On the other hand, tiled arrays able to actually (re)sequence fragments of interest can be even more descriptive[67] but, at the same time, of lower overall scope or, alternatively, extremely massive, and thus requiring more sample or elaborate incubation protocols. This specific array type became irrelevant once massively efficient sequencing chemistries became available.

Finally, there is always the issue of reading and scoring an array. The development of uninstrumented formats for point-of-care and field use with almost real-time readings (following the example of pregnancy tests) is a prospective advantage over other technologies (Table 7.1). Alternatively, mobile compact instruments of more elaborate formats requiring processing is within the technology level and may be developed in ruggedized forms with few requirements in terms of infrastructure such as stability, power, cooling and IT interfaces, and environmental control.

7.5 METAGENOMICS: THE FUTURE IS HERE AND NOW

To alleviate the occasional failure of implementing Koch's principles due to fastidious microorganisms or to inherently complicated samples, metagenomic analysis was developed initially as a spin-off of pooled samples' processing, only to evolve to shotgun—full or partial—genome readout. The novelty was the direct molecular processing without prior culture enrichment, which allowed less bias against fastidious organisms, minority populations, and slow growers. The method had been devised primarily to make way with the infrastructure necessary for live microbe processing and to tackle complex/mixed samples. Metagenomics allows either multiagent resolution or agnostic application, but not both simultaneously, mainly—but not exclusively—because the latter requires invariably shotgun metagenomics, while the former may be better approached by consensus single-target PCR, although conditionally. For samples suspected for totally novel, naturally evolving or engineered pathogens, metagenomic analysis cannot be fully implemented as a one-off assay due to the lack of established methods and protocols designed to subtract in silico partial-sequence reads of known genomes from a metagenomic sequence mix. The concept is not unheard of: background depletion for optimization of microbiome sequence reads is easier[78] and is performed routinely in quite a number of metagenomic applications, especially in studies of microbiomes hosted in bio-compartments of macroorganisms. Thus the fully agnostic approach in shotgun metagenomics is practically limited to cases where a background-sequence compilation is available as a previous record (proactively acquired) or as an ad hoc event (reactively acquired) so as to compare and single out new signatures and contacts.

Metagenomics is currently implemented through NGS, which is responsible for much of the accompanying glamour. As mentioned earlier, NGS is not a prerequisite, and direct molecular protocols allowed diagnosis and detection in various samples, since the 1990s, in both clinical and environmental contexts. On the contrary, in the shotgun metagenomic approach, NGS allows the detection of novel, hitherto unchartered agents, in practically every sample type (Table 7.1). Once an unidentifiable genomic signature is compiled by software from the wet reads, it can be compared against entries submitted to depositories and databases for either identification or, should that fail in the case of a totally novel genome, phylogenetic resolution and classification. Thus migrational, evolutionary, or malevolent engineering and release events may be substantiated[20] conditional to the volume of beforehand processed and recorded samples, which would compile the default microbiome picture[20,79,80] in terms of metagenomes.

Obviously, the end usefulness of (meta)genomics in public health for surveillance and resolution of outbreaks is directly proportional to the volume of available sequences. Should the multitude of microbiomic databases curated worldwide become accessible under cooperation agreements, Global Health Security Agenda,[2,4,10,81] the Big Data, and Internet of Things operating principles could work miracles (see also Chapter 15). Their contribution would be the optional utilization of the data collected by the wet-end instrumentation practically throughout the world to resolve incidents and enact surveillance, implementing partially the One Health–One Humanity effort, which pinpointed risk factors of potentially global impact due to failed healthcare institutions in a number of states.[82]

One of the catches is the temporal dimension: shotgun metagenomics is not fast (Table 7.1) and depends on, among others, processing resources, connections/communications, and availability and quality of repository resources, such as databases. Moreover, metagenomics as such remains singularly unsuited when artificial agents are concerned.[20] The algorithms of current IT pipelines align, bin, and assemble raw sequencing data according to patterns, comparisons, and logical assumptions most probably intentionally irrelevant in custom-engineered genomes intended for nefarious ends, causing failure of detection or misidentification due to erroneous assemblies.

7.5.1 NEXT-GENERATION SEQUENCING: A ROBUST AND FLEXIBLE TOOL

NGS, if applied to abundant and essentially pure samples in a whole-genome sequencing (WGS) context, shows impressive potential.[83] It may classify every possible specimen harvested in sampling events; it also deciphers encoded virulence factors of different origins by resorting to trans-omics resources, such as libraries and databases of transcriptomics, proteomics, toxinogenomics, and metabolomics, even if applied in more restrictive, but faster and more affordable, whole-exome sequencing (WES) contexts.

In WGS formats, NGS allows elucidation of a whole genome in a medical, research, or industrial context with affordability and relevant, though not low, TAT. Improved chemistry, innovative instruments, and enhanced IT resources take the credit.[83] Just some years ago, the price for the resolution of a human genome and accessing the assembled sequence and processed information through a smartphone application had dropped below $1000.[84] Novel approaches have already furthered expectations, in terms of flexibility and user-friendliness, from compact benchtop formats with modest requirements in utilities, logistics, expertise, and preparation, which allow mobile lab applications,[20] to man-portable formats, such as the Oxford Nanopore (http://nanoporetech.com) or other portable devices (see also Chapter 4).

The use of WGS in direct surveillance and screening is restricted mainly by cost as it entails processing of large numbers of samples collected in a wide, long, deep context, possibly more than once. Moreover, WGS runs are slow and their cost and TAT differ according to the genomes involved, but it comes into its own when presented with pure samples of even artificial microbiomes, an occasion definitely within the scope of the state of the art of synthetic genomics,[2,5,6] along with the dispersion of the ability to resurrect eradicated pathogens, such as smallpox or polio, among facilities throughout the world.[2,85]

WGS is perhaps the only way to characterize a novel pathogen. Generic, unbiased priming can ensure multiple sequences amplified simultaneously while the depth of sequencing provides a filter against in vitro or, more importantly, in vivo contaminants. In resolving complex, tainted samples while seeking novel pathogens, sampling is of paramount importance to furnish as abundant and pure samples as possible. Segregation strategies, such as flow cytometry, might furnish single cells before scrutinizing with WGS, thus tackling the issue of contaminants.

In a public health context, where both clinical and environmental samples must be processed—although most probably in different events—a flexible and sustainable method is required (Fig. 7.2). It seems that the long-range sequencing spin-off is meant to be the answer to respective prayers. It follows, in the evolutionary line, the versions of Sanger dideoxy termination (either dyed by radioactive residues or fluorescents and run in anything between glass plates and capillaries) principle and the today's reigning NGS formats.

NGS, the current standard, which requires presequencing amplification and is characterized by massively parallel sequence processing, was made possible by advanced chemistries and dedicated bioinformatics tools. These formats are mostly—but not exclusively—operating by some microarray principle, either positional or self-assembled, but they use it not as the actual readout method but as an enabler (on-array sequencing). The Genome Analyzer, Solexa, Hiseq2000 and their updates, presented by Illumina and the RS of Pacific Biosciences NGS platforms use spin-offs of planar arrays.[86] At the same time the SOLiD platform of Applied Biosystems, the Ion platforms of Ion Torrent Systems Inc., and the GS FLX 454 of Roche are based on suspension/self-assembled arrays.[87]

The operating name of long-range sequencing is self-explanatory: extremely long-sequence runs are produced due to improved chemistries, not necessarily performing matrix-guided synthesis. Longer runs, which require fewer assembly rounds, fewer overlapping sequences, and thus sequencing depth, are easier to score unique alignments and as a result lower computational power and less DNA matrix are required. In turn, amplification of extracted DNA may become redundant, which drops the cost and time but most importantly removes a source of considerable bias in "sequenceable" DNA chunks, thus reducing the background and the necessary computational power and further improving accuracy, as minority genomes are not shadowed by majority ones, or at least not as much as with mandatory amplification steps, whether they are isothermal or not.

Optimistic projections of achieving very high−speed reads with unprecedented affordability may prove unsubstantiated and overbearing. Still, the long-range sequencing platforms are developed for individual and not collective, centralized deployment, thus not only reestablishing the self-sufficiency offered by Sanger sequencing instruments, but also allowing several steps further into operational flexibility: truly man-portable devices, which with full support packages (processing and networking equipment, power management, and consumables for some rounds of operating) weight less than a standard hand luggage and occupy a fraction of the latter's volume. Such qualities and virtues allow deployable, mobile, and thus responsive formats with optimized dispersion of the inventory to increase coverage in temporal and spatial terms, streamlined support pipelines, manage contingency provisions and backups, and, ultimately, near real-time response for outbreak classification and containment.[88]

Lastly, the "tracker effect" must be taken into account: WGS reads and records a novel genome so as to inform the development of respective NAATs for specific signatures, rendering it detectable and possibly identifiable in both clinical diagnostics and environmental surveillance contexts, affordably and promptly, like with other, existing NAAT assays.[20]

7.6 CULTUROMICS: TOO LITTLE TOO LATE BUT OCCASIONALLY INDISPENSABLE

Culturomics looks like an oddity when time sensitivity is supreme, and CIDTs are developed with high cost and investment in labor, cash, and intellect. The methodology features labor-intensiveness, unwieldy logistical footprint in total resources, and unimpressive TAT compared to CIDT and NAATs in particular.[79,80] On the contrary, this may be a biased view of its potential, as in the cases of environmental, impure, multiagent samples, possible suspicion for the presence of artificial microbiomes, culturomics is the only option allowing multiple agent resolution, so that

WGS may be applied to segregated isolates. Thus it constitutes a valuable asset, which may assist genomic approaches in extracomplex scenarios and allow more stratified preparation for projected risk management.

7.7 EXPANDING THE HORIZON

A three-dimensional integration seems to have become the new standard in public-health infectiomics and interactomics. The first dimension is the integration in terms of geography (One Planet–One Health), as the speed and volume of transportation sensu lato nullifies border-based containment efforts. Thus surveillance must be adaptable, expendable in space and time, and responsive, as funds will never be enough to cover the planet with standardized equipment.

The second dimension is the integration among different host types (multisectorial and transdisciplinary surveillance) as pathogens break host-species barriers and adapt onto new and very different hosts. The opportunists that infect both plants and humans, for example, some fungi, such as *Fusarium* and *Aspergillus* genera, are the perfect examples.

The third dimension is temporal, which defines the evolutionary traits of host and symbionts, such as virulence, resistance, and immunity. Response measures will be developed, procured, stockpiled, distributed, and fielded more rapidly and affordably if *multitasking* is suggested, encouraged, or enforced.[89] This means preferable development and fielding of countermeasures (materials, methods, and skills) applicable to multiple needs and challenges. In this way, research and development and production costs may be dissipated, a prospect vital for the viability of such endeavors.[36,90]

Since this approach contrasts the current, high-cost, high-expertise diagnostic services, a heated skepticism is bound to appear, but at the end of the day, it will be a matter of policy: business enjoying priority over politics, as happens today, or vice versa, to counter a global threat of unknown kinetics but disastrous dynamics.

To detect dispersion and identify infected environment and subjects, monitor the success of containment and treatment, and introduce routine surveillance protocols for future outbreaks, IA and NAAT are needed. Their development takes more than the time frames usually acknowledged, namely, 4–6 weeks for monoclonal antibody-based IAs,[48] 2 weeks for NAAT.[20]

ACKNOWLEDGMENTS

We acknowledge support for this work by the project "Synthetic biology: from omics technologies to genomic engineering (OMIC-ENGINE)" (MIS 5002636), which is implemented under the action "Reinforcement of the Research and Innovation Infrastructure," funded by the Operational Program "Competitiveness, Entrepreneurship and Innovation" (NSRF 2014-2020) and cofinanced by Greece and the European Union (European Regional Development Fund).

REFERENCES

1. Frinking E, Sweijs T, Sinning P. *The Increasing Threat of Biological Weapons*. The Hague Center for Strategic Studies; 2016.
2. Koblentz GD. The de novo synthesis of horsepox virus: implications for biosecurity and recommendations for preventing the reemergence of smallpox. *Health Secur*. 2017;15:620–628.
3. Connell ND. The challenge of global catastrophic biological risks. *Health Secur*. 2017;15:345–346.
4. Wolicki SB, Nuzzo JB, Blazes DL, Pitts DL, Iskander JK, Tappero JW. Public health surveillance: at the core of the global health security agenda. *Health Secur*. 2016;14:185–188.
5. Jackson RJ, Ramsay AJ, Christensen CD, Beaton S, Hall DF, Ramshaw IA. Expression of mouse interleukin-4 by a recombinant *Ectromelia virus* suppresses cytolytic lymphocyte responses and overcomes genetic resistance to mousepox. *J Virol*. 2001;75:1205–1210.
6. Smith HO, Hutchison III CA, Pfannkoch C, Venter JC. Generating a synthetic genome by whole genome assembly: phiX174 bacteriophage from synthetic oligonucleotides. *Proc Natl Acad Sci USA*. 2003;100:15440–15445.
7. Casadevall A. Don't forget the fungi when considering global catastrophic biorisks. *Health Secur*. 2017;15:341–342.
8. Casadevall A, Pirofski LA. The weapon potential of human pathogenic fungi. *Med Mycol*. 2006;44:689–696.
9. Khalil AT, Shinwari ZK. Threats of agricultural bioterrorism to an agro dependent economy; what should be done? *J Bioterror Biodef*. 2014;5:127–134.
10. Millett P, Snyder-Beattie A. Human agency and global catastrophic biorisks. *Health Secur*. 2017;15:335–336.
11. Schoch-Spana M, Cicero A, Adalja A, et al. Global catastrophic biological risks: toward a working definition. *Health Secur*. 2017;15:323–328.
12. Nuzzo JB. Improving biosurveillance systems to enable situational awareness during public health emergencies. *Health Secur*. 2017;15:17–19.
13. Meyer W, Aanensen DM, Boekhout T, et al. Consensus multi-locus sequence typing scheme for *Cryptococcus neoformans* and *Cryptococcus gattii*. *Med Mycol*. 2009;47(6):561–570.
14. Taylor JW, Fisher MC. Fungal multilocus sequence typing—it's not just for bacteria. *Curr Opin Microbiol*. 2003;6(4):351–356.
15. Arabatzis M, Kambouris M, Kyprianou M, et al. Polyphasic identification and susceptibility to seven antifungals of 102 *Aspergillus* isolates recovered from immunocompromised hosts in Greece. *Antimicrob Agents Chemother*. 2011;55:3025–3030.
16. Salunke S, Lal DK. Multisectoral approach for promoting public health. *Indian J Public Health*. 2017;61(3):163–168.
17. The White House. National biodefense strategy. Available at: <https://www.whitehouse.gov/wp-content/uploads/2018/09/National-Biodefense-Strategy.pdf>; 2018.
18. Kambouris ME, Kantzanou M, Arabatzis M, Velegraki A, Patrinos GP. Rebooting bioresilience: a multi-OMICS approach to tackle global catastrophic biological risks (GCBRs) and next generation biothreats. *OMICS-JIB*. 2018;22:35–51.
19. Velsko S, Bates T. A conceptual architecture for national biosurveillance: moving beyond situational awareness to enable digital detection of emerging threats. *Health Secur*. 2016;14:189–201.
20. Doggett NA, Mukundan H, Lefkowitz EJ, et al. Culture-independent diagnostics for health security. *Health Secur*. 2016;14:122–142.
21. Kauchak M. *Letting Sensors Do the Work*. Military Medical Technology; 2006.

22. Ludovici GM, Gabbarini V, Cenciarelli O, et al. A review of techniques for the detection of biological warfare agents. *Def S&T Tech Bull.* 2015;8:17–26.
23. Primmerman CA. Detection of biological agents. *Lincoln Lab J.* 2000;12:3–31.
24. Halász L, Pintér I, Solymár-Szocs A. Remote sensing in the biological and chemical reconnaissance. *AARMS.* 2002;1:39–56.
25. Joshi D, Kumar D, Maini AK, Sharma RC. Detection of biological warfare agents using ultra violet-laser induced fluorescence LIDAR. *Spectrochim Acta A Mol Biomol Spectrosc.* 2013;112:446–456.
26. Pak T. *A Wireless Remote Biosensor for the Detection of Biological Agents.* Serving Nanoscale Science, Engineering & Technology. National Nanotechnology Infrastructure Network, 2008 Research Accomplishments, Stony Brook University; 2008:24–25.
27. Deshpande A, Mcmahon B, Daughton AR, et al. Surveillance for emerging diseases with multiplexed point-of-care diagnostics. *Health Secur.* 2016;14:111–121.
28. Petrovick MS, Harper JD, Nargi FE, et al. Rapid sensors for biological-agent identification. *Lincoln Lab J.* 2007;17:63–84.
29. Wang HY, Luo M, Tereshchenko IV, et al. A genotyping system capable of simultaneously analyzing >1000 single nucleotide polymorphisms in a haploid genome. *Genome Res.* 2005;15:276–283.
30. Mayboroda O, Gonzalez Benito A, Sabate Del Rio J, et al. Isothermal solid-phase amplification system for detection of *Yersinia pestis. Anal Bioanal Chem.* 2016;408:671–676.
31. Rohrman BA, Richards-Kortum RR. A paper and plastic device for performing recombinase polymerase amplification of HIV DNA. *Lab Chip.* 2012;12:3082–3088.
32. Ramage JG, Prentice KW, Depalma L, et al. Comprehensive laboratory evaluation of a highly specific lateral flow assay for the presumptive identification of *Bacillus anthracis* spores in suspicious white powders and environmental samples. *Health Secur.* 2016;14:351–365.
33. Bartholomew RA, Ozanich RM, Arce JS. Evaluation of immunoassays and general biological indicator tests for field screening of *Bacillus anthracis* and ricin. *Health Secur.* 2017;15:81–96.
34. Kambouris ME, Pavlidis C, Skoufas E, et al. Culturomics: a new kid on the block of OMICS to enable personalized medicine. *OMICS.* 2018;22:108–118.
35. Wieser A, Schneider L, Jung J, Schubert S. MALDI-TOF MS in microbiological diagnostics-identification of microorganisms and beyond (mini review). *Appl Microbiol Biotechnol.* 2012;93:965–974.
36. Millett P, Snyder-Beattie A. Existential risk and cost-effective biosecurity. *Health Secur.* 2017;15:373–383.
37. Ema SG, Mwadwo KS, Paul BP. Integrated microsystem for multiplexed genosensor detection of biowarfare agents. *World News Nat Sci.* 2016;5:20–32.
38. Jakupciak JP, Colwell RR. Barcoding methodology and applications: biological agent detection technologies. *Mol Ecol Resour.* 2009;9:51–57.
39. Song L, Ahn S, Walt DR. Detecting biological warfare agents. *Emerg Infect Dis.* 2005;11:1629–1632.
40. Sapsford KE, Bradburne C, Delehanty JB, Medintz IL. Sensors for detecting biological agents. *Mater Today.* 2008;11:38–49.
41. Madad SS. Bioterrorism: an emerging global health threat. *J Bioterror Biodef.* 2014;5.
42. Emami K, Nelson A, Hack E, et al. MALDI-TOF mass spectrometry discriminates known species and marine environmental isolates of *Pseudoalteromonas. Front Microbiol.* 2016;7:104.
43. Ferreira L, Sanchez-Juanes F, Garcia-Fraile P, et al. MALDI-TOF mass spectrometry is a fast and reliable platform for identification and ecological studies of species from family *Rhizobiaceae. PLoS One.* 2011;6:e20223.
44. Schroeder Jr HW, Cavacini L. Structure and function of immunoglobulins. *J Allergy Clin Immunol.* 2010;125:S41–S52.

45. Vojdani A, Cooper EL. Identification of diseases that may be targets for complementary and alternative medicine (CAM). In: Cooper EL, Yamaguchi N, eds. *Complementary and Alternative Approaches to Biomedicine Springer Science & Business Media*. Springer Science & Business Media; 2004:90.
46. Vojdani A. Reaction of monoclonal and polyclonal antibodies made against infectious agents with various food antigens. *J Clin Cell Immunol*. 2015;6:359.
47. Morel N, Volland H, Dano J, et al. Fast and sensitive detection of *Bacillus anthracis* spores by immunoassay. *Appl Environ Microbiol*. 2012;78:6491–6498.
48. Ouisse LH, Gautreau-Rolland L, Devilder MC, et al. Antigen-specific single B cell sorting and expression-cloning from immunoglobulin humanized rats: a rapid and versatile method for the generation of high affinity and discriminative human monoclonal antibodies. *BMC Biotechnol*. 2017;17:3.
49. Velegraki-Abel A. *Mycotoxins*. Athens [In Greek]; 1986.
50. McCutcheon S, Langeveld JP, Tan BC, et al. Prion protein-specific antibodies that detect multiple TSE agents with high sensitivity. *PLoS One*. 2014;9:e91143.
51. Velegraki A, Kambouris M, Kostourou A, Chalevelakis G, Legakis NJ. Rapid extraction of fungal DNA from clinical samples for PCR amplification. *Med Mycol*. 1999;37:69–73.
52. Velegraki A, Kambouris ME, Skiniotis G, Savala M, Mitroussia-Ziouva A, Legakis NJ. Identification of medically significant fungal genera by polymerase chain reaction followed by restriction enzyme analysis. *FEMS Immunol Med Microbiol*. 1999;23:303–312.
53. Kambouris ME, Reichard U, Legakis NJ, Velegraki A. Sequences from the aspergillopepsin PEP gene of *Aspergillus fumigatus*: evidence on their use in selective PCR identification of *Aspergillus* species in infected clinical samples. *FEMS Immunol Med Microbiol*. 1999;25:255–264.
54. Kambouris ME. Staged oligonucleotide design, compilation and quality control procedures for multiple SNP genotyping by multiplex PCR and single base extension microarray format. *e-JSTech*. 2009;4:21–40.
55. Kambouris ME. Integrated real-time PCR formats: methodological analysis and comparison of two available industry options. *e-JSTech*. 2010;5:33–40.
56. Lin Z, Cui X, Li H. Multiplex genotype determination at a large number of gene loci. *Proc Natl Acad Sci USA*. 1996;93:2582–2587.
57. Pastinen T, Raitio M, Lindroos K, Tainola P, Peltonen L, Syvanen AC. A system for specific, high-throughput genotyping by allele-specific primer extension on microarrays. *Genome Res*. 2000;10:1031–1042.
58. Kambouris ME. Population screening for hemoglobinopathy profiling: is the development of a microarray worthwhile? *Hemoglobin*. 2016;40:240–246.
59. CDC. *Federal Select Agents Program*. 2017. Available at: <https://www.selectagents.gov/selectagentsandtoxinslist.html>.
60. DoD. *Department of Defense, Chemical and Biological Defense Program Annual Report to Congress'2006*. <https://fas.org/irp/threat/cbdp2006.pdf>; 2006:F33 Accessed 18.11.15.
61. Liu Y, Sam L, Li J, Lussier YA. Robust methods for accurate diagnosis using pan-microbiological oligonucleotide microarrays. *BMC Bioinformatics*. 2009;10(suppl 2):S11.
62. Palacios G, Quan P, Jabado OJ, et al. Panmicrobial oligonucleotide array for diagnosis of infectious diseases. *Emerg Infect Dis*. 2007;13:73–81.
63. Ozanich RM, Colburn HA, Victry KD, et al. Evaluation of PCR systems for field screening of *Bacillus anthracis*. *Health Secur*. 2017;15:70–80.
64. Euler M, Wang Y, Heidenreich D, et al. Development of a panel of recombinase polymerase amplification assays for detection of biothreat agents. *J Clin Microbiol*. 2013;51:1110–1117.
65. Heller MJ. DNA microarray technology: devices, systems, and applications. *Annu Rev Biomed Eng*. 2002;4:129–153.

66. Lin B, Wang Z, Vora GJ, et al. Broad-spectrum respiratory tract pathogen identification using resequencing DNA microarrays. *Genome Res.* 2006;16:527–535.
67. Malanoski AP, Lin B, Wang Z, Schnur JM, Stenger DA. Automated identification of multiple microorganisms from resequencing DNA microarrays. *Nucleic Acids Res.* 2006;34:5300–5311.
68. Wang Z, Daum LT, Vora GJ, et al. Identifying influenza viruses with resequencing microarrays. *Emerg Infect Dis.* 2006;12(4):638–646.
69. Gardner SN, Jaing CJ, McLoughlin KS, Slezak TR. A microbial detection array (MDA) for viral and bacterial detection. *BMC Genomics.* 2010;11:668.
70. Chou C-C, Lee T-T, Chen C-H, et al. Design of microarray probes for virus identification and detection of emerging viruses at the genus level. *BMC Bioinformatics.* 2006;7:232.
71. Huguenin A, Moutte L, Renois F, et al. Broad respiratory virus detection in infants hospitalized for bronchiolitis by use of a multiplex RT-PCR DNA microarray system. *J Med Virol.* 2012;84:979–985.
72. Chiu CY, Alizadeh AA, Rouskin S, et al. Diagnosis of a critical respiratory illness caused by *Human metapneumovirus* by use of a pan-virus microarray. *J Clin Microbiol.* 2007;45:2340–2343.
73. Hong B-X, Jiang L-F, Hu Y-S, Fang D-Y, Guo H-Y. Application of oligonucleotide array technology for the rapid detection of pathogenic bacteria of foodborne infections. *J Microbiol Methods.* 2004;58:403–411.
74. Sakai K, Trabasso P, Moretti ML, Mikami Y, Kamei K, Gonoi T. Identification of fungal pathogens by visible microarray system in combination with isothermal gene amplification. *Mycopathologia.* 2014;178:11–26.
75. Sato T, Takayanagi A, Nagao K, et al. Simple PCR-based DNA microarray system to identify human pathogenic fungi in skin. *J Clin Microbiol.* 2010;48:2357–2364.
76. Chizhikov V, Rasooly A, Chumakov K, Levy DD. Microarray analysis of microbial virulence factors. *Appl Environ Microbiol.* 2001;67:3258–3263.
77. Kambouris ME, Gaitanis G, Manoussopoulos Y, et al. Humanome versus microbiome: games of dominance and pan-biosurveillance in the omics universe. *OMICS.* 2018;22:528–538.
78. Nakamura S, Maeda N, Miron IM, et al. Metagenomic diagnosis of bacterial infections. *Emerg Infect Dis.* 2008;14:1784–1786.
79. Bardet L, Cimmino T, Buffet C, et al. *Microbial Culturomics Application for Global Health: Noncontiguous Finished Genome Sequence and Description of* Pseudomonas massiliensis *Strain CB-1T sp. nov. in Brazil.* OMICS; 2017.
80. Lagier JC, Armougom F, Million M, et al. Microbial culturomics: paradigm shift in the human gut microbiome study. *Clin Microbiol Infect.* 2012;18:1185–1193.
81. Forzley M. Global health security agenda: joint external evaluation and legislation—a 1-year review. *Health Secur.* 2017;15:312–319.
82. Barbeschi M. A global catastrophic biological risk is not just about biology. *Health Secur.* 2017;15:349–350.
83. Voelkerding KV, Dames SA, Durtschi JD. Next-generation sequencing: from basic research to diagnostics. *Clin Chem.* 2009;55:641–658.
84. Regalado A. *Rewriting Life: For $999, Veritas Genetics Will Put Your Genome on a Smartphone App.* MIT Technology Review. Available at: <https://www.technologyreview.com/s/600950/for-999-veritas-genetics-will-put-your-genome-on-a-smartphone-app/>; 2016.
85. Cello J, Paul AV, Wimmer E. Chemical synthesis of poliovirus cDNA: generation of infectious virus in the absence of natural template. *Science.* 2002;297:1016–1018.
86. Quail MA, Smith M, Coupland P, et al. A tale of three next generation sequencing platforms: comparison of Ion Torrent, Pacific Biosciences and Illumina MiSeq sequencers. *BMC Genomics.* 2012;13:341.

87. Mardis ER. The impact of next-generation sequencing technology on genetics. *Trends Genet.* 2008;24:133–141.
88. Heather JM, Chain B. The sequence of sequencers: the history of sequencing DNA. *Genomics.* 2016;107(1):1–8.
89. Cameron EE. Emerging and converging global catastrophic biological risks. *Health Secur.* 2017;15:337–338.
90. Lipsitch M. If a global catastrophic biological risk materializes, at what stage will we recognize it? *Health Secur.* 2017;15:331–334.

CHAPTER

GENOME INFORMATICS PIPELINES AND GENOME BROWSERS

8

Evaggelia Barba[1], Evangelia-Eirini Tsermpini[1], George P. Patrinos[1,2,3] and Maria Koromina[1]

[1]Department of Pharmacy, School of Health Sciences, University of Patras, Patras, Greece [2]Department of Pathology, College of Medicine and Health Sciences, United Arab Emirates University, Al Ain, United Arab Emirates [3]Zayed Center of Health Sciences, United Arab Emirates University, Al Ain, United Arab Emirates

ABBREVIATIONS

BAM	Binary Alignment Map
BED	binary PED
BLAST	Basic Local Alignment Search Tool
CNV	copy number variant
EBI	European Bioinformatics Institute
ENA	European Nucleotide Archive
FASTA	Fast All
GMOD	Generic Model Organism Database
GEO	Gene Expression Omnibus
GFF	general feature format
GTF	Gene transfer format
MGI	Mouse Genome Informatics
NCBI	National Center for Biotechnology Information
NGS	next-generation sequencing
SAM	Sequence Alignment Map
SGD	Saccharomyces Genome Database
SMRT	single-molecule real time
SNP	single-nucleotide polymorphism
SRA	Sequence Read Archive
TAIR	The Arabidopsis Information Resource
UCSC-GB	University of California Santa Cruz Genome Browser
VCF	Variant Call Format
WES	whole-exome sequencing
WGS	whole-genome sequencing
ZMWs	zero-mode waveguides
1KGP	1000 Genomes Project

Applied Genomics and Public Health. DOI: https://doi.org/10.1016/B978-0-12-813695-9.00008-X

8.1 INTRODUCTION

Research on the genetic background of common and rare diseases of humans has progressed over the last decade leading to a massive volume of data deposited. Moreover, electronic health-care records have been integrated into the modern medical practice in order to decipher the genetic changes leading to inherited diseases. Since genetics now play a vital role in the application of clinical medicine, it is of utmost importance to educate "modern health-care professionals" in understanding the scientific depth of medical genetics.

Progress in the field of molecular biology, and more precisely in the analysis and detailed mapping of genomes of thousands of organisms, has led to accumulation of a multitude of biological data that provide useful information in determining protein-encoding genes, RNA genes, structural motifs, and regulatory regions. Therefore it is necessary to find a way to effectively and efficiently manage all these data by developing an interdisciplinary field (bioinformatics), which is an alloy of biology and computer science. At the same time, there is an urgent need for a strong knowledge base for genotype–phenotype information in the context of translational medicine.

Next-generation sequencing (NGS) is a powerful, innovative technology that fully inspires the field of gene research. Nowadays, genomics research uses multiple NGS platforms, and NGS studies produce a massive amount of data. Big data analysis is an essential aspect of the genomics study, and bioinformatics plays a vital role in the analysis and interpretation of genomic data. NGS formats and powerful bioinformatics tools strengthen and support translational research in genomics by gathering individual or population data from genomic analysis at the level of individuals or populations and gradually integrating them into the therapeutic and preventative guidelines of modern health-care systems. NGS is essentially the matrix of many techniques that can be used to assess a wide range of genetic information within a biological system. As expected, several computational barriers and challenges arise from analyzing a massive volume of genetic data produced through NGS technologies. Such examples include sequence data, annotation data and quantitative data, and read alignments.

8.2 BIG DATA IN GENOMICS

Newly developed high-throughput sequencing technologies produce in most cases a large generation of datasets as their output. Thus sequencing technologies with high-throughput performance in terms of accuracy and depth of reading are used in a wide range of applications during the human lifespan,[1] from prenatal diagnostics and newborn screening to rare diseases diagnosis.[2–5] The Big Data revolution brought us to the doorstep of data-driven, evidence-based, precision medicine.

The interdisciplinary field of bioinformatics combines biology and computer science and is thus endowed in analyzing, interpreting, and finally translating into applications, biological data. Moreover, bioinformatics has already significantly contributed to handling and exploiting omics-related Big Data.

8.2.1 NEXT-GENERATION SEQUENCING

The pioneering work that enabled the first entire genomes to be sequenced took place in 1976. As a result of decades of laborious scientific bench work, the number of genomes sequenced continued to rise throughout the years, which enabled the worldwide expansion of genomics research on the human. NGS is a blanket term used to collectively address diverse methods of high-throughput DNA sequencing, which arose in the previous 15 years.[6] NGS is a powerful high-throughput technology that has created unprecedented opportunities for analyzing whole genomes.[7] It is, therefore, a useful tool that provides valuable information for further understanding of biological functions. NGS revolutionizes fields, such as personalized medicine, genetic diseases, and clinical diagnosis, while offering a high-performance capability, such as multiperson sequence at the same time. Interpretation of data, as results from DNA sequence analysis, becomes impossible without specific tools and databases. An interdisciplinary field of science, bioinformatics, has a crucial role in analyzing and interpreting genomic data. NGS formats and powerful bioinformatics tools strengthen and support translational research in genomics.

The newly developed NGS technologies consist of several methods and strategies broadly grouped as steps of (1) template generation, (2) sequencing and imaging, and (3) data analysis.[8] A particular combination of these steps distinguishes between each NGS manufacturer and sequencing platform and determines the type and quality of the produced data.[6,8]

8.2.2 DATA SOURCES

There are three prominent institutions around the globe, which contain and process extensive databases for raw sequencing data. These are the DNA Data Bank of Japan, the GenBank in the United States, and the European Nucleotide Archive (ENA) in the United Kingdom. These three organizations joined efforts and formed the International Nucleotide Sequence Database Collaboration[9] in 2011, which was a milestone event for genetics. The primary purpose of the agreement is to synchronize and integrate the sequencing data that are separately organized and handled by each of these institutions. For this reason the Sequence Read Archive (SRA) is the most proper database platform.[10] Each institution maintains its separate instance of SRA, and all cases are synchronized sporadically.

SRA began as a repository for short reads primarily for RNA-seq for expression profiling and ChIP-seq research for protein–DNA interaction analysis, and it has evolved to encompass all arrays of NGS experiment types. Apart from raw sequencing data, SRA holds library preparation protocols, sample description, used sequencing machines and bundles, everything in an intuitive online user interface. SRA offers a vast amount of resources (SRA toolkit) that process beforehand, download, and convert data to different kinds of formats. The most widely available and accessible public repository for functional genomics is the Gene Expression Omnibus (GEO) database[11] that contains a vast amount of data regarding gene expression. GEO incorporates microarrays, genotypic arrays, real time (RT)-polymerase chain reaction (PCR), and NGS data in addition to using SRA for storing and managing raw NGS data. This GEO–SRA synergy is just a small portion of a larger scheme comprised a network of large and closely connected databases from the NCBI (National Center for Biotechnology Information). These databases encompass the full spectrum of genomic information from raw sequencing data to discoveries of medical significance.[12]

Within the same context as the NCBI, the European Bioinformatics Institute (EBI) offers a vast assortment of highly connected tools and databases that refer to the full spectrum of genomic information. The ArrayExpress–ENA for EBI, which is similar to NCBI's GEO–SRA synergy,[13] is a database for operational genomics data with the ENA[14] functioning as a part of the SRA database.

One more encompassing set of sequencing data comes from the 1000 Genomes Project (1KGP). 1KGP investigated human genetic variation by gathering data from an extensive populous primarily by using sequencing techniques. At its latest stage the 1KGP encompassed data of 2,504 people from 26 various populations thus covering a large portion of the world.[15] The most considerable contribution of 1KGP is the access to all data from raw sequencing reads in FASTQ format to final variant calls in VCF (Variant Call Format). The most common format for storing sequencing reads is FASTQ. Every entry in a FASTQ file contains both the sequence and a quality score of each nucleotide. The quality score, that is, typically used is Phred, which is the most common metric used to assess the accuracy of sequencing.[16]

8.2.3 DATA FORMATS

In general, the raw data produced by NGS experiments are short reads of sequences. Typically, these reads do not exceed 200 base pairs, although new promising techniques that produce longer reads of high quality are in the rise. An NGS experiment can provide billion of sequences that result in many gigabytes of data.

Sequences that represent assembled genomes are stored in the FASTA (Fast All) format. For example, a FASTA file may contain the complete genome of an organism, a chromosome, or single gene. FASTA files contain lines starting with ">," which consequently contain information for the genome (name of the gene, or the version of the genome assembly). These lines are followed by usually very long lines of genome sequences that contain the letters "A", "C", "G," and "T," which correspond to the nitrogenous bases of DNA. The most crucial tool for FASTA file exploration is the Integrative Genomics Viewer (IGV).[17]

After obtaining a set of raw DNA sequences in FASTQ format, their alignment is followed in a reference genome (encoded in FASTA format). This process is called alignment and is the first computational resourceful step of most NGS pipelines. The purpose of this step is to assign the best matching genomic position to most of the reads in the FASTQ files. SAM stands for Sequence Alignment Map. SAM files contain alignment information for each read in a collection of FASTQ data. In contrast to FASTQ and FASTA files, SAM files have a wide variety of fields that contain very useful metainformation regarding the performed experiment. A BAM file is a binary version of a SAM file, containing exactly the same information, but in smaller size. The most used data format for aligned reads is SAM/Binary Alignment Map (BAM).[18]

Specific formats exist for the annotation of genomic sequences. Entries on these files consist of a declaration of a genomic region accompanied with the annotation of this region. Annotation files are a valuable accompaniment to NGS pipelines to prioritize in specific regions or give insights into the biological significance of a specific locus. These formats are GFF, Gene transfer format (GTF), and binary PED (BED). Finally, BED file (.bed) is a tab-delimited text file that defines a feature track. The GTF is a file format used to hold information about gene structure. Also, the general feature format (gene-finding format, generic feature format, GFF) is a file format, which

Table 8.1 Summary of Data Sources and Databases for Sequencing Data.

Data Sources	Platforms	Data Formats
ENA	INSDC	FASTQ
GenBank	SRA	FASTA
DNA Data Bank	GEO	SAM/BAM
	NCBI	GFF
	EBI	GTF
	1KGP	BED

The first column lists the three primary data sources for sequencing data; the second column lists a couple of databases/platforms also available for sequencing data, while the last column shows the different formats for raw sequencing data.
1KGP, 1000 Genomes Project; *BAM,* Binary Alignment Map; *BED,* Binary PED; *EBI,* European Bioinformatics Institute; *ENA,* European Nucleotide Archive; *FASTA,* Fast All; *GEO,* Gene Expression Omnibus; *GFF,* general feature format; *GTF,* Gene transfer format; *INSDC,* International Nucleotide Sequence Database Collaboration; *NCBI,* National Center for Biotechnology Information; *SRA,* Sequence Read Archive; *SAM,* Sequence Binary Alignment.

describes genes and other features of nucleic acids and protein sequences. Finally, BED file (.bed) is a tab-delimited text file that defines a feature track.

The final analysis steps of most NGS pipelines generate files that contain genomic variants. These are sequences that differ at a significant level from the reference genome. The most common format for describing these variants is VCF.[19] It is a simple text, a tab-delimited format where each line contains the position, the alternative sequence, the reference sequence, various quality metrics, and most importantly the genotype of all samples in an experiment on these locations. A summary listing the different types of data sources, data formats, and platforms is provided inTable 8.1.

8.2.4 NEXT-GENERATION SEQUENCING PLATFORMS

NGS technologies are a number of different modern sequencing technologies, but the most frequently used platforms include Illumina (Solexa) (https://www.illumina.com), Applied Biosystems SOLiD System (www.appliedbiosystems.com), Ion Torrent sequencing (https://www.iontorrent.com), Pacific Biosciences (https://www.pacificbiosciences.com), and Oxford nanopore (https://www.nanoporetech.com).[20,21] Roche 454 pyrosequencing technology is a sequencing technology, which has been widely used in the previous decade, although nowadays it has been discontinued (https://www.454.com). NGS platforms can be categorized into second and third-generation sequencing technologies based on their launch year. Among the NGS technologies as listed earlier, the following two platforms, Pacific Biosciences and Oxford Nanopore, are the most updated technologies, and they are classified as third-generation sequencing technologies with the remaining NGS technologies classified as second-generation technologies.[22] Third-generation NGS technologies are based on sequencing individual DNA molecules without a previous amplification step, which is the single long molecule sequencing or clonally amplification. In contrast, second-generation NGS platforms rely on PCR to grow clusters of a given DNA template, which is one of their disadvantages as sequencing methods.[23]

8.2.4.1 Roche 454 pyrosequencing

The first parallel next-generation DNA sequencing was based on the pyrosequencing method, developed in 1996 by the Stockholm Royal Institute of Technology and launched in 2005 by "454." Roche then acquired it in 2007.[24,25] The workflow of the platform begins with the creation of the library, which is completed either by fragmenting DNA into smaller fragments or by multiplying the segments with multiplex PCR. Sequence adapters finally surround the DNA fragments generated by either of these methods. Adapters play a key role in both library enhancement and sequencing. Pyrosequencing is a technique based on a "sequencing by synthesis" (SBS) method. More precisely, pyrosequencing is based on detection and quantification of DNA polymerase activity, which is carried out using the enzyme luciferase, thus leading to light production. This bioluminescence signal is proportional to the amount of pyrophosphate produced, which is directly proportional to the number of added nucleotides. DNA fragments are first attached to microbeads and clonally amplified in an emulsion PCR. The beads are consequently spread into individual picoliter-sized wells and subjected to pyrosequencing. Roche/454 sequencing platform was sequenced 580,069 bp of *Mycoplasma* genitalia genome at 96% coverage and 99.96% accuracy in a single run for the first time.

This system was the first NGS technology to sequence a complete human genome by producing 400 Mb per run with the maximum of 450 bp read length at the beginning and then increased up to 700 bp.[26] The technique continued to evolve and to overcome the pitfalls encountered in the beginning, so the length of sequence reads as well as the output (yield per run) was constantly increasing (from kilo to megabase pairs). The latest instruments overcame the initial limitation of NGS short reads, achieving up to 1 kb read lengths.[27] These long-read lengths enabled better alignment to reference genomes, thus making 454 pyrosequencing NGS the method of choice in metagenomic studies for years,[28] alongside with sequencing of conserved marker gene tags, now commonly referred to as "pyrotag" sequencing.[29] Roche/454 sequencing was also successful for both confirmatory sequencing and de novo sequencing. However, one of the limitations of pyrosequencing laid in its inaccuracy in the sequencing of strings of repeated nucleotides (homopolymers); more than five identical nucleotides could not be detected efficiently.[26] Therefore other NGS platforms are being evolved, such as Illumina sequencings, which have successfully coped with some other weaknesses of the pyrosequencing as well, such as the high error rate, low yield, and high cost per bp.[30]

8.2.4.2 Illumina (Solexa) sequencing

One of the most widely used NGS technologies at the moment is, undoubtedly, Illumina, which was launched in 2006. Illumina has contributed a lot to the advancement of sequencing platforms in terms of simplicity, flexibility, and capacity, so that it can be applied in the field of humans and animals genomics research projects. Its goal is to use innovative sequencing technologies to the analysis of genetic variation and function. Initially developed as a Solexa method and subsequently acquired by Illumina, it is based on the SBS principle as well. The principle of Illumina (Solexa) sequencer was based on SBS chemistry concept that enabled the identification of single bases as they are introduced into DNA strands. Solexa sequencing—branded fluorescently labeled modified nucleotides and a particular DNA polymerase enzyme to sequence the millions of clusters present on the flow cell surface.[31] The high data yield has been one of the hallmarks of Illumina sequencing technology. Currently, Illumina generates more than 90% of the world's DNA sequencing data.

Before the introduction of Mi-Seq and Hi-seq platforms which can sequence up to 15 and 600 Gbps, Illumina purchased Solexa in 2007 which was released in 2005.[32] The first Solexa sequencer, the Genome Analyzer, which was launched in 2006, yielded an astonishing 1 Gb of data in a single run. The technology has skyrocketed since then, with sequencers in 2014 outputting more than 1000 times more data, that is, 1.8 Tb in a single run.[33] Furthermore, low error rate makes this technology preferable in metagenomic studies, where the detection of single-nucleotide variants in bacterial genomes is vital as these changes could lead to significant changes in phenotypic characters, such as antibiotic resistance and virulence.[34]

8.2.4.3 Sequencing by Oligonucleotide Ligation and Detection

Life Technologies developed sequencing by oligonucleotide ligation and detection (SOLiD) and acquired by Applied Biosystems. Similar to Illumina, Roche/454, and Ion Torrent platforms, SOLiD applies a clonal amplification through emulsion PCR and optical detection system.[35] The sequencer, launched in 2007, adopts the sequencing-by-ligation technology. First, template DNA fragments are merged with adapters, which allow their hybridization to individual beads. For bead-based preparation the method begins by applying amplified DNA fragments to microbeads. Beads are then deposited on a glass slide in which DNA fragments can be fixed. The sequencing process begins with the annealing of a universal sequencing primer onto the adapter sequence of the template DNA fragment.[36,37] Then, a specially designed probe hybridizes to the fragment next to the primer, and the ligase seals phosphate backbone between the probe and the primer. The glass slides can be segmented up to eight chambers to facilitate upscaling of the number of analyzed samples. The 8-mer oligonucleotides with a fluorescent label at the end are sequentially ligated to DNA fragments. The resulting product is then removed and the process repeated for five more cycles with hybridized primers. Two bases are interrogated in each ligation reaction providing increased specificity; the primer is comprised of five independent rounds of extension improving the signal the system; each base is interrogated twice in separate primer rounds; four dyes are used to encode 16 possible two-base enables error checking capacity. A single run using this sequencing technology generates tens of millions of up to 85 bp reads. The distinctive two-base encoding principle and the 8-bp labeled probe enable high quality and accuracy of 99.9%.[38] SOLiD technology may not be the best choice when palindromic sequences need to be sequenced. However, it is used for whole-genome resequencing, targeted resequencing, transcriptome research (including gene expression profiling, small RNA analysis, and whole-transcriptome analysis), and epigenome research (such as ChIP-seq and methylation).[36,39]

8.2.4.4 Ion Torrent sequencing

Life Technologies (Thermo Fisher) commercialized the Ion Personal Genome Machine sequencing platform in 2010. This sequencing platform is one of the sequencing technologies, which contributed to large-scale transcriptome studies in the last decade.[40] Similar to the previously mentioned NGS technologies, 454 pyrosequencing, and Illumina, Ion Torrent sequencing also utilizes a SBS approach. Ion Torrent sequencing platform employs a similar technique as pyrosequencing, but it does not apply, and optic fluorescent labeled other second-generation technologies.[40,41] It detects the release of hydrogen ions (H^+), a by-product of nucleotide incorporation, as quantitated changes in pH through a novel coupled silicon detector.[42] The resulting change in pH is detected by an integrated complementary metal-oxide semiconductor and an ion-sensitive field-effect transistor.

Detected pH is imperfectly proportional to the number of nucleotides identified and converted into a voltage signal, which is equivalent to the number of nucleotides incorporated.[43] The platform has several different types of chips and instruments to improve its performance. A throughput of these chips ranges from 50 Mb to 15 Gb with runtimes between 2 and 7 hours.[43] Nowadays, a new system and new chips are being developed, which will aim to produce high-yield regions of the entire genome sequence. Ion Torrent launched its follow-up on the system, Ion Proton in 2012, which will allow larger chips with higher densities necessary for whole-exome sequencing (WES) and whole-genome sequencing (WGS) (https://www.thermofisher.com). The most important advantage of this technology is that it allows rapid sequencing alongside the actual detection phase.

Moreover, Ion Torrent sequencing provides the lowest run cost due to the failure of the cost of incorporation of modified bases. A compromise between the low throughput of 454 pyrosequencing and ultrahigh-throughput of Illumina is achieved by running the length of up to 400 bp and a yield of up to 15 Gb.[29,44] However, similar with 454 pyrosequencing, Ion Torrent is more error prone in the interpretation of homopolymer sequences due to the loss of signal as multiple identical deoxynucleotide triphosphates (dNTPs) are incorporated.

8.2.4.5 Nanopore single-molecule sequencing

Nanopore single-molecule sequencing is the most contemporary third-generation technology released by UK-based company Oxford Nanopore Technologies in 2014.[45] This company is developing and distributing nanopore sequencing products (including the mobile DNA sequencing, MinION) for the linear, electronic analysis of single molecules (http://www.nanoporetech.com). MinION was first introduced at the Advances in Genome Biology and Technology conference in Florida in February 2012; however, it was not publicized until an early access program under the name MinION Access Program that commenced in April 2014.[46] MinION is an affordable, small (powered from the USB port of a laptop computer), mobile, high-throughput sequencing device that generates real-time data. The apparatus provides multikilobase reads and a streamed modus operandi that enables handling of reads as they are produced. This technology is useful in producing very long reads (about 50,000 bp) with the smallest amount of the sample preparation.[47]

Nanopore sequencing is applied in solving restrictions of short-read sequencing technologies and allows sequencing of large DNA molecules in a short time from effortlessly prepared libraries. A nanopore is simply a small hole, and its internal diameter is 1 nm. A nanopore is integrated into an electrically resistant artificial membrane, and a voltage is applied across the membrane. DNA molecules are developed based on one of the standard library preparation protocols that include connecting a leader adapter and motor protein to one strand of DNA.[48] Throughout the sequencing process, an ion current passes through the hole that is obstructed by the nucleotide.[49] If every nucleotide that passes produces a certain current, the documentation of the particular DNA sequence[50] meaning alterations in the electric current shows which base is present. Nanopore DNA sequencing provides a new prospective edge over the other short-read sequencing technologies incorporating sensitive detection from the limited starting material, ultralong reads, fast time results, low cost, and small footprint (https://nanoporetech.com). In spite of all these, an advantageous high error rate (15%–40%) problem brought to question the previous versions of nanopore DNA-sequencing technology.[51] In order to overcome high error rate difficulties of the earlier readers, a new version of MinION ultralong single-molecule read named R9.4 was developed with an expansion of median accuracy up to 92% and much-grown yield 127,000–217,000 reads per flow cell, four flow cells sequenced.[52]

8.2.4.6 Pacific Biosciences single-molecule real-time sequencing

Single-molecule real-time (SMRT) sequencing is a parallelized SMRT sequencing procedure created by Pacific Biosciences (PacBio) of California, Inc. and classified as a third-generation sequencing technique.[53] Contrary to the majority of other sequencing technologies, PacBio does not require clonal amplification of DNA by utilizing the SMRT sequencing technology (SMRT). It sequences single molecules. The first commercially accessible long read single-molecule platform was the PacBio RS II system that was publicly available in 2011.[54] PacBio RS II system can be implemented in WGS, targeted sequencing, complex population analysis, RNA sequencing, and epigenetics characterization. This series of systems such as RS II and Sequel systems aim to fix low-quality reads created by PacBio RS,[55] and it is the most extensively utilized third-generation sequencing technology. PacBio SMRT sequencing reaction on silicon dioxide chips is named zero-mode waveguides (ZMWs).[36,56] The platform relies on SBS approach and real-time identification of integrated fluorescently labeled nucleotides, as they are discharged.

ZMWs make use of the properties of light passing through gaps with a diameter smaller than its wavelength. Each ZMW consists of a DNA polymerase attached to their end and the target DNA fragment for sequencing. The fluorescent dye of the integrated nucleotide can be observed while on normal speed reverses strand synthesis. The identification of the labeled nucleotides allows determination of the DNA sequence. Compared with other sequencing technologies described by Roberts et al.,[57] crucial advantages of SMRT sequencing platform contain long-read lengths (for de novo assemblies of novel genomes), direct measurement of separate molecules. In addition, templates can be prepared without PCR amplification; the system notes the kinetics of every nucleotide incorporation reaction, making simple and enhancing genomic assembly and comprehension of disease heritability. However, PacBio SMRT technology has a restriction of highly elevated error rates.[58]

8.3 BIOINFORMATICS METHODS FOR ANALYZING GENOMIC DATA

High-throughput sequencing analysis provides different quality and quantity of data. Therefore it is essential to apply bioinformatics workflows for the accurate study and analysis of data produced by NGS technologies. The standard procedure for marker gene analysis consists of the following steps: preprocessing of raw sequencing reads, binning (grouping similar sequences into units that represent taxonomic clusters), taxonomic classification, and statistical analysis.[59,60] In the case of shotgun approach, the standard workflow involves preprocessing, the assembly of short sequences into longer continuous or semicontinuous genome fragments (contigs and scaffolds), binning (grouping reads/contigs/scaffolds into clusters of related fragments), the annotation of genetic variation, the prediction of the putative role of genes and proteins, as well as the prediction of proteins' domains, functions, and pathways.[60,61]

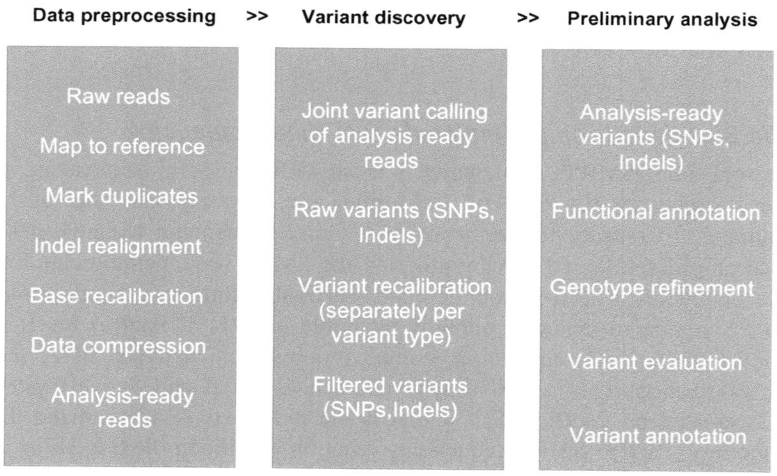

FIGURE 8.1

Schematic diagram of the steps of a typical NGS pipeline. The pipeline mainly consists of "data preprocessing," "variant discovery," and "preliminary analysis." Each one of these steps is then split into different "sub"-steps. Raw reads can be obtained from sequencing, online, or from software that generates artificial datasets. Alignment, visualization of results, and annotation are steps included in all NGS pipelines. Quality control, which is not shown in this diagram, is a crucial step in all NGS pipelines and should never be omitted. *NGS*, Next-generation sequencing.

8.3.1 NEXT-GENERATION SEQUENCING PIPELINES

A common NGS pipeline is constructed from the steps as illustrated in the following subsections. These steps are also illustrated diagrammatically in Fig. 8.1.

8.3.1.1 Read alignment

Aligning a DNA sequence to a reference sequence has been one of the central problems of genetics, which has attracted much interest from computer science researchers. Starting with the Needleman–Wunsch algorithm,[62] which was introduced in 1970, computers could efficiently perform exact sequence alignment (or else global alignment). This was followed by the introduction of the Smith–Waterman algorithm,[63] introduced in 1981, which is a variation of Needleman–Wunsch that allowed sequence mismatches (local alignment). The Smith–Waterman algorithm also allowed the computational detection of sequence variants, mainly insertions and deletions. Interestingly, the demanding time requirements of these algorithms gave rise to the prominent FASTA[64] and BLAST[65] algorithms (introduced at 1985 and 1990, respectively) that are approximately 50 times faster, although they produce suboptimal results. Nowadays, these algorithms are available either as online web services or programs implemented in low-level programing languages that make optimal utilization of the underlying computer architecture. These algorithms and their variations are the foundational components of modern NGS techniques. There are more than 90 tools that perform read alignment, the vast majority (~80%) of which have been developed over the last decade.[66] Choosing the optimal tool requires careful consideration of many criteria regarding our experimental setup.[67]

8.3.1.2 Variant calling

The alignment of reads in a specific genomic reference allows to creation of a SAM/BAM file. Assuming that all quality control steps for eliminating biases in these files have been applied, the next analysis step is variant calling. This step has been the subject of thorough scrutiny from researchers since many options are available each implementing a different analytic approach. According to the results of Hwang et al.,[68] the task of single-nucleotide polymorphism (SNP) calling was performed better with the combination of Burrows-Wheeler Alignment (BWA)–MEM[69] for alignment and SAMtools[18] for a variant call. Another software for variant calling, Freebayes,[70] had similar performance when combined with any alignment tool. For the task of Indel calling, the combination tools of BWA–MEM and GATK (Genome Analysis Toolkit)[71] had better performance. Given that all of the presented tools being improved continuously, readers should always consult comparisons that include the latest versions before making decisions.

There are two types of possible computational pipelines for variant calling. The first is for germline mutation, and the second is for somatic mutations. Somatic mutations happen in any cell of somatic tissue, and since these mutations are not present in germline, they are not passed to the progenies of the organism.

The primary use of NGS pipelines is to acquire a partial (e.g., WES) or a complete (WGS) image of an individual's DNA. In both cases the last step of the pipeline (variant calling) can give insights of already known or unknown mutations. NGS can also be applied to perform a typical genotyping experiment, where the final result is the assessment of alleles in a given set of genotypes. Typical NGS pipelines contain tools that focus on SNPs and copy number variations (CNVs) of relatively small size (usually up to 1 kbp). Although WGS allows identification of SNPs and CNVs to a satisfactory level, the nature of WES experiments limits the ability of many tools to identify CNVs significantly.[72] Special tools that attempt to tackle these limitations are ERDS-exome,[73] CODEX,[74] CNVnator,[75] and PennCNV-seq.[76]

8.3.1.3 Downstream analysis

Downstream analysis refers to possible computation pipelines that can be applied after variant calling. Usually, the first step after variant calling is variant annotation. This step adds information from existing genomic databases to all of the identified variants. This information often includes allele frequency distribution of the variants across various general populations, known clinical implications, known or predicted functional consequences. Useful sources for the allele frequency distribution of the identified variant are the 1KGP, the Exome Aggregation Consortium, or the Genome Aggregation Database (http://exac.broadinstitute.org). Tools that can perform variant annotation are ANNOVAR (ANNOtate VARiation) (http://annovar.openbioinformatics.org/en/latest/) and the VEP tool in Ensembl browser (variant effect predictor) (https://www.ensembl.org/info/docs/tools/vep/index.html). Both tools facilitate fast and easy variant annotations, including gene-based, region-based, and filter-based annotations on a VCF file generated from human genomes. VEP allows variant annotation in other types of formats and not only in VCF files.

Annotation can be performed not only on mutations identified after variant calling but also on the sequences themselves. For example, large sequences that represent genes can be annotated with information such as conservation scores [i.e., Genomic Evolutionary Rate Profiling (GERP) or Phylop scores].

8.4 GENOME BROWSERS

One of the problems of repositories is that the available information is usually not organized in a user-friendly way to aid researchers search, browse information, and draw conclusions. To overcome problems that limit the use of data, developers are trying to create tools, such as genomic browsers,[77] which present the information in an understandable and friendly way to meet the needs of users. A genome browser provides a graphical interface for users to browse, search, retrieve, and analyze the genomic sequence and annotation data. Several genome browsers with multiple species genomic information as well as species-specific genome browsers can be found online.

All the sequence data as well as the annotations generated via completed or ongoing genome projects are collated in the genome databases and are publicly available through web portals, such as the NCBI genome portal (http://www.ncbi.nlm.nih.gov/genome/) and the EBI genome database website (http://www.ebi.ac.uk/Databases/genomes.html). Genome browser provides a unique platform for molecular biologists to search, retrieve, and analyze these genomic data efficiently and conveniently. Furthermore, genome browser helps users to extract and summarize information intuitively from a massive amount of raw data via a graphical interface. An additional advantage is the ability to integrate different types of annotations from multiple sources into a genome browser so that users can effectively analyze data between different data providers. Users can navigate the whole genome using the same genomics coordinate system of the uniform interface and make comparative analysis across different lineages.

8.4.1 WEB-BASED GENOME BROWSERS

Web-based genome browsers are useful in promoting biological research owing to their data quality, flexible accessibility, and high performance. First, it is worth mentioning that the web-based genome browsers are installed on high-performance servers and can support more complex and more extensive scale data types and applications. High-quality annotation data into web-based genome browsers are collected and integrated by organizations, thus providing full, up-to-date information for the community. Users can then access this data anywhere with a standard web browser without requiring any additional effort to configure the local environment for application installation and data preparation.

Nowadays, there are two types of web-based genome browsers. The first type is the multiple-species genome browsers implemented in, among others, the UCSC (University of California Santa Cruz) (http://genome.ucsc.edu/) genome database,[78] the Ensembl (European Molecular Biology Laboratory) (http://www.ensembl.org) project,[79] and the NCBI Map Viewer website (http://www.ncbi.nlm.nih.gov/mapview/).[80] These genome browsers integrate sequence and annotations for a large number of organisms and also promote further comparative species analysis. Some elements included in the web genome browsers are abundant commentary, covering the gene model, transcription elements, expression profiles, regulatory DAT, and genomic conversation. Each set of

Table 8.2 Summary of the Most Commonly Used Web-Based Genome Browsers in Multiple Species and Also Some Species-Specific Genome Browsers.

Multiple Species		Species Specific	
Resource	**Link**	**Resource**	**Link**
UCSC	http://genome.ucsc.edu/cgi-bin/hgGateway	Xenbase	http://www.xenbase.org/fgb2/gbrowse/
Ensembl	http://genome.ucsc.edu/cgi-bin/hgGateway	MGI	http://gbrowse.informatics.jax.org/cgi-bin/gbrowse/
Map Viewer	http://www.ncbi.nlm.nih.gov/mapview/	Flybase	http://flybase.org/cgi-bin/gbrowse/
Phytozome	http://www.phytozome.net/cgi-bin/gbrowse/	Wormbase	http://www.wormbase.org/db/gb2/gbrowse/
Gramene	http://www.gramene.org/genome_browser/	SGD	http://browse.yeastgenome.org/
VISTA	http://pipeline.lbl.gov/cgi-bin/gateway2/	TAIR	http://www.arabidopsis.org/browse/
Genome Projector	http://www.g-language.org/g3/		
Annmap	http://annmap.picr.man.ac.uk/		

Interestingly, the majority of the species-specific genome browsers are implemented based on the GBrowse framework.
MGI, Mouse Genome Informatics; *NCBI*, National Center for Biotechnology Information; *SGD*, Saccharomyces Genome Database; *TAIR*, The Arabidopsis Information Resource; *UCSC*, University of California Santa Cruz.

precomputed annotation data is called a track in genome browsers. No genome browser is "better" than others, but each browser has the advantage of performing specific tasks depending on what each function. For example, the UCSC browser aims to be a comprehensible, fast, and easy to use browser. However, it hosts fewer genomes than Ensembl and does not allow for some of the advanced searches as performed in Ensembl browser.

The second type of browsers is the species-specific genome browsers, and they mainly focus on one model organism. Moreover, these browsers may provide a higher number of annotations for a particular species. Dozens of open-source software tools are also powered by the Generic Model Organism Database (GMOD) project (http://gmod.org/) in order to create and manage genome biological databases. In addition, the GBrowse[29] framework is one of the most popular tools found within the GMOD project. Most of these implemented species-specific genome browsers are based on the GBrowse framework, such as the Mouse Genome Informatics, the FlyBase, the WormBase, the Saccharomyces Genome Database, and The Arabidopsis Information Resource (Table 8.2).

8.4.2 GENOME BROWSER FRAMEWORKS

The large volume of sequencing data from different species has led to an increasing demand for building genome browsers. This way, researchers are able to view and analyze this data in a more intuitive way. Nevertheless, building a web-based genome browser from scratch is as time-consuming and labor. Therefore well-designed genome browser frameworks could be useful in this aspect. Users can locally install, configure, and customize their annotations on some web-based genome browser frameworks. In this direction, Ensembl and the UCSC systems are also released as software packages for local installation. GBrowse is the most popular genome browser framework[81] and has been widely used in modeling organism projects for data visualization. The new genome browsers, JBrowse,[82] ABrowse,[83] and Anno-J,[84] support Google-Map-like navigation, while LookSeq[85] is designed for raw sequencing reads the presentation. In addition, there are some genome browsers for comparative visualization of several species, with separated genome coordinates.

8.4.3 FUNCTIONALITIES AND FEATURES

8.4.3.1 Visualization

High-throughput sequencing technology and high-performance computing resources contribute to the availability of high volumes of genomic annotation data. Collecting different types of annotation data and integrating them into an abstract graphical view are some of the essential functions of a genome browser.

A web genome browser usually provides a centralized database or a set of databases for storing different types of annotation data from different organizations. Presenting this massive volume of information appropriately for different genomic scales is a significant challenge for generic genome browsers. The server and the network are substantially burdened if a large amount of information needs to be embedded in the image. Another issue is that many dense and complex details can disturb the user's attention. The UCSC genome browser tries to solve the problem by providing multiple views for one piece.[78]

To allow for easier comparison between studies, such as finding the conserved or rapidly involved elements between different genomes, it is necessary to display genome alignments within or between species. Examples of these types of browsers are the Generic Synteny Browser,[86] Sybil (http://sybil.sourceforge.net/), SynBrowse,[87] SynView,[88] and VISTA.[89] The alignments could only be organized under the coordinates of one of the genomes. Nevertheless, some of the general genome browsers could also display genomic alignments. Most of the Synteny browsers may perform better arrangements of each segment of the alignment under the coordinate of its genome while piling them together. This type of genome browsers often focuses on sequence alignment of DNA or protein sequences rather than other types of annotation data.

Users can find genome structure variations with the help of the NGS raw data viewer, which aims to provide graphical views for the short sequences aligned with the genome. LookSeq provides a simple graphical representation for paired sequence reads, which aims in revealing potential insertions and deletions.[85] Users can also have a view of high-depth original readings and manipulate them manually to digest information at different levels of resolution.

8.4.3.2 Data retrieval and analysis

A genome browser also allows data retrieval and analysis. The majority of the current genome browsers support search functions to locate genomic regions with the use of keywords, coordinates, or sequences. In addition, some genome browsers use a system to recover bulk data. For example, the UCSC system offers the Table Browser in order to retrieve specified datasets,[90] while the

Ensembl, Gramene, and ABrowse projects employ the BioMart system[91,92] in order to deal with large data queries.

Multiple data access approaches for analysis tools are supported aiming to retrieve data from genome browsers and to facilitate further data analysis. Galaxy's genome browser, Trackster, supports analysis by integrating tools in the same platform and connecting data manipulation with visualization tightly. Users can visualize the data in the genome browser and improve the results conveniently and efficiently.

Machine-oriented data retrieval is becoming even more important for large-scale data analysis.[93] Currently, the web service has contributed significantly to the exchange of structured information between different data resources.[81] ABrowse supports native standard Simple Object Access Protocol (SOAP)-based web service for underlying data access,[83] which is also supported by BioMart[91,92] and employed by the Ensembl and Gramene projects. Moreover, BioDAS[94] has been widely used in some genome browsers for data exchange[78,81,95] of distributed platforms.

8.4.3.3 Customization

A genome browser based on a framework is much easier to build. The majority of the frameworks have configuration files for users to customize local data. Various popular data formats, such as GFF, BED, SAM, and wiggle format (WIG), are available for incorporation annotations into general genome browsers. However, one problem that often occurs is that the data format is incompatible with the genome browser. It is often difficult to convert data formats to meet system requirements by the different users. A few browser frameworks, such as GBrowse and ABrowse, provide plug-in mechanisms or application programming interface (API) to extend new data types.[81,83]

The genome browser is rapidly becoming a collaboration platform for researchers promoting remote cooperation among a group of scientists. Particularly, most genome browsers provide a facility for end users to upload, create, and share their annotation data, thus creating a collaborative platform. The user annotation comments can also be attached for selected items on-the-fly[83,95] and shared with specified users and groups,[95] or the entire research community.[83] In addition, users can save any critical analysis status as bookmarks[81,83,95] or sessions[96] and share them efficiently among researchers.

8.5 CONCLUSION

Nowadays, genome sequencing is becoming a commonly used method of screening for health care and research.[97] DNA sequencing combined with WES and WGS technologies improves and furthers our understanding of the association of genetic variation with complex human diseases. The perspectives on this technology exceed our grasp, achieving population-wide sequencing, portable sequencers, and even in vivo DNA editing. Since this research field is gaining more ground, researchers and clinicians need to familiarize themselves with recent developments and new tools as they are constantly being upgraded. Moreover, it is important for researchers to consider that there is still much space for improvement in these technologies. For example, although theoretically variant discovery through RNA-seq should be equally efficient with NGS pipelines, there is still a considerable discrepancy.

Recent advances in bioinformatics and particularly in workflow management systems allow the rapid automation and pipelining of these procedures. This chapter provides details about common steps in NGS pipelines alongside with a list of different publicly available genome browsers. The majority of the NGS pipelines are available, as ready-to-run computational scripts that can be combined at will in sophisticated analysis protocols.

It is crucial to consider that genetic research in modern times is more than just simple tool composition and experimentation. More specifically, it requires a social culture, cooperative skills, and the ability to construct analytical methods that are directly verifiable, reusable, and beneficial to a wide and diverse community.

REFERENCES

1. Morganti S, Tarantino P, Ferraro E, et al. Complexity of genome sequencing and reporting: next generation sequencing (NGS) technologies and implementation of precision medicine in real life. *Crit Rev Oncol Hematol*. 2019;133:171–182. Available from: https://doi.org/10.1016/j.critrevonc.2018.11.008.
2. Andjelkovic M, Minic P, Vreca M, et al. Genomic profiling supports the diagnosis of primary ciliary dyskinesia and reveals novel candidate genes and genetic variants. *PLoS One*. 2018;13:e0205422. Available from: https://doi.org/10.1371/journal.pone.0205422.
3. Komazec J, Zdravkovic V, Sajic S, et al. The importance of combined NGS and MLPA genetic tests for differential diagnosis of maturity onset diabetes of the young. *Endokrynol Pol*. 2019;70:28–36. Available from: https://doi.org/10.5603/EP.a2018.0064.
4. Skakic A, Djordjevic M, Sarajlija A, et al. Genetic characterization of GSD I in Serbian population revealed unexpectedly high incidence of GSD Ib and 3 novel *SLC37A4* variants. *Clin Genet*. 2018;93:350–355. Available from: https://doi.org/10.1111/cge.13093.
5. Stojiljkovic M, Klaassen K, Djordjevic M, et al. Molecular and phenotypic characteristics of seven novel mutations causing branched-chain organic acidurias. *Clin Genet*. 2016;90:252–257. Available from: https://doi.org/10.1111/cge.12751.
6. Rizzo JM, Buck MJ. Key principles and clinical applications of "next-generation" DNA sequencing. *Cancer Prev Res*. 2012;5:887–900. Available from: https://doi.org/10.1158/1940-6207.CAPR-11-0432.
7. Schuster SC. Next-generation sequencing transforms today's biology. *Nat Methods*. 2008;5:16–18. Available from: https://doi.org/10.1038/nmeth1156.
8. Metzker ML. Sequencing technologies—the next generation. *Nat Rev Genet*. 2010;11:31–46. Available from: https://doi.org/10.1038/nrg2626.
9. Karsch-Mizrachi I, Nakamura Y, Cochrane G, International Nucleotide Sequence Database Collaboration. The International Nucleotide Sequence Database Collaboration. *Nucleic Acids Res*. 2012;40:D33–D37. Available from: https://doi.org/10.1093/nar/gkr1006.
10. Leinonen R, Sugawara H, Shumway M, International Nucleotide Sequence Database Collaboration. The Sequence Read Archive. *Nucleic Acids Res*. 2011;39:D19–D21. Available from: https://doi.org/10.1093/nar/gkq1019.
11. Edgar R, Domrachev M, Lash AE. Gene Expression Omnibus: NCBI gene expression and hybridization array data repository. *Nucleic Acids Res*. 2002;30:207–210. Available from: https://doi.org/10.1093/nar/30.1.207.
12. NCBI Resource Coordinators. Database Resources of the National Center for Biotechnology Information. *Nucleic Acids Res*. 2017;45:D12–D17. Available from: https://doi.org/10.1093/nar/gkw1071.

13. Kolesnikov N, Hastings E, Keays M, et al. ArrayExpress update—simplifying data submissions. *Nucleic Acids Res*. 2015;43:D1113–D1116. Available from: https://doi.org/10.1093/nar/gku1057.
14. Leinonen R, Akhtar R, Birney E, et al. The European Nucleotide Archive. *Nucleic Acids Res*. 2011;39: D28–D31. Available from: https://doi.org/10.1093/nar/gkq967.
15. Sudmant PH, Rausch T, Gardner EJ, et al. An integrated map of structural variation in 2,504 human genomes. *Nature*. 2015;526:75–81. Available from: https://doi.org/10.1038/nature15394.
16. Ewing B, Hillier L, Wendl MC, Green P. Base-calling of automated sequencer traces using Phred. I. Accuracy assessment. *Genome Res*. 1998;8:175–185.
17. Thorvaldsdottir H, Robinson JT, Mesirov JP. Integrative Genomics Viewer (IGV): high-performance genomics data visualization and exploration. *Brief Bioinform*. 2013;14:178–192. Available from: https://doi.org/10.1093/bib/bbs017.
18. Li H, Handsaker B, Wysoker A, et al. The sequence alignment/Map format and SAMtools. *Bioinformatics*. 2009;25:2078–2079. Available from: https://doi.org/10.1093/bioinformatics/btp352.
19. Danecek P, Auton A, Abecasis G, et al. The variant call format and VCFtools. *Bioinformatics*. 2011;27:2156–2158. Available from: https://doi.org/10.1093/bioinformatics/btr330.
20. Henson J, Tischler G, Ning Z. Next-generation sequencing and large genome assemblies. *Pharmacogenomics*. 2012;13:901–915. Available from: https://doi.org/10.2217/pgs.12.72.
21. Morey M, Fernández-Marmiesse A, Castiñeiras D, Fraga JM, Couce ML, Cocho JA. A glimpse into past, present, and future DNA sequencing. *Mol Genet Metab*. 2013;110:3–24. Available from: https://doi.org/10.1016/j.ymgme.2013.04.024.
22. Pareek CS, Smoczynski R, Tretyn A. Sequencing technologies and genome sequencing. *J Appl Genet*. 2011;52:413–435. Available from: https://doi.org/10.1007/s13353-011-0057-x.
23. Khodakov D, Wang C, Zhang DY. Diagnostics based on nucleic acid sequence variant profiling: PCR, hybridization, and NGS approaches. *Adv Drug Deliv Rev*. 2016;105:3–19. Available from: https://doi.org/10.1016/j.addr.2016.04.005.
24. Margulies M, Egholm M, Altman WE, et al. Genome sequencing in microfabricated high-density picolitre reactors. *Nature*. 2005;437:376–380. Available from: https://doi.org/10.1038/nature03959.
25. Pillai S, Gopalan V, Lam AK-Y. Review of sequencing platforms and their applications in phaeochromocytoma and paragangliomas. *Crit Rev Oncol Hematol*. 2017;116:58–67. Available from: https://doi.org/10.1016/j.critrevonc.2017.05.005.
26. Author C, Mohammad Mazumdar R, Chowdhury A, Hossain N, Mahajan S, Islam S. Pyrosequencing—a next generation sequencing technology. *World Appl Sci J*. 2013;24:1558–1571. Available from: https://doi.org/10.5829/idosi.wasj.2013.24.12.2972.
27. Siqueira JF, Fouad AF, Rôças IN. Pyrosequencing as a tool for better understanding of human microbiomes. *J Oral Microbiol*. 2012;4:10743. Available from: https://doi.org/10.3402/jom.v4i0.10743.
28. Harrington CT, Lin EI, Olson MT, Eshleman JR. Fundamentals of pyrosequencing. *Arch Pathol Lab Med*. 2013;137:1296–1303. Available from: https://doi.org/10.5858/arpa.2012-0463-RA.
29. Bragg L, Tyson GW. Metagenomics using next-generation sequencing. *Methods Mol Biol (Clifton, N.J.)*. 2014;183–201. Available from: https://doi.org/10.1007/978-1-62703-712-9_15.
30. De Mandal S, Panda AK. Microbial ecology in the era of next generation sequencing. *J Next Gener Seq Appl*. 2015;. Available from: https://doi.org/10.4172/2469-9853.S1-001. 01.
31. Heather JM, Chain B. The sequence of sequencers: the history of sequencing DNA. *Genomics*. 2016;107:1–8. Available from: https://doi.org/10.1016/j.ygeno.2015.11.003.
32. Barba M, Czosnek H, Hadidi A. Historical perspective, development and applications of next-generation sequencing in plant virology. *Viruses*. 2014;6:106–136. Available from: https://doi.org/10.3390/v6010106.

33. Park ST, Kim J. Trends in next-generation sequencing and a new era for whole genome sequencing. *Int Neurourol J*. 2016;20:S76–S83. Available from: https://doi.org/10.5213/inj.1632742.371.
34. Schmieder R, Edwards R. Quality control and preprocessing of metagenomic datasets. *Bioinformatics*. 2011;27:863–864. Available from: https://doi.org/10.1093/bioinformatics/btr026.
35. Levy SE, Myers RM. Advancements in next-generation sequencing. *Annu Rev Genomics Hum Genet*. 2016;17:95–115. Available from: https://doi.org/10.1146/annurev-genom-083115-022413.
36. Ambardar S, Gupta R, Trakroo D, Lal R, Vakhlu J. High throughput sequencing: an overview of sequencing chemistry. *Indian J Microbiol*. 2016;56:394–404. Available from: https://doi.org/10.1007/s12088-016-0606-4.
37. Mardis ER. The impact of next-generation sequencing technology on genetics. *Trends Genet*. 2008;24:133–141. Available from: https://doi.org/10.1016/j.tig.2007.12.007.
38. Liu L, Li Y, Li S, et al. Comparison of next-generation sequencing systems. *J Biomed Biotechnol*. 2012;2012:1–11. Available from: https://doi.org/10.1155/2012/251364.
39. Valouev A, Ichikawa J, Tonthat T, et al. A high-resolution, nucleosome position map of *C. elegans* reveals a lack of universal sequence-dictated positioning. *Genome Res*. 2008;18:1051–1063. Available from: https://doi.org/10.1101/gr.076463.108.
40. Rothberg JM, Hinz W, Rearick TM, et al. An integrated semiconductor device enabling non-optical genome sequencing. *Nature*. 2011;475:348–352. Available from: https://doi.org/10.1038/nature10242.
41. Salipante SJ, Kawashima T, Rosenthal C, et al. Performance comparison of Illumina and Ion Torrent next-generation sequencing platforms for 16S rRNA-based bacterial community profiling. *Appl Environ Microbiol*. 2014;80:7583–7591. Available from: https://doi.org/10.1128/AEM.02206-14.
42. Quail M, Smith ME, Coupland P, et al. A tale of three next generation sequencing platforms: comparison of Ion torrent, pacific biosciences and Illumina MiSeq sequencers. *BMC Genomics*. 2012;13:341. Available from: https://doi.org/10.1186/1471-2164-13-341.
43. Goodwin S, McPherson JD, McCombie WR. Coming of age: ten years of next-generation sequencing technologies. *Nat Rev Genet*. 2016;17:333–351. Available from: https://doi.org/10.1038/nrg.2016.49.
44. Besser J, Carleton HA, Gerner-Smidt P, Lindsey RL, Trees E. Next-generation sequencing technologies and their application to the study and control of bacterial infections. *Clin Microbiol Infect*. 2018;24:335–341. Available from: https://doi.org/10.1016/j.cmi.2017.10.013.
45. Lee H, Gurtowski J, Yoo S, et al. Third-generation sequencing and the future of genomics. *bioRxiv*. 2016;. Available from: https://doi.org/10.1101/048603. 048603.
46. Timp W, Comer J, Aksimentiev A. DNA base-calling from a nanopore using a Viterbi algorithm. *Biophys J*. 2012;102:L37–L39. Available from: https://doi.org/10.1016/j.bpj.2012.04.009.
47. Wang Y, Yang Q, Wang Z. The evolution of nanopore sequencing. *Front Genet*. 2015;5:449. Available from: https://doi.org/10.3389/fgene.2014.00449.
48. Magi A, Semeraro R, Mingrino A, Giusti B, D'Aurizio R. Nanopore sequencing data analysis: state of the art, applications and challenges. *Brief Bioinform*. 2017;19:1256–1272. Available from: https://doi.org/10.1093/bib/bbx062.
49. Diaz-Sanchez S, Hanning I, Pendleton S, D'Souza D. Next-generation sequencing: the future of molecular genetics in poultry production and food safety. *Poult Sci*. 2013;92:562–572. Available from: https://doi.org/10.3382/ps.2012-02741.
50. Derrington IM, Butler TZ, Collins MD, et al. Nanopore DNA sequencing with MspA. *Proc Natl Acad Sci USA*. 2010;107:16060–16065. Available from: https://doi.org/10.1073/pnas.1001831107.
51. Magi A, Giusti B, Tattini L. Characterization of MinION nanopore data for resequencing analyses. *Brief Bioinform*. 2016;18:940–953. Available from: https://doi.org/10.1093/bib/bbw077.
52. Leggett RM, Clark MD. A world of opportunities with nanopore sequencing. *J Exp Bot*. 2017;68:5419–5429. Available from: https://doi.org/10.1093/jxb/erx289.

53. Shin SC, Ahn DH, Kim SJ, et al. Advantages of single-molecule real-time sequencing in high-GC content genomes. *PLoS One*. 2013;8:e68824. Available from: https://doi.org/10.1371/journal.pone.0068824.
54. Nakano K, Shiroma A, Shimoji M, et al. Advantages of genome sequencing by long-read sequencer using SMRT technology in medical area. *Hum Cell*. 2017;30:149–161. Available from: https://doi.org/10.1007/s13577-017-0168-8.
55. Ardui S, Ameur A, Vermeesch JR, Hestand MS. Single molecule real-time (SMRT) sequencing comes of age: applications and utilities for medical diagnostics. *Nucleic Acids Res*. 2018;46:2159–2168. Available from: https://doi.org/10.1093/nar/gky066.
56. Eid J, Fehr A, Gray J, et al. Real-time DNA sequencing from single polymerase molecules. *Science*. 2009;323:133–138. Available from: https://doi.org/10.1126/science.1162986.
57. Roberts RJ, Carneiro MO, Schatz MC. The advantages of SMRT sequencing. *Genome Biol*. 2013;14:405. Available from: https://doi.org/10.1186/gb-2013-14-6-405.
58. Carneiro MO, Russ C, Ross MG, Gabriel SB, Nusbaum C, DePristo MA. Pacific biosciences sequencing technology for genotyping and variation discovery in human data. *BMC Genomics*. 2012;13:375. Available from: https://doi.org/10.1186/1471-2164-13-375.
59. Jünemann S, Kleinbölting N, Jaenicke S, et al. Bioinformatics for NGS-based metagenomics and the application to biogas research. *J Biotechnol*. 2017;261:10–23. Available from: https://doi.org/10.1016/j.jbiotec.2017.08.012.
60. Oulas A, Pavloudi C, Polymenakou P, et al. Metagenomics: tools and insights for analyzing next-generation sequencing data derived from biodiversity studies. *Bioinform Biol Insights*. 2015;9:75–88. Available from: https://doi.org/10.4137/BBI.S12462. BBI.S12462.
61. Roumpeka DD, Wallace RJ, Escalettes F, Fotheringham I, Watson M. A review of bioinformatics tools for bio-prospecting from metagenomic sequence data. *Front Genet*. 2017;8:23. Available from: https://doi.org/10.3389/fgene.2017.00023.
62. Needleman SB, Wunsch CD. A general method applicable to the search for similarities in the amino acid sequence of two proteins. *J Mol Biol*. 1970;48:443–453.
63. Smith TF, Waterman MS. Identification of common molecular subsequences. *J Mol Biol*. 1981;147:195–197.
64. Lipman DJ, Pearson WR. Rapid and sensitive protein similarity searches. *Science*. 1985;227:1435–1441.
65. Altschul SF, Gish W, Miller W, Myers EW, Lipman DJ. Basic local alignment search tool. *J Mol Biol*. 1990;215:403–410. Available from: https://doi.org/10.1016/S0022-2836(05)80360-2.
66. Otto C, Stadler PF, Hoffmann S. Lacking alignments? The next-generation sequencing mapper segemehl revisited. *Bioinformatics*. 2014;30:1837–1843. Available from: https://doi.org/10.1093/bioinformatics/btu146.
67. Fonseca NA, Rung J, Brazma A, Marioni JC. Tools for mapping high-throughput sequencing data. *Bioinformatics*. 2012;28:3169–3177. Available from: https://doi.org/10.1093/bioinformatics/bts605.
68. Hwang S, Kim E, Lee I, Marcotte EM. Systematic comparison of variant calling pipelines using gold standard personal exome variants. *Sci Rep*. 2016;5:17875. Available from: https://doi.org/10.1038/srep17875.
69. Li H. *Aligning Sequence Reads, Clone Sequences and Assembly Contigs With BWA-MEM*. 2013.
70. Marth GT, Korf I, Yandell MD, et al. A general approach to single-nucleotide polymorphism discovery. *Nat Genet*. 1999;23:452–456. Available from: https://doi.org/10.1038/70570.
71. McKenna A, Hanna M, Banks E, et al. The Genome Analysis Toolkit: a MapReduce framework for analyzing next-generation DNA sequencing data. *Genome Res*. 2010;20:1297–1303. Available from: https://doi.org/10.1101/gr.107524.110.
72. Tan R, Wang Y, Kleinstein SE, et al. An evaluation of copy number variation detection tools from whole-exome sequencing data. *Hum Mutat*. 2014;35:899–907. Available from: https://doi.org/10.1002/humu.22537.

73. Tan R, Wang J, Wu X, et al. ERDS-exome: a Hybrid Approach for Copy Number Variant Detection from Whole-exome Sequencing Data. *IEEE/ACM Trans Comput Biol Bioinform*. 2018;1. Available from: https://doi.org/10.1109/TCBB.2017.2758779.
74. Jiang Y, Oldridge DA, Diskin SJ, Zhang NR. CODEX: a normalization and copy number variation detection method for whole exome sequencing. *Nucleic Acids Res*. 2015;43:e39. Available from: https://doi.org/10.1093/nar/gku1363.
75. Abyzov A, Urban AE, Snyder M, Gerstein M. CNVnator: an approach to discover, genotype, and characterize typical and atypical CNVs from family and population genome sequencing. *Genome Res*. 2011;21:974–984. Available from: https://doi.org/10.1101/gr.114876.110.
76. de Araújo Lima L, Wang K. PennCNV in whole-genome sequencing data. *BMC Bioinform*. 2017;18:383. Available from: https://doi.org/10.1186/s12859-017-1802-x.
77. Schattner P. *Genomes, Browsers and Databases*. Cambridge: Cambridge University Press; 2008. Available from: https://doi.org/10.1017/CBO9780511754838.
78. Karolchik D, Baertsch R, Diekhans M, et al. The UCSC genome browser database. *Nucleic Acids Res*. 2003;31:51–54. Available from: https://doi.org/10.1093/nar/gkg129.
79. Hubbard T, Barker D, Birney E, et al. The Ensembl genome database project. *Nucleic Acids Res*. 2002;30:38–41. Available from: https://doi.org/10.1093/nar/30.1.38.
80. Wolfsberg TG. Using the NCBI map viewer to browse genomic sequence data. *Curr Protoc Hum Genet*. 2011;69:18.5.1–18.5.25. Available from: https://doi.org/10.1002/0471142905.hg1805s69.
81. Stein LD, Mungall C, Shu S, et al. The generic genome browser: a building block for a model organism system database. *Genome Res*. 2002;12:1599–1610. Available from: https://doi.org/10.1101/gr.403602.
82. Skinner ME, Uzilov AV, Stein LD, Mungall CJ, Holmes IH. JBrowse: a next-generation genome browser. *Genome Res*. 2009;19:1630–1638. Available from: https://doi.org/10.1101/gr.094607.109.
83. Kong L, Wang J, Zhao S, Gu X, Luo J, Gao G. ABrowse – a customizable next-generation genome browser framework. *BMC Bioinform*. 2012;13:2. Available from: https://doi.org/10.1186/1471-2105-13-2.
84. Lister R, O'Malley RC, Tonti-Filippini J, et al. Highly integrated single-base resolution maps of the epigenome in Arabidopsis. *Cell*. 2008;133:523–536. Available from: https://doi.org/10.1016/j.cell.2008.03.029.
85. Manske HM, Kwiatkowski DP. LookSeq: a browser-based viewer for deep sequencing data. *Genome Res*. 2009;19:2125–2132. Available from: https://doi.org/10.1101/gr.093443.109.
86. McKay SJ, Vergara IA, Stajich JE. Using the Generic Synteny Browser (GBrowse_syn). *Curr Protoc Bioinformatics*. 2010;. Available from: https://doi.org/10.1002/0471250953.bi0912s31. Chapter 9, Unit 9.12.
87. Pan X, Stein L, Brendel V. SynBrowse: a synteny browser for comparative sequence analysis. *Bioinformatics*. 2005;21:3461–3468. Available from: https://doi.org/10.1093/bioinformatics/bti555.
88. Wang H, Su Y, Mackey AJ, Kraemer ET, Kissinger JC. SynView: a GBrowse-compatible approach to visualizing comparative genome data. *Bioinformatics*. 2006;22:2308–2309. Available from: https://doi.org/10.1093/bioinformatics/btl389.
89. Frazer KA, Pachter L, Poliakov A, Rubin EM, Dubchak I. VISTA: computational tools for comparative genomics. *Nucleic Acids Res*. 2004;32:W273–W279. Available from: https://doi.org/10.1093/nar/gkh458.
90. Karolchik D, Hinrichs AS, Furey TS, et al. The UCSC Table Browser data retrieval tool. *Nucleic Acids Res*. 2004;32:493D–496D. Available from: https://doi.org/10.1093/nar/gkh103.
91. Haider S, Ballester B, Smedley D, Zhang J, Rice P, Kasprzyk A. BioMart Central Portal—unified access to biological data. *Nucleic Acids Res*. 2009;37:W23–W27. Available from: https://doi.org/10.1093/nar/gkp265.
92. Smedley D, Haider S, Ballester B, et al. BioMart – biological queries made easy. *BMC Genomics*. 2009;10:22. Available from: https://doi.org/10.1186/1471-2164-10-22.

93. Sen TZ, Harper LC, Schaeffer ML, et al. Choosing a genome browser for a Model Organism Database: surveying the Maize community. *Database*. 2010;2010:baq007. Available from: https://doi.org/10.1093/database/baq007.
94. Dowell RD, Jokerst RM, Day A, Eddy SR, Stein L. The distributed annotation system. *BMC Bioinform*. 2001;2:7.
95. Flicek P, Aken BL, Beal K, et al. Ensembl 2008. *Nucleic Acids Res*. 2007;36:D707–D714. Available from: https://doi.org/10.1093/nar/gkm988.
96. Karolchik D, Kuhn RM, Baertsch R, et al. The UCSC Genome Browser Database: 2008 update. *Nucleic Acids Res*. 2007;36:D773–D779. Available from: https://doi.org/10.1093/nar/gkm966.
97. Shendure J, Balasubramanian S, Church GM, et al. DNA sequencing at 40: past, present and future. *Nature*. 2017;550:345–353. Available from: https://doi.org/10.1038/nature24286.

CHAPTER

TRANSLATIONAL TOOLS AND DATABASES IN GENOMIC MEDICINE

9

Evaggelia Barba[1], Maria Koromina[1], Evangelia-Eirini Tsermpini[1] and George P. Patrinos[1,2,3]

[1]*Department of Pharmacy, School of Health Sciences, University of Patras, Patras, Greece* [2]*Department of Pathology, College of Medicine and Health Sciences, United Arab Emirates University, Al-Ain, United Arab Emirates* [3]*Zayed Center of Health Sciences, United Arab Emirates University, Al-Ain, United Arab Emirates*

ABBREVIATIONS

CPIC	Clinical Pharmacogenetics Implementation Consortium
dbSNP	single-nucleotide polymorphism database
EHR	electronic health record
GVDs	general variation databases
LSDBs	locus-specific databases
PGx	pharmacogenomics
PharmGKB	Pharmacogenomics Knowledgebase

9.1 INTRODUCTION

The concept of personalized and genome-based medicine can be traced back in ancient times with Hippocrates quoting that "It's far more important to know what person the disease has than what disease the person has." Although enormous progress has been made in drug development, prescribing the appropriate drug, in the appropriate and right dose, to the right patient is still to be implemented in a routine clinical framework.

Pharmacogenomics (PGx) studies all genetic factors that could potentially influence drug response, and it is a research field that has greatly advanced over the last decade. PGx aims to revolutionize drug therapy by "dissecting" associations between genetic factors and response to drug treatment. In this direction, genome-wide association studies (GWAS) have contributed to many genetic and biological discoveries by revealing a considerable amount of disease-associated loci and providing insights into the allelic architecture of complex traits.[1]

Therefore designing the appropriate tools, pipelines, and databases, which would allow the integration of both genetic and clinical information, is an emerging necessity. Such tools and databases could translate the genetic information into a useful clinical guideline and, consequently, could assist in improving clinical practice.[2]

In the previous chapter, we have given an overview of the various genomic data analysis pipelines that are most frequently used in genomic medicine applications and discovery. Here, we

Applied Genomics and Public Health. DOI: https://doi.org/10.1016/B978-0-12-813695-9.00009-1

discuss the closely related issue of big data analysis in genomics. In particular, we focus on the various translational tools in PGx and genomic medicine applications and the closely related discipline of genomic databases that serve as auxiliary resources in genomic medicine.

9.2 TRANSLATIONAL TOOLS IN GENOMIC MEDICINE

Discovering and validating actionable genomic variants is not a straightforward process, especially when studying complex traits, such as drug metabolism. Numerous PGx studies have shown that most of the genetic variability in drug response involves several genes encoding proteins involved in multiple pathways of drug metabolism and disposition. In this chapter, we discuss current and future initiatives in the field of genomic medicine, mainly through developing and optimizing informatics tools and databases, which will assist in translating the genetic association information into better clinical practice and outcomes in individual patients.

9.2.1 PHARMACOGENOMICS AND GENOME INFORMATICS

As mentioned earlier, the term PGx refers to the study of the association of a set of genes with the response of patients to pharmacotherapy. It also refers to the adaptation of the modes of medical treatment to the individual genetic characteristics, needs, and preferences of each patient. Owing to a documented lack of PGx knowledge from clinicians,[3] there is a need to develop informatics solutions and tools to translate genomic information into a clinically meaningful format with an emphasis in individualization of drug-treatment modalities. To this end, appropriate genome informatics platforms and services should be designed and developed in order to achieve an accurate analysis of the resulting PGx resequencing, perform secure storage of the huge volume of genomic information from the PGx resequencing, and provide an important and clinician-friendly report that can be easily utilized in different clinical frameworks. To deal with this challenge, novel approaches utilizing information from large and complex PGx resources have to be explored and developed. The scope of the task has two sides: first, to appropriate and enhance recognition and fact-based documentation of (existing or newly reported) PGx gene/variant–drug–phenotype associations and second, to translate and transfer the well-documented PGx knowledge to clinical application with focusing on both rationalizing and individualizing drug-treatment modes. The following sections focus on this particular challenge.

Since the Internet has become a valuable tool for biomedical researchers, genomic information overload was inevitable. Although there are many genomic and biological databases, which often create indecision for users in means of which might be the best to investigate a given biological question, the amount of the existing databases that are directly tied to PGx is limited in number.[4]

The main question about the exploitation of PGx knowledge and its use in clinical practice relates to the heterogeneity and the lower degree of connectivity between the different PGx information. A large quantity of raw data is difficult to handle, and, consequently, PGx biomedical researchers and stakeholders are often unable to retain all the information regarding the genomic variation and its association with fluctuating drug response. Consequently, the design of an incorporated web information system aiming to merge the diverse PGx information resources into a singular portal is still a challenge.[5]

9.2.2 THE CONCEPT OF INTEGRATED PHARMACOGENOMICS ASSISTANT SERVICES

The PGx data overflow challenge requires specialized informatics services focusing on understanding and unifying the vast quantities of molecular and clinical data. Such a multidimensional effort requires both translational and clinical bioinformatics approaches that may not only lead to analytical and interpretational methods to enhance the transformation of ever-increasing biomedical data into proactive, predictive, preventive, and participatory (4Ps) medicine[6] but also, at the same time, enable the clinical utilization of discovery-driven bioinformatics methods to understand the molecular mechanisms underlying complex disease phenotypes and boost the search for possible therapies for human diseases.[7] An electronic PGx assistant platform could be used as the PGx bench-to-bedside enabling medium to aid to this direction. The foundational components that underpin the innovative nature of such a platform are based on its ability to offer translational services, accumulated from interrelating genotypic to phenotypic (metabolizer status) information. This platform is an essential tool for clinicians as well as biomedical researchers, by (1) assisting toward informed decisions based on state-of-the-art PGx data, and (2) providing an "one-stop solution" where knowledge can be acquired to create an understanding of interindividual differences in drug effectiveness, toxicity, and pharmacokinetics (PK), as well as propelling the findings of new PGx variants.

To this scope the aim is to provide a web-based platform, which will make the handling, assimilation, and sharing of PGx knowledge more comfortable, as well as facilitate the accumulation of different PGx stakeholders' standpoints. The platform should make the use of interoperable and adjustable bioinformatics and advanced fact-handling elements that can provide personalized treatment advice based on reliable genomic evidence, while simultaneously cutting down on health-care expenses by increasing drug effectiveness and minimizing adverse drug reactions.

Such a system can be improved by focusing on (1) retrieval of PGx information concerning ADMET genes, their corresponding variants, and drugs; (2) updating the database/system with information regarding recently discovered PGx variants; and (3) the ability to present personalized medical care recommendations established on personalized PGx profiles.

9.2.3 DEVELOPMENT OF AN ELECTRONIC PHARMACOGENOMICS ASSISTANT

Several external data sources are designed to extract and accumulate PGx information. Therefore the concept of a data warehouse may be the primary and most suitable data model to encircle the distinct requirements for database technology. Distinct types of data extraction tools, such as application programming interface, web services, JSON/XML, or text parsers, are employed to deliver and transform data from the different heterogeneous data sources, such as Pharmacogenomics Knowledgebase (PharmGKB), National Center for Biotechnology Information databases [e.g., single-nucleotide polymorphism database (dbSNP) and PubMed], and genotyping platform(s) (e.g., Affymetrix) annotations, into the primary data warehouse, following an extraction–transform–load procedure. Typical ontologies and nomenclatures are used to represent the diverse data and information of PGx modules uniformly (e.g., gene–variant nomenclatures, gene ontology, and Internal Classification of Diseases for disease classification and encoding).

Regarding the management of people's genotype/SNP profiles, an electronic health-care documentation [electronic health records (EHR)] solution can be adopted. To achieve these, state-of-the-art directions and data models associated with genetic tests and their interpretation can be implemented.[5] Standard ontologies and data models could be implemented for the rendition of genotype profiles, such as the Genetic Variant Format www.sequenceontology.org/recourses/gvf.html and the Logical Observation Identifiers Names and Codes (https://loinc.org). The efficacy of linking genotype data to EHRs is vital for the translation and transmittance of PGx knowledge into the clinic. The method is both low-cost and time-saving as there is no need to engage actively and collect samples from a study population, since cases and controls are easily accessible and constantly identified from EHRs and the linked genetic samples. The eMERGE Consortium (http://emerge.mc.vanderbilt.edu) has already successfully utilized this alternative.[8]

Based on the previous concept, there are few such electronic solutions and services that have been generated to satisfy this need. Abomics PGx (www.abomicspgx.com) is a service that interprets the complex PGx data and offers laboratories, clinicians, and patients an easy access to reliable and highly updated PGx information. Abomics PGx interpretation service, coupled with the corresponding decision-support database, is a comprehensive solution to accessing and utilizing genetic information obtained through PGx testing, providing a medical professional–approved interpretation of the data, which is easily accessible to the patient, laboratory, and treating clinician. A simple yet comprehensive report covers the genetic tests with explanations, a list of drugs with genetic variation, and dosing recommendations for each drug. The reports are approved by various medical professionals accepting the responsibility for accuracy of the information. The reports are easy to read and give specific recommendations based on the genetic information acquired with genetic tests.

Similar to the Abomics solution, the YouScript precision prescribing software (www.youscript.com) integrates PGx testing with comprehensive drug–gene and drug–drug interaction information to assess the cumulative impact of a patient's genetics and drug regimen, and their risk for developing adverse drug reactions. Information, alerts, and recommended medication alternatives are presented in an easy-to-understand, clinically actionable format.

A similar, with the previously described, solution is the Medication Safety Code (MSC; www.safety-code.org) that captures the results for PGx test to make them available whenever needed during medical care. MSCs can be printed on personalized plastic cards (like a credit card) so that patients can carry them in their wallet, or the cards can be incorporated in paper-based lab reports. After scanning the QR code with a standard smartphone, the user is led to a website that displays drug-dosing recommendations extracted from PGx guidelines that are highly relevant for the patient based on his or her PGx test results.

The Pharmacogenomics Clinical Annotation Tool (PharmCAT; www.pharmcat.org) is a software tool to extract guideline variants from a genetic dataset (represented as a .vcf file; see previous chapter), interpret the variant alleles, and generate a report with genotype-based prescribing recommendations, which can be used to inform treatment decisions. These recommendations are retrieved from the Clinical Pharmacogenetics Implementation Consortium (CPIC; www.cpicpgx.org) that are surrounding gene–drug pairs that can and should lead to treatment modifications based on genetic variants. These guidelines are used for the initial version of PharmCAT, and other sources of PGx information and guidelines will be included in the future. In June 2019 this tool was still under development and not officially released.

Several crucial ethical issues surface regarding the assembly, stockpiling, and management of individuals' genotype profiles, as well as the current concerns about depositing data from genetic tests and ethical issues regarding translational research.[9] These comprise public genetic comprehension and genomic literacy, physicians' awareness of genomics, managing genomic information in and beyond the clinic, and online direct-to-consumer PGx testing, with the corresponding arguments to be extremely polarized.[10] Moreover, all the pertinent ethical, privacy protection, and security matters should be taken to consideration, and special caution should be applied when designing the guidelines by performing critical assessment of the effect of genetics and PGx on society. In addition to this, an effort should be made toward improving the understanding of the general public, health-care professionals, and biomedical scientists to PGx and personalized medicine.

The suggested PGx assistant will boost PGx research by easing the discovery of new PGx variants. Discovering genetic variants associated with drug response, incorporating GWAS, expression analysis, and even whole-genome sequencing (WGS), are already present. The suggested PGx assistant presents reporting and analytical services to the user with the use of a secure and user-friendly interface, while it assists in unveiling hidden relations between genes, variants, and drugs, thus steering the discovery of candidate genomic regions of interest. Furthermore, as understanding about drug–gene and drug–drug interactions is improved, the suggested freely available system, which is further paired with upgraded literature-mining characteristics and updatable elements, becomes even more advantageous to the research community and society.

Undoubtedly, the primary concept in the proposed PGx assistant platform is the notion of personalization. The inclusion of a personalized PGx translational component in this platform is based on the idea that "clinical high-throughput and preemptive genotyping will eventually become common practice and clinicians will increasingly have patients' genotypes available before a prescription is written."[9] The personalized translation component would aim toward the (1) automated correspondence to patients' genotype profiles with confirmed and/or newly uncovered genetic variants/alleles, established on the customization of an intricate allele-matching algorithm; (2) deduction of particular phenotypes (e.g., metabolized profiles), and (3) delivery of pertinent and updated clinical interpretations and (drug) dosing suggestions.

9.2.4 PERSONALIZED PHARMACOGENOMICS PROFILING USING WHOLE GENOME SEQUENCING

It is already highlighted in previous sections that the development of next-generation sequencing (NGS) technologies has created unprecedented opportunities to analyze whole genomes.[11] The main hypothesis is that individuals may harbor rare and novel variants of functional significance in well-established pharmacogenes, which may render an individual intermediate or poor metabolizer to certain drugs, and which may go undetected when using a genetic screening assay.[12] Over recent years, WGS has been one of the drivers of PGx research with many exciting findings,[13] with the majority of variants that impact drug response remaining to be identified.[12] As GWAS may not detect all risk biomarkers,[14] since GWAS are genetic screening rather than full scanning approach, the identification of unknown (possibly very rare or even novel) variants from WGS may provide indicative associations between specific genotypes and (rare) adverse drug reactions. Whole exome sequencing (WES) and WGS can now be easily performed utilizing several commercially available or proprietary platforms, in order to analyze genome variation comprehensively and accurately at a reasonable cost.[15]

Several consortia, such as the Pharmacogenomics Research Network (http://pgrn.org) and the eMERGE Consortium (http://emerge.mc.vanderbilt.edu) are developing whole pharmacogenes resequencing platforms, which focus on the most significant (i.e., actionable) fraction of pharmacogenes. As such, relevant pipelines aiming to integrate information from NGS analysis into a clinical PGx workflow from deep sequencing to PGx consultation, while adhering to the current clinical guidelines and recommendations needs to be improvized. To this end, designing a computational tool, which will classify and calibrate the different types of genetic variation in (pharmaco) genes, may be beneficial in NGS implementations and in clinical PGx.

It is evident that since WGS-based PGx testing is becoming extensively accessible, interpretation of this genomic information into clinically major guidelines will be necessary. With regards to PGx the PGx clinical scenarios are very intricate in a real-life situation, which often constitute notable dilemmas to the medical professionals concerning the selection of a most suitable treatment procedure and drug dose. This complication emerges mainly owing to the substantial translation gap in moving PGx (as with the other -omics) scientific discoveries toward successful changes that can be implemented in the clinic. This gap arises owing to the lack of conceptualization of information-based PGx innovation as an ecosystem that conveys innovation factors, such as pharmacology, PGx molecular biology, and genetics scientists and innovation narrators.[16] It is important to implement a multidisciplinary approach focused on a portfolio of interoperating translational or clinical genome information elements and their adjustment to modern information engineering and processing techniques. Such methods should focus on devising (1) a PGx knowledge assimilator that effortlessly links multiple PGx knowledge sources, and (2) knowledge-extraction systems capable of recognizing functional genotype-to-phenotype connections and knowledge from these origins.

Furthermore, the identified PGx genotype-to-phenotype associations should be analyzed according to their PK and pharmacodynamic (PD) background. This kind of research could be enhanced by the elaboration of the proper PK/PD modeling, which is a technique combining the two traditional pharmacologic disciplines of PK, as well as PD. There is a variety of PK/PD modeling approaches to describe exposure–response relationships. PK/PD simulation models assist in evaluating PGx connection's covariance in artificially devised populations, for example, following the example of SimCYP (www.simcyp.com) and NONMEM (www.iconplc.com/technology/products/nonmem) virtual simulation commercial packages, as well as using free, open-source PK modeling software tools, such as PK report (cran.r-project.org/web/packages/PKreport) and WFN (wfn.sourceforge.net/wfnxpose.htm) R-packages.

In order to translate PGx knowledge into clinical decision, the linkage and seamless integration of established PGx resources (e.g., PharmGKB, CPIC, and DruGeVar), literature, and other genomic databases [e.g., PubMed, dbSNP, dbGAP, ClinVar, and Frequency of Inherited Disorders database (FINDbase)] is crucial. This will allow the elaboration and operationalization of the standard PGx/clinical ontologies and data models. Furthermore, one should incorporate (1) literature mining/natural language processing for extraction of putative disease–drug–gene/variant–phenotype associations from PGx resources, (2) published literature, as well as, (3) a virtual population PK simulator for testing putative variant–phenotype associations and assessing relevant genotype-to-phenotype covariance statistics in virtual populations. Moreover, a collaborative environment is also essential in order to enable communication and collaboration between various PGx players toward the formation, validation, and evidential assessment of such associations.

Moreover, two additional components, such as an electronic health-care genotype and a genotype-to-phenotype translation, and respective services are required to align and harmonize such a platform. The first component is readily compatible with the general EHR, so as to service the management of patients' genotype profiles. The second component aims to service the automated matching of patients' genotype profiles with established and newly discovered gene/variant alleles, assess the respective phenotypes, and deliver up-to-date, relevant clinical annotations, and respective (drug) dosing guidelines.

In this direction, three additional fundamental hurdles need to be overcome, which are (1) ensuring that all the necessary consents are provided by the patients, (2) safeguarding sensitive personal data to avoid the inappropriate leaking of genetic information, which may lead to a person's stigmatization, and (3) enhancing the genetics awareness and genetics education of health-care professionals.[17]

9.3 HUMAN GENOMIC DATABASES

The rapid increase in data production combined with the urgent need to understand how variations in the human genome sequence affect human health has led to the development of many genomic databases. Genomic databases are repositories containing all or a large part of the genomic DNA sequence of one or more organisms and are widely used alongside with NGS (WGS and WES) data. Furthermore, the genomic databases contain additional information in the form of "comments" describing the characteristics and properties of the sequences or biological organisms.[18] They usually have a graphical user interface based on web technologies, which allows all the sequence information to be visible in the form of comments. These environments are known as "genome browsers."

Genomic databases serve as repositories of publicly available genomic variants that refer to either one or more genes or a particular population or race. Furthermore, genomic databases are of great importance not only for diagnosis but also for clinicians and researchers.

Victor McKusick was the innovator in this field in 1966, when he first attempted to bridge the gap between DNA variations and their clinical interpretation. Then, the Mendelian Inheritance in Man (MIM) was issued as a paper compendium of information in genetic disordered and genes.[19] Nowadays, this issue is distributed electronically [Online MIM (OMIM), http://www.omim.org] and updated regularly.[20] Moreover, the first "locus-specific database" collecting genomic variants from a single gene was published in 1976, including 200 mutations from the globin gene in a book at that time. This database was then formatted and then evolved to the HbVar database for hemoglobin variants and thalassemia mutations.[21–23] A decade later, David Cooper started mapping variants in genes to decide which one was the most common.[24] Richard Cotton, so as to perform "mutation analysis" for different genetic domains,[25] initially founded the Human Genome Organization-Mutation Database Initiative in the mid-1990s. This later evolved into the Human

Genome Variation Society (HGVS; http://www.hgvs.org). Nowadays, the main target of HGVS is "...to foster discovery and characterization of genomic variations, including population distribution and phenotypic associations."

This research field is quickly growing with numerous databases of genomic variants being available on the Internet. Nevertheless, not all genomic databases meet quality requirements, whereas others have been created by researchers "on the side" for their use.

9.3.1 DATABASE MANAGEMENT

The term "databases" refers to organized, discrete collections of related, electronically and digitally stored data to the software that manages such collections (Database Management System, or DBMS). In addition to its inherent ability to store data, the database provides the ability to quickly retrieve and refresh data through the design and prioritization of data.

Nowadays, relational databases are the most frequently used ones, since they are very effective than any other database types in dealing with large volumes of information. Relational database is a data collection organized in correlated tables, which simultaneously provides a mechanism for reading, writing, modifying, or for even more complex data processes. The database aims to organize information storage and the possibility of extracting this information, in a more organized form, according to queries placed in the relational database. The data can be reorganized in many different ways, inconceivable tables, without having to reorganize the physical tables that store them. The relational database was invented in 1970 by Codd.[26] Although interest in this model was initially confined to academia, subsequently, relational databases became the dominant type for high-performance applications because of their productivity and friendly user environment. Virtually, all relational database systems use SQL (Structured Query Language) for querying and maintaining the database. By executing queries the user creates, modifies and deletes data on the database, or retrieves information by multiple search criteria depending on his/her rights. There are different SQL types, among which the main ones are Microsoft SQL (http://www.microsoft.com/sql), MySQL (http://www.mysql.com), and postgreSQL (http://www.postgresql.org).

Information on genomic variants was provided in plain text websites in the first genomic databases; in other words, the most straightforward "database" format. The advantages of this model were the development and maintenance simplicity, since no specific software was required. However, there were no valid data-querying options, apart from the standard searching tool provided by the respective Internet browser, while the database was quite complicated to maintain in the case of expanded datasets. Moreover, flat-file databases were the simplest database types, especially for small-scale datasets and simple applications. These databases had modest querying capacity and could accommodate small-to-moderately big datasets, while their development required average computing skills, even though they were based on simple software. The first version of the ETHNOS software[27] was developed by using the flat-file database model.

9.3.2 GENOMIC DATABASE TYPES

The various depositories that fall under the banner of "genomic databases" can be categorized into three main types: general (or central) variation databases (GVDs), locus-specific databases (LSDBs), and national/ethnic genomic databases (NEGDBs; Table 9.1). It is widely accepted that

Table 9.1 Types and Characteristic Examples of Human Genomic Databases (See Also Text for Details)

Database	URL
General (Central) Variation Databases	
ClinVar	http://www.ncbi.nlm.nih.gov/clinvar
HGMD	www.hgmd.org
Varsome	www.varsome.com
Locus-Specific Databases	
LOVD	www.lovd.nl
HbVar	http://globin.bx.psu.edu.hbvar
National/Ethnic Genomic Databases	
Finnish disease database	www.findis.org
FINDbase	www.findbase.org

FINDbase, Frequency of Inherited Disorders Database; *HGMD,* Human Gene Mutation Database.

LSDBs and NEGDBs are increasingly becoming valuable tools in translational medicine. In brief, GMDs, LSDBs, and NEGDBs share the same primary purpose of representing DNA variations that have a definitive or likely phenotypic effect, in other words of clinical interest. They tackle this goal from very different perspectives, and there is a need for these three types of resource in various disciplines of human genetics and genomics, particularly, genetic testing.

Even though there have been noticeable improvements in both qualitative (data uniformity and database quality) and quantitative terms (an increase of the number of LSDBs and NEGDBs), there are certain restrictions in the level of interconnection of these resources to comprehend all that are known and being discovered about pathogenic DNA variants. The leading cause for this deficiency is that modern research ethos fails to produce significant initiatives and to encourage researchers to build and curate new existing databases.

9.3.2.1 General (or central) variation databases

As previously mentioned, GVDs include basic genotype and related phenotypic information of a large number of genomic variants that have been described for every gene locus. The most characteristic examples of GVDs are ClinVar and Human Gene Mutation Database (HGMD) (www.hgmd.org).

ClinVar database (http://www.ncbi.nlm.nih.gov/clinvar) was launched in 2012,[28] and the first dataset included variations from OMIM and GeneReviews databases, as well as affiliated laboratories. Nowadays, ClinVar[29] is a public resource serving as an archive of reports of genotype–phenotype relationships with supporting evidence of clinical utility. ClinVar reports genetic associations stemming from data linking between human genomic variation and observed health status, as well as the history of that interpretation. ClinVar records include various supporting data. More precisely, each ClinVar record represents the submitter, the variation and the phenotype, assessments of their clinical significance, that is, the unit that is assigned an accession of the format SCV000000000.0. The submitter can update the submission at any time, in which case a new version is attached. A unique feature of ClinVar is that it archives previous versions of submissions, meaning that when submitters update their records, the previous version is retained for review. In cases, where there are multiple submissions about the same variant/condition relationship, these are aggregated within ClinVar's data flow and reported as a reference accession of the format RCV000000000.0. As a result, one variant allele may be included in multiple RCV accessions whenever different phenotypes may be reported for that variant allele. It is worth mentioning that the alleles described in submissions are mapped to reference sequences and reported according to the HGVS nomenclature.

ClinVar supports data submission of different levels of complexity. In other words the submission may be as simple as a representation of an allele and its interpretation (sometimes termed a variant-level submission), or as detailed as providing multiple types of structured observational (case-level) or experimental and functional evidence about the effect of the particular genomic variant on phenotype. The purpose is to achieve a high level of confidence in the accuracy of variant calls and assertions of clinical significance. This can only be achieved when the user himself/herself has access to this information; for this reason the supporting evidence is collected and made visible to users. The fairly common problem is that the availability of supporting evidence may vary. For this reason the archive accepts submissions from multiple groups, and aggregates related information, to reflect both consensus and conflicting assertions of clinical significance transparently.

ClinVar currently includes clinical assertions for variants identified through clinical testing, where clinical significance is reported as part of the genetic testing process in CLIA-certified or ISO 1589–accredited laboratories. In research, for variations identified in human samples as part of a research project, as well as extraction from the literature, reporting of the phenotype is extracted directly from the literature without modification of authors' statements.

The information documented in ClinVar is freely available to users and organizations to ensure the broadest utility to the Medical Genetics community. Clinical laboratories can integrate the information available from ClinVar into their workflow by submitting variants and related clinical claims and using the information available to determine the clinical relevance of the already documented variants.

Moreover, attribution is essential to identify the source of variants and assertions, to facilitate communication, and to give due credit to data submitters. Each submitter is explicitly acknowledged to facilitate communication and collaboration within the scientific community.

In the same context, Varsome is a new database (https://varsome.com) aiming to provide useful information and annotation for identified variants, either inserted manually in the database or by directly uploading a .vcf file.

9.3.2.2 Locus-specific databases

In contrast to GVDs, LSDBs focus on one or more specific genes[30,31] usually associated with a disease entity. The purpose of LSDBs is to be labeled highly scrutinized repositories of published and unpublished mutations so as to provide a much-needed complement to GVDs. LSDBs can facilitate molecular diagnosis of inherited diseases in different ways. For instance, LSDBs can assist in gathering whether a DNA variation is indeed causative, bringing about a genetic condition, or benign.

Likewise, some high-quality LSDBs supply detached phenotypic information that is connected to disease-causing genomic variants. Ultimately, LSDBs can aim toward optimizing the "mutation-detection" strategies. An increasing number of LSDBs are generated by often using downloadable, LSDB management system with the issue of tackling the data-content heterogeneity existing in the early 2000s. Mitropoulou et al.[31] proposed an LSDB-based structure, which could potentially contribute to a federated genetic variation browser. In spite of the shortage of specific disease and phenotypic info for each variant, combined with available clinical data information, LSDBs may be allowed to be further developed and be useful for clinical genetics communities.

The fact that several LSDBs are available on the Internet, makes the choice of the "best" LSDB even more difficult. Moreover, there is often more than one LSDB per gene locus, so it is hard to determine which is the "reference" and best-curated LSDB. In 2010 Mitropoulou et al.[31] conducted a detailed domain analysis of the 1188 existing LSDBs in an attempt to thoroughly map data models and ontology alternatives. It is also worth noting that the research field should further evolve in order to offer advice toward the application of LSDBs for use in a clinical and genetic laboratory environment. This attempt came as a follow-up of the comparative analysis of Claustres et al.,[30] dictated by the quick expansion of LSDBs and the extensive data content heterogeneity that distinguishes the field. These LSDBs were evaluated for a total of 44 content criteria concerning an overall presentation, locus-specific information, database structure, data collection, variant information table, and database querying.

The use of an LSDB can provide researchers with valuable information regarding the variant's causative role. In addition, a comprehensive LSDB includes both the reference sequence of the gene as well as a description of structural domains and data about interspecies conservation for each protein residue. Since many recent publications include large datasets, it is often possible to observe errors or use of wrong variant nomenclature, due to reference to an old sequence. The use of LSDBs may be useful for this purpose. Several LSDBs include an automatic nomenclature system, based on a reference sequence.[32] Interestingly, Mutalyzer, which is a dedicated module to automatically produce any sequence variation nomenclature (http://www.lovd.nl/mutalyzer),[33] provides unambiguous and correct sequence variant descriptions to avoid mistakes and uncertainties that may lead to undesired errors in clinical diagnosis. Mutalyzer follows the current HGVS recommendations. In addition, a couple of LSDBs include data-presentation tools to visualize their content in a graphical display. VariVis is a generic visualization toolkit for variation databases that can be employed by LSDBs to generate graphical models of gene sequence with corresponding variations and their consequences.[34] This tool can be integrated into generic DBMSs used for LSDBs development.

Also, since LSDBs are more than simple inert repositories, as they incorporate proofing tools, which utilize computing capacity to solve complex queries, such as phenotypic heterogeneity and genotype/phenotype correlations, the vast majority of them, especially all LOVD-based LSDBs, offer phenotypic interpretations in a conceptual format. Phenotypic descriptions are remarkably more comprehensive in many custom-built databases, such as HbVar. All LSDBs make an effort to impose controlled vocabulary to ease straightforward data querying. An exciting project was attempted to interrelate human phenotype and clinical data in various LSDBs with data on genome sequences, evolutionary history, and function from the ENCODE project and other resources in genome browsers. PhenCode yields a seamless, bidirectional relation between LSDBs and ENCODE data at genome browsers, which allows users to effortlessly explore phenotypes related to functional elements and search for genomic data that could describe clinical phenotypes. Hence, PhenCode is not only useful for clinicians for diagnostics but also for biomedical scientists by integrating numerous types of information and facilitating the creation of testable hypotheses. In this way, our understanding of both the functions of genomic DNA and the mechanisms, by which it attains those functions, is enhanced and a better understanding of difficult clinical phenotypes may be achieved.

Lastly, since LSDBs gather all published and unpublished genomic variants, they can be implemented to determine optimal genetic screening strategy, mainly when targeted resequencing is required. Therefore an outline of the dispersal of variants at the exonic level can assist to focus on specific exons, where most of the variants are located. Soussi and Béroud[35] describe a strategy that is cost-effective but points out that particular attention should be paid, in the case of a negative result and complete scanning should be performed to avoid bias. Similarly, a summary list of variations documented in LSDBs will assist in choosing the best experiment approach. Specific LSDBs might include further information about primers and technical conditions to assist new research groups or diagnostic laboratories to construct their diagnostic strategies, such as those for the several genes involved in muscular dystrophies (http://www.dmd.nl) and thalassemias (http://goldenhelix.org/xprbase).[36] In conclusion, coupling LSDBs with NEGDBs resources (see also next section), if convenient would additionally facilitate these attempts. Nonetheless, utmost care should be taken in the event of a negative outcome that will require absolute mutation scanning to avoid bias.

9.3.2.3 National/ethnic genomic databases

National databases of genomic variants are repositories of information per nation or population and document the genetic defects that lead to various hereditary disorders and their frequencies calculated on the basis of population. NEGDBs help to process the demographic history of population groups but are also prerequisite for optimizing national DNA diagnostics. These resources have lately surfaced and are mainly driven by the necessity to record the varying mutation spectrum discerned for any gene (or multiple genes) linked to a genetic disorder, among a different population and ethnic groups.[37]

In general, NEGDBs can be divided into two subcategories:

- The "National Genetic" (or disease mutation) databases: The first databases that appeared online record the extant genetic background of a population or ethnic group but with limited or no description of mutation frequencies. The first NEGDB to come online was the Finnish database (http://www.findis.org);[38] in spite of being rich in information, it provided limited querying capacity, particularly for allelic frequencies.
- The "National Mutation Frequency" databases: These databases provide comprehensive information only for inherited, mostly monogenic, disorders whose disease-causing variants spectrum is well defined. The Hellenic and Cypriot NEGDBs[39,40] introduced a specialized DBMS for NEGDBs that enabled both basic query formulation and restricted-access data entry, so that all records are manually curated to ensure high and consistent data quality.[41]

The first NEGDBs were released in a very primitive structure back in 2003, mainly focusing on data from specific populations and ethnic groups. Then, ETHNOS software (http://ethnos.findbase.org) was developed aiming to maintain a homogeneous data content and to allow the development and curation of NEGDB. In 2006 FINDbase was created (http://www.findbase.org), which is a global database aspiring to comprehensively document the prevalence of clinically relevant genomic variation allele frequencies in various populations and ethnic groups worldwide. In 2010 FINDbase migrated to the new version of the ETHNOS software, which included new data and querying and visualization tools in order to exploit the expanded FINDbase data collection further.[42] The data querying and visualization tools were built around Microsoft's PivotViewer software (http://www.getpivot.com) and were based on Microsoft Silverlight technology (http://www.silverlight.net). Microsoft Silverlight technology provides an elegant, web-based multimedia interface for population-based genetic variation data collection and retrieval. All FINDbase data records were converted to a set of files on a server with the format of CXML and Deep Zoom-formatted images. The Hellenic National Mutation database was set up within the context of ETHNOS. It is a standard database that aims to provide high-quality information on the genetic heterogeneity of hereditary diseases in the Greek population.

In the same context, FINDbase aims at global and population-based frequency recording of mutations and variants of clinical significance (e.g., PGx markers). FINDbase offers a user-friendly query interface, providing direct access to the lists. The frequencies of different mutations and query results can be exported either in a tabular or in a graph format. In particular, the FINDbase data querying and visualization of environment enable the user to perform simple and complex queries, visualize and sort, organize and categorize data dynamically and discover trends across all items using different views. In 2012 additional visualization tools were implemented[43,44] on the basis of the Flare visualization toolkit (http://flare.prefuse.org), which provided two additional types of data query and visualization outputs: the Gene and Mutation Map and the Mutation Dependency Graph. The Gene and Mutation Map was based on a tree map, which is an easy way of analyzing large datasets (http://www.cs.umd.edu/hcil/treemap). When using the FINDbase, population names were placed along a circle, and these populations were clustered on the basis of the presence and the frequency of a particular genomic variant. A link between populations indicates that these populations have the same genomic variant in common and by clicking on a specific population, the user gets information about all the relevant dependencies of that population concerning the selected genomic variant.

Apart from the documentation of causative genomic variants, which lead to inherited disorders, the PGx biomarker module comprehensively documents the occurrence of PGx biomarkers in different populations.[42] This is a necessary addition, since population- and ethnic group-specific allele frequencies of PGx markers are poorly documented and not systematically collected in structured data repositories.

The efficacy and utility of NEGDBs can be improved via an integration of the associated content in both LSDBs and GVDs,[27] as well as with other means and tools (i.e., VarioML[39]). Moreover, extensive links to supplementary external information (e.g., to OMIM and different types of genome sequence annotation) would preferably be supplied as part of the mandatory tying together of the growing network of genomic databases.

NEGDBs can be beneficial in various ways in translational medicine. First, NEGDBs may be used in optimizing national molecular diagnostic resources by providing crucial reference information for the outlining and application of regional or national-wide mutation screening efforts. Second, NEGDBs can assist clinicians, bioscientists, and the general public by providing information for the most common genetic disorders among certain populations and ethnic groups. More importantly, these databases can also assist in the interpretation of diagnostic test outcomes in countries with heterogeneous populations, especially, where the interpretation of test results in minority ethnic groups might be debatable or problematic.[45]

In addition to their significance in a molecular diagnostic setting, NEGDBs can also be important in identifying populations' origin and migration. History of a defined population is tightly linked with the history of its allele(s). Therefore NEGDBs and especially those containing data from several population groups can be used as the platform for comparative genomic studies that can collectively provide useful information, such as the demographic history of human populations, patterns of their migration and admixture, and gene/mutation flow.

Recently, FINDbase has been refurbished using a new data-querying interface that allows queries to be performed and results to be delivered in a much faster pace than before. Also, the database has been enriched with a new interactive map interface that indicates, using a color code, the populations for which allelic frequency data are available. This upgrade, along with the addition of new data on population-specific PGx biomarkers allele frequencies, will further enrich this useful resource.

9.3.2.4 *Other genomic database types*

Apart from the aforementioned main database types, DNA variations are also recorded in various databases, such as those provided from the National Center of Biotechnology Information, namely, dbSNP and dbGAP (http://www.ncbi.nlm.nih.gov/projects/SNP)[46] and the HapMap Data-Coordination Center (http://www.hapmap.org).[47] These resources are useful in completing the picture for any gene or region of interest, by summarizing all the variants that are typically not included in GVDs, LSDBs, and NEGDBs.

9.4 DISCUSSION

In the era of the post–genomic revolution, there is a rise of massively parallel NGS technologies, which has consequently led to the correlation of specific genomic variants with disease predisposition and other clinical features, such as response to commonly prescribed drugs. As genomic medicine and personalized drug treatment are slowly gaining ground in clinical practice, the use of WGS may be the most useful approach, since it covers all possible genetic alterations for different ethnic groups.[48]

The design and development of advanced informatics solutions that fill in the gap between PGx research findings and clinical practice remains a significant challenge. As soon as the operational requirements and design specifications of a PGx electronic assistant are first addressed, it can then be developed and act as a means of communication between the various PGx communities. The goal of the PGx assistant should be to provide effective solutions to address all the challenges associated with the PGx annotation of whole genomes.[49]

It is widely accepted that LSDBs and NEGDBs are slowly becoming valuable tools in translational medicine. However, although they are constantly improving, limitations on the degree of interconnection of these resources continue to exist. The biomedical community must first recognize the overwhelming demand for further improvement of the genetic/mutation databases, and then, perhaps, the satisfactory solution will follow. Owing to the development of new web and

Internet technologies, the concept of genome browser is becoming a collaboration platform for researchers. This allows exchange of data and knowledge by promoting collaboration between scientists in different locations.

ACKNOWLEDGMENT

This work was partly funded by a Greek General Secretariat of Research and Technology grant [eMoDiA (ΣYN11_0415)] and European Commission grants [GEN2PHEN (FP7-200754), RD-Connect (FP7-305444)] to GPP.

REFERENCES

1. Lippert C, Listgarten J, Davidson RI, et al. An exhaustive epistatic SNP association analysis on expanded Wellcome Trust data. *Sci Rep*. 2013;3:1099. Available from: https://doi.org/10.1038/srep01099.
2. Dunnenberger HM, Crews KR, Hoffman JM, et al. Preemptive clinical pharmacogenetics implementation: current programs in five US medical centers. *Annu Rev Pharmacol Toxicol*. 2015;55:89–106. Available from: https://doi.org/10.1146/annurev-pharmtox-010814-124835.
3. Mai Y, Mitropoulou C, Papadopoulou XE, et al. Critical appraisal of the views of healthcare professionals with respect to pharmacogenomics and personalized medicine in Greece. *Per Med*. 2014;11:15–26. Available from: https://doi.org/10.2217/pme.13.92.
4. Potamias G, Lakiotaki K, Katsila T, et al. Deciphering next-generation pharmacogenomics: an information technology perspective. *Open Biol*. 2014;4:140071. Available from: https://doi.org/10.1098/rsob.140071.
5. Squassina A, Manchia M, Manolopoulos VG, et al. Realities and expectations of pharmacogenomics and personalized medicine: impact of translating genetic knowledge into clinical practice. *Pharmacogenomics*. 2010;11:1149–1167. Available from: https://doi.org/10.2217/pgs.10.97.
6. Wang X, Liotta L. Clinical bioinformatics: a new emerging science. *J Clin Bioinform*. 2011;1:1. Available from: https://doi.org/10.1186/2043-9113-1-1.
7. Butte AJ. Translational bioinformatics: coming of age. *J Am Med Inform Assoc*. 2008;15:709–714. Available from: https://doi.org/10.1197/jamia.M2824.
8. Shoenbill K, Fost N, Tachinardi U, Mendonca EA. Genetic data and electronic health records: a discussion of ethical, logistical and technological considerations. *J Am Med Inform Assoc*. 2014;21:171–180. Available from: https://doi.org/10.1136/amiajnl-2013-001694.
9. Vayena E, Prainsack B. Regulating genomics: time for a broader vision. *Sci Transl Med*. 2013;5:198ed12. Available from: https://doi.org/10.1126/scitranslmed.3005797.
10. Denny JC. Chapter 13: Mining electronic health records in the genomics era. *PLoS Comput Biol*. 2012;8: e1002823. Available from: https://doi.org/10.1371/journal.pcbi.1002823.
11. Schuster SC. Next-generation sequencing transforms today's biology. *Nat Methods*. 2008;5:16–18. Available from: https://doi.org/10.1038/nmeth1156.
12. Katsila T, Patrinos GP. Whole genome sequencing in pharmacogenomics. *Front Pharmacol*. 2015;6:61. Available from: https://doi.org/10.3389/fphar.2015.00061.
13. Mizzi C, Peters B, Mitropoulou C, et al. Personalized pharmacogenomics profiling using whole-genome sequencing. *Pharmacogenomics*. 2014;15:1223–1234. Available from: https://doi.org/10.2217/pgs.14.102.

14. Gurwitz D, McLeod HL. Genome-wide studies in pharmacogenomics: harnessing the power of extreme phenotypes. *Pharmacogenomics*. 2013;14:337–339. Available from: https://doi.org/10.2217/pgs.13.35.
15. Pareek CS, Smoczynski R, Tretyn A. Sequencing technologies and genome sequencing. *J Appl Genet*. 2011;52:413–435. Available from: https://doi.org/10.1007/s13353-011-0057-x.
16. Georgitsi M, Patrinos GP. Genetic databases in pharmacogenomics: the Frequency of Inherited Disorders Database (FINDbase). *Methods Mol Biol (Clifton, NJ)*. 2013;1015:321–336. Available from: https://doi.org/10.1007/978-1-62703-435-7_21.
17. Kampourakis K, Vayena E, Mitropoulou C, et al. Key challenges for next-generation pharmacogenomics: science & society series on science and drugs. *EMBO Rep*. 2014;15:472–476. Available from: https://doi.org/10.1002/embr.201438641.
18. Schattner P. *Genomes, Browsers and Databases*. Cambridge: Cambridge University Press; 2008. <https://doi.org/10.1017/CBO9780511754838>.
19. McKusick VA. Mendelian inheritance in man and its online version, OMIM. *Am J Hum Genet*. 2007;80:588. Available from: https://doi.org/10.1086/514346.
20. Amberger JS, Bocchini CA, Schiettecatte F, Scott AF, Hamosh A. OMIM.org: Online Mendelian Inheritance in Man (OMIM®), an online catalog of human genes and genetic disorders. *Nucleic Acids Res*. 2015;43:D789–D798. Available from: https://doi.org/10.1093/nar/gku1205.
21. Giardine B, van Baal S, Kaimakis P, et al. HbVar database of human hemoglobin variants and thalassemia mutations: 2007 update. *Hum Mutat*. 2007;28:206. Available from: https://doi.org/10.1002/humu.9479.
22. Hardison RC, Chui DHK, Giardine B, et al. HbVar: a relational database of human hemoglobin variants and thalassemia mutations at the globin gene server. *Hum Mutat*. 2002;19:225–233. Available from: https://doi.org/10.1002/humu.10044.
23. Patrinos GP, Wajcman H. Recording human globin gene variation. *Hemoglobin*. 2004;28:v–vii.
24. Cooper DN, Ball EV, Krawczak M. The human gene mutation database. *Nucleic Acids Res*. 1998;26:285–287. Available from: https://doi.org/10.1093/nar/26.1.285.
25. Cotton RG, McKusick V, Scriver CR. The HUGO mutation database initiative. *Science*. 1998;279:10–11.
26. Codd EF. A relational model of data for large shared data banks. *Commun ACM*. 1970;13:377–387. Available from: https://doi.org/10.1145/362384.362685.
27. Patrinos G, Brookes A. DNA, diseases and databases: disastrously deficient. *Trends Genet*. 2005;21:333–338. Available from: https://doi.org/10.1016/j.tig.2005.04.004.
28. Landrum MJ, Lee JM, Riley GR, et al. ClinVar: public archive of relationships among sequence variation and human phenotype. *Nucleic Acids Res*. 2014;42:D980–D985. Available from: https://doi.org/10.1093/nar/gkt1113.
29. Landrum MJ, Lee JM, Benson M, et al. ClinVar: public archive of interpretations of clinically relevant variants. *Nucleic Acids Res*. 2016;44:D862–D868. Available from: https://doi.org/10.1093/nar/gkv1222.
30. Claustres M, Horaitis O, Vanevski M, Cotton RGH. Time for a unified system of mutation description and reporting: a review of locus-specific mutation databases. *Genome Res*. 2002;12:680–688. Available from: https://doi.org/10.1101/gr.217702.
31. Mitropoulou C, Webb AJ, Mitropoulos K, Brookes AJ, Patrinos GP. Locus-specific database domain and data content analysis: evolution and content maturation toward clinical use. *Hum Mutat*. 2010;31:1109–1116. Available from: https://doi.org/10.1002/humu.21332.
32. den Dunnen JT, Antonarakis SE. Nomenclature for the description of human sequence variations. *Hum Genet*. 2001;109:121–124.
33. Wildeman M, van Ophuizen E, den Dunnen JT, Taschner PEM. Improving sequence variant descriptions in mutation databases and literature using the Mutalyzer sequence variation nomenclature checker. *Hum Mutat*. 2008;29:6–13. Available from: https://doi.org/10.1002/humu.20654.

34. Smith TD, Cotton RG. VariVis: a visualisation toolkit for variation databases. *BMC Bioinform.* 2008;9:206. Available from: https://doi.org/10.1186/1471-2105-9-206.
35. Soussi T, Béroud C. Assessing TP53 status in human tumours to evaluate clinical outcome. *Nat Rev Cancer.* 2001;1:233–239. Available from: https://doi.org/10.1038/35106009.
36. Giardine B, Riemer C, Hefferon T, et al. PhenCode: connecting ENCODE data with mutations and phenotype. *Hum Mutat.* 2007;28:554–562. Available from: https://doi.org/10.1002/humu.20484.
37. Patrinos GP. National and ethnic mutation databases: recording populations' genography. *Hum Mutat.* 2006;27:879–887. Available from: https://doi.org/10.1002/humu.20376.
38. Sipilä K, Aula P. Database for the mutations of the Finnish disease heritage. *Hum Mutat.* 2002;19:16–22. Available from: https://doi.org/10.1002/humu.10019.
39. Kleanthous M, Patsalis PC, Drousiotou A, et al. The Cypriot and Iranian National Mutation Frequency Databases. *Hum Mutat.* 2006;27:598–599. Available from: https://doi.org/10.1002/humu.9422.
40. Patrinos GP, van Baal S, Petersen MB, Papadakis MN. Hellenic National Mutation Database: a prototype database for mutations leading to inherited disorders in the Hellenic population. *Hum Mutat.* 2005;25:327–333. Available from: https://doi.org/10.1002/humu.20157.
41. van Baal S, Zlotogora J, Lagoumintzis G, et al. ETHNOS: a versatile electronic tool for the development and curation of national genetic databases. *Hum Genomics.* 2010;4:361–368.
42. Georgitsi M, Viennas E, Antoniou DI, et al. FINDbase: a worldwide database for genetic variation allele frequencies updated. *Nucleic Acids Res.* 2011;39:D926–D932. Available from: https://doi.org/10.1093/nar/gkq1236.
43. Papadopoulos P, Viennas E, Gkantouna V, et al. Developments in FINDbase worldwide database for clinically relevant genomic variation allele frequencies. *Nucleic Acids Res.* 2014;42:D1020–D1026. Available from: https://doi.org/10.1093/nar/gkt1125.
44. Viennas E, Gkantouna V, Ioannou M, et al. Population-ethnic group specific genome variation allele frequency data: a querying and visualization journey. *Genomics.* 2012;100:93–101. Available from: https://doi.org/10.1016/j.ygeno.2012.05.009.
45. Zlotogora J, van Baal S, Patrinos GP. Documentation of inherited disorders and mutation frequencies in the different religious communities in Israel in the Israeli National Genetic Database. *Hum Mutat.* 2007;28:944–949. Available from: https://doi.org/10.1002/humu.20551.
46. NCBI Resource Coordinators. Database resources of the National Center for Biotechnology Information. *Nucleic Acids Res.* 2016;44:D7–D19. Available from: https://doi.org/10.1093/nar/gkv1290.
47. The International HapMap Project. *Nature.* 2003;426:789–796. Available from: https://doi.org/10.1038/nature02168.
48. Drmanac R. The ultimate genetic test. *Science.* 2012;336:1110–1112. Available from: https://doi.org/10.1126/science.1221037.
49. Altman RB, Whirl-Carrillo M, Klein TE. Challenges in the pharmacogenomic annotation of whole genomes. *Clin Pharmacol Ther.* 2013;94:211–213. Available from: https://doi.org/10.1038/clpt.2013.111.

CHAPTER

GENETIC TESTING

10

Kariofyllis Karamperis[1], Sam Wadge[1], Maria Koromina[2] and George P. Patrinos[2,3,4]

[1]*Department of Twin Research and Genetic Epidemiology, King's College London, London, United Kingdom* [2]*Department of Pharmacy, School of Health Sciences, University of Patras, Patras, Greece* [3]*Department of Pathology, College of Medicine and Health Sciences, United Arab Emirates University, Al Ain, United Arab Emirates* [4]*Zayed Center of Health Sciences, United Arab Emirates University, Al Ain, United Arab Emirates*

10.1 INTRODUCTION

From the discovery of the deoxyribonucleic acid (DNA) double helix structure by Watson and Crick in 1953 to the present day, our knowledge of human genetics is based on, at most, 2%–3% of the entire sequence of the genome, and thus on about 2%–3% of all human genes. An innovative project known as the Human Genome Project (HGP) was designed to complete the entire sequence before 2005, thus expanding our knowledge of human genes from 2% to 3% up to 100% within the next 7 years.[1,2]

In recent years, decoding of the human genome has empowered scientists to provide new ideas and strategies with the aim of finding new potential treatments. The term "genome" is defined as the entire genetic material found in a cell or carried within an individual. According to the HGP, humans have between 20,000 and 25,000 genes [National Human Genome Research Institute (NHGRI)]. In biology, a gene is a nucleotide sequence in DNA or RNA encoding a functional molecule, such as a protein. As a result of multiple factors and conditions, a gene can acquire variants in their sequence, leading to different variants. If the variant occurs inside a gene, even an error of one nucleotide would cause an erroneous reading of each subsequent codon in the message, resulting in a nonfunctional protein with an altered amino acid sequence. In that case, depending on the significance and the rate of protein function can lead to a genetic disease.[3]

It is now becoming obvious that the ability to analyze the genome provides us with sufficient knowledge to identify these variants that will help us understand better the pathogenesis and characterization of diseases. In addition, genome analysis provides a human genome mapping that will allow us to detect/predict possible variants that cause genetic diseases, as well as design new drugs and therapies by using molecular approaches, underlining the importance of the HGP and defining a new era in medicine, new concepts, and fields such as personalized medicine and genetic testing.[4]

Applied Genomics and Public Health. DOI: https://doi.org/10.1016/B978-0-12-813695-9.00010-8

10.2 THE HISTORICAL CONTEXT OF HUMAN GENOME MAPPING—THE HUMAN GENOME PROJECT

The NHGRI was created to launch this research project, known as the Human Genome Project, in short HGP.[4] The HGP is defined as the international effort by distinguished scientists, such as James Watson and Francis Collins, to determine the DNA sequence of the entire human genome, by providing us with a comprehensive map of all genes, from both physical and functional standpoints (NHGRI).

Eventually, in 1990, human genome mapping formally launched and at the time represented the largest biological experiment to have been conducted, with an initial implementation deadline of 15 years. Within a short period of time, the program became international with the participation of many researchers (more than 2000) from many countries, such as England, France, Germany, Italy, Canada, and Japan.[4]

However, this idea was also approached by various competitors because of its prequalification. In 1998 a parallel project by Celera Genomics (business unit of Applera, US biotech firm) was formally launched and, in relation to US National Center for Human Genome Research, was intended to be implemented within 3 years.[4]

In 2001 both research teams finally announced the end of the experimental process simultaneously, since by then the reading and decoding of this "rough draft" was completed. Celera Genomics did not produce a final version of the project compared to the US NHGRI. In 2003 researchers successfully proof read the sequence and the Human Research Project was published including mapping of 3000,000,000 DNA base pairs. More precisely, the finished sequence covers about 99% of the total genome including gene-containing regions to an accuracy of 99.9%, less than 1 error for every 100,000 base pairs (National Human Genome Research Institute, 2012).

Sequence mapping of the human genome played a decisive role as it sets new horizons and directions compared to the existing technology. In the early years, it was mainly implemented in the field of research and biotechnology (sequencing and decoding of other species) but in a short period of time, due to its high potential, it was also introduced in medicine by improving diagnostic and therapeutic methods. Since 2000, the NHGRI has been involved in launching more than 25 important research projects presenting new challenges to biomedical research, focused on different population groups and societal issues by sharing and analyzing vast data sets; this project is known as ethical, legal, and social implications (ELSI).[4] In recent years, a similar action such as ELSI or HGP has been launched, a project known as the 1000 Genomes Project providing genomic information of 5 major population groups and 26 specific population groups.

Furthermore, this large-scale project laid the foundations for the creation of new branches of medicine based on the genome of each person (personalized medicine) and drug treatment (pharmacogenomics) leading to a dynamic new era of medicine with several benefits, which are summarized as follows:

- Detection and study of genes for monogenic and polygenic subsets.
- Prevention of genetic diseases or their diagnosis at preclinical level.
- Detection of SNPs to determine whether a genetic variant is associated with a disease or trait.
- Implementation of new therapeutic methods (pharmacogenomics, gene therapy).
- Study of proteins and their interaction at the cellular level (proteomics).
- Genome analysis of other living species.

10.3 GENETIC TESTING

In recent years, mapping of the human genome has offered significant advances in the understanding of the genetic basis of inherited disorders and the correlations between mutant genotypes and clinical phenotypes, both for monogenic and polygenic subsets. Over the years, multiple techniques have been developed to provide health-care services in clinical diagnosis, such as molecular genetics and cytogenetic testing. The use of these methods to date is valuable; however, technology is being evolved with new upcoming innovative methods. For example, high-throughput genetic analysis is based on a new concept thus providing high susceptibility and efficacy with a rapid development in the industrial field and also producing different types of "genetic testing."[5,6]

Genetic testing is defined as "the analysis of human DNA, RNA, chromosomes, proteins and certain metabolites to identify or detect changes in genotype, phenotype or karyotype." More precisely, it is a specific medical test that can be used to diagnose a disease, predict future disease, predict risk or susceptibility to disease, direct clinical management, identify carriers of genomic variants, and establish prenatal or clinical diagnosis or prognosis in individuals, families, groups, or populations. Genetic tests can be used for both research (research testing) and/or clinical purposes (clinical genetic testing). Nowadays, there are more than 1000 genetic tests in use, and more are being developed[7,8] [(US National Library of Medicine, National Institutes of Health (NIH)].

The various types of genetic testing are as follows:

1. Molecular genetic tests, or gene tests, are used to study single genes or short fragments of DNA to identify variations that lead to a genetic disorder.
2. Chromosomal genetic tests analyze entire chromosomes or large fragments of DNA (karyotyping) capable of detecting chromosomal differences.
3. Biochemical genetic tests study the quantity or the activity level of enzymes or proteins that may be unusual; abnormalities in either can indicate on DNA that may result in a genetic disorder.

10.4 GENETIC TESTING IN CLINICAL DIAGNOSIS

Genetic testing is an important tool in the screening and diagnosis of many disease conditions. The HGP and advances in molecular genetics have helped in the identification of a couple of genetic disorders for which DNA testing is available from about 10 to over 1000. Technology continues to evolve and the new major step in the medical action was carried out from reliance to next-generation sequencing (NGS) technology, a versatile tool that will be further analyzed later with the possibility of detecting even a single variant of a nucleotide.[7,8] The current chapter is focused on the usage of genetic testing in clinical diagnosis and medicine.

10.4.1 GENETIC TESTING SERVICES

In the postgenomic era, adding new technologies in laboratories or health service providers, such as genetic testing, has boosted health services. First, an important prerequisite for their implementation is the selection of the appropriate and proficient staff including primary-care physicians, medical geneticists, pathologists, genetic counselors, and genetic nurses.[6,9] Second, it is necessary for the person concerned to provide the appropriate instructions, such as providing support and genetic counseling to patients (familiarization with the purpose of genetic tests such as limitations and expectations) and assistance with the interpretation and management of clinical treatment.[10]

The use of genetic tests should be considered as a medical examination, which provides relevant information. There is no substantial difference in laboratory services compared to routine tests regarding the management of biological samples. Their main difference concerns the management and provision of medical information. It is worth noting that any laboratory that performs these kinds of tests must comply with the federal regulatory standards called the Clinical Laboratory Improvement Amendments.[11,12]

First, the individual must be informed about the testing procedure, the benefits, the limitations, and the possible consequences of the test results; this process of educating is called "informed consent." Before performing the genetic test, it is important to choose the appropriate parameters regarding the validity (if this test provides an accurate result) and the usefulness of the test. The characteristics that determine the accuracy of the test are (1) analytical validity, (2) clinical validity, and (3) clinical utility. More precisely, analytical validity refers to how well a test performs in the laboratory, clinical validity refers to the accuracy with which a test predicts the presence or absence of a clinical condition or predisposition, and clinical utility refers to the usefulness of the test and the value of information to the person being tested. In addition, the Food and Drug Administration (FDA) requires information about clinical validity and success rates for some genetic tests.[7,10–12] In some clinical cases the use of genetic testing is undoubtedly recommended owing to the high success rates and precision in the detection of variant of interest. In contrast, the rate of success may vary in some cases depending on what part of the genome is being analyzed (single genes, multiple genes, exomes, or an entire genome), and the techniques used such as microarrays or sequencing for instance.[12] For all laboratories that perform health-related testing it is necessary to provide also this information.[11]

Genetic tests and services have expanded the diagnostic and predictive capabilities of clinical laboratories, whilst replacing in some cases the other existing diagnostic test because of their high performance. However, genetic tests and services face many challenges as they become integrated into the health-care system, such as ensuring sufficient evidence and strong reliability. Careful data management is consistent with the ethical, legal, psychosocial, and financial implications of genetic testing.[9]

10.4.2 COST AND ETHICAL ISSUES OF GENETIC TESTING

The rapid development of genetic tests linked with the high competition in the market has established a low-cost rate, based on the dogma of marketing (see also Chapter 19). Overall, the cost ranges from under $100 to more than $2000, depending on the test. Costs are higher in instances where a family member must be tested to obtain a meaningful result. Overall, according to recent studies, the highest cost of genetic testing concerns prenatal testing.[10,12]

Regarding the sequencing technologies, the total cost is still relatively high, approximately US $1000 for a whole-genome sequencing (WGS) test. The abundant information provided makes it difficult to further reduce costs. It should be noted that a few years ago, the use of this technology was not even accessible to the provision of health services because of the extremely high cost.

As mentioned previously, there has been a lot of debating about the ethical issues that may arise with genetic testing. Genetic tests can provide a sense of relief from uncertainty for individuals, to manage their health care knowing the required genetic information. In contrast, a positive result aims to properly guide a person by providing options, such as prevention, monitoring, and treatment.[10] Ethical issues concern different social groups (infants, minors, relatives) and populations. Briefly, some of the ethical issues must be taken into account, such as the severity (morbidity and mortality) of a disease, the actionability, health risks, phycological risks (anxiety), social risks (like financial costs and discrimination), the age of onset, and the psychiatric/somatic distinction.[13,14]

Nowadays, platforms have been developed to take that into account, in addition to the characteristics of a genetic test (analytical validity, clinical validity, and clinical utility) and moral, legal, and social implications. One of the most used platforms is the ACCE framework.[14] Existing and developing technologies offer tremendous benefits to the society and public health but deciphering human existence, even interfering with and modifying a person's genetic information, will always be a major moral issue and an effort to resolve. The results of a genetic test affect greatly a person's life both presently and in the future, since predicting a genetic disease could bring psychological and social destabilization to an individual.[10,11,14]

10.5 CLASSIFICATION OF GENETIC TESTING METHODS

Thousands of genetic tests are now available, and others are expected to enter into the market soon. As a general rule, genetic tests are classified based on the implemented technology and their serving purpose (i.e., disease diagnosis). Regarding the technology classification, genetic tests are classified into cytogenetic testing, biochemical testing, and molecular testing, whilst regarding the serving purpose, they can be classified as diagnostic testing, prenatal testing, predictive testing, preimplantation testing, pharmacogenomic testing, carrier testing, and newborn screening. Depending on the technology, the majority of genetic tests that are widely used in both the private and the public sectors are based on molecular genetic testing technology.

10.5.1 DIAGNOSTIC TESTING

Diagnostic testing is used to confirm or rule out the possibility of a known or suspected genetic disorder in a symptomatic person. Methods, such as creatine kinase measurement and muscle biopsy, are inconclusive when detecting myotonic dystrophy (caused by variant on chromosome 19), compared with DNA testing which is less expensive, more accurate, and offers less risk to the patient[8,15,16] (US National Library of Medicine).

10.5.2 PREDICTIVE TESTING

Predictive testing is used to clarify the genetic status and history (genealogy) of an asymptomatic family member at risk for a genetic disorder. A basic condition for the use of predictive testing is the identification of specific disease mutation(s) in an affected family member. Predictive testing can be divided into two different types *presymptomatic* and *predispositional*. Huntington's disease is a typical example of presymptomatic testing. If there is a known genetic condition in individual's family, and the faulty gene that causes that condition is known, then predictive test is a necessary option to confirm the heredity of this mutant gene. Furthermore, it is used to provide genetic information if (1) the condition can be prevented or its symptoms effectively treated or (2) if the condition can be neither prevented, nor its symptoms effectively treated. In instances where a disease has no cure, a positive result can be used for life planning, reproductive planning, and treatment[15] (Department of Health and Human Services-USA, 2006).

Genetic testing of the *BRCA1* gene for breast and ovarian cancer susceptibility is the most common example of predispositional testing; the findings of a particular genomic variant predispose the occurrence of the breast cancer. It is worth noting that predispositional testing does not indicate a 100% risk of developing the condition. It differs essentially from presymptomatic testing as it informs individuals about the risk developing the condition; however, the degree of certainty is unknown[8] (US National Library of Medicine).

A proper discussion between physician and patient is highly recommended before using predictive testing, since early diagnosis may result in medical interventions because of the reduction of morbidity or mortality. In addition, patients need to provide their full consent about the risks and the benefits of knowing the results of testing[15,17–19] (US National Library of Medicine Department of Health and Human Services-USA). Till now, there are many ethical and practice standards that need to be developed. An extensive reference will be made later on the ethical issues.

10.5.3 CARRIER TESTING

Carrier testing is used to identify individuals carrying a genomic variant for a disorder inherited in an autosomal recessive or X-linked recessive manner. In this case the carrier does not have usually any symptoms and the carrier is informed about possible reproduction problems. Carrier testing aims to help couples make reproductive decisions. Despite the differences from predictive testing, pretest and posttest counseling are equally important given the social and personal implications that may occur following test results. Depending on the genetic condition, other additional options are also highly recommended to be discussed between physician and carrier, such as pregnancy without prenatal testing and preimplantation diagnosis[15,18] (US National Library of Medicine).

10.5.4 PRENATAL TESTING

The use of prenatal testing is to identify the genetic status of a pregnancy to detect any possible risks for a genetic disorder. Some of the risk factors are advanced maternal age, family history, and ethnicity. There are several prenatal options including maternal serum screening, amniocentesis, and chorionic villus sampling, which are widely invasive. Furthermore, placental biopsy and periumbilical blood sampling are more highly specialized techniques. However, prenatal testing may pose risks for the fetus; genetic counseling and the importance of informed consent are key to this process.[15,18]

10.5.5 PREIMPLANTATION TESTING

Preimplantation testing or preimplantation genetic diagnosis (PGD) is the genetic profiling of embryos and is used to select early embryos for implantation that have been conceived by in vitro fertilization (IVF). It has been observed amongst women of advanced maternal age undergoing IVF that embryos can exhibit numerical chromosomal abnormalities. With preimplantation testing, the embryos that were identified as abnormal are discarded and embryos with normal genetic constitutions are selected for implantation.[20] Preimplantation genetic testing was primarily developed as an alternative form of prenatal diagnosis which is carried out on embryos. Originally used for diagnosis in partners who are at high risk of single-gene genetic disorders, such as cystic fibrosis and Huntington's disease; preimplantation genetic testing has most frequently been employed in assisted reproduction. In recent years, major improvements have been seen in PGD analysis with the addition of new molecular tools, such as DNA microarrays and NGS, and replacing older methods, such as the fluorescence in situ hybridization (FISH) analysis.[18,21]

PGD analysis is performed at specialized centers and targeted to a small number of disorders due to the high cost and lack of health insurance.[21,22]

10.5.6 PHARMACOGENOMIC TESTING

This type of testing is mainly used to individualize patient care by applying genotype results for selected drug-metabolizing enzymes, transporters, and/or targets to inform a person about their response to a particular drug and/or the possibility of having any adverse drug reactions.[6,23] For over a decade, this form of testing has been rapidly evolved and it is expected to be established in the clinical practice due to its significant benefits. Up to this day, there are approximately 200 medications with pharmacogenomic information in the FDA-approved labeling and more are being developed.[24] Pharmacogenomic tests seem to contribute greatly to areas of medicine that lack such services, psychiatry being an example. An interesting survey has shown that the majority of psychiatrists (over 90%) believe that pharmacogenomic testing will become a standard of care for medication treatment in psychiatry in the future (see also Chapter 6). For example, a pharmacogenomic test is a useful clinical tool helping mental health professionals prescribe psychiatric medications. Cytochrome 450 testing determines an individual's required dosage of identified drugs that are metabolized by the CYP2C19 (generic name: Rabeprazole) and CYP2D6 (generic name: Aripiprazole) enzymes[24–27] (Department of Health and Human Services-USA, 2006).

How is the method behind pharmacogenomics testing any different from the aforementioned? The usage may be for a different purpose. Shall we introduce a sentence somewhere about it?

10.5.7 NEWBORN SCREENING

Newborn screening involves a series of tests of a newborn to control certain disorders and conditions that can prevent their normal development, thus avoiding significant morbidity and mortality. Ideally, screening should be conducted in the first week of a baby's life to ensure treatment initiation before the age of 4 weeks. From simple blood to urine screening tests, new technologies have now empowered and led to the development of a more comprehensive and complex screening system by taking only a simple heel prick, allowing this way detection of over 50 different conditions that are treatable, but not clinically evident in the immediate postnatal period. Some of the diseases that can be identified include hereditary diseases (hemoglobinopathies, for instance), immunodeficiencies, congenital heart defects, inborn errors of organic acid metabolism, hearing loss, etc., which are likely to be genetic in origin. Screening tests are often performed on every infant regardless of the parents' health insurance or economic status.[15] Cost of these tests depends on the screening and treatment, and of course the range varies in each country, although in most cases, newborn screening is carried out by laboratories that are government funded (US National Library of Medicine).

Nowadays, a new approach has been demonstrated, known as newborn genomic sequencing, which is based on the technology of NGS. The purpose of this method is to collect and analyze large amounts of DNA sequence data in the newborn period. Up to this day, it is not meant to replace standard newborn screening, as the ethical, social, and legal issues need to be considered before the technology is widely adopted. However, it is a promising method that is expected to offer multiple benefits in the future and some organizations, such as NIH (National Institutes of Health), which has already sponsored several studies to gain further information and knowledge.[28]

10.6 TYPES OF DIAGNOSTIC GENETIC TESTING

In this section, we describe how standard genetic methods can serve as tools used in clinical diagnosis, bearing in mind that these methods can be also used for research purposes as well.

10.6.1 CYTOGENETICS

Cytogenetics is the study of chromosomes present in cell and tissues, a branch of genetics that investigates how chromosomes relate to cell behavior (particularly during mitosis and meiosis phases). It is widely used in the field of biology and medicine to understand better genetic disorders, such as cancer, sickle cell anemia, cystic fibrosis, and Down's syndrome, and generally cytogenetic approaches to studying chromosomes and their relationship to human disease. Recently, a new cytogenetic branch, molecular cytogenetics, has been incorporated, which contributes significantly to new approaches to medicine, such as personal medicine, targeted cancer treatment, and pharmacogenomics. The techniques that are mainly used in cytogenetics are karyotyping, FISH, and array-based comparative genomic hybridization (aCGH) as well as molecular cytogenetics.[29]

10.6.1.1 Karyotyping

This method is used to detect any numerical and structural chromosomal abnormalities in newborns (such as the loss or gain of upward of 5 Mb of genetic material) with the aim of predicting the occurrence of a genetic disease that can cause deficits in growth and development. There are different banding methods, with G-banding being the most widely used method for cytogenetic analysis. This technique has been used for many years but over time, new techniques have emerged to give more efficient results and accuracy, whilst reducing the time from diagnosis to treatment. Up to this day, it is used in clinical practice; however, it is recommended to be combined with techniques such as chromosomal microarrays (CMAs) in order to identify additional clinically significant cytogenetic information, such as aneuploidies and unbalanced rearrangements.[30,31] Moreover, in prenatal diagnosis, several studies comparing different chromosomal alteration analysis techniques in high-risk patients report that the diagnostic frequencies are between 2.5% and 4.2% with karyotyping, a low percentage compared to methods such as comparative genomic hybridization.[30]

10.6.1.2 Fluorescence in situ hybridization

FISH is a macromolecule recognition technology based on the complementary nature of DNA or DNA/RNA double strands. In 1953 James Watson and Francis Crick described the extensive network of hydrogen bonds that hold together the two antiparallel strands in the double helix.[32] If the hydrogen bonds that hold the helix together are ruptured (by heat or chemical), the helix can reform, if conditions permit. This ability of the DNA helix to reform, or renature, provides the basis for molecular hybridization.[33,34]

In the early 1980s the technique has been developed to detect and localize the presence or absence of specific RNA sequences of interest, as well as to localize the target sequences to specific cells or chromosome sites.[35] More precisely, selected DNA strands incorporated with fluorophore-coupled nucleotides can be used as probes to hybridize onto the complementary sequences in tested cells and tissues and then visualized via a fluorescence microscope or an imaging system.[33] In the beginning, this technology was meant to be used as a physical mapping tool to delineate genes and polymorphic loci within chromosomes (specifically onto metaphase chromosomes) in order to construct a physical map of the human genome. Owing to its high analytical resolution to a single gene level and high specificity and susceptibility, it is also used for genetic diagnosis of constitutional common aneuploidies, such as microdeletion/microduplication syndromes, and subtelomeric rearrangements.[34]

FISH technology enabled the detection of an increased spectrum of genetic disorders from chromosomal abnormalities to submicroscopic copy number variants (CNVs) and extended the cell-based analysis from metaphases to interphases.[33] The diagnostic applications of FISH technology are chromosomal gene mapping, characterizing genetic abnormalities, identifying genetic abnormalities related to genetic disease or neoplasmic disorders, and detecting viral genomes.[33,34,36]

Briefly, FISH technology demonstrates three major advantages; high sensitivity and specificity in recognizing DNA or RNA sequences of interest, direct application to both metaphase chromosomes and interphase nuclei as mentioned, and visualization of hybridization signals at the single-cell level. However, there are some limitations. FISH can only detect known genetic aberrations (only in metaphase and interphase), using a specific probe which must be hybridized to the specimen in order to indicate the presence or absence of that specific genetic aberration alone. In addition to this, analysis with locus-specific probes or chromosome-specific DNA libraries is restricted to the targeted chromosome or chromosomal subregion which means FISH cannot be used as a screening test for chromosomal rearrangements, since most of the FISH techniques are able to detect only known imbalances.[33]

Despite the aforementioned limitations, FISH is considered as an essential tool in cytogenetics and it has become a powerful tool in molecular cytogenetics and an important method for genetic testing as a diagnostic and discovery tool in the fight against cancer.[34] FISH is constantly evolving and it is used in combination with other techniques, such as RNA FISH (smRNA-FISH), CASH FISH, multiplex FISH, and comparative genomic hybridization (CGH) in cytogenetics.[33,34] It has evolved to allow the screening of the whole genome by utilizing multicolor whole-chromosome probe techniques, such as multiplex FISH or spectral karyotyping. These FISH techniques, such as CASFISH, oligopaint-FISH, and smRNA-FISH, have been developed mainly to be implemented in genetic research applications and they have also been introduced to assess the spatiotemporal changes of intranuclear genomic organization and cytoplasmic RNA profiling.[34,37]

10.6.1.3 Comparative genomic hybridization (CGH)

Comparative genomic hybridization (CGH) has been developed as a molecular test for chromosomal analysis and it is used in prenatal diagnosis, pediatric patients, or adults with specific indications. CGH detects microdeletions and microduplications that are more than 500 base pairs and they are not detected by karyotype. In 2010 a consensus document and an economic analysis were published suggesting that CGH should be considered as the first diagnostic test, replacing karyotyping in patients with neurological problems, autism, and cognitive deficits and in newborns with congenital anomalies of unknown etiology.[30] Compared to the karyotyping (diagnostic frequencies: 2.5%–4.2%) in high-risk patients, the diagnostic frequencies are between 5.3% and 15%. Indeed, this detection can be significantly higher for CGH (9.3%–39%) when fetal anatomic defects are indicated.[36]

Despite the high cost, it should be highlighted that aCGH is used widely in prenatal diagnosis, and in most of the cases the additional cost is borne by health insurance as well as by countries or states where abortion is permitted.[38]

10.6.2 DEOXYRIBONUCLEIC ACID SEQUENCING

A human genome contains a complete set of nucleic acid sequences, encoded as DNA. Any method or technology that is used to determine the nucleic acid sequence is called DNA sequencing.[39]

In the 1970s and 1980s the basic DNA sequencing methods used were the Maxam–Gilbert sequencing and Sanger sequencing. In 1977 the first DNA sequenced from the bacteriophage φX174 and in 1985 the Epstein–Barr virus. Subsequently, in the 1990s several microorganisms have been decrypted as shifting to faster, automated sequencing methods allowed the sequencing of whole genomes. In 1995 the Institute for Genomic Research (TIGR) published in the journal *Science* the use of whole-genome shotgun sequencing, the first complete genome of a free-living organism, the bacterium *Haemophilus influenzae*. Thereafter, analysis of the genome of other organisms was made, such as *Saccharomyces cerevisiae* (1996), *Caenorhabditis elegans* (1998), and subsequently the entire human genome (2000) by use of automated capillary sequencers by Celera Genomics and Applied Biosystems (now called Life Technologies), for the HGP.[39]

Although capillary sequencing was the first successful approach to sequence a nearly full human genome, it is a long-term procedure and too expensive for commercial purposes. Since the completion of the HGP, various techniques have been developed leading to creation of a modern DNA sequencing era and surpassing the limitations of the prior methods (Sanger sequencing etc.). These techniques are known as NGS or high-throughput sequencing.[39,40]

The high-throughput-NGS technologies are currently the hottest topic in the field of human and animal genomics research, which can produce over 100 times more data compared to the most sophisticated capillary sequencers based on the Sanger method.[41]

10.6.2.1 Next-generation sequencing

NGS technologies have been established in different fields of life sciences, including functional genomics, transcriptomics, oncology, evolutionary biology, forensic sciences, and medicine.[41] In contrast with the previous sequencing techniques, NGS technology is highly scalable, allowing sequencing of the entire genome of interest at once (known as massively parallel sequencing) in an automated process. This is achieved by fragmentation of the genome into smaller pieces; random sampling for a fragment and sequencing it by using one of the many technologies accomplish this.[42]

Furthermore, NGS methods have many advantages, such as detection of genetic variation in patients at high accuracy and the reduced cost, thus offering the promise of fundamentally altering medicine.[38] NGS can produce over 100 times more data compared to the most sophisticated capillary sequencers based on the Sanger method.[41]

Still, there are some limitations not only on the performance of genetic tests but also on the complexity of choosing the appropriate genetic test. Therefore the combination of different laboratory methods is highly recommended for the verification and accuracy of the results.

NGS technology has been rapidly adapted to clinical testing and it is radically changing the dogma of clinical diagnostics. The widespread use of NGS in clinical laboratories has allowed an incredible amount of progress in the genetic diagnostics of several inherited disorders. Nowadays, there are numerous molecular tests available and different NGS panels including single-gene tests, gene panels, and exome or genome sequencing, depending on the genetic condition. As a result, physicians face the conundrum of selecting the best diagnostic tool/method for their patients with a variety of genetic conditions. It should be highlighted that NGS-based gene panel testing is usually complemented with array comparative genomic hybridization and other ancillary methods, thus yielding a comprehensive and feasible approach for heterogeneous disorders.[43] Based on recent evidence and marketing data, single tests are frequently used since they are suitable for conditions with distinct clinical features and minimal site heterogeneity; however, these data are constantly changing due to the rapid development of new technologies in genetic testing.[12]

10.6.2.2 Single-gene panel testing

Many genetic diseases are known to be caused by variants in a single gene (known as single-gene disorders). Briefly, single-gene testing is performed to identify changes or variants in a single gene and the association between the disorder and this specific gene. The type of testing and who will require testing will depend on the inheritance pattern and type of variant being investigated. Single-gene disorders are divided into categories depending on how they are inherited. The most common inheritance patterns are dominant, recessive, and X-linked. A typical example of a single-gene disorder is cystic fibrosis, which is caused by variants in the *CFTR* gene. The high clinical sensitivity and the minimal possibility of discovering multiple confounding variants of unknown clinical significance make single gene tests valuable and depend on each genetic condition and usage. For example, the *FGFR3* gene is the only one known to be associated with achondroplasia. Single-gene testing of the *FGFR3* gene can detect genomic variants in 99% of the gene and is therefore the most efficient approach in terms of both cost and time.[43]

10.6.2.3 Multigene panel testing

In contrast to monogenic diseases, genomic variants can be identified within multiple genes (polygenic subsets), especially in genetic diseases such as cancer. Cancer can result from either germ line (inherited) or somatic (acquired) variant in the DNA. Based on the expected gene-variant interplay, multigene panel testing is considered an effective detection method with the potential to perform parallel sequencing of multiple prespecified genes to identify pathologic DNA variations.[9] In addition, a different number of preselected genes can be analyzed, ranging from 5 to more than 60, and genes with differing levels of penetration (penetrance is associated with disease susceptibility) may also be analyzed. For example, multigene panel testing is widely used to detect variations for breast cancer (*BRCA1*, *BRCA2*, etc.), although penetrance of genes can vary in other cancer types.[9]

The integration of multigene panel testing into clinical application allowed the use of different panel approaches by genetic counselors and selection of the most suitable method such as (1) a single-gene test using Sanger sequencing, (2) a cancer- or syndrome-specific high-penetrant gene panel, (3) a cancer- or syndrome-specific high- and moderate-penetrant gene panel, or (4) a comprehensive cancer gene panel.[9]

Compared to single-gene panel testing, multigene panel testing is not considered as a superior option, the use of the latter will always depend on each genetic condition and the gene/genes of interest. In general, multigene panel testing is a more efficient and/or cost-effective option for patients with a single inherited cancer syndrome.[38]

10.6.2.4 Whole-genome and whole-exome sequencing in diagnostic testing

As previously mentioned, NGS allows us sequencing of many genes of interest through massively parallel sequencing. Furthermore, NGS can be used for the identification of unknown genetic variants and the management of incidental findings.[42] This indicates that whole-exome sequencing (WES) and WGS can be implemented in the field of diagnostic genetic testing.

Based on the central dogma of biology by Francis Crick (1957), DNA contains all the important genetic information and instructions for making proteins and molecules essential for our growth, development, and health.[44] These regions provide instructions for the proteins that are known as exons which comprise only 1% of the genome.[38] All the exons in an individual's genome are collectively called the exome. Our current understanding is that most of the genomic variants that cause disease occur in exons; WES facilitates efficient analysis of the coding regions in the genome and identifies possible disease-causing genomic variants[38] (US National Library of Medicine). However, there are some variations that can be found outside of the exons, by affecting gene activity and gene production. WGS detects variations in any part of genome, including the coding and noncoding regions (estimates almost all the euchromatic human genome). Both WGS and WES are valuable methods in the fields of research and personalized medicine. Although these methods are providing a huge amount of data and information for interpretation of the results, this is a labor-intensive procedure, while some identified variants may be reported to be associated with different genetic disorders and their exact role in disease pathogenesis is yet to be fully defined. Currently, there are several clinical laboratories worldwide that can perform WES and WGS.[38,43]

10.6.3 MICROARRAYS

Microarray technology is another powerful tool used in a variety of fields in research such as genomics, proteomics, toxicogenomics, and oncology. It is also used widely in medicine, pharmacy (for drug design and treatment), and clinical diagnostics, where it provides valuable information about the molecular mechanisms that lead to the emergence of diseases. It is a high-performance technology used to analyze the gene expression of thousands of genes simultaneously, even the entire genome, of an organism. This allows rapid comparative studies of a large number of samples in different tissues. In recent years, CMA technique has been used widely as a genetic test.

CMA is a molecular genetic test which detects CNV abnormalities; deletions or duplication in chromosomes that range in size from approximately 1 kb to multiple megabases (Mb) resulting a loss or gain of an entire chromosome. It is worth noting that the clinical significance of CNVs may be important (pathogenic or benign) or unknown.

The most common types of CMA include oligonucleotide array comparative genomic hybridization (oligo aCGH), single-nucleotide polymorphism (SNP) genotyping array or genome-wide SNP panels (SNP array), phenotype-specific SNP panels, and oligo aCGH/SNP combination array.[38,45] Genome-wide SNP panels offer an efficient platform to detect structural anomalies and to assess the risk of common genetic disorders in one test. Nowadays, multiple panels are being designed for clinical diagnosis, such as panels for retinal degeneration.[38] In addition, these panels can be used via a multiplex platform as a predictive and presymptomatic testing (even as pharmacogenetic tests) for a variety of conditions such as specific cancers, as well as ophthalmologic, cardiac, renal, and neurological disorders.

As mentioned previously, CMA can identify deletions and duplications across the genome or in any region of interest. Compared to karyotype analysis, CMA is more sensitive at detecting CNVs; most CMA can detect up to 100 kb, even a small single exon (with microarray oligo aCGH). Using a high-resolution karyotype analysis, one can detect deletions as small as 3–5 Mb and duplications larger than 5 Mb.[38] CMA is recommended as an important clinical diagnostic test (first-line test) for individuals with developmental delay, intellectual disability, multiple congenital anomalies, and/or autism spectrum disorders.[31] However, there are some limitations in which arrays cannot be deterministic of identifying a wide range of variants as there is a limited scope for analysis in the genome (preselected points for analysis of specific variants).

It should be highlighted that the rapid developments in the field of human genomic technologies have led to a plethora of available genetic diagnostic tests. Based on recent evidence and reports, it is estimated that more than 700,000 genetic tests are performed in Europe each year and this number is expected to grow even more in the next few years. The implementation of each method depends on multiple factors such as the genetic condition, the purpose (predictive, diagnostic) of the analysis, cost, and the accuracy that this method attributed to the genomic variant under study. Based on these criteria, different techniques and types of genetic tests can be used. The use of these different approaches or even the combination of methods should be considered before performing a diagnostic test; to summarize, the decision always comes down to the geneticist and the corresponding clinical laboratory.

10.7 ALLOWANCE AND COSTS OF GENETIC TESTS

Besides the dubious results of the genetic tests, the cost and coverage by government agencies is a very important issue that must be considered by each patient/customer. The cost of genetic testing varies from hundreds to several thousand dollars depending on the genetic variation targets to be analyzed. So far, some of the costs are being covered by the government, although this is not a general rule for all countries. The current procedure, although time consuming and complex, is to apply for insurance, since the applicant must adhere to some comparative criteria, and some may even require genetic counseling services in order to be accepted.[10,38,46]

Insurance coverage of genetic testing is highly variable and depends on the patient's insurer, the insurance package, and the test indication (e.g., diagnostic or predictive). Genetic tests are used to be covered by public services regarding rare diseases and specific conditions, such as hereditary cancer testing, cystic fibrosis, Tay–Sachs disease, and hereditary hemochromatosis. Compared with diagnostic genetic testing, predictive genetic testing is particularly variable, since the majority of individuals are asymptomatic and therefore testing is not being conducted to inform about the specific active medical problem.

Regarding other types of genetic tests, such as chromosomal abnormalities, prenatal and neonatal diagnosis, and PGD in certain situations (e.g., advanced maternal age and suspected fetal anomaly), the coverage of insurance varies, and multiple parameters need to be taken into account.[17,46]

10.8 DISCUSSION

Despite the differences and arguments in the entire research and medical community, and the ethical and economic issues, genetic testing is still a promising and powerful tool with multiple possibilities. This chapter assessed both the positives and negatives of genetic testing to fully understand and update the current views about genetic testing and its implementation.

Nowadays, there is a wide variety of public entities and private companies offering a broad range of antenatal and postnatal molecular genetic testing services for monogenic and multigene disorders, classical and molecular cytogenetics for chromosomal rearrangements, pharmacogenomic testing, and even predictive genomics for genetic disorders.[47,48] Apart from the public health services, there are hundreds of companies that offer ancestry and lifestyle information known as "direct-to-consumer genetic tests." Through a simplified and fast-tracked procedure, consumers are able to access information based on their genetic profile without the involvement of a physician. This includes the patients' ancestry or genealogy (ethnicity, origin, etc.) and also lifestyle factors, such as nutrition, weight loss, or even their wine preferences.[6,49]

This increasingly unrestricted access to human genomes may be the key point to predict or treat several common and genetic diseases, preventing the burden on health systems globally. In some cases it is an ideal method for detecting or for predicting specific variants. According to scientific findings, celiac disease is up to 87% attributable to genetics, for instance. Genetic tests look for specific versions of genes (a set of genes on chromosome 6 that encode the protein HLA-DQ). Other typical examples concern genetic tests for breast and ovarian cancer. According to the National Cancer Institute, more than 200,000 breast cancer diagnoses given to American women each year occur in disorders with no known family history of the disease. In most of the cases, *BRCA1* and *BRCA2* (up to three genomic variants in each gene) are the responsible genes for the development of disease. In recent years, in combination with NGS, a multigene (panel) test has been created, with a high efficacy and sensitivity against these harmful genomic variants.[9]

A few years ago, completion of the HGP offered significant clinical advances with tremendous impact in the field of medicine. New approaches and methods coupled with next-generation technology have become more accurate and continue to be quite promising in the near future. For example, a new comprehensive custom panel was designed in 2017 for routine hereditary cancer testing, which improved diagnostic sensitivity and detection by resolving the issue of uncertain clinical diagnosis.[50]

Despite the advantages and their integration to clinical diagnostics, genetic tests are not able to provide any "gold standards" autonomously. The complexities and difficulties for the interpretation of tests, the current gaps in methodology, information needs, policies, and systems are some of the key issues that are highly recommended to be carefully considered, before the usage of genetic tests.[7]

According to multiple scientific reports, the appropriate genetic test and analysis method plays a crucial role in the correct diagnosis and interpretation of results. Depending on the genetic case, there are specific guidelines that are provided worldwide. However, each decision must be made by a physician or genetic counselor, a choice which has raised several doubts given the lack of experience and knowledge in the management of new technologies. Apart from the necessary experience that a genetic counselor has to gain, a better management of a genetic case can be achieved through an assessment framework, such as ACCE, a model that was developed in the United States. This framework (ACCE) is derived from the following three evaluated components: analytical validity, clinical validity, and the ELSI of genetic testing. Based on the selected ACCE program, several frameworks have been created afterward, such as the United Kingdom Genetic Testing Network Gene Dossier evaluation framework, with a common goal to emphasize the key components of a genetic test evaluation.[7,10,17]

In spite of the constraints and the difficulties mentioned in the previous paragraphs, the complex nature of the genetic tests does not inhibit their rapid growth and expansion in the clinical practice. According to Srtatistics MRC, the Global Genetic Testing market is accounted for $7.749 billion in 2017 and is expected to reach $25.948 billion by 2026.

As described earlier, the process of compensating for genetic testing takes place through a complex and time-consuming process that often forces insurers to cover the cost of genetic testing themselves. In most cases, however, the insurance company will cover the costs of the tests; however, this depends on the type of the genetic test. According to recent evidence though, as many genetic tests as possible should be included in the reimbursement list based on their significance and predictive value.[10]

To summarize, genetic tests have now been established in modern medicine and clinical practice by offering a number of advantages and potential. Advances in omics technologies, such as genomics, transcriptomics, proteomics, and metabolomics, have begun to enable personalized medicine at an extraordinarily detailed molecular level. However, as mentioned previously, it is necessary to clarify that more data, experience, and knowledge are needed in particularly in the use of new technologies for the overall benefit of public health. According to global health services, each technology individually cannot capture the entire biological complexity of most human diseases.

Integration of multiple technologies has emerged as an approach to provide a more comprehensive aspect of the biology and the etiology of the disease. For this reason the combination of technologies and methods, depending on the genetic disease situation, is often recommended for a comprehensive and thorough diagnosis. Indeed, the use of algorithms and different models seems to be a meaningful solution to understand, diagnose, and inform disease treatment of modalities.[43] For example, following clinical genetics evaluation, there are different action points suggested in case clinical presentation of the patient is suggestive of a specific genetic condition or not. In the former case, targeted genetic testing, using single-gene, panel testing or cytogenetic testing is sought, and if the result is negative, then CMA or NGS is recommended. In the latter case, when the clinical presentation of the patient is not specific or clinical testing is not available for this disorder, then CMA and/or NGS is selected for the identification of the genetic etiology of the disease. In case where clinical presentation is not consistent with a genetic disorder, then no further genetic evaluation is recommended (see also Ref. [51]).

In addition, a substantive approach to ethical, legal, psychosocial, and financial issues through the integration of these powerful molecular tools is yet to be designed. Public provision of genetic information (e.g., particularly without genetic counseling) and the impact on the patient's life are major issues that need to be considered in detail.

ACKNOWLEDGMENTS

The authors wish to acknowledge the valuable help of Mr. Nikolaos Patronas with graphic design.

REFERENCES

1. Cleeren E, Van der Heyden J, Brand A, Van Oyen H. Public health in the genomic era: will Public Health Genomics contribute to major changes in the prevention of common diseases? *Arch Public Health.* 2011;69:8. Available from: https://doi.org/10.1186/0778-7367-69-8.
2. Evans GA. The Human Genome Project: applications in the diagnosis and treatment of neurologic disease. *Arch Neurol.* 1998;55:1287–1290.
3. B. Alberts, A. Johnson, J. Lewis, M. Raff, K. Roberts and P. Walter, From RNA to protein. Molecular Biology of the Cell, 2002, 4th ed. Garland Science, New York.
4. Green ED, Watson JD, Collins FS. Human Genome Project: twenty-five years of big biology. *Nature.* 2015;526:29. Available from: https://doi.org/10.1038/526029a.
5. Katsanis N. Point: treating human genetic disease one base pair at a time: the benefits of gene editing. *Clin Chem.* 2018;64:486–488. Available from: https://doi.org/10.1373/clinchem.2017.278309.
6. Patrinos GP, Baker DJ, Al-Mulla F, Vasiliou V, Cooper DN. Genetic tests obtainable through pharmacies: the good, the bad, and the ugly. *Hum Genomics.* 2013;7:17. Available from: https://doi.org/10.1186/1479-7364-7-17.
7. Zimmern RL, Kroese M. The evaluation of genetic tests. *J Public Health (Oxf).* 2007;29:246–250. Available from: https://doi.org/10.1093/pubmed/fdm028.
8. McPherson E. Genetic diagnosis and testing in clinical practice. *Clin Med Res.* 2006;4:123–129.
9. Grissom AA, Friend PJ. Multigene panel testing for hereditary cancer risk. *J Adv Pract Oncol.* 2016;7:394–407.

10. Becker F, van El CG, Ibarreta D, et al. Genetic testing and common disorders in a public health framework: how to assess relevance and possibilities. *Eur J Hum Genet*. 2011;19:S6–S44. Available from: https://doi.org/10.1038/ejhg.2010.249.
11. Evans BJ, Burke W, Jarvik GP. The FDA and genomic tests—getting regulation right. *N Engl J Med*. 2015;372:2258–2264. Available from: https://doi.org/10.1056/NEJMsr1501194.
12. Phillips KA, Deverka PA, Hooker GW, Douglas MP. Genetic test availability and spending: where are we now? Where are we going? *Health Aff (Millwood)*. 2018;37:710–716. Available from: https://doi.org/10.1377/hlthaff.2017.1427.
13. Suarez CJ, Yu L, Downs N, Costa HA, Stevenson DA. Promoting appropriate genetic testing: the impact of a combined test review and consultative service. *Genet Med*. 2017;19:1049–1054. Available from: https://doi.org/10.1038/gim.2016.219.
14. Bunnik EM, Schermer MH, Janssens ACJW. The role of disease characteristics in the ethical debate on personal genome testing. *BMC Med Genomics*. 2012;5:4. Available from: https://doi.org/10.1186/1755-8794-5-4.
15. Pagon RA, Hanson NB, Neufeld-Kaiser W, Covington ML. Genetic testing. *West J Med*. 2001;174:344–347.
16. Botkin JR. Ethical issues in pediatric genetic testing and screening for current opinion in pediatrics. *Curr Opin Pediatr*. 2016;28:700–704. Available from: https://doi.org/10.1097/MOP.0000000000000418.
17. Vozikis A, Cooper DN, Mitropoulou C, et al. Test pricing and reimbursement in genomic medicine: towards a general strategy. *Public Health Genomics*. 2016;19:352–363. Available from: https://doi.org/10.1159/000449152.
18. Bioethics C. Ethical issues with genetic testing in pediatrics, on*Pediatrics*. 2001;107:1451–1455. Available from: https://doi.org/10.1542/peds.107.6.1451.
19. Bentley C. Integrating oral health into social marketing campaigns: program development and description for San Francisco Children's Oral Health Strategic Plan. *Pac J Health*. 2018;1.
20. Traeger-Synodinos J. Pre-implantation genetic diagnosis. *Best Pract Res Clin Obstet Gynaecol*. 2017;39:74–88. Available from: https://doi.org/10.1016/j.bpobgyn.2016.10.010.
21. Stern HJ. Preimplantation genetic diagnosis: prenatal testing for embryos finally achieving its potential. *J Clin Med*. 2014;3:280–309. Available from: https://doi.org/10.3390/jcm3010280.
22. Cheng THT, Jiang P, Teoh JYC, et al. Noninvasive detection of bladder cancer by shallow-depth genome-wide bisulfite sequencing of urinary cell-free DNA for methylation and copy number profiling. *Clin Chem*. 2019;65. Available from: https://doi.org/10.1373/clinchem.2018.301341.
23. Cicali EJ, Weitzel KW, Elsey AR, et al. Challenges and lessons learned from clinical pharmacogenetics implementation of multiple gene-drug pairs across ambulatory care settings. *Genet Med*. 2019; (in press). Available from: https://doi.org/10.1038/s41436-019-0500-7.
24. Dressler LG, Bell GC, Abernathy PM, Ruch K, Denslow S. Implementing pharmacogenetic testing in rural primary care practices: a pilot feasibility study. *Pharmacogenomics*. 2019;20. Available from: https://doi.org/10.2217/pgs-2018-0200.
25. Gross T, Daniel J. Overview of pharmacogenomic testing in clinical practice. *Ment Health Clin*. 2018;8:235–241. Available from: https://doi.org/10.9740/mhc.2018.09.235.
26. Vogenberg FR, Isaacson Barash C, Pursel M. Personalized medicine. *P T*. 2010;35:560–576.
27. Pestka EL, Hale AM, Johnson BL, Lee JL, Poppe KA. Cytochrome P450 testing for better psychiatric care. *J Psychosoc Nurs Ment Health Serv*. 2007;45:15–18.
28. Berg JS, Agrawal PB, Bailey DB, et al. Newborn sequencing in genomic medicine and public health. *Pediatrics*. 2017;139. Available from: https://doi.org/10.1542/peds.2016-2252.
29. Pickard BS, Millar JK, Porteous DJ, Muir WJ, Blackwood DHR. Cytogenetics and gene discovery in psychiatric disorders. *Pharmacogenomics J*. 2005;5:81. Available from: https://doi.org/10.1038/sj.tpj.6500293.

30. Saldarriaga W, García-Perdomo HA, Arango-Pineda J, Fonseca J. Karyotype versus genomic hybridization for the prenatal diagnosis of chromosomal abnormalities: a metaanalysis. *Am J Obstet Gynecol.* 2015;212(330):e1–e10. Available from: https://doi.org/10.1016/j.ajog.2014.10.011.
31. Wapner RJ, Martin CL, Levy B, et al. Chromosomal microarray versus karyotyping for prenatal diagnosis. *N Engl J Med.* 2012;367:2175–2184. Available from: https://doi.org/10.1056/NEJMoa1203382.
32. Watson JD, Crick FHC. Reprint: molecular structure of nucleic acids. *Ann Intern Med.* 2003;138:581–582. Available from: https://doi.org/10.7326/0003-4819-138-7-200304010-00015.
33. Cui C, Shu W, Li P. Fluorescence in situ hybridization: cell-based genetic diagnostic and research applications. *Front Cell Dev Biol.* 2016;4:89. Available from: https://doi.org/10.3389/fcell.2016.00089.
34. Ratan ZA, Zaman SB, Mehta V, Haidere MF, Runa NJ, Akter N. Application of fluorescence in situ hybridization (FISH) technique for the detection of genetic aberration in medical science. *Cureus.* 2017;9: e1325. Available from: https://doi.org/10.7759/cureus.1325.
35. Jansen ME, Lister KJ, van Kranen HJ, Cornel MC. Policy making in newborn screening needs a structured and transparent approach. *Front Public Health.* 2017;5:53. Available from: https://doi.org/10.3389/fpubh.2017.00053.
36. Bishop R. Applications of fluorescence in situ hybridization (FISH) in detecting genetic aberrations of medical significance. *Biosci Horiz.* 2010;3:85–95. Available from: https://doi.org/10.1093/biohorizons/hzq009.
37. Lai L-T, Meng Z, Shao F, Zhang L-F. Simultaneous RNA-DNA FISH. *Methods Mol Biol.* 2016;1402:135–145. Available from: https://doi.org/10.1007/978-1-4939-3378-5_11.
38. Katsanis SH, Katsanis N. Molecular genetic testing and the future of clinical genomics. *Nat Rev Genet.* 2013;14:415–426. Available from: https://doi.org/10.1038/nrg3493.
39. Shendure J, Balasubramanian S, Church GM, et al. DNA sequencing at 40: past, present and future. *Nature.* 2017;550:345–353. Available from: https://doi.org/10.1038/nature24286.
40. Verma M, Kulshrestha S, Puri A. Genome sequencing. In: Keith JM, ed. *Bioinformatics: Volume I: Data, Sequence Analysis, and Evolution, Methods in Molecular Biology.* New York: Springer; 2017:3–33. <https://doi.org/10.1007/978-1-4939-6622-6_1>.
41. Pareek CS, Smoczynski R, Tretyn A. Sequencing technologies and genome sequencing. *J Appl Genetics.* 2011;52:413–435. Available from: https://doi.org/10.1007/s13353-011-0057-x.
42. Di Resta C, Galbiati S, Carrera P, Ferrari M. Next-generation sequencing approach for the diagnosis of human diseases: open challenges and new opportunities. *EJIFCC.* 2018;29:4–14.
43. Xue Y, Ankala A, Wilcox WR, Hegde MR. Solving the molecular diagnostic testing conundrum for Mendelian disorders in the era of next-generation sequencing: single-gene, gene panel, or exome/genome sequencing. *Genet Med.* 2015;17:444–451. Available from: https://doi.org/10.1038/gim.2014.122.
44. Cobb M. 60 years ago, Francis Crick changed the logic of biology. *PLoS Biol.* 2017;15:e2003243. Available from: https://doi.org/10.1371/journal.pbio.2003243.
45. Wallace SE, Bean LJ. *Educational Materials—Genetic Testing: Current Approaches.* Seattle: University of Washington; 2018.
46. Uhlmann WR, Schwalm K, Raymond VM. Development of a streamlined work flow for handling patients' genetic testing insurance authorizations. *J Genet Couns.* 2017;26:657–668. Available from: https://doi.org/10.1007/s10897-017-0098-3.
47. Marzuillo C, De Vito C, D'Andrea E, Rosso A, Villari P. Predictive genetic testing for complex diseases: a public health perspective. *QJM.* 2014;107:93–97. Available from: https://doi.org/10.1093/qjmed/hct190.
48. Sagia A, Cooper DN, Poulas K, Stathakopoulos V, Patrinos GP. Critical appraisal of the private genetic and pharmacogenomic testing environment in Greece. *Per Med.* 2011;8:413–420. Available from: https://doi.org/10.2217/pme.11.24.

49. Allyse MA, Robinson DH, Ferber MJ, Sharp RR. Direct-to-consumer testing 2.0: emerging models of direct-to-consumer genetic testing. *Mayo Clin Proc.* 2018;93:113–120. Available from: https://doi.org/10.1016/j.mayocp.2017.11.001.
50. Castellanos E, Gel B, Rosas I, et al. A comprehensive custom panel design for routine hereditary cancer testing: preserving control, improving diagnostics and revealing a complex variation landscape. *Sci Rep.* 2017;7:39348. Available from: https://doi.org/10.1038/srep39348.
51. Shashi V, McConkie-Rosell A, Rosell B, et al. The utility of the traditional medical genetics diagnostic evaluation in the context of next-generation sequencing for undiagnosed genetic disorders. *Genet Med.* 2014;16:176–182. Available from: https://doi.org/10.1038/gim.2013.99.

FURTHER READING

Du L, Becher SI. Genetic and genomic consultation: are we ready for direct-to-consumer telegenetics? *Front Genet.* 2018;9:550. Available from: https://doi.org/10.3389/fgene.2018.00550.

Grier S, Bryant CA. Social marketing in public health. *Ann Rev Public Health.* 2005;26:319–339. Available from: https://doi.org/10.1146/annurev.publhealth.26.021304.144610.

Jensen E. Technical review: in situ hybridization. *Anat Rec.* 2014;297:1349–1353. Available from: https://doi.org/10.1002/ar.22944.

Karczewski KJ, Snyder MP. Integrative omics for health and disease. *Nat Rev Genet.* 2018;19:299–310. Available from: https://doi.org/10.1038/nrg.2018.4.

Londhe BR. Marketing mix for next generation marketing. *Procedia Econ Finance.* 2014;11:335–340. Available from: https://doi.org/10.1016/S2212-5671(14)00201-9.

Other Resources

Allied Market Research. <https://www.alliedmarketresearch.com/cytogenetics-market/purchase-options>.

CDC Centers for Disease Control and Prevention. <https://www.cdc.gov>.

Concert Genetics. *The Current Landscape of Genetic Testing*; 2018 ed.

Department of Health and Human Services-USA. *Coverage and Reimbursement of Genetic Tests and Services.* Report of the Secretary's Advisory Committee on Genetics, Health and Society; 2006.

Genetics Home Reference, US National Library of Medicine. <https://ghr.nlm.nih.gov/>.

National Health Service (NHS England). <https://www.england.nhs.uk/publication/national-genomic-test-directories/>.

National Human Genome Research Institute (NHGRI). <https://www.genome.gov/human-genome-project/What>

NIH (National Institutes of Health). <www.nih.gov>.

Stratistics Market Research Consultant. <https://www.strategymrc.com/report/genetic-testing-market/description>; 2018.

PART

PERSONALISED MEDICINE AND PUBLIC HEALTH

II

CHAPTER

ASSESSING THE STAKEHOLDER LANDSCAPE AND STANCE POINT ON GENOMIC AND PERSONALIZED MEDICINE

11

Christina Mitropoulou[1], Konstantina Politopoulou[2], Athanassios Vozikis[3] and George P. Patrinos[2,4,5]

[1]The Golden Helix Foundation, London, United Kingdom [2]Department of Pharmacy, School of Health Sciences, University of Patras, Patras, Greece [3]Economics Department, University of Piraeus, Piraeus, Greece [4]Department of Pathology, College of Medicine and Health Sciences, United Arab Emirates University, Al-Ain, United Arab Emirates [5]Zayed Center of Health Sciences, United Arab Emirates University, Al-Ain, United Arab Emirates

11.1 INTRODUCTION

Genomic medicine aims to optimize the overall medical decision-making process and to rationalize drug prescription to the benefit of both the patient and the national health-care system, by means of exploiting an individual's unique genomic profile. This enables health-care professionals to make tailor-made disease and treatment risk assessments based on a patient's unique pharmacogenomic profile,[1] hence individualizing conventional therapeutic interventions.[2,3]

The genomic medicine environment is complex, with a plethora of key players and stakeholders with varying levels of genetics awareness and education. Previous studies have attempted to shed light on the level of genetics awareness of the general public and have investigated the genetics education level of health-care professionals, as well as their views on ethical, legal, and social issues pertaining to genomics toward adopting certain policies and perform the necessary steps that would facilitate integration of genomics into health care. These studies, conducted in various countries in Europe and the United States,[4–8] have shown that the level of general public awareness of genomic medicine and its impact on society is often rather low, which constitutes a major barrier to expediting the implementation of genomic medicine. In addition, the lack of proper mapping of the opinions of various stakeholders leads to an inadequately regulated environment in the field.

In order to better understand and analyze the genomic and personalized medicine policy environment in any given country and health-care system, one needs to pursue a stepwise approach, so that all the key stakeholders and their role in the field can be adequately identified, together with the main opportunities and obstacles for policy implementation in the genomic and personalized medicine environment. Also, in order to better plan the undertaking of various measures and

Applied Genomics and Public Health. DOI: https://doi.org/10.1016/B978-0-12-813695-9.00011-X

prioritize policymaking in genomic health care, the various stakeholders' interests should also be precisely mapped and analyzed.

This chapter aims to address this emerging but at the same time important concept in the field of genomic and personalized medicine. In particular, we are going to summarize the key stakeholders involved in the implementation of genomic and personalized medicine, their interests and power of intervention and the possible involvement in taking genomic and personalized medicine from the bench to the bedside. Also, we are going to allude to earlier conducted work to assess such an environment, which could be used as a paradigm for other health-care systems, with the aim of understanding the challenges and pitfalls in relation to pharmacogenomics and genomic medicine and to identify ethical, legal, and regulatory deficiencies that need to be rectified.

11.2 IDENTIFYING STAKEHOLDERS IN GENOMIC AND PERSONALIZED MEDICINE

The key stakeholders in the field of genomic and personalized medicine can be identified by expert consultation, extensive literature review, and, most importantly, previously published reports. In particular, the key stakeholders in the field of genomic and personalized medicine policymaking environment are the following:

1. Academic and research organizations, such as state and private universities, colleges, and research institutes.
2. Private and public genetic laboratories, including those affiliated with hospital clinics.
3. Physicians, including geneticists, in particular, but also physicians from other medical specialties.
4. Payers, including both the main public health insurance fund and also the various private health insurance companies, where applicable.
5. Genetics and genomics professional associations, including clinical and laboratory geneticists and genetic counseling professionals.
6. Pharmaceutical and biotechnology corporate entities, especially those with previous involvement in the field of genomic and personalized medicine.
7. National Medicines Organization that regulates the pharmacogenomics environment in the respective country.
8. Ministry of Health.
9. National Bioethics Council, where applicable.
10. Various religious organizations and the church.
11. Public and private providers in the field of genomic and personalized medicine, such as resellers of reagents, service providers, and manufacturers.
12. Pharmacy practices.
13. Consumers and citizens.
14. Press and the media.

11.3 METHODOLOGY OF ANALYZING THE STAKEHOLDERS' VIEWS AND OPINIONS

An important aspect of getting the views and opinions of all the stakeholders outlined earlier is to get them interviewed either by structured interviews and/or questionnaires that can be subsequently analyzed using specific analysis tools. According to our previous experience, it seems that the *PolicyMaker* method is the most adequate for collecting and organizing important policy information for genomic and personalized medicine and public health genomics in general.

PolicyMaker is a political mapping tool that serves as a database for assessments of the policy's content, the major stakeholders, their power of intervention, interests and the policy positions, and the networks and coalitions that interconnect them.[9,10] This research method aims to help policymakers managing the processes of reform and promote strategic programming as well as strategic thinking.[11] In other words, *PolicyMaker* is a logical and formal procedure to provide practical advice on how to manage the political aspects of public policy, hence assisting decision-makers to improve the political feasibility of their policy.

Analysis through *PolicyMaker* is carried out through the following five main analytic steps:

1. *Policy content*: It helps defining and analyzing the policy content, identifying the major goals of the policy, specifying a mechanism that is intended to achieve each goal, and determining whether the goal is already on the agenda.
2. *Players/Stakeholders*: It aims to identify the most important stakeholders and analyze their positions, power of intervention, and interests and assess the consequences of the policy for them. Also, it aims to analyze the networks and coalitions among the players.
3. *Opportunities and obstacles*: It helps assessing the opportunities and obstacles that affect the feasibility of a given policy, by analyzing conditions within specific organizations and in the broader political environment.
4. *Strategies*: It facilitates the design strategies to improve the feasibility of a given policy, by using expert advice provided in the program. Then, evaluate your strategies, and create alternative strategy packages as potential action plans.
5. *Impact of strategies*: Finally, the tool estimates the impact of any given strategies on the positions, power, and number of mobilized stakeholders, which constitute the three main elements that affect the feasibility of a policy.

The analysis outcomes in *PolicyMaker* are presented in a series of tables and diagrams that systematically organize essential information about the policy in question. These tables and diagrams can be used in strategic planning for policy formulation and implementation, to assist in improving the political feasibility of this policy (Fig. 11.1). The results can help with the

1. analysis of the political circumstances faced by a given policy;
2. rapid identification of bottlenecks and definition of obstacles;
3. policymaking process, by assisting in communication among different organizations;
4. data organization for storing, tracking, and analyzing positions, power, and other aspects of a political question;
5. implementation strategies, by proposing new ideas and strategies, helping policymakers to evaluate their consequences, and to track down their implementation; and
6. assessment of the overall impact of a policy.

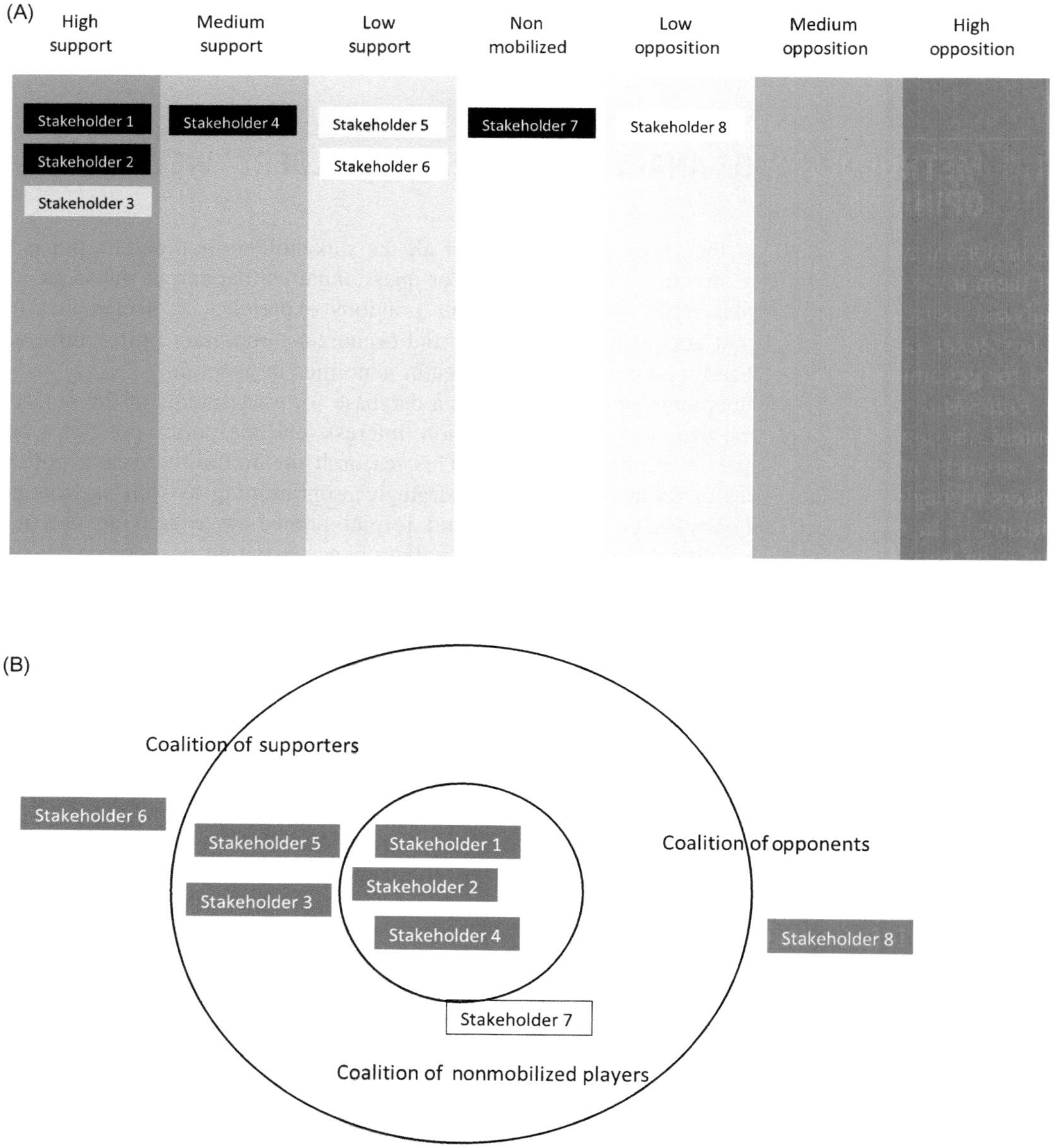

FIGURE 11.1

Schematic drawing of the *PolicyMaker* analysis output. (A) The current position map, in which each stakeholder is grouped as supporter (high, medium, and low support, in different shades of green), opponent (high, medium, and low opposition, in different shades of *red*) and nonmobilized (in *white*). The power of intervention of each stakeholder is depicted in black (high), gray (medium), and white (low). (B) The coalition map, depicting the stakeholders' initial position, various interests, and clustering, namely, coalition of supporters (*blue*; in the left), coalition of nonmobilized players (*white*; in the center), and coalition of opponents (*red*; in the right). Stakeholders are positioned in the same-centered circles, according to their power of intervention (the higher the power of intervention, the more centrally the stakeholder is placed). The coalition map is always evaluated jointly with the current position map.

The output graphs and tables can be stored and compared to comparatively analyze current against previous positions, as outlined in the position and feasibility graphs to show the impacts of an undertaken strategy (Fig. 11.2).

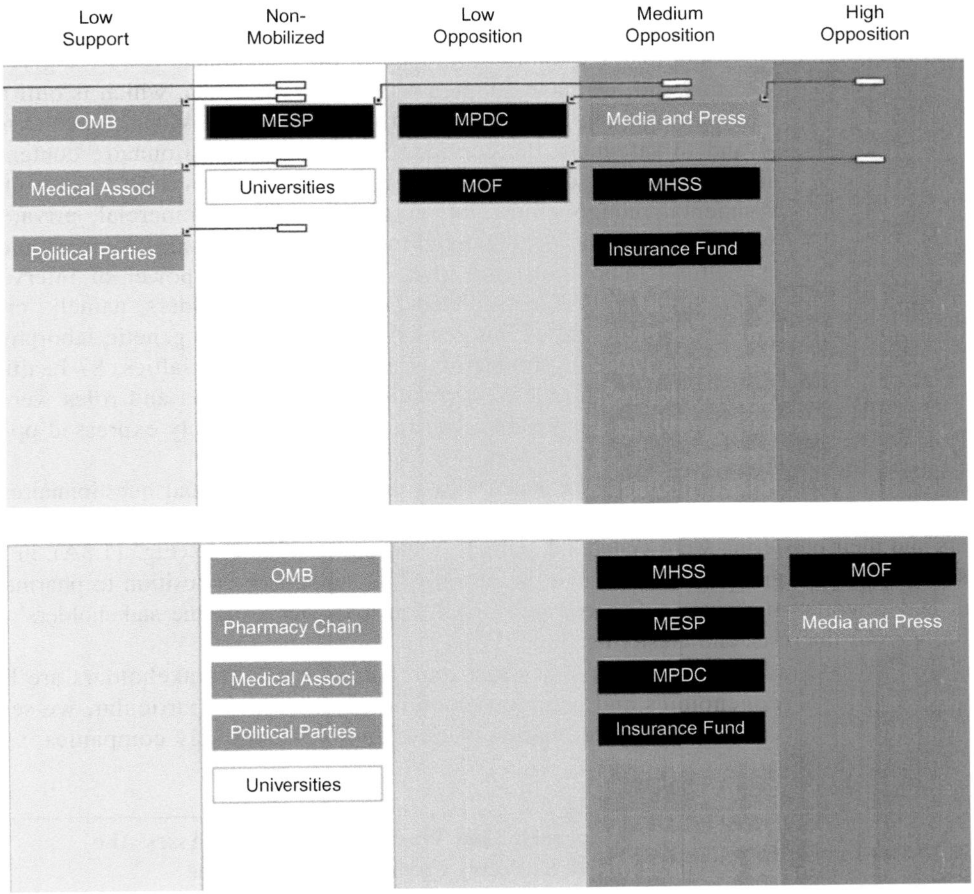

FIGURE 11.2

Comparative analysis of the current position map against the previous positions of the same stakeholders (lower part of the figure). Note that the "OMB," "medical associ," and "political parties" stakeholders have been relocated from the "nonmobilized" to the "low-support" group, the "MESP" stakeholder has been relocated from the "medium opposition" to the "nonmobilized" group, the "MPDC" stakeholder has been relocated from the "medium-opposition" to the "low-opposition" group, the "MOF" stakeholder has been relocated from the "high-opposition" to the "low-opposition" group, and the "media and press" stakeholder has been relocated from the "high-opposition" to the "medium-opposition" group. The latter findings depict a shift of various stakeholders toward a more supportive stance to the new policy under evaluation.

11.4 AN EXAMPLE OF STAKEHOLDER ANALYSIS IN GENOMIC AND PERSONALIZED MEDICINE: PRELIMINARY ASSESSMENT OF THE GENOMIC AND PERSONALIZED MEDICINE LANDSCAPE IN GREECE

In order to assess the genomic and personalized medicine environment in Greece, we have decided to use the *PolicyMaker* tool (see previous paragraph), which is one the very few tools currently available to precisely map the opinions and views of stakeholders toward a given policy and/or innovation. To this end, we have adopted a three-tier approach, which is outlined in Table 11.1, where all the participating entities and individuals have provided their consent to ensure their anonymity and to safeguard the confidentiality of the questionnaire content. For every different stakeholder, we have obtained its current territorial level (be it national or regional), sector (governmental, nongovernmental, political, media, commercial, private, and social), its position toward genomic and personalized medicine (namely, high support, medium support, nonmobilized, medium opposition, and high opposition), and power of intervention (namely, low, medium, and high). Also, for selected groups of stakeholders, namely, citizens and consumers (including 1717 members of the general public[7]), private genetic laboratories[12] and physicians and pharmacists [704 physicians of various medical specialties, 87 health-care professionals (other than physicians), and 86 pharmacists[7,8]], their views and roles were also identified mainly from our previous studies and also through their publicly expressed opinions in the media and conferences.

Based on these published findings and stated views during interviews and questionnaires, two graphical representations depicting the key stakeholders' initial position, including their various interests and their clustering were generated, namely, a current position map (Fig. 11.3A), in which each stakeholder was grouped according to the extent of its support or opposition to pharmacogenomics and genomic medicine in Greece, and a coalition map, depicting the stakeholders' initial position, various interests, and clustering (Fig. 11.3B).

Based on these findings, it is clear that more than half of the key stakeholders are highly supportive of pharmacogenomics and genomic medicine in Greece. In particular, we see that among the supportive stakeholders are pharmaceutical and biotechnology companies, as well

Table 11.1 A Three-Tier Stepwise Approach That Was Implemented to Assess the Pharmacogenomics and Genomic Medicine Policy Environment in Greece

Step 1	Policy • Policy content analysis, including goals and mechanisms
Step 2	Players • Identify the key players • Analyze their positions, power, interest, and interrelationships • Assess feasibility of current policy
Step 3	Opportunities and threats • Assess opportunities and threats that impact on the feasibility of the policy

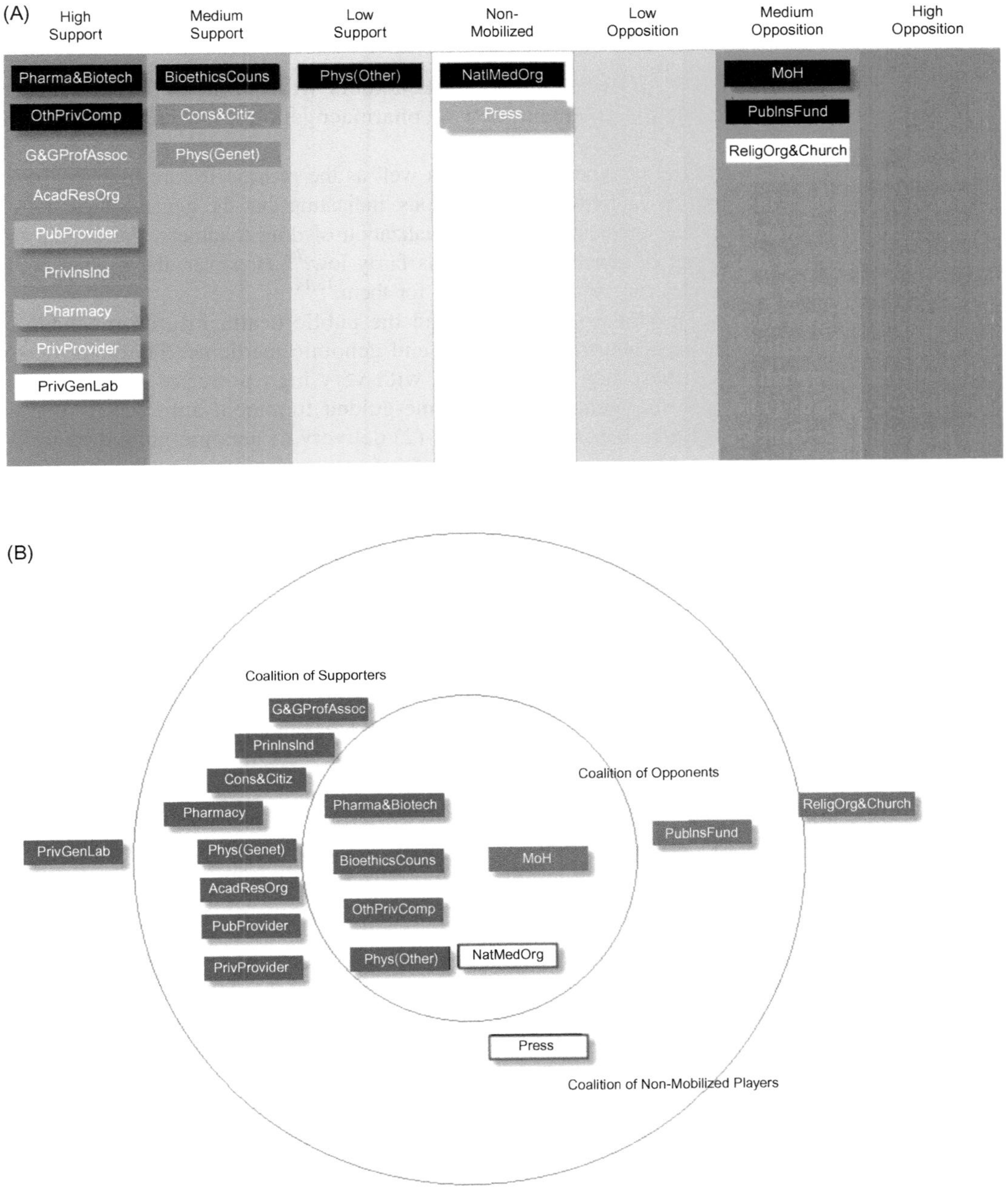

FIGURE 11.3

(A) Current position map, depicting the stakeholders' current position on pharmacogenomics and genomic medicine. Black boxes depict "high power" of the stakeholders to intervene, gray boxes "medium power," and white boxes "low power," respectively. (B) Coalition map, providing a more comprehensive graphical presentation of the key stakeholders' current position, but also of the homogeneity of their interests and their grouping. For abbreviations, please refer to Table 11.2.

Reproduced with permission from Mitropoulou et al. [19].

as molecular diagnostics laboratories, which are stakeholders with strong influence and are driving forces to support clinical implementation of pharmacogenomics from a technology-driven perspective.

We found that citizens, geneticists, other physicians as well as the pharmacies are highly supportive. These findings are in line with our previous observations, indicating that the general population is in general positive toward genomic medicine and individualization of drug treatment, despite the fact that they professed that their level of genetics awareness is fairly low.[7,8] However, the general public may get confused as to which tests can be truly beneficial for them.[13,14]

Interestingly, both the Greek Ministry of Health and the public health insurance funds display a medium opposition toward pharmacogenomics and genomic medicine. The main reason for the medium opposition of these key stakeholders, with very high power of intervention is most likely twofold: (1) cost-effectiveness of a genome-guided treatment approach is not yet fully proven in the Greek health-care environment and (2) delivery of genome-guided treatment modalities requires capacity building, both in terms of infrastructure and experienced human capital. As both stakeholders have a very strong financial interest and responsibility (Table 11.2), this could partly explain this finding, particularly since public health insurance funds lack the evidence on how genetic testing reimbursement could lead to a steady decrease of the overall health-care expenditure, by reducing adverse drug reactions. The latter is also supported by the professed opinion of over 75% of the physicians who think that the costs of genetic testing services should be reimbursed by insurance companies.[8] Obviously, and given the time period where the study was performed, namely, within the peak of the Greek financial crisis, this would constitute a major challenge for an economy for which the gross domestic product (GDP) contracted by almost 20% in 4 years (2010–13), unemployment rate increased by 15 percentage points to almost 24%,[15] and for a health-care system struggling to rationalize licensing, pricing, and reimbursement systems for health-care services, medicines, and medical devices. Quite surprisingly, the position of the private health insurance companies is highly supportive, that is at the opposite direction compared to that of the public health insurance funds, a fact that warrants further investigation and possibly exploitation in order to convince the latter funds to also adopt a supportive attitude toward this emerging trend of genomic medicine.

The current lack of proper legislation to oversee the operation of private genetic testing laboratories[13,16] could also explain the medium opposition of these stakeholders. This, however, contradicts the positions of both the National Medicines Organization (nonmobilized) and the National Bioethics Council (highly supportive), particularly since the latter has issued an opinion regarding direct-to-consumer genetic testing services (http://www.bioethics.gr/images/pdf/GNOMES/OPINION%20DTC%20genetic%20tests-Final-GR.pdf; report in Greek). Lastly, the church displays a medium opposition, although the power to intervene is lower than that of the other stakeholders.

The media and the press currently hold a neutral position on genomic medicine, which if changed to a medium-to-high support and present objective opinions and facts by academics, qualified professionals, and regulatory bodies, it would significantly facilitate and expedite adoption of pharmacogenomics and genomic medicine and also alter the position of governmental organizations that currently hold an opposition stance toward genomic medicine (see earlier paragraphs).

Table 11.2 Key Stakeholders, Their Respective Sector, Interests, and Power to Intervene in Pharmacogenomics and Genomics Medicine in Greece

Stakeholder	Abbreviations	Sector	Interest	Power
Academic and research organizations	AcadResOrg	Local nongovernmental	Scientific and financial	Medium
Greek National Bioethics Council	BioethicsCouns	Local nongovernmental	Humanitarian	High
Private genetic laboratories	PrivGenLab	Private	Financial and professional	Low
Religious organizations and church	ReligOrg&Church	Religious	Religious	Low
Consumers and citizens	Cons&Citiz	Social	Self-interest and financial	Medium
Pharmaceutical and biotechnology companies	Pharma&Biotech	Private	Financial	High
Genetics and genomics professional associations	G&GProfAssoc	Local nongovernmental	Professional and scientific	Medium
Ministry of Health	MoH	Governmental	Financial	High
Payers (private health insurance industry)	PrivInsInd	Private	Financial	Medium
Payers (public health insurance funds)	PubInsFund	Governmental	Financial	High
Other private companies[a]	OthPrivComp	Private	Financial, professional, and scientific	High
Pharmacies	Pharmacy	Private	Financial and professional	Medium
Physicians (geneticists)	Phys (Genet)	Private	Professional, scientific, and financial	Medium
Physicians (others)	Phys (others)	Private	Professional, scientific, and financial	High
Press and media	Press	Media	Ideological and political	Medium
Private providers	PrivProvider	Private	Financial	Medium
Public providers	PubProvider	Governmental	Financial	Medium
Greek National Medicines Organization	NatlMedOrg	Governmental	Scientific	High

[a]*In the field of genomic medicine and/or pharmacogenomics, including pharmaceutical and biotechnology companies.*

11.5 DEFINING OPPORTUNITIES AND THREATS WHEN IMPLEMENTING GENOMIC AND PERSONALIZED MEDICINE

The stakeholders' view and standpoint analysis study, described in the earlier paragraph, has identified several opportunities and threats in the genomic and personalized medicine policymaking environment in Greece that are described in the following paragraphs.

11.5.1 OPPORTUNITIES

According to the professed opinion of the interviewed stakeholders, one of the most obvious and at the same time important opportunities that arise from the implementation of genomic medicine is the ability of treating diseases in a personalized manner. In other words, not only can personalized medicine help toward genome-guided treatment rationalization, hence minimizing adverse drug reactions and improving the overall quality of life, but can also contribute toward better disease prevention, allowing patients to optimize their own health plans and decisions. Similarly, drug treatment individualization can also help to reduce the overall health-care expenditure at a national level (see also Ref. [17]), since the lower incidence of adverse drug reactions reduces the number and duration of the hospitalizations and reciprocally the number of deaths.

From the technology perspective, implementation of pharmacogenomics and genomic medicine could help improving the next-generation sequencing technology and help adopting novel next-generation sequencing approaches in a clinical setting. This could be also accompanied by companion technologies, such as array-on-demand diagnostics approaches for genetic diseases, noninvasive prenatal diagnosis, pharmacogenomics, and so on, particularly in developing countries and low-resource settings, where the means for establishing many clinical centers with routine next-generation sequencing capability are limited.[18] Of equal importance the interviewed stakeholders believe that implementation of genomic medicine and pharmacogenomics would lead to a very rapid development of the genomics and biotechnology industry and create opportunities for start-up establishment, especially for start-ups that adopt a more multidisciplinary product or service, for example, combining an innovative technology with an adjacent analysis software.

Lastly, continuous strengthening of national and international cooperation and interactions of various corporate and academic key players in the genomic medicine arena, biobanking area, is yet another important parameter that the stakeholders conceive as an important opportunity that arises from the implementation of genomic medicine.

11.5.2 OBSTACLES AND THREATS

Along with the elements that the stakeholders defined as opportunities that arise from the implementation of genomic medicine, they also defined obstacles and threats that if not addressed properly and timely, they could hold the field back.

First of all, the low genetic awareness of the general public, combined with the poor and/or incomplete genomics knowledge of health-care professionals, is a major hurdle to the broad adoption of genomic medicine. If coupled to the incomplete information, lack of proper evidence, and of necessary translational tool to assist physicians selecting the appropriate therapy based on the patients' genetic profile, then this obstacle becomes a serious threat that discourage physicians to implement genomic medicine and at the same time deprive the general public from the benefits of genome-guided treatment.

Ethics also poses a serious barrier to the implementation of genomic medicine, according to the interviewed stakeholders. First of all, the available legislation for safeguarding personal data is currently very limited in Greece, which does not guarantee personal data protection. To this end, stakeholders believe that insurance companies could deny insurance to a group of patients, based on their genetic profile, which leads to patient stigmatization, while health-care reforms to rectify this deficiency is likely to elicit opposition from many powerful and well-organized interest groups.[19] Similarly, lack of the necessary legislation to control the provision of genetic services via direct-to-consumer or over-the-counter approaches, especially for those tests that lack scientific evidence, such as nutrigenomic tests and genetic test for multifactorial genetic disease with a strong environmental component, is also conceived as a threat, especially combined with the lack of genetic awareness from the general public, where a patient cannot tell the difference between a scientifically sound and a scientifically flawed genetic test.[13,16] If the lack of a stable health-care environment and of a consistent national strategy for genetics and genomics is added to the equation, then a serious threat arises that could hold back the field and weaken the contribution of genomic medicine interventions in the eyes of the main beneficiaries, which are the patients and the treating physicians.

Lastly, the interviewed stakeholders consider the lack of funding and resources for capacity building along with the lack of reimbursement of genetic tests from the public and private insurance companies to be an obstacle of equal importance for the implementation of genomic medicine (see also Chapter 17).

11.6 CONCLUDING REMARKS

In this chapter, we have provided the basis for performing stakeholder analysis in an effort to determine the policy environment and to identify the role, the interests, and the position of the key stakeholders toward a given policy. We also provided a summary of the findings from our earlier work implementing stakeholder analysis to assess the views of the key stakeholders related to pharmacogenomics and personalized medicine in Greece.

These data underline that the majority of the key stakeholders are favorably viewing the implementation of genomic medicine, despite the fact that very few but key stakeholders, such as the Ministry of Health and the public health insurance funds, are currently opposing this new trend in medical practice. It is anticipated that once some tangible benefits from the implementation of pharmacogenomics become available, such as additional scientific evidence or economic evaluation studies indicating that the genomic medicine interventions are cost-effective and hence reimbursable by payers; the overall position of these key stakeholders are likely to change to a more favorable one. Also, these data underline the fact that the majority of the stakeholders seem to unveil their financial interest at a high priority. Most of the professional key players also express their scientific and professional interest, while the consumers highly prioritize their self-interest to access high quality and affordable health services.

These findings will be valuable to adopt the necessary steps and measures not only to maintain the overall positive attitude of most stakeholders toward genomic medicine but most importantly to shift the remaining stakeholders from a neutral-to-negative opinion into a more supportive position. Such studies could not only be replicated in the same country in the near future in order to acquire further insights as to how the views of the same stakeholders develop over time but also in other countries to be used as means of comparison of the views and attitudes of the same stakeholders in order to harmonize national policies toward the establishment of the genomic medicine landscape into future medical practice.

COMPETING INTERESTS

The authors declare that they have no competing interests.

ACKNOWLEDGMENTS

This work was partly funded by the University of Patras research budget and a Golden Helix Foundation research grant to GPP. This study was encouraged by the Genomic Medicine Alliance Public Health Genomics Working Group.

REFERENCES

1. Guttmacher AE, McGuire AL, Ponder B, Stefánsson K. Personalized genomic information: preparing for the future of genetic medicine. *Nat Rev Genet*. 2010;11(2):161–165.
2. Cooper DN, Chen JM, Ball EV, et al. Genes, mutations, and human inherited disease at the dawn of the age of personalized genomics. *Hum Mutat*. 2010;31(6):631–655.
3. Squassina A, Artac M, Manolopoulos VG, et al. Translation of genetic knowledge into clinical practice: the expectations and realities of pharmacogenomics and personalized medicine. *Pharmacogenomics*. 2010;11(8):1149–1167.
4. Makeeva OA, Markova VV, Roses AD, Puzyrev VP. An epidemiologic-based survey of public attitudes towards predictive genetic testing in Russia. *Per Med*. 2010;7(3):291–300.
5. Pavlidis C, Karamitri A, Barakou E, et al. Analysis and critical assessment of the views of the general public and healthcare professionals on nutrigenomics in Greece. *Per Med*. 2012;9(2):201–210.
6. Hietala M, Hakonen A, Aro AR, Niemelä P, Peltonen L, Aula P. Attitudes toward genetic testing among the general population and relatives of patients with a severe genetic disease: a survey from Finland. *Am J Hum Genet*. 1995;56(6):1493–1500.
7. Mai Y, Koromila K, Sagia A, et al. A critical view of the general public's awareness and physicians' opinion of the trends and potential pitfalls of genetic testing in Greece. *Per Med*. 2011;8(5):551–561.
8. Mai Y, Mitropoulou C, Papadopoulou XE, et al. Critical appraisal of the views of healthcare professionals with respect to pharmacogenomics and personalized medicine in Greece. *Per Med*. 2014;11(1):15–26.
9. Reich RM. Applied political analysis for health policy reform. *Curr Iss Public Health*. 1996;2:186–191.
10. Reich MR, Cooper DM. *PolicyMaker: computer-assisted political analysis. Software and Manual*. Newton Centre, MA: PoliMap; 1996.
11. Mintzberg H. *The Rise and Fall of Strategic Planning*. New York: Free Press; 1994.
12. Sagia A, Cooper DN, Poulas K, Stathakopoulos V, Patrinos GP. A critical appraisal of the private genetic and pharmacogenomic testing environment in Greece. *Per Med*. 2011;8(4):413–420.
13. Patrinos GP, Baker DJ, Al-Mulla F, Vasiliou V, Cooper DN. Genetic tests obtainable through pharmacies: the good, the bad, and the ugly. *Hum Genomics*. 2013;7:17.
14. Kampourakis K, Vayena E, Mitropoulou C, et al. Key challenges for next-generation pharmacogenomics. *EMBO Rep*. 2014;15(5):472–476.
15. European Commission (E.C.). *The Second Economic Adjustment Programme for Greece: Third Review. European Economy, Occasional Papers 159*, European Commission, Directorate-General for Economic and Financial Affairs. Brussels: Publications; 2013.

16. Kechagia S, Yuan M, Vidalis T, Patrinos GP, Vayena E. Personal genomics in Greece: an overview of available direct-to-consumer genomic services and the relevant legal framework. *Public Health Genomics*. 2014;17(5-6):299–305.
17. Patrinos GP, Mitropoulou C. Measuring the value of pharmacogenomics evidence. *Clin Pharmacol Ther*. 2017;102(5):739–741.
18. Mitropoulos K, Cooper DN, Mitropoulou C, et al. Genomic medicine without borders: which strategies should developing countries employ to invest in precision medicine? A new "fast-second winner" strategy. *OMICS*. 2017;21(11):647–657.
19. Mitropoulou C, Mai Y, van Schaik RH, Vozikis A, Patrinos GP. Documentation and analysis of the policy environment and key stakeholders in pharmacogenomics and genomic medicine in Greece. *Public Health Genomics*. 2014;17(5–6):280–286.

FURTHER READING

Balck F, Berth H, Meyer W. Attitudes toward genetic testing in a German population. *Genet Test Mol Biomarkers*. 2009;13(6):743–750.

Reydon TA, Kampourakis K, Patrinos GP. Genetics, genomics and society: the responsibilities of scientists for science communication and education. *Per Med*. 2012;9(6):633–643.

CHAPTER

HEALTH-CARE PROFESSIONALS' AWARENESS AND UNDERSTANDING OF GENOMICS 12

Konstantina Papaioannou[1] and Kostas Kampourakis[2]

[1]Independent Scholar, London, United Kingdom [2]Section of Biology and IUFE, University of Geneva, Geneva, Switzerland

12.1 INTRODUCTION

Advances in genomic technologies are transforming health care, but genomic medicine is a relatively new concept in the area of public health. Genomic medicine comprises several areas such as pharmacogenomics and personalized medicine, including genetic screening (GS), and precision medicine. All of these require DNA screening using next-generation sequencing (NGS) technologies, such as whole-genome sequencing (WGS) or whole-exome sequencing. During the last 15 years or so the cost of these technologies has been continuously decreasing so that genomic analyses have become affordable for many people. As a result, it is possible that genomic medicine could become a vital part of the routine health-care practice in the future.

Despite the rising significance of genomic analyses, it has been observed that physicians and other health-care professionals (HCPs) lack the required knowledge and understanding (see[1] for an overview) to consider and incorporate genomics in their everyday practice and, most importantly, the confidence and ability to explain them to laypeople. Not only delays in medical diagnosis but also errors during patient–doctor counseling can cause patient confusion and psychological distress, reduce the medical and scientific validity, and lead to huge health-care costs for patients, for their families, as well as for medical institutions.[2] Therefore patients need to be informed about the actual benefits and the uncertainties of genomic technologies, which in turn requires well-informed HCPs.

A number of studies across the world, focusing on different areas and specialties, have attempted to document the level of knowledge, expertise, confidence, and educational needs related to genomics that are necessary for HCPs across all medical fields. The aim of this chapter is to provide an overview of the existing literature. Several broad as well as more focused studies have been conducted in recent years in order to figure out which kinds of genomics knowledge are missing from clinical practice. In this chapter, we present key findings in order to highlight some of the educational gaps. The conclusions of the various studies are critically evaluated in an attempt to raise awareness of future educational needs.

As this field is rapidly expanding, in this chapter, we have mainly considered studies conducted since 2013. Although there exist relevant research published before 2013, we think that looking at more recent studies provides a more up-to-date view of the situation. Some related

Applied Genomics and Public Health. DOI: https://doi.org/10.1016/B978-0-12-813695-9.00012-1

research until 2013 is summarized in Kampourakis.[3] We have decided to present the studies discussed in this chapter thematically, following a chronological order for studies within the same area.

12.2 RESEARCH ON HEALTH-CARE PROFESSIONALS' KNOWLEDGE AND UNDERSTANDING OF GENOMICS

An extensive literature review was conducted focusing on studies since 2013. In Sections 12.2–12.4, empirical findings are summarized thematically (Table 12.1). Three main themes were identified including first researches focusing on specific countries or differences between regions within and outside Europe, America, and Asia. The second theme explores oncology studies with research primarily focusing on oncologists' genomics knowledge, application in their fields, and familiarity. The third and concluding sections in this chapter examine nurses' perceptions and preparedness toward adopting genomics to their practice.

12.2.1 STUDIES IN INDIVIDUAL COUNTRIES

Selkirk et al.[4] recruited 260 primary care physicians (PCPs), specialists, and surgeons from the Northshore University Hospital, in Illinois, United States, in order to evaluate their experiences with genomics, pharmacogenomics, and genetic testing (GT). The study focused on their expertise and confidence levels of including GT in their practice. This is important as during a 1-year period 41% of physicians had at least one patient who had undergone GT. The study assessed physicians' knowledge of genomics concepts, including genetics of complex diseases, pharmacogenomics, genome-wide association studies (GWAS), and basic genetics principles. It also determined participants' expectations for future opportunities and potential challenges in effectively bringing genomics into clinical practice. The researchers found that 72% of the participants had no or limited understanding and knowledge regarding GWAS, whereas 42% had above-the-average or expert knowledge on basic concept of genetics. A general observation was that specialists knew more than the generalists, whereas PCPs were less confident in decision-making of when and how to use medical genomics. When participants were asked to describe their level of confidence in discussing the benefits and drawbacks of GT with their patients, a significantly higher number reported that they were *not very* or were *not at all* comfortable (41%) comparing to those who felt *highly* comfortable (16%). Participants who had graduated after 1990 were more likely to highlight their lack of knowledge compared to older ones. This may stem from the fact that genomic medicine is a relatively new topic that was not included in the curricula of older HCPs. Younger HCPs are thus more likely than older ones to be aware that it exists and that they lack sufficient knowledge to discuss and use GT or pharmacogenomic testing (PT) with their patients.

Marzuillo et al.[5] conducted a genomics study with HCPs in Italy. A large-scale survey of 797 randomly chosen respondents registered in the Italian Society of Hygiene, Preventing Medicine and Public Health took part in the study. Results were quite promising as, overall, the participants expressed a positive attitude as well as exhibited an adequate educational background that supported the introduction and application of predictive GT for chronic diseases. In fact, the majority

Table 12.1 Researches' Main Features and Key Demographics

HCP Types	*N* (%)	Region	*N* (%)	Hospital/Institution/Project	Source
Studies in and Between Individual Countries					
PCPs, specialists, and surgeons	260	Illinois, United States	–	NorthShore University HealthSystem	[4]
General HCPs	797 (–)	Italy	–		[5]
PCPs	11 (55)	United States	–	Medseq project	[6]
Cardiologists	9 (45)				
Total	**64**	China	–	Peking Union Medical College Hospital	[7]
Gynecologist/Obstetrician	4 (6)				
Pediatrician	2 (3)				
Other	58 (91)				
Study 1 total	**164**	United Kingdom	–	–	[8]
Undergraduate students of optometry	60 (74)				
Eye care professionals	35 (35)				
General public	69 (34)				
Study 2 total	**219**		–	–	
Undergraduate students of optometry	127 (47)				
Primary eye care professionals	22 (57)				
General public	70 (87)				
General HCPs	70 (–)	Austria	13 (18.6)	The European U-PGx project	[9]
		Great Britain	14 (20.0)		
		Greece	13 (18.6)		
		Italy	9 (12.9)		
		The Netherlands	9 (12.9)		
		Slovenia	1 (1.4)		
		Spain	11 (15.7)		
Studies in Oncology					
Oncologists	215 (–)	Africa	4 (1.9)	–	[10]
		Asia	12 (5.6)		
		Europe	151 (70.2)		
		North America	7 (3.3)		
		South America	22 (10.2)		
		Oceania	16 (7.4)		
		Missing	3 (1.4)		

(*Continued*)

Table 12.1 Researches' Main Features and Key Demographics *Continued*

HCP Types	*N* (%)	Region	*N* (%)	Hospital/Institution/Project	Source
Total	**51**	Ontario and Alberta, Canada	–	–	[11]
Family physician	30 (59)				
Registered nurse	11 (21)				
Nurse practitioner	2 (4)				
Physician assistant	1 (2)				
Family medicine resident	4 (8)				
Medical student	1 (2)				
Other	2 (4)				
Physicians and oncologists	160 (–)	Boston, MA, United States	–	Profile research study at Dana-Farber Cancer Institute, Brigham, and Women's Hospital study	[12]
Pediatric oncologists	52 (–)	Memphis, TN, United States	–	St. Jude Children's Research Hospital, Part of the Genomes for Kids study	[13]
Studies Focusing on Nurses' Perceptions and Knowledge					
Nurses	175 (–)	Istanbul, Turkey	–	University Hospital in Istanbul	[14]
Total	**54**	Sao Paulo, Brazil	–	–	[15]
Nurses	30 (55.6)				
Physicians	24 (44.4)				
Total	**102**	Bologna, Italy	–	Two public hospitals	[16]
Nurses	59 (58)				
Midwives	43 (42)				
Multivariate professions including doctors, teachers, university lecturers, lawyers, and office workers	5404	78 Countries	–	–	[17]

HCPs, Health-care professionals; *PCPs*, primary care physicians; *U-PGx*, Ubiquitous Pharmacogenomics.

of respondents were able to describe the role of predictive GS in identifying individuals showing a higher predisposition for developing chronic diseases. More than 70% of participants highlighted the importance of genetic counseling; however, only 50% of them were familiar with the guidelines and counseling recommendations available for genetic tests. Highly positive views were recorded toward the use of predictive GT in clinical settings, although cost-effectiveness issues in the health-care practice were mentioned as potential obstacles. Despite the positive attitudes and the high rate of general awareness regarding predictive GS, when participants were asked more specific questions, about 90% failed to answer all of them correctly. In addition, lack of more specialized knowledge was strongly correlated with previous experience with GT during postgraduate studies and continuous professional development (CPD). In addition, the majority of public health practitioners described their genomics knowledge for chronic diseases as inadequate, expressed the need to improve their knowledge, and suggested that additional training modules should be developed to ensure continuous education on the fast-paced field of genomics and testing.[5]

Data from the MedSeq project in the United States have also been analyzed in an effort to understand the preparedness of HCPs for WGS use in the clinics, as well as identify future educational needs.[6] This qualitative study involved PCPs and cardiologists and revealed their knowledge and attitude differences. Cardiologists had more professional experience with the use of genetics than PCPs. In fact, 56% of cardiologists stated that they discuss genetic information with their patients, compared to only 9% of the PCPs. The study assessed participants' genetics literacy, preparedness, and personal motivations about upskilling themselves in the area of genetics. Indeed, PCPs felt less confident in discussing results with patients, mostly due to lack of specialized knowledge when data revealed oncological clinical areas. In addition, PCPs considered their knowledge in genomics as "weak" and "limited" and felt that even though medical schools introduced genetics as a course, they were not adequately prepared for WGS. In contrast, stronger confidence and feeling of obligation were noted among cardiologists who felt that it is their responsibility to discuss WGS results related to their specialty. However, concerns were raised when incidental findings, unrelated to cardiac-related matters, were seen in the WGS results. A lack of accountability in responsibly reporting all WGS data was mentioned by many cardiologists. Both groups felt motivated to become proficient in WGS, raising the need for additional support. Collaboration with genetic counselors was a recurrent theme that could potentially alleviate much of the anxiety of both groups, increasing guidance and support on data handling and patient discussions. Both groups felt that it is their responsibility to self-educate themselves by either meeting experts in the field or devoting more time to acquiring information before appointments with their patients. In addition, both groups agreed that familiar resources were the most convenient way to keep up with recent development, with cardiologists mentioning genetics-specific resources such as GeneTests (www.genetests.org) and the PCPs turning to scientific journals (e.g., *The New England Journal of Medicine*) or general medical resources.

Another study conducted in China assessed the knowledge, confidence, and experience of different physicians about GT and revealed significant educational needs.[7] In total, 64 physicians from the Peking Union Medical College Hospital, working in different specialties including pediatrics and gynecologists, took part in the study. Li et al.[7] assessed the routine use of family history collection and HCPs' experiences with GT. In fact the majority of HCPs (88%) had never or rarely used pedigree drawing as part of family history collection. Key reasons for GT use included help with diagnosis (51%), treatment (23%), risk assessment (17%), and prenatal diagnosis (9%).

Interestingly, among the physicians who had previously ordered a GT, less than 60% were able to acknowledge and name the tests and techniques used. Furthermore, the majority of HCPs had no or limited knowledge of NGS, single nucleotide polymorphism (SNP) microarray, and Sanger sequencing. Younger HCPs were more likely than older ones to exhibit knowledge, understanding, and confidence levels toward themes in genetics. In addition, the need for further educational resources was more prominent among recent graduates. The authors attributed these results to the lack of knowledge of older respondents. Overall, 84% of HCPs felt that they need more educational support and resources to strengthen their knowledge in genomics, highlighting the inadequate education on this topic provided in China.[7] The greater confidence of younger people in China and better understanding of younger HCPs in earlier study by Selkirk, Weissman, and Anderson[4] show a correlation between young age and genomic knowledge.

Other studies have focused on particular diseases. Inherited eye disease is a major cause of blindness worldwide and particularly in the middle-income, highly industrialized regions.[18] Approximately 15% of working adults in the United Kingdom suffer from visual impairment or blindness due to genetic factors, while this number rises up to 30% for children in the United Kingdom.[19,20] However, HCPs, medical students', and the public's understanding of inherited eye diseases remains elusive. Ganne et al.[8] looked at the level of general and eye-genetics knowledge as well as the perceptions and views about GT and gene therapy of three groups: (1) primary eye care professionals, (2) first-, second-, and third-year undergraduate students of optometry, and (3) members of the general public in the United Kingdom in two separate studies. The first study explored the perspectives and understanding of genetic eye diseases and other genetics concepts among 164 participants while the second study looked at the attitudes of a total of 219 respondents toward gene therapy and gene testing. As expected, students and eye care professionals' understandings of general genetics, including topics such as the number of chromosomes, the impact of genes, and the nature of mutations and gene therapy, were higher than the general public. Nonetheless, nearly 50% of eye care professionals and 24% of students did not know that genes exert some kind of control in the body, or that all cells in the human body in principle have exactly the same genetic information. In fact, 13% of students and eye care professionals together did not know the number of chromosomes per human cell. Less than 20% of the respondents in each group knew the genetic basis of age-related macular degeneration. Despite the extremely low proportion of correct responses, nearly 80% of all respondents expressed their positive views about GT, while 70% of them agreed that they would choose GT and gene therapy should they were diagnosed with a genetic eye disease. Nonetheless, it must be noted that 5% of participants from all groups felt that gene therapy and testing should be completely forbidden. Finally, approximately 76% of all participants failed to identify and explain the genetic centers and facilities available in the United Kingdom, while more than 50% of participants lacked any knowledge regarding genetic counseling.

12.2.2 THE UBIQUITOUS PHARMACOGENOMICS: A EUROPEAN INITIATIVE

Just et al.[9] looked specifically at the medical education of pharmacogenomics as part of the European Ubiquitous Pharmacogenomics (U-PGx) project. This is a European project focusing on the study of genetic variability among individual patients and the potential different reactions to different types of medication. The project aims to increase awareness, knowledge, and routine application of PGx in clinical practice.[21] The study focused on assessing the knowledge and attitude toward PGx in the seven European implementation sites where U-PGx was run and potentially propose future strategies to enhance confidence and improve knowledge related to PGx. HCPs participating at the U-PGx's implementation sites were asked to complete online questionnaires. In total, 70 questionnaires were obtained from Austria, Great Britain, Greece, Italy, The Netherlands, Slovenia, and Spain. Although a high proportion of HCPs were familiar with general genetic concepts, pharmacology, and the drug metabolizer phenotypes, their knowledge was more limited in the context of PGx and interpretation of the respective results. The study reported that the most common sources of learning were universities, whereas additional resources from post-graduate education, conferences, and scientific journals were less popular with less than 30% of responses commenting on these. In addition to their knowledge and experience the study empirically assessed respondents' performance on several scientific topics. Surprisingly, 41% of the questions were answered correctly, whereas 36.6% of them were not. The survey revealed that nearly half of the HCPs who participated in the study were not able to identify and determine whether or not PGx testing is necessary before administration of a drug. Furthermore, more than one-third of the participants admitted that they lacked the knowledge to effectively interpret results and take advantage of them in order to modify and adjust treatment to patients accordingly.

In the question "What knowledge would you need to utilize pharmacogenomics for adjustment of therapy?", a wealth of interesting responses was recorded as follows:

- More in-depth knowledge on drug metabolism (67.1%) and general PGx (60%).
- Stronger empirical evidence on the use of PGx on improving clinical outcomes and drug effectiveness (55.7%).
- More information and education on genetics (52.9%) and pharmacology (51.4%).
- Approximately half of the participants mentioned the need to improve their practical skills and knowledge in order to apply PGx in practice.
- More than half of HCPs were positive toward attending accredited learning courses, CPD workshops, and modules in order to enhance their knowledge and confidence among HCPs with respects to the clinical routine use of PGx. In addition, e-learning and other scientific articles were also mentioned as favorable education sources.

12.3 STUDIES WITH ONCOLOGISTS

A study carried out by Gingras et al.[10] looked at oncologists' attitudes and opinions regarding the integration and use of genomics services in the clinics for the management of breast cancer patients. Overall, 215 oncologists were recruited from different countries, the majority (70%) from Europe with Italy, Belgium, and France being the most frequent with 32 (16%), 25 (12%), and 22 (10%) participants, respectively. Most common countries outside Europe included Asia (5.6%) and South America (10.2%). All participants were highly experienced with breast cancer patients, as most of them (more than 90%) were seeing at least six new breast cancer patients every month. Surprisingly though, low levels of consensus were noted when participants were asked to describe when tumor genome sequencing should be requested. The lack of mutual agreement was exemplified by the fact that only 19% of respondents reported that there are available guidelines for multiplex tumor genome sequencing at their institutions. In fact, out of the 215 participants, only 82 (38%) had an experience with genome sequencing for their patients at least once a month in the past. Confidence level responses in analyzing tumor genome sequencing results ranged from *highly* confident (27% of participants) to *somewhat* confident (52%) to *poorly* confident (21%). However, 44% of participants pointed out the lack of support in interpreting genome sequencing results. These findings are actually in line with the research conducted by Gray et al.[12] who looked at oncologists' perceptions toward multiplex somatic genomic testing in a single cancer center before the onset of a genomic multiplex testing initiative. Overall, it was found that tumor genome sequencing is sometimes, but not always, used as part of the management of breast cancer patients in clinics. Stronger literature awareness, better access to clinical trials, and opportunities for modern and more efficient sequencing platforms were described as the key factors in encouraging oncologists' involvement in medical genomics. The development of tumor boards and educational resources could also dramatically increase oncologists' confidence in data analysis and hence in incorporating tumor genome sequencing in their clinical practice. Overall, oncologists from all countries felt unprepared to use genomics in their day-to-day medical practice. However, it should be noted that differences in funding and access to resources were associated with the frequency of use of tumor genome sequencing. In fact, the procedure was shown to be used more frequently in Asia than Europe; however, data validity is questionable given the low number of participants outside Europe.

Personalized medicine was described as the *medicine of the future* in a qualitative research study conducted with PCPs from various specialties.[11] Carroll et al. observed that PCPs had limited experience with personalized medicine and involvement on some clinical areas, including hereditary breast cancer but not colorectal cancer, and some cases with prenatal care. Interestingly, it was reported that patients were more informed than the physicians about the availability of GS. A significant or almost complete lack of knowledge regarding personalized medicine, the potentials and availability of tests, and direct-to-consumer GT (DTC-GT) as well as a lack of collaboration between PCPs and genetic counselors was highlighted during the study. PCPs' lack of knowledge regarding personalized medicine was again reported. The most urgent need mentioned during the interviews was obtaining genetics information. A profound education gap was recognized, especially in the areas of DTC-GT where patients' questions made PCPs uncomfortable. Educational knowledge, including the genetic causes of cancer, was the topic of interest, whereas new guidelines for screening, the suitability of available tests and referrals, proof of benefits of testing, and preventing treatment were among the areas that PCPs mostly felt incompetent to deal with. When respondents were asked to recommend potential changes in the clinical institutions to increase the use of personalized medicine practices and their confidence, a stronger collaboration with genetics counselor emerged as a potential and effective solution. PCPs described a significant lack of contact with experts in the field which often "causes frustration". Additional resources ranged from more reliable and up-to-date information, training, and workshops from experts in the field, e-learning resources such as websites as well as more personalized mobile application, point of care tools (e.g., tools and guidelines to help them in their decision-making), and finally more knowledgeable pool of people with expertise in personalized medicine and genomics to be available in the hospitals.

Gray et al.[12] studied the attitudes of medical physicians and oncologists toward the use of multiplex tumor genomic testing and their confidence levels in using and interpreting the results as part of the diagnosis process. The project was part of a Profile research study at Dana-Farber Cancer Institute in Brigham and Women's Hospital study, in Boston, MA, United States, with 160 participating physicians. The confidence levels reported varied among physicians, with researchers with high baseline testing and medical oncologists recording higher degrees of certainty than others. Again, there was consistency between the study's results and other research as it confirmed, once again, that a strong association between the physicians' confidence levels in supporting and recommending GT to their patients related to the actual use of tumor GT to test their patients' susceptibility. Furthermore, they also looked at the variation of the scientific language and terms used by all physicians to describe multiplex testing to their patients. Popular terms ranged from tumor and molecular testing (77% and 72%, respectively), tumor profiling (66%), and GT (62%). Interestingly, the use of open-ended questions generated another 28 terms that participants referred to when describing the procedure to the general public. Significant language discrepancies and lack of universal language standardization in physicians explain the multiplex tumor testing procedure to their patients calls for a vital need to revisit the doctor–patient communication and the scientific clarity about tumor testing. The use of an inconsistent language is likely to confuse patients and decrease their confidence in undergoing tumor testing.[22] Finally, the study highlighted the urgency in assessing the disparities in the genomic confidence among groups of physicians across different geographical regions, supporting the need for an enriched genetic education program.

A smaller scale, and more focused, study assessed the confidence and genomics knowledge level of pediatric oncologists and the potential use and integration of NGS into pediatric oncology clinical care.[13] The survey was part of the Genomes for Kids study whose purpose was to assess and facilitate the integration of clinical genome and exome sequencing (CGES) using germ line and tumor cells from patient children in St. Jude Children's Research Hospital, Memphis, TN, United States. When participants were asked to comment on their confidence of analyzing CGES results, less than half felt confident in interpreting results from somatic cells and germ line test results. In addition, oncologists were asked about the role and participation of a genetic counselor (GC) in the process of addressing and disclosing the germ line test results. Almost all physicians (93%) expressed the desire to consult a GC before disclosing the results, whereas 78% wished they had a GC present during the disclosure discussion with the patient. Follow-up questions were made in assessing the ways in which respondents would use the test results for their patients. Of those who used somatic test outcomes, 93% said that they use the results in order to engage the patient in a new study. In addition, 93% said that they would use the results in order to change or add a new medication to the therapy schedule of a patient, while 79% said that the results provide guidance in adjusting the current treatment.

12.4 NURSES' PERCEPTIONS AND UNDERSTANDING OF GENOMICS

A focused group in a University hospital in Istanbul, Turkey, investigated the knowledge of nurses in topics such as genetics, medical genomics, and gene sequencing in order to evaluate their confidence levels and highlight educational needs in the field.[14] Out of the 175 nurses who participated in the study, 142 stated that the genetics was not included in their undergraduate studies (81%). This finding is in line with previous studies in other regions, such as Italy and Taiwan, which reported that the majority of the nurses did not have access to any specific genomics-related course as part of their degree.[16,23] Moreover, 70.3% of nurses expressed a strong interest in increasing their genomics knowledge by accessing more educational resources. Most popular topics include prenatal and newborn screening or familial cancer. Interestingly, a small proportion of nurses regarded genetics health care as a priority for increasing their confidence, including genetics centers in Istanbul, family history collection, and drawing pedigree as key topics. The study assessed the genetic knowledge of nurses by asking them to identify genetic diseases and other genetic concepts such as mutations, mitosis, meiosis as well as X-linked inheritance patterns and autosomal dominant, and recessive inheritance patterns. Most nurses managed to describe and define the basic concepts of mitosis, meiosis, and mutations correctly; however, most failed to describe inheritance patterns and X-linked disorders. The average genetic score was 6.47 (out of 11). Themes that were commonly understood were the relationship between genetic risk and disease and more specifically breast cancer (92%) and ovarian cancer (90.9%). However, further relations and issues were less commonly understood among the respondents including hereditary breast cancer syndromes (12.6%), patterns of inheritance of Down syndrome (13.1%), and mutations in *BRCA 1–2* genes and their associations with breast and ovarian cancer (20%). This indicates that although genetic factors leading to cancer and familial relationships are well understood concepts in Turkey, nurses failed to report understanding that the likelihood cause of ovarian and breast cancer can stem from the same gene mutation. In addition to molecular understanding, the knowledge of genetic and health services was assessed. Interestingly, only 20% of nurses knew the location of genetic services in Istanbul, whereas 51.4% lacked any knowledge about the availability of health services provided by genetic centers.

Similar were the results of a cross-sectional study in Brazil where Lopes-Júnior et al.[15] examined the genetics education, understanding, and practical experiences among nurses and physicians in the primary health-care sector in Sao Paulo. The majority of respondents (79.6%) had not performed a GC, whereas 62.9% were not aware of genetic counseling institutions in Brazil. This implied nurses' underestimation of the number of genetic centers available in Brazil, particularly in the south regions.[24] In contrast the percentage of physicians who had used a pedigree was 62.5%. These findings strongly correlate with the knowledge of the genetic centers available, as most PCPs were aware of at least one genetic center. Less than half of the nurses and physicians who took part in the study had used a pedigree to record patients' family history, and 60% of nurses had never used a pedigree in clinical practice. Overall, the significant lack of training on GS observed among the respondents (90.7%) correlates with the limited use of GC in the clinics and the low confidence in assessing genetic cases of their patients. Genetic counseling is considered an effective way to apply a risk assessment strategy in managing the risks related to genetic disorders. GC provides information and assessment risk strategies to both the HCPs and the patients involved, offering them the necessary information to interpret and understand the medical and family implications of a genetic disorder. When participants were asked about the importance of genetic counseling in health care, 90.7% agreed that the development of educational resources and opportunities for improving HCPs' knowledge and understanding is highly needed. This is in agreement with the study carried out by Seven et al.[14] where nurses expressed their preferences to receive further educational resources in genomics and their clinical applications. Nurses who had previously attended genetics courses as part of their undergraduate programs expressed stronger genetic confidence and strongly wished to expand their knowledge on the subject. This is also in accordance with earlier studies where the level of genetic knowledge was positively correlated with the willingness for further knowledge development.[25]

Similar were the findings of a survey conducted by Godino et al.[16] in Italy. More specifically, more than half of the nurses who took part in the study (61%) felt that GC should be used for informative and advisory purposes, while 53.9% failed to explain the reasons and identify patient groups who might be eligible for a GC. In addition, when asked about the role of nurses and their responsibility to genomics communication, 62% felt that there was no relevance. This study highlights a lack of genomics accountability to the nursing profession as well as points out the incomplete knowledge of nurses with regard to genetic counseling, health-care use of genetics, and clinical implications for the future. The lack of understanding that ultimately leads to responsibility denial increases the need to develop training and educational resources to upskill professionals in the clinics.

12.5 A MULTIDEMOGRAPHIC PERSPECTIVE

A large-scale, multidemographic study conducted by Chapman et al.,[17] recruiting 5404 participants from 78 countries, looked at the level of genetic knowledge and differences across countries, professions, education levels, religious affiliations, and political associations. The research examined participants' perceptions of genetics, considering not only HCPs but also the well-educated general public. Overall, genetic literacy was shown to be poor with an average score of 65.5%, with respondents revealing significant gaps in their understanding. Less than half of participants responded correctly about the number of genes in human DNA, whereas about 30% regarded complex conditions including schizophrenia and autism to be a result of single genetic mutations despite a plethora of research studies showing otherwise.[17] The inadequate genetic understanding was evident across different professions such as medical doctors, teachers, lawyers, and university lecturers, raising the critical role that these groups have in increasing literacy with teachers and lecturers in the education and medical doctors in the health context. Despite the lack of genetic understanding, more than 85% of respondents confirmed that they would be willing to undergo GT in order to increase health-care benefits. Cross-country differences were also evident, and in some cases, it was found that genetic knowledge and confidence are influenced by factors such as country policies, legislation, and curricula.[17] This research is in line with the aforementioned studies, advocating that continuous education with updated curricula is key to raise awareness to the general public as well as prepare HCPs to integrate genomics into their practice. In addition the authors argued that the responsibility lies to other bodies and institutions as well as media and communications strategies whose purpose should be to provide accurate genetic information to reduce misconceptions and provide clarity.[17]

12.6 EDUCATIONAL CHALLENGES IN IMPLEMENTING GENOMIC MEDICINE

It should not be assumed that even health educators, whose main responsibilities include the identification of future educational needs, are competent enough to promote and encourage the integration of genomics to the medical arena. In fact, one of the first studies surveying a total of 835 health educators in the United States revealed very limited knowledge, education, and training among participants.[26] More specifically, more than 80% of respondents felt that they had no or very limited genomics training, implying a significant lack of exposure to genomics and its potential application in the public health care. The majority of participants felt they needed genomics education, whereas 94.1% expressed their positive attitudes toward promoting genomics in health care. In addition to surveying health educator's knowledge and experience levels, Chen and Kim[26] also looked at their respondents' preferred genomics areas for training and suitable educational means. Areas focusing on genomic diseases or disorders, genetic risk predisposition, and the association between genomics knowledge and health education promotion were highlighted as the more important and highly interested educational ones. In addition, continuing professional development, online training, workshops, and conferences were considered the most preferred ways of enhancing training prospects within the health education profession. Empirical evidence therefore prompts us to assume that despite the central role of health educators in promoting medical advancements in public health care, they should be the first to develop the skills, knowledge, and competencies necessary to further raise awareness of medical genomics in their fields.

A multiregional study in Southeast Europe assessed the education of pharmacogenomics in all bachelor and master-level Health Sciences degrees. More specifically, the study looked at any potential differences and deficiencies among university curricula in Cyprus, Greece, Italy, Turkey, Albania, Serbia, Malta, Bulgaria, Montenegro, Bosnia and Herzegovina, and Croatia in order to highlight the importance of smooth integration of pharmacogenomics in the routine medical practice.[27] The authors revealed significant discrepancies in the level and depth of pharmacogenomics education between Southeast European countries and grouped countries into three main categories: (1) universities that do not provide any undergraduate or graduate pharmacogenomics subject include Bosnia and Herzegovina, Cyprus, and Malta; (2) those that include some optional or compulsory pharmacogenomics modules as part of their graduate and/or undergraduate degrees, namely, Bulgaria, Albania, Croatia, Serbia, and Turkey; and (3) universities that offer substantially more and in-depth courses on pharmacogenomics, such as Greece and Italy. The findings stand in contrast to studies conducted in other European countries, namely, United Kingdom, Germany, and The Netherlands where pharmacogenomics is taught more extensively, and there is a variety of courses available for students. When the level of pharmacogenomics education was assessed and compared among students, residents, and specialists in Cagliari, where pharmacogenomics was extensively embedded in students' education, significant deficiencies were recorded. In particular, almost all undergraduates and residents had attended introductory courses on pharmacogenomics, yet, the number was significantly lower among specialists (55%). The majority of general and specialized physicians appreciated the utilization of pharmacogenomics tests in the clinical practice and highlighted the need to enhance education on pharmacogenomics and personalized medicine. The results revealed extremely low confidence levels in ordering a pharmacogenomics test and in interpreting the results for their patients, with only 16% of specialists and 7% of residents reporting confident enough to do so. Despite the lack of confidence, a very high percentage of students, residents, and specialists expressed positive views about enhancing pharmacogenomics education in medicine surgery courses.[27]

Studies have also looked at the confidence level of different HCP specialties in teaching genomics concepts to university students.[28,29] Read and Ward[28] used the Genomic Nursing Concept Inventory to examine the level of understanding in genomics knowledge among members of the nursing faculty in the United States. Out of the 495 nurses who took part in the study, about 70% described their proficiency with the genetics area as *poor* while only 33% of them had pursued additional educational genomics programs. The results signify the importance of increasing the variety and accessibility of education materials for nursing faculties in educational settings. Being in the academic sector, the nursing department ought to be accountable for updating their knowledge, expertise, and competence in the fast-paced field of genomics and transferring their knowledge in the educational settings and university programs.

As evidenced, numerous surveys reveal the reluctance and unwillingness of many specialties of HCPs and other physicians to apply genomics in their clinical practice. Given the rapid progress, the ongoing development of sequencing technologies, and the unraveling of complex genetic conditions, it is time to call for an urgent and uniform strategy toward improving the global genomics education. The Inter-Society Coordinating Committee for Physicians Education in Genomics (ISCC) initiative was established in 2013 by the collaboration of the National Human Genome Research Institute along with 23 professional societies, 15 institutions at the National Institutes of Health collaborated, and other medical educational organizations. The ISCC aims to "improve genomic literacy of physicians and other practitioners and to enhance the practice of genomic medicine through sharing of educational approaches and joint identification of educational needs."[30] As part of ISCC work, a standardized and widely applicable framework with suggested guidelines was developed, highlighting the essential skills, knowledge, and competencies toward a successful genomics intervention in the clinical practice.[30] The framework is considered a starting point of providing guidance and listing key competencies necessary for the genomics integration into the clinical practice. The committee offers this basic, starting model in an endeavor to encourage other relevant institutions and professional group in formulating, more specialized frameworks, contingent on HCP's specialty, and level of care.[31]

In a descriptive review, Slade et al.[32] also discussed the complex and interlinked setting for genomics education in the United Kingdom, highlighting the power of a plethora of organizational bodies and institutions in directing genomics education, development, and upskilling of HCPs. Their review supports a stronger focus on genomics education and curricula in the postgraduate degrees by the General Medical Council, royal colleges, and other training boards and their role in medical CPD. However, the fast-paced progress in genomics should be considered during the design of educational courses. The media developed to facilitate this continuous education should support mechanisms to ensure that HCPs are kept up-to-date with new diagnostic and therapeutic strategies. This would ensure sufficient knowledge and competency across the medical workforce in adopting some of these advancements into their clinical practice. The potential, independent role in HCP training and education of organizational bodies aiming to improve medical training in the United Kingdom by refining the structure and delivery of medical genomics education is also described. A clear case in point is the Royal Society of Medicine which streamlines cross-specialty education and integration of genomics medicine into practice. Independent or other advisory boards such as the Public Health Genomics Foundation also aim to bridge the gap between scientific advances and health system interventions into the health-care practice. The framework and interlinked net of various bodies, institutions, and other health organizations and their critical role in promoting the use of genomics into the clinical practice and upskilling of the current and future medical workforce is discussed. In line with other research, the authors also arguably raised the risk of lack of benefits and the need to demonstrate stronger and wider scale evidence whilst raising the need of increasing collaboration and better integration to achieve synergies and avoid duplication of effort and resources.[32]

We argue that for a successful genomics education, specific guidelines, training programs, and CPD courses should target each of the different specialties of physicians and other HCPs as well as nurses and pharmacists. Further challenges in improving genetic literacy stem from the lack of funding and availability of resources (i.e., materials and manpower) and the diffusion rate of genetic technology application to the clinics.[33] A dilemma between improving the global genetic literacy and the currently available infrastructure and network capable of supporting these educational activities arise, arguing that variations between different regions in the world are likely to result again in a substantial training inequality.[33]

Education of HCPs in substantially improving genomic literacy in health-care faces its own challenges across the scientific and medical community and the society in general. In a survey conducted in 2010, it was revealed that out of the 850,000 physicians in the United States and the District of Columbia, 47% were 50 years old or older, while 22% were more than 60.[34] Therefore this substantial proportion of aging medical workforce generates more challenges in improving genomics literacy, as many completed their medical education before the era of human genome and DNA sequencing. Therefore it is not surprising that the majority of empirical studies discussed outline stronger genomics awareness and higher confession rates of genomic educational needs among younger physicians and other HCPs comparing to older ones. HCPs are required to be kept up-to-date with a range of fast-paced and technological advances in their respective fields and specialties including conventional therapeutic and diagnostic areas. For many of them, medical genomics education is an additional topic in their dense "to-do" list, further constraining genomics integration in their medical practices.[35]

12.7 CONCLUSION

The advent of new high-throughput sequencing technologies has dramatically decreased application costs, turnaround times, while accuracy and validity of results are continuously increasing.[35] Despite the remarkable progress, it is argued that the costs of running large-scale genomics diagnostics still outweigh the beneficial implications in the medical arena.[36] Institutions and governments in many countries are struggling to maintain medical care costs at the lowest level possible while educational organizations strive to secure additional funding every year. The health economic evaluations currently available should be optimized to incorporate elements and cost centers of the genomics medical use[37] (see also Chapter 16). In fact, it is time to critically assess the short-term costs of medical genomics application and the long-term benefits realized by the lower rate of medical errors, misdiagnoses, trial-and-error treatment strategies, drug toxicities, and so on.

The aim of this chapter was to critically examine the level of genomics literacy among general physicians, specialists, and other relevant HCPs by drawing upon available empirical findings. By bringing together research across different types of HCPs, social contexts, and regions across the world, we have aimed to provide a more holistic, critical overview of the level of genomics education and HCP understanding toward genomics application into their medical practices. Despite different medical institutions and educational programs across the world, common themes were identified. First, the lack of basic genomics knowledge across all types of HCPs presents a major barrier in the progress of clinical application of genomics to the public health-care systems. The majority of HCP respondents reported low levels of confidence and skills in understanding basic genomic concepts and translating genetic results into appropriate clinical recommendations. Their lack of confidence triggers the mis- or underusage of GT in performing disease diagnosis and recommending target treatments to patients. Furthermore, HCPs feel underprepared in discussing GT and benefits of medical genomics with their patients, further widening the gap between the medical society, general public, and medical genomics. In fact, even though that the majority of the general public in the so-called developed countries possess a fundamental understanding of genetic concepts, empirical evidence suggests that there is a clear misinterpretation of the underlined gene implications to human health including gene–environment interactions and the inaccurate, commonly used frame of "gene for traits."[1] It is therefore time to highlight common misconceptions about the certainty of genomic tests and the underlying assumptions these hold when communicating test results to patients. Modern society and social influences always tend to distort information accuracy, thus increasing the general public's confusion while restraining their positive attitudes toward genomic testing. For this reason, urgent but also effective educational awareness of general HCPs and other specialized physicians could improve patient–HCP interaction and communication, reducing ambiguity and raising patient awareness as well. The findings from science education research might also be relevant.[38]

The chapter critically analyses empirical findings and the educational strategies that could be employed in order to improve general genetic concept understanding as well as confidence levels across all HCPs in the medical field ranging from general practitioners, primary care and specialist physicians, nurses, and health educators in the public health care (Fig. 12.1). In specific, short-term educational strategies in upskilling today's registered HCPs worldwide range from online or face-to-face training courses, genetic conferences, or symposiums with workshops, personalized educational materials fitted to individual's medical practice and specialty care as well as CPD modules are some of the few preferred educational means described by the majority of participants in these studies. Although these methods could be effective in enhancing the current medical workforce, additional long-term educational strategies should be implemented across the globe. We have highlighted significant educational discrepancies among universities across different regions including countries in the European Union where institutions in some regions such as Greece, Germany, United Kingdom, The Netherlands, and Italy have structured genomics courses embedded into the students' curricula of all medically related degrees versus others (e.g., Croatia or Turkey) that completely lack or provide limited genetics education to their students. This is expected to create differing attitudes, competencies, and confidence levels among the future medical graduates, leading to the formation of a viscous cycle of continuous challenges and slow progress in achieving our optimum end objective in creating global, systemic genomics integration in the medical care. For this reason, global initiatives in integrating the field of genomics into the higher education systems will ultimately result in the development of more a competent and up-to-date medical workforce in the future.

Finally, we also appreciate the multivariate stakeholder involvement in the public and HCP genomics literacy. Besides, direct contribution of the medical community, localized and global governmental, educational, political, research, and economic institutions play a central role in this field. It is argued that the costs of running global medical genomics projects are immense, thus restraining their official roll out into health-care systems. Health economic evaluations should be modified in order to better assess the long-term benefits of educational genomics and their practical implications of their integration into the medical arena. A global collaboration in upskilling current

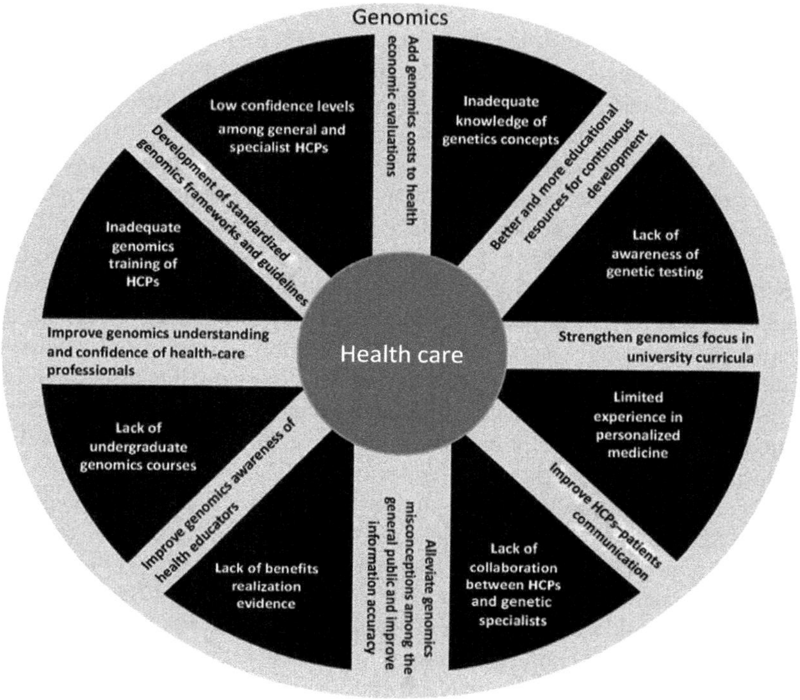

FIGURE 12.1

Focal challenges (*black circle*) that prevent genomics integration into the health-care practice and key recommendations (*gray boxes*) for best practice implementation in the future.

and future workforce, generating evidence of the value of genomics for patients and health-care systems as well as developing global guidelines fit for each specialty and optimized to each regional health-care system could potentially create synergies, reduce costs, and increase overall genomics application and stakeholder confidence in employing these genomics tools.[39]

REFERENCES

1. Kampourakis K. *Making Sense of Genes*. Cambridge: Cambridge University Press; 2017.
2. Brierley KL, Blouch E, Cogswell W, et al. Adverse events in cancer genetic testing: medical, ethical, legal, and financial implications. *Cancer J*. 2012;18(4):303–309.
3. Kampourakis K. Public understanding of genetic testing and obstacles to genetics literacy. In: Patrinos GP, Danielson PB, Ansorge WJ, eds. *Molecular Diagnostics*. 3rd ed., Elsevier, 2017:469–477.
4. Selkirk CG, Weissman SM, Anderson A, Hulick PJ. Physicians' preparedness for integration of genomic and pharmacogenetic testing into practice within a major healthcare system. *Genet Test Mol Biomarkers*. 2013;17(3):219–225.

5. Marzuillo C, De Vito C, D'Addario M, et al. Are public health professionals prepared for public health genomics? A cross-sectional survey in Italy. *BMC Health Serv Res*. 2014;14(239):1–8.
6. Christensen KD, Vassy JL, Jamal L, et al. Are physicians prepared for whole genome sequencing? A qualitative analysis. *Clin Genet*. 2016;89(2):228–234.
7. Li J, Xu T, Yashar BM. Genetics educational needs in China: physicians' experience and knowledge of genetic testing. *Genet Med*. 2014;17(9):757–760.
8. Ganne P, Garrioch R, Votruba M. Perceptions and understanding of genetics and genetic eye disease and attitudes to genetic testing and gene therapy in a primary eye care setting. *Ophthalmic Genet*. 2015;36 (1):50–57.
9. Just KS, Steffens M, Swen JJ, Patrinos GP, Guchelaar HJ, Stingl JC. Medical education in pharmacogenomics—results from a survey on pharmacogenetic knowledge in healthcare professionals within the European pharmacogenomics clinical implementation project Ubiquitous Pharmacogenomics (U-PGx). *Eur J Clin Pharmacol*. 2017;73(10):1247–1252.
10. Gingras I, Sonnenblick A, De Azambuja E, et al. The current use and attitudes towards tumor genome sequencing in breast cancer. *Sci Rep*. 2016;6(22517):1–8.
11. Carroll JC, Makuwaza T, Manca DP, et al. Primary care providers' experiences with and perceptions of personalized genomic medicine. *Can Fam Physician*. 2016;62(10):e626–e635.
12. Gray SW, Hicks-Courant K, Cronin A, Rollins BJ, Weeks JC. Physicians' attitudes about multiplex tumor genomic testing. *J Clin Oncol*. 2014;32(13):1317.
13. Johnson LM, Valdez JM, Quinn EA, et al. Integrating next-generation sequencing into pediatric oncology practice: an assessment of physician confidence and understanding of clinical genomics. *Cancer*. 2017;123(12):2352–2359.
14. Seven M, Akyüz A, Elbüken B, Skirton H, Öztürk H. Nurses' knowledge and educational needs regarding genetics. *Nurse Educ Today*. 2015;35(3):444–449.
15. Lopes-Júnior LC, Carvalho Júnior PM, Faria Ferraz VE, Nascimento LC, Van Riper M, Flória-Santos M. Genetic education, knowledge and experiences between nurses and physicians in primary care in Brazil: a cross-sectional study. *Nurs Health Sci*. 2017;19(1):66–74.
16. Godino L, Turchetti D, Skirton H. Genetic counseling: a survey to explore knowledge and attitudes of Italian nurses and midwives. *Nurs Health Sci*. 2013;15(1):15–21.
17. Chapman R, Likhanov M, Selita F, Zakharov I, Smith-Woolley E, Kovas Y. New literacy challenge for the twenty-first century: genetic knowledge is poor even among well educated. *J Community Genet*. 2018;10(1):73–84.
18. WHO. Prevention of blindness and visual impairment. In: *Priority Eye Diseases* [Online]. Available from: <http://www.who.int/blindness/causes/priority/en/>; 2018 Accessed 09.04.18.
19. Bunce C, Wormald R. Leading causes of certification for blindness and partial sight in England & Wales. *BMC Public Health*. 2006;6(58):1–7.
20. Rahi JS. Childhood blindness: a UK epidemiological perspective. *Eye*. 2007;21(10):1249–1253.
21. Ubiquitous Pharmacogenomics (U-PGx). U-PGx & study [Online]. Available from: <http://upgx.eu/>; 2018 Accessed 15.04.18.
22. Gray SW, Hicks-Courant K, Lathan CS, Garraway L, Park ER, Weeks JC. Attitudes of patients with cancer about personalized medicine and somatic genetic testing. *J Oncol Pract*. 2012;8(6):329–335.
23. Hsiao CY, Lee SH, Chen SJ, Lin SC. Perceived knowledge and clinical comfort with genetics among Taiwanese nurses enrolled in a RN-to-BSN program. *Nurse Educ Today*. 2013;33(8):802–807.
24. Horovitz DDG, de Faria Ferraz VE, Dain S, Marques-de-Faria AP. Genetic services and testing in Brazil. *J Community Genet*. 2013;4(3):355–375.
25. Kim MY. The nurses' knowledge and perception of their role in genetics. *J Korean Acad Nurs*. 2003;33 (8):1083–1092.

26. Chen LS, Kim M. Needs assessment in genomic education: a survey of health educators in the United States. *Health Promot Pract*. 2014;15(4):592–598.
27. Pisanu C, Tsermpini EE, Mavroidi E, Katsila T, Patrinos GP, Squassina A. Assessment of the pharmacogenomics educational environment in Southeast Europe. *Public Health Genomics*. 2014;17 (5–6):272–279.
28. Read CY, Ward LD. Faculty performance on the Genomic Nursing Concept Inventory. *J Nurs Scholarsh*. 2016;48(1):5–13.
29. Donnelly MK, Nersesian PV, Foronda C, Jones EL, Belcher AE. Nurse faculty knowledge of and confidence in teaching genetics/genomics: implications for faculty development. *Nurse Educ*. 2017;42 (2):100–104.
30. National Human Genome Research. Inter-Society Coordinating Committee for Physicians Education in Genomics (ISCC) [Online]. Available from: <https://www.genome.gov/27554614/intersociety-coordinating-committee-for-practitioner-education-in-genomics-iscc/>; 2018 Accessed 25.04.18.
31. Korf BR, Berry AB, Limson M, et al. Framework for development of physician competencies in genomic medicine: report of the Competencies Working Group of the Inter-Society Coordinating Committee for Physician Education in Genomics. *Genet Med*. 2014;16(11):804–809.
32. Slade I, Subramanian DN, Burton H. Genomics education for medical professionals—the current UK landscape. *Clin Med*. 2016;16(4):347–352.
33. De Abrew A, Dissanayake VH, Korf BR. Challenges in global genomics education. *Appl Trans Genomics*. 2014;3(4):128–129.
34. Young A, Chaudhry HJ, Rhyne J, Dugan M. A census of actively licensed physicians in the United States, 2010. *J Med Regul*. 2011;96(4):10–20.
35. Passamani E. Educational challenges in implementing genomic medicine. *Clin Pharmacol Ther*. 2013;94 (2):192–195.
36. Phimister EG, Feero WG, Guttmacher AE. Realizing genomic medicine. *N Engl J Med*. 2012;366:757–759.
37. Buchanan J, Wordsworth S, Schuh A. Issues surrounding the health economic evaluation of genomic technologies. *Pharmacogenomics*. 2013;14(15):1833–1847.
38. Stern F, Kampourakis K. Teaching for genetics literacy in the post-genomic era. *Stud Sci Educ*. 2017;53 (2):193–225.
39. Manolio TA, Abramowicz M, Al-Mulla F, et al. Global implementation of genomic medicine: we are not alone. *Sci Transl Med*. 2015;7(290):1–20.

FURTHER READING

National Commission for Health Education Credentialing (NCHEC), 2018National Commission for Health Education Credentialing (NCHEC). Responsibilities and competencies for health education specialists [Online]. Available from: <https://www.nchec.org/responsibilities-and-competencies>; 2018 Accessed 25.04. 18.

CHAPTER

"GENETHICS" AND PUBLIC HEALTH GENOMICS

13

Emilia Niemiec and Heidi Carmen Howard

Centre for Research Ethics and Bioethics, Uppsala, Sweden

13.1 INTRODUCTION TO ETHICAL ISSUES IN PUBLIC HEALTH

The World Health Organization defines public health as "the art and science of preventing disease, prolonging life and promoting health through the organized efforts of society."[1,2] Awofeso distinguished six public health eras in history, each of which is characterized by a different primary focus.[3] The earliest public health approach of "health protection" can be recognized in antiquity; it was based on following cultural and religious rules, which had a health-protecting function. The other approaches include the 19th century "sanitary movement" focused on addressing unsanitary conditions; "contagion control" characterized by efforts to interrupt the transmission of diseases; "preventive medicine" developed in the mid-20th century and focused on preventive measures applied to specific groups (e.g., elderly, pregnant women, and schoolchildren); "primary health care" with its approach of "health for all" and emphasis on equity in health care; and the currently dominant "health promotion", which has the key tasks to "build healthy public policy, to create supportive environments, to strengthen community action, to develop personal skills, and to reorient health services."[3] These six approaches to public health have been intertwined to some extent and all of them appear to various degrees in current public health practices.

Importantly, public health in any region or country possesses specific features and needs which should be addressed in order to improve the health of a given population. As Awofeso puts is

> ... public health is essentially an expression of the ways different societies address questions of social order and nationhood. By first addressing the structures of power and socioeconomic development within the history of national and regional cultures, the suitability of implementing specific public health paradigms might become clearer.
>
> **Awofeso.**[3]

The way health care is organized in a given country or region, whether there are public or private health-care services, the scope of such services, and the quality and competences of health professionals depend, to a great extent, on the economic situation of a given population, its government's attitudes to these issues and the values accepted in a given society. Moreover, the question of how society, including public health, should be organized may be answered from various philosophical perspectives, including utilitarianism, liberalism, and communitarianism.[4] Each of these approaches uses a different basis on which to find the answer to "what should we do?" in

Applied Genomics and Public Health. DOI: https://doi.org/10.1016/B978-0-12-813695-9.00013-3

a given situation, a question addressed by a branch of philosophy called ethics.The first approach, utilitarianism, dominant in the realm of public health policies, aims at "the greatest happiness of the greatest number"[5]; in the context of public health, this goal can be framed as the best health for the greatest number of people. The second ethical position, liberalism, ascribes the greatest value to individual's rights to decide about how to live and would leave choices related to health care to each individual. In the communitarian approach, everyone should aim to live a virtuous life defined by moral norms upheld in a given society (derived, e.g., from religion).[4] Each of these approaches raises questions, one of which is "what is health". In 1948 the World Health Organization extended the classical negative definition of health (i.e., the absence of disease) by proposing a contentious formulation: "a state of complete physical, mental and social well-being and not merely the absence of disease or infirmity."[6] One may rightly ask what is this state of *complete* well-being and how can it be achieved? Subjective utilitarians would argue that well-being should be determined according to personal experience; for example, citizens may be asked to assess what their gain is from a certain public health program. Meanwhile, objective utilitarians use measurements, such as quality-adjusted life years calculated on the basis of existing evidence.[4] Additional questions may be posed when discussing various ethical approaches, for example, in liberalism: which rights should take precedence if these are conflicting? In the public discourse the tensions between these three approaches and their variations are often tangible; these are also present in publichealth genomics.

In this chapter, we first present ethical, legal, and social issues in genetics and genomics, with particular attention to public health and the historical perspective, which can help us understand the current approaches and ethical problems in (public health) genetics and genomics. We then introduce the field of publichealth genomics and discuss selected seemingly particularly important ethical issues in this area.

13.2 INTRODUCTION TO ETHICAL, LEGAL, AND SOCIAL ISSUES IN GENETICS AND GENOMICS

The beginnings of genetics, the science of heredity, can be traced back to ancient times, when philosophers pondered on questions of how traits are passed on to offspring.[7] Yet, the first organized attempts to use the knowledge of human heredity to improve health of a given population seem to appear much later. In the beginning of the 20th century, eugenics, defined by Francis Galton as "science of improving stock" aiming to "give to the more suitable races or strains of blood a better chance of prevailing speedily over the less suitable than they otherwise would have had"[8] became a part of public health programs in Nazi Germany, the United States, and Scandinavian countries, among others.[a10] The eugenic approaches in public health differed in each country and involved, among others, compulsory sterilizations and murders (in Nazi Germany) of those who were considered "unfit[10] One example of how eugenics was practiced in the United States is the infamous case of Buck v. Bell in Virginia, in which the court ruling allowed for compulsory sterilization of Carrie

[a]For the relation between eugenics and human genetics, see the presendential address to the American Society of Human Genetics by Dunn.[9]

Buck, on the basis of her alleged "feeble-mindedness," which as it turned out later on, was not supported by any evidence.[10] These are examples of harmful and unacceptable programs based on a misunderstanding of science and representing at least in part utilitarian thinking. For example, in the aforementioned case of Buck v. Bell, the ruling was justified on the grounds of public welfare, understood in a particular way:

> We have seen more than once that the public welfare may call upon the best citizens for their lives. (. . .) It is better for all the world, if instead of waiting to execute degenerate offspring for crime, or to let them starve for their imbecility, society can prevent those who are manifestly unfit from continuing their kind. The principle that sustains compulsory vaccination is broad enough to cover cutting the Fallopian tubes.
>
> **Justia Legal Resources.**[11]

In the years following the World War II (WWII), such practices were condemned, for example, Nazi doctors who conducted "Euthanasia Program" were found guilty of war crimes and crimes against humanity in the Nuremberg doctors' trial.[12] Post-war human genetics in the United Kingdom and the United States was mostly against eugenics.[10] Notwithstanding, many German eugenicists were reinstated in their positions after WWII, and there have been examples of clearly articulated eugenic ideas by prominent scientists in the years since.[b10]

In 1963 the first population screening program was introduced in the United States to detect a genetic disease, phenylketonuria in newborns, which allowed for early diagnose of this disease and dietary therapy.[13] In 1966 the first prenatal diagnosis using cytogenetics was conducted by Steele and Breg.[14] Around that time, abortion was legalized in many countries—if a genetic disease was detected in a fetus, the pregnancy could be terminated.[10] With time, diagnostic and screening methods allowing the detection of more genetic diseases became available and along with that the practice of genetic counseling developed having nondirectiveness as its principle. Obtaining informed consent for genetic and genomic testing, both in research and clinical contexts, as well as securing confidentiality of health-related information became other seemingly cardinal elements of ethical practice of medical genetics and genetic research.[15] Implementation of these principles into practice was meant to secure that there is no coercion, patients can make informed decisions and the access to their genetic information is controlled.

Ethical aspects of the developments in genetics received attention also in the context of the Human Genome Project (HGP), a 13-year-long endeavor funded by the US Congress, which involved efforts by teams in the United States and other countries (such as France and the United Kingdom). The HGP resulted in the sequencing of all human DNA—the human genome, as well as the sequencing of genomes of model organisms.[16] The HGP was accompanied by the Ethical, Legal, and Social Implications (ELSI) Program, which initially investigated topics, such as privacy, fairness in the use of genetic information, safety and effectiveness in the implementation of

[b]For example, Linus Pauling, Nobel prize laureate said, "I have suggested that there should be tattooed on the forehead of every young person a symbol showing possession of the sickle-cell gene or whatever other similar gene, such as the gene for phenylketonuria, that he has been found to possess in single dose. If this were done, two young people carrying the same seriously defective gene in single dose would recognise this situation at first sight, and would refrain from falling in love with one another. It is my opinion that legislation along this line, compulsory testing for defective genes before marriage, and some form of public or semi-public display of this possession, should be adopted."[10]

genetics in clinical care, education of public and professionals in genetics, and was meant to be informative to policy developments.[17,18] The successful accomplishment of the HGP fueled further genetic,and from this point also *genomic* studies,which involve large-scale genomic sequencing using high-throughput technologies and the collection of massive amounts of genomic data.[17,19] The fast pace of genomic research, which, among others, has resulted in new tools and approaches for medical practice, has also raised new ethical, legal, and social challenges. Therefore to ensure timely reflection on these issues, the ELSI research has continued to be funded by the US National Institutes of Health in the form of grants for researchers, impacting how genomic research is performed and its results are implemented in medical practice.[20] Importantly, similar research projects have been initiated in other countries, and the term "ELSI research" started being used to name a distinctive area of studies.[20,21]

13.3 CAN GENOMICS IMPROVE PUBLIC HEALTH?

As described earlier, genomics is a relatively new approach in health care, yet, some genomic technologies, such as genomic sequencing, have already started to be implemented into clinical care in some countries, such as in the United States, the United Kingdom, and The Netherlands.[22] It is important to keep in mind that in many other countries, particularly with lower incomes, the use of genomics in health care may not even be considered, as there may be much more pressing (health care) needs to be addressed.

The main genomic technology that is currently used in some health-care systems is next-generation sequencing (i.e., high-throughput massively parallel sequencing). Next generation sequencing can be applied to sequence selected parts of genome (targeted sequencing), whole genome, and whole exome. These approaches may be used for different purposes: diagnostic testing, predictive testing, screening, pharmacogenomic testing; and in different life stages: in embryos, during the prenatal, newborn, childhood, or adult periods. Each different use and (potential) combination with life stage raises overlapping and distinct ELSI. Another emerging field of applications of genomics is pathogen medicine, in particular, infection diagnosis, monitoring of microbial resistance, and virulence prediction using next-generation sequencing.[23,24] Furthermore, impact on public health can be considered when discussing genedrive technologies, which may be usedto eradicate a population of a disease vector, for example, mosquitoes carrying malaria[25].

Public health genomics has been defined as "The responsible and effective translation of genome-based knowledge and technologies for the benefit of population health."[26,27] The feature of public health genomics which makes it distinct from genomic medicine is the focus on the health of a population rather than of an individual. To answer the question whether genomics can improve public health, that is, whether it can be helpful in efforts to improve health of a given population (e.g. of a country or region), evidence of the impact (and its effectiveness) of a given genomic intervention (e.g. whole genome sequencing and analysis of genes associated with cancer) on a given group of individuals is needed (e.g. healthy individuals or adults with familial history of cancer). Such evidence together with the characteristics of a given community (e.g. prevalence of a given genotype) should be taken into account when considering introduction of the intervention to improve health of that community in question. Since public health structures usually operate on limited resources, the costs of an intervention also play a role in such deliberations. Furthermore, as stated in the definition, genomic knowledge and technologies should be translated responsibly, which entails that related ELSI should be considered. In the following sections we focus on selected applications of genomic technologies and related ethical issues relevant to public health. We firstly provide insights into available evidence of the impact of genomic sequencing results on diagnoses, its use for screening purposes, and discuss the problem of the definition of clinical utility. In the subsequent sections we focus on specific applications of genomic technologies: sequencing and genome editing in the context of reproductive health.

13.3.1 GENOMIC SEQUENCING IN DIAGNOSIS

The use of genomic technologies is relatively new, and the question of whether genomic sequencing can improve and/or be a staple of public health has yet to be answered. Whole exome sequencing (WES) was shown to have a diagnostic rate of 26% in a group of 814 undiagnosed patients with symptoms of developmental delay, ataxia, and others.[28] Whole genome sequencing (WGS) applied as a first-tier test in pediatric patients increased diagnostic yield compared to conventional genetic testing (i.e., chromosomal microarray analysis), which is currently offered as a standard of care in that patients.[29–31] Moreover, Farnaes et al. found that the diagnoses led to changes in treatment of 31% of 42 infants who underwent WGS, indicating clinical utility of the sequencing in that group.[29] Other studies showed diagnostic successes of genomic sequencing in specific groups of patients, such as those affected by neurological Joubert syndrome,[32] galactosialidosis,[33] and hypercholesterolemia.[34] The impact of genomic sequencing on diagnosis, selection of targeted therapy, and treatment when a patient ceases to respond to a targeted therapy due to a resistance mutation was demonstrated in oncology.[35] In many other diseases, evidence for such clinical utility is still missing, together with larger scale data on cost-effectiveness of genome sequencing in specific contexts.[36]

13.3.2 GENOMIC SCREENING

Evidence for benefits of genomic sequencing is also limited in another area of applications for which WGS has been suggested, that is, screening of a healthy population,[37] newborn screening,[38,39] and so-called opportunistic screening for a set of diseases, which would be offered every time diagnostic WGS is conducted.[40] Screening is an approach aiming to detect disease risk factors (e.g., presence of a genetic variant) or disease at the early stage of development in asymptomatic, potentially at-risk population, and unlike diagnostic testing, it usually does not provide a diagnosis, but a suspicion of a disease and indication for a confirmatory diagnostic test. In 1968 Wilson and Jungner proposed principles for screening for diseases, which became the basis of policies for many screening programs.[41] They require, among others, "... (2) There should be an accepted treatment for patients with recognized disease. (3) Facilities for diagnosis and treatment should be available. (4) There should be a recognizable latent or early symptomatic stage. (5) There should be a suitable test or examination."[41] As previously mentioned, genetic screening was used for the first time in newborns in the 1960s to detect a metabolite, which indicates a genetic disease, phenyloketonuria.[13] Over time, other diseases have been added in newborn screening programs. Furthermore, carrier screening, that is, screening of couples to detect those heterozygous for recessive diseases, such as Tay–Sachs was introduced in certain populations; such heterozygous couples are at 25% risk of having an affected child.[10] Moreover, screening of fetuses (prenatal screening) and more recently screening of embryos created in in vitro fertilization (IVF) became available (see the following section). Importantly, the development of next-generation sequencing and its increased accessibility prompted suggestions for its routine use for various screening purposes previously mentioned and expansion of the number of diseases for which a given population would be screened, which supposedly would allowed for earlier treatment and/or prevention of genetic diseases. Furthermore, the American College of Medical Genetics (ACMG) in 2013 suggested active searching for a list of secondary variants, or in other words opportunistic screening, which is the analysis of a set of medically actionable variants every time a clinical WGS is offered.[43] As mentioned earlier, the evidence for benefits or clinical utility of opportunistic screening and WGS approaches in screening at various stages of life and various purposes is limited.[37–40] Furthermore, it has been questioned whether such proposals for using WGS for screening, for example, of newborns comply with the established principles of Wilson and Jungner and if not, whether such genomic screening should be considered.[42] Meanwhile, the risks and limitations of such potential screening applications of WGS are numerous and weighty. Adams et al., who conducted an analysis of algorithms for the selection of variants, which could be considered for the screening of general adult population, capture many caveats of genomic screening in the conclusion of their study:

> (...) we must be cognizant of our limited ability to interpret the pathogenicity of rare genomic variations, the implications this has for the standards used to analyze and report genomic findings, and the downstream consequences of screening among the general population. Otherwise, great harm could result. Much additional research is required to assess the overall cost of screening efforts, including the downstream effects of false-positive tests, in order to evaluate the overall economic impact of genomic screening. A host of factors need to be studied, including cost-effectiveness, adverse consequences of interventions, insurance impact, education and consent materials that address the complexity of potential harms and benefits of screening, implications for reproductive choices, potential psychosocial impacts, and the benefits and harms of familial cascade testing.
>
> **Adams et al.**[37]

Many of these limitations of genomic screening in various contexts have been discussed at length.[42–45] Genomic screening and diagnostic tests have been offered also by companies that advertise directly to consumers, which raises additional set of ethical considerations relating to a lack of involvement of a health-care professional, unbalanced or misleading information provision, and impact of such offers on public health services, among others.[46] WGS/WES approaches have been evaluated also in government-funded study programs.[22] These evaluations are often challenged by difficulties in assessing or quantifying impacts of downstream factors, such as costs of analysis, storage and interpretation of genomic data, and needs for training of health-care professionals.[47]

13.3.3 CLINICAL UTILITY

The evaluation of impact or utility of genomic sequencing is further complicated if we consider additional potential benefits and risks of genomic sequencing, which are not related to the treatment of a disease yet may be considered affecting utility of sequencing. In 2015 the ACMG stated

> We submit that the clinical utility of genetic testing and services should take into account effects on diagnostic or therapeutic management, implications for prognosis, health and psychological benefits to patients and their relatives, and economic impact on health-care systems.
>
> **ACMG Board of Directors.**[48]

This broadly understood utility of sequencing may include providing comfort to patients and families by offering diagnosis (even if it does not change the treatment), possibility of enrollment in clinical trials following the diagnosis, and impact of the findings on family members (e.g., indications of increased risk of a disease for which preventive measures can be taken).[48] Yet, important questions arise when considering this expanded definition of clinical utility from public health perspective and when trying to answer the question if genomic sequencing can improve public health. Which exactly is the clinical utility of genomic sequencing and how should it be measured? What importance should be placed on psychological aspects (or benefits) given that there may be more pressing medical needs (e.g., saving lives) to address in a population? Should we focus on potential short- or long-term impacts (such as related to enabling research in which sequencing results are used)? Such reflection on what "clinical utility" is relates to the problems with defining "health" and "well-being," and what constitutes "risks" and "benefits" as well as problems of prioritization of interventions and distribution of goods if we discuss publicly funded health-care systems. The answers to such issues impact the decisions on policies about which technologies or approaches should be used and how they should be implemented; for example, which findings should be returned to patients or whether/when screening should be offered. The definitions, measuring of costs and risks, and importance of values in medical practice are even more important when we consider the applications of genomic sequencing in the context of reproduction.

13.3.4 GENOME SEQUENCING IN THE CONTEXT OF REPRODUCTION

The use of genomics in the context of human reproduction, that is, for preconceptional or carrier (in prospective parents), preimplantational (in embryos in vitro), and prenatal testing and screening (of fetuses during pregnancy) raises specific issues, which are, to a great extent, overlapping with those related to "traditional" genetic testing in these circumstances. However, due to the huge amount of information contained in genomic data, some concerns are amplified when using high-throughput sequencing in these contexts. The results of screening/testing of (prospective) parents may result in a choice to undergo IVF procedures with embryo testing in order to select those that do not carry a given disease-causing genetic variant(s) and have these implanted to establish a pregnancy. When prenatal testing/screening is performed, the findings may be used to decide for abortion performed within the (public) health-care system (in the countries where the legislation allows for that). These practices may recall eugenics as their aim is to ensure that the babies are born healthy, by the elimination of embryos/fetuses that carry disease-causing genes. What distinguishes the current offer of genetic and genomic testing/screening from past eugenic approaches is, first, that it is not based on false understanding of mechanisms of heredity which were underlying the eugenic practices in the first half of the 20th century, and, second, there is no coercion; as discussed previously, one of the principles in genetic counseling is nondirectiveness to facilitate informed choice of a patient. Yet, the practice shows that the application of our knowledge of human genetics to interpret genetic and genomic findings and the task of facilitating informed choices do not remain without challenges. The cases of misdiagnosis based on genetic test have been reported, also in the context of prenatal testing.[49,50] Interpretation of genetic variants is challenging, and sometimes the link between the genetic makeup and phenotypic manifestations of diseases is more complex than was initially understood. Furthermore, genetic findings often only provide a probability of the occurrence of a disease in the future. A question may be posed: do the patients [in this context (prospective) parents] understand the results and their implications, and, if it is the case, their probabilistic nature? How much and which information should be given to parents about the implications of the test results so that we are sure that their choice is informed and free? Informed consent is problematized further if prenatal testing/screening is offered by companies advertising directly to consumers (i.e., pregnant women in this case); it was shown that some of these companies presented inadequate information about the testing on their websites.[51] Even if both the scientific accuracy and informed choice were always secured (indeed, one may doubt if it is ever truly possible), a fundamental question may still be posed whether the practices of selective abortions (which are often offered within public health-care services) and embryo selection are ethical. Should the presence of genetic variants be a reason for choosing abortion and why? Which genetic variants/disease should allow for choosing abortion and why? Where do we draw the line? What are (social) implications of using subjective criteria, such as "normalcy" and "serious disability?" Importantly, what healthy people may consider a low quality of life for a disabled person may not be perceived as such by that affected individual; on the contrary, studies have shown that many people with serious disabilities report their quality of life as good or excellent.[52] There are many more questions surrounding the use of genomics in the reproductive context. The central question yet seems to relate to the value of human life, and whether it can be arbitrarily decided that some lives are worth less because of being affected by a disease. The use of genetics and genomics in the context of reproductive choice has been expanding; public prenatal screening programs using noninvasive screening testing have been implemented in a few countries. Furthermore, proposals to use WGS in prenatal care were made. In this context, the articulation of values, which (should) guide the practice of public health genomics, limitations of genomic technologies, and sober evaluation of the impacts for individuals and society are particularly pertinent. This reflection reminds us that health of a population should not be pursued at any cost; there are other values, such as human life, which are more important than achieving the goal of a healthy population.

13.3.5 GERMLINE GENOME EDITING

CRISPR-Cas9 is a novel technology, which emerged a few years ago, allowing faster, easier, and cheaper modification of DNA than was possible with previously used methods. Currently, CRISPR-Cas9 and other CRISPR-Cas systems are widely used in human genetics research, both in cell lines and (less often) embryos, to investigate function of genes and to develop the treatment approaches for genetic diseases.[53] Currently, there are many clinical trials running involving somatic genome editing, the goal of which is to cure/treat disease via somatic cells (i.e., the majority of body cells that will not become gametes, such as skin and blood cells); such approaches are being developed for blood and neuromuscular disorders, among others.[54] Furthermore, despite calls for caution and moratoriums, a case of germline genome editing (GLGE) on embryos, which developed into children, was reported in China in November 2018.[55]

When discussing the impact of genome editing on public health, we should pose the question: can genome editing improve health of a given population? As mentioned, in the case of somatic genome editing, this question is being answered by collecting evidence from clinical trials to evaluate safety and efficacy of the genome editing approaches.[54] Meanwhile, potential applications of GLGE and their evaluation raise more problems and their prospective impact on health, including many possible harms and uncertainties, is the subject of vivid debates. Herein, we will ponder over the questions of the impact of potential clinical application of GLGE on (public) health and related ethical aspects.

In the following sections, we discuss three situations in which GLGE has been suggested to be used. For all these situations, the aim of GLGE would be *a genetically related child* not affected by certain disease(s) or possessing certain features. The procedure would involve IVF using parents' gametes and modification of DNA of created embryos to correct disease-causing gene(s) or other genes to achieve certain features. Descendants of such "genome-edited" persons would inherit their modified DNA and consequently would not be affected by the disease in question (caused by the "corrected" gene) or would possess certain feature introduced by genome editing (whichever goal was the intent). In all three scenarios outlined next, there exist alternatives to GLGE, albeit in some cases not all the parents' wishes may be delivered (e.g., child will not be biologically related in the case of adoption).

13.3.5.1 Context 1: Using germline genome editing to avoid having a child with disease in a situation where there is no chance parents will not pass on disease to all biological offspring

The first situation occurs when a couple carries gene variants in configuration, which will certainly cause a monogenic disease in *all their offspring*; that is, where both prospective parents are homozygous for a recessive disorder (and thus both are affected by that disease, e.g., sickle-cell disease, and phenylketonuria) or when one parent is homozygous for a dominant disorder (e.g., Huntington's disease and familial hypercholesterolemia). Importantly, such situations are rare; we can only speculate how many such couples exist and if they would be interested in pursuing GLGE.[56] Indeed, in such situations, preimplantation genetic diagnosis (PGD) would not work, since no embryos would be expected to be free from the disease. In such case, in order to have a healthy child who is genetically related to parents, one would need to consider GLGE. An existing alternative is adoption; while the child would not be biologically related, parents would avoid passing a disease on to their offspring.

13.3.5.2 Context 2: Using germline genome editing to avoid having a child with a disease in a situation where, in theory, some parents' gamete combinations should be disease free

In the second situation, a couple would carry gene–variants combination, which would cause a disease *in part of their offspring* (e.g., parents being heterozygous for a recessive disease). As in the previous scenario, in theory, GLGE could be considered if such couples wished to have genetically related offspring, without passing on disease and were willing to undergo IVF. Importantly, at present PGD may be used to detect disease-causing variants and select embryos which do not carry them. Some argue that PGD performed along with GLGE could increase the number of embryos eligible for transfer to uterus to establish pregnancy, improving pregnancy rates,[57] and therefore, it would be an advantageous approach over PGD alone. Yet, such arguments are highly speculative, there are currently no data on how efficient IVF combined with GLGE would be.[56] Clinical uses of GLGE in such cases have not been reported. A few studies on embryos aiming to correct disease-causing genes, for example, for cardiomyopathy, have been published.[57–60]

13.3.5.3 Context 3: Using germline genome editing to prevent disease/enhance

The first clinical application of CRISPR-Cas9 in the germline was conducted for a purpose different from those previously outlined in contexts 1 and 2 and one that may be qualified as enhancement or prevention (the line between these two is arguably not clear cut). He Jiankui, a Chinese scientist, reported that he inactivated the gene *CCR5*, which encodes a receptor used by HIV to enter cells, in embryos that were subsequently used to establish pregnancy. The goal of that DNA modification was to obtain a child genetically related to a given couple and who would have reduced risk of acquiring AIDS[55] Related to such application, yet distinct and rather far-fetched are the proposals of using GLGE to improve nonhealth-related traits, such as intelligence.

Could GLGE improve health if used in one of the contexts listed earlier? To answer this question, we should specify how we understand health and whose health we are considering (see Section 13.1). Considering the currently initial stage of development of the technology, technical problems related to the occurrence of off-target effects (i.e., unintended modifications of the genome caused by genome editing) and mosaicisms (when genome editing occurs only in part of the cells of an embryo) when GLGE is performed, the usage of the technology entails serious risks. Therefore at this point, based on what we know about actual problems with the use of GLGE, it is unlikely to improve health (neither in narrow nor in broad understanding) of any of the actors involved or of a population as a whole.

Yet, we may consider a very optimistic scenario in which GLGE technology is perfected and the safety risks are eliminated or reduced significantly. Of note, in reality, the first clinical GLGE would *necessarily* involve risks and uncertainties and would be of experimental nature, since the effects of this technology on a developing organism are simply unknown and are impossible to fully predict.

So, if we imagined that such an experimental stage was successful and did not cause any side effects or only minor ones, could the abovementioned applications of GLGE improve health of a child-to-be, its parents, or a population as a whole? Importantly, GLGE would take place in the context of IVF; the recent study shows that injection of CRISPR-Cas9 ingredients together with sperm into oocytes can minimize mosaicism in embryos; therefore most likely gene-editing system would be introduced into an oocyte at the moment of fertilization.[57] From this perspective, GLGE (coupled with IVF) is an approach of creating embryos, which does not *treat* or *cure* an embryo, an individual, or their parents; it rather aims to create an individual with desired characteristics (e.g., not being affected by a disease) at request of (prospective) parents. Assuming the broad definition of health as "complete well-being" (see Section 13.1), one may argue that GLGE may improve such understood health of a given couple by fulfilling their desire of having genetically related offspring not affected by a given disease or possessing a given trait. There is also a chance that GLGE could improve the health of a (sub)population, again, assuming that the technique would work well. However, given that monogenic diseases tend to be rare, it is questionable that any significant difference would be experienced in a population as whole. Furthermore, we should consider at what costs these potential benefits of satisfied parents and so-called healthier population would be achieved. Many risks and (ethical) problems can be considered when answering this question, such as health risks both to mother (e.g., related to oocyte extraction) and future child (including psychological aspects), costs for a health-care system (if GLGE was funded by public money), destruction of human embryos, potential decrease in care and "tolerance" for people who are different and/or who suffer from disease. Due to limited space, we do not have the opportunity to discuss all these aspects here; however, we would like to highlight one of the issues, the problem of instrumentalization of embryos/persons.

This problem was raised in the context of IVF; Pacholczyk stated, "It (IVF) dehumanizes embryonic children, treating them as objects to be frozen, manipulated, abandoned, or destroyed."[61] Indeed, in IVF, human life and its beginnings are in the hands of third parties, the doctors, who work to fulfill the desire of a couple; this process involves freezing and/or discarding human embryos; human life is therefore used as a means to satisfy someone. If we add to this picture that the procedure (IVF coupled with GLGE) would aim to create a child with given traits, this may recall eugenic practices to some.[62] Interestingly, Cussins and Lowthorp state that IVF "has a widely unacknowledged legacy of eugenics" considering the relationships of their pioneers with eugenics movement; the current developments related to nuclear genome transfer and gene editing may be seen as related to this legacy.[63] The same authors point out:

> We cannot talk about NGT or gene editing in embryos as tools that would only mitigate disease propensity. There is significantly more social baggage than that. Ethical concerns about children's right to an open future, and for the parent/child relationship not to be reduced to an overt commercial transaction, do not hinge on intended use of modification technologies.[63]

Currently, reflection on these ethical aspects, clarity on the definitions (e.g., of "treatment" and "health"), and values guiding public health practices are very desirable, especially given that the first clinical application of GLGE already has taken place in China with violation of number of ethical principles and recommendations.[64]

13.4 CONCLUSION

In this chapter, we first reflected on the approaches to public health, ethical positions on public health issues, and definitions of health. We subsequently presented ethical, legal, and social issues in genetics and genomics, with particular attention to public health and the historical perspective. We then discussed the issues related to implementation of genomic technologies in public health with focus on genomic sequencing used for diagnosis, screening, its applications in the context of reproductive technologies, and the potential use of genome editing technology in the germline. We did not aim to provide a whole picture of public health genomics but rather to highlight aspects which seem to be most pressing given their current development and/or use and ethical dimensions.

As much as genomics opens new opportunities of improving health, it introduces also risks of abuses. In the introduction, we aimed to shed light on the past of genetics and genomics, related ethical problems and misuses in applying the knowledge about heredity, which to a great extent shaped the current principles guiding research and clinical genetics and genomics. The understanding of the origin of these ethical norms, and overall, the history of genetics may help to apply these ethical principles more consciously and carefully, which may prevent future potential abuses. Indeed, for this cautious and responsible approach to be used, stakeholders who simultaneously guide the use of the technologies and have invested a lot in the development of research and clinical uses of genomics will have to slow down enough to fully understand and acknowledge these potential risks. In particular, conflicts of interests will have to be more closely followed, identified, and highlighted as more and more special groups are being created to "guide" the use of new genomic technologies.

To conclude, we would like to quote the reflection and advice of Peter Harper:

> It seems unlikely that the character of scientists has changed greatly over the years; elements of opportunism and naïveté, along with a conviction of the importance of one's own research, can all be seen today in the context of pressure to apply new genetic findings, though open discussion and peer review should help to limit their misuse.
>
> (...)
>
> a continuation of vigilance, scientific openness, humility, skepticism of extravagant claims, and valuing of the individual, will all be important in preventing major eugenic abuse in the future. And this will be easier to achieve if the disasters produced in the name of eugenics during the first half of the 20th century are not forgotten.
>
> **Harper.**[10]

The open discussion and peer review mentioned, to be effective, should be vigilant and transparent about the limitations of technologies, risks, and uncertainties they involve, stakeholders and their interests and the most vulnerable groups who may be affected by a given technology. Apart from discussion, the education and research on ethical aspects of genetics and genomics, the history of genetics and genomics, their social impacts, and values upheld in genomic practices should be undertaken.

ACKNOWLEDGMENTS

The chapter is partly based upon the 2.1 and 2.4 reports of the SIENNA project (Stakeholder-informed ethics for new technologies with high socio-economic and human rights impact)—which has received funding under the European Union's H2020 research and innovation programme under grant agreement No 741716. The chapter and its contents reflect only the views of the authors and do not intend to reflect those of the European Commission. The European Commission is not responsible for any use that may be made of the information it contains.

These authors have also been partly supported by a grant funded by the Swedish Research Council (2017–01710) "ethical, legal, and social issues of gene editing."

In this chapter, we include content originally written for the introduction and conclusion sections of the doctoral thesis of Dr. Emilia Niemiec titled "Ethical, legal and social issues related to the offer of whole genome and exome sequencing" (2018).

REFERENCES

1. Acheson ED. *Public Health in England. Report of the Committee of Enquiry into the Future Development of the Public Health Function*. London: Department of Health; 1988.
2. World Health Organization. Public health services. From <http://www.euro.who.int/en/health-topics/Health-systems/publichealth-services/publichealth-services>; 2019 Retrieved 26.04.19.
3. Awofeso N. What's new about the new public health. *Am J Public Health*. 2004;94(5):705–709.
4. Roberts MJ, Reich MR. Ethical analysis in public health. *Lancet*. 2002;359:1055–1059.
5. Bentham J. *An Introduction to the Principles of Morals and Legislation*. Oxford: Clarendon Press; 1907.
6. United Nations World Health Organization Interim Commission. *Official Records of the World Health Organization*. Vol. 2. 1948.
7. Aristotle. *History of Animals. In Ten Books*. London: George Bell & Sons; 1883.
8. Galton F. *Inquiries Into Human Faculty and Its Development*. London: Macmillan; 1883.
9. Dunn LC. Cross currents in the history of human genetics. *Am J Hum Genet*. 1962;14(1):1–13.
10. Harper PS. *A Short History of Medical Genetics*. New York: Oxford University Press; 2008.
11. Justia Legal Resources. Buck v. Bell, 274 U.S. 200 (1927). From <https://supreme.justia.com/cases/federal/us/274/200/#207>; 2019 Retrieved 26.04.19.
12. Germany (Territory under Allied Occupation, U.S. Zone). *Trials of War Criminals Before the Nuernberg Military Tribunals under Control Council Law No. 10*. Vol. 1. Washington, DC: Government Printing Office; 1949.
13. Guthrie R, Susi A. A simple phenylalanine method for detecting phenylketonuria in large populations of newborn infants. *Pediatrics*. 1963;32:338–343.
14. Steele MW, Breg WR. Chromosome analysis of human amniotic-fluid cells. *Lancet*. 1966;287(743):383–385.
15. Nuffield Council. *Genetics Screening – Ethical Issues*. 1993.
16. National Human Genome Research Institute. Human genome project results. From <https://www.genome.gov/human-genome-project/results>; 2019.
17. Collins F, Galas D. A new five-year plan for the U.S. Human Genome Project. *Science*. 1993;262(5130):43–46.
18. Drell D, Adamson A. *DOE ELSI Program Emphasizes Education, Privacy A Retrospective (1990–2000)*. 2001.
19. Green ED, Guyer MS, National Human Genome Research Institute. Charting a course for genomic medicine from base pairs to bedside. *Nature*. 2011;470(7333):204–213.
20. McEwen JE, Boyer JT, Sun KY, Rothenberg KH, Lockhart NC, Guyer MS. The ethical, legal, and social implications program of the National Human Genome Research Institute: reflections on an ongoing experiment. *Annu Rev Genomics Hum Genet*. 2014;15:481–505.
21. Walker RL, Morrissey C. Bioethics Methods in the Ethical, Legal, and Social Implications of the Human Genome Project Literature. *Bioethics*. 2014;28(9):481–490.
22. Stark Z, Dolman L, Manolio TA, et al. Integrating genomics into healthcare: a global responsibility. *Am J Hum Genet*. 2019;104(1):13–20.
23. Centers for Disease Control and Prevention. Pathogen genomics. From <https://www.cdc.gov/genomics/pathogen/>; 2019.

24. Gwinn M, MacCannell D, Armstrong GL. Next-generation sequencing of infectious pathogens. *JAMA*. 2019;321(9):893–894.
25. Kyrou K, Hammond AM, Galizi R, et al. A CRISPR-Cas9 gene drive targeting doublesex causes complete population suppression in caged Anopheles gambiae mosquitoes. *Nat Biotechnol*. 2018;36 (11):1062–1066.
26. Genome-based Research and Population Health. Report of an expert workshop held at the Rockefeller Foundation Study and Conference Center, Bellagio, Italy, 14–20 April, 2005.
27. Burke W, Khoury MJ, Stewart A, Zimmern RL. The path from genome-based research to population health: Development of an international collaborative public health genomics initiative. *Genet Med*. 2006;8:451–458.
28. Lee H, Deignan JL, Dorrani N, et al. Clinical exome sequencing for genetic identification of rare Mendelian disorders. *JAMA*. 2014;312(18):1880–1887.
29. Farnaes L, Hildreth A, Sweeney NM, et al. Rapid whole-genome sequencing decreases infant morbidity and cost of hospitalization. *NPJ Genom Med*. 2018;3:10.
30. Lionel AC, Costain G, Monfared N, et al. Improved diagnostic yield compared with targeted gene sequencing panels suggests a role for whole-genome sequencing as a first-tier genetic test. *Genet Med*. 2018;20(4):435–443.
31. Stavropoulos DJ, Merico D, Jobling R, et al. Whole genome sequencing expands diagnostic utility and improves clinical management in pediatric medicine. *NPJ Genom Med*. 2016;1:1–9.
32. Tsurusaki Y, Kobayashi Y, Hisano M, et al. The diagnostic utility of exome sequencing in Joubert syndrome and related disorders. *J Hum Genet*. 2013;58(2):113–115.
33. Prada CE, Gonzaga-Jauregui C, Tannenbaum R, et al. Clinical utility of whole-exome sequencing in rare diseases: galactosialidosis. *Eur J Med Genet*. 2014;57(7):339–344.
34. Futema M, Plagnol V, Whittall RA, et al. Use of targeted exome sequencing as a diagnostic tool for familial hypercholesterolaemia. *J Med Genet*. 2012;49(10):644–649.
35. Gagan J, Van Allen EM. Next-generation sequencing to guide cancer therapy. *Genome Med*. 2015;7 (1):80.
36. Douglas MP, Ladabaum U, Pletcher MJ, Marshall DA, Phillips KA. Economic evidence on identifying clinically actionable findings with whole-genome sequencing: a scoping review. *Genet Med*. 2016;18 (2):111–116.
37. Adams MC, Evans JP, Henderson GE, Berg JS. The promise and peril of genomic screening in the general population. *Genet Med*. 2016;18(6):593–599.
38. Ceyhan-Birsoy O, Murry JB, Machini K, et al. Interpretation of genomic sequencing results in healthy and ill newborns: results from the BabySeq Project. *Am J Hum Genet*. 2019;104(1):76–93.
39. Bodian DL, Klein E, Iyer RK, et al. Utility of whole-genome sequencing for detection of newborn screening disorders in a population cohort of 1,696 neonates. *Genet Med*. 2016;18(3):221–230.
40. Burke W, Antommaria AH, Bennett R, et al. Recommendations for returning genomic incidental findings? We need to talk!. *Genet Med*. 2013;15(11):854–859.
41. Wilson JMG, Jungner G. *Principles and Practice of Screening for Disease*. Geneva: World Health Organization; 1968.
42. Howard HC, Knoppers BM, Cornel MC, et al. Whole-genome sequencing in newborn screening? A statement on the continued importance of targeted approaches in newborn screening programmes. *Eur J Hum Genet*. 2015;23(12):1593–1600.
43. Green RC, Berg JS, Grody WW, et al. ACMG recommendations for reporting of incidental findings in clinical exome and genome sequencing. *Genet Med*. 2013;15(7):565–574.
44. Pinxten W, Howard HC. Ethical issues raised by whole genome sequencing. *Best Pract Res Clin Gastroenterol*. 2014;28(2):269–279.

45. Borry P, Chokoshvili D, Niemiec E, Kalokairinou L, Vears DF, Howard HC. Current ethical issues related to the implementation of whole-exome and whole-genome sequencing. In: Schneider S, Brás J, eds. *Movement Disorder Genetics*. Springer, Cham; 2015.
46. Niemiec E, Kalokairinou L, Howard HC. Current ethical and legal issues in health-related direct-to-consumer genetic testing. *Per Med*. 2017;14(5):433–445.
47. Delaney SK, Hultner ML, Jacob HJ, et al. Toward clinical genomics in everyday medicine: perspectives and recommendations. *Expert Rev Mol Diagn*. 2016;16(5):521–532.
48. ACMG Board of Directors. Clinical utility of genetic and genomic services: a position statement of the American College of Medical Genetics and Genomics. *Genet Med*. 2015;17(6):505–507.
49. Manrai AK, Funke BH, Rehm HL, et al. Genetic misdiagnoses and the potential for health disparities. *N Engl J Med*. 2016;375(7):655–665.
50. Yong E. Clinical genetics has a big problem that's affecting people's lives. *The Atlantic*. 2015.
51. Skirton H, Goldsmith L, Jackson L, Lewis C, Chitty LS. Non-invasive prenatal testing for aneuploidy: a systematic review of Internet advertising to potential users by commercial companies and private health providers. *Prenat Diagn*. 2015;35(12):1167–1175.
52. Albrecht GL, Devlieger PJ. The disability paradox: high quality of life against all odds. *Soc Sci Med*. 1999;48:977–988.
53. Knott GJ, Doudna JA. CRISPR-Cas guides the future of genetic engineering. *Science*. 2018;361:866–869.
54. Maeder ML, Gersbach CA. Genome-editing technologies for gene and cell therapy. *Mol Ther*. 2016;24(3):430–446.
55. Cyranoski D, Ledford H. International outcry over genome-edited baby claim. *Nature*. 2019;563:607–608.
56. Lander ES, et al. Adopt a moratorium on heritable genome editing. *Nature*. 2019;567:165–168.
57. Ma H, Marti-Gutierrez N, Park SW, et al. Correction of a pathogenic gene mutation in human embryos. *Nature*. 2017;548(7668):413–419.
58. Liang P, Ding C, Sun H, et al. Correction of beta-thalassemia mutant by base editor in human embryos. *Protein Cell*. 2017;8(11):811–822.
59. Liang P, Xu Y, Zhang X, et al. CRISPR/Cas9-mediated gene editing in human tripronuclear zygotes. *Protein Cell*. 2015;6(5):363–372.
60. Tang L, Zeng Y, Du H, et al. CRISPR/Cas9-mediated gene editing in human zygotes using Cas9 protein. *Mol Genet Genomics*. 2017;292(3):525–533.
61. Pacholczyk T. Gene-edited babies and the runaway train of IVF. From <https://www.ncbcenter.org/files/5215/4817/5871/MSOB162_Gene-Edited_Babies_.pdf>; 2018.
62. Dance A. Better beings? *Nat Biotechnol*. 2017;35(11):1006–1011.
63. Cussins J, Lowthorp L. Germline modification and policymaking: the relationship between mitochondrial replacement and gene editing. *New Bioeth*. 2018;24(1):74–94.
64. Krimsky S. Ten ways in which He Jiankui violated ethics. *Nat Biotechnol*. 2019;37(1):19–20.

CHAPTER

LEGAL ASPECTS OF GENOMIC AND PERSONALIZED MEDICINE

14

Zoe Kordou[1], Stavroula Siamoglou[1] and George P. Patrinos[1,2,3]

[1]Department of Pharmacy, School of Health Sciences, University of Patras, Patras, Greece [2]Department of Pathology, College of Medicine and Health Sciences, United Arab Emirates University, Al Ain, United Arab Emirates [3]Zayed Center of Health Sciences, United Arab Emirates University, Al Ain, United Arab Emirates

14.1 INTRODUCTION

According to the WHO definitions, "Genetics is the study of heredity," while "Genomics is defined as the study of genes and their functions, and related techniques."[1] According to this definition, "the main difference between genomics and genetics is that genetics scrutinizes the functioning and composition of the single gene, whereas genomics addresses all genes and their interrelationships in order to identify their combined influence on the growth and development of the organism" (WHO definitions of genetics and genomics).[2] According to the European Bioinformatics Institute, "Genomics is the study of whole genomes of organisms and incorporates elements from genetics, and it differs from 'classical genetics' in that it considers an organism's full complement of genetic material, rather than one gene or one gene product at a time".[3]

Many jurisdictions expressly forbid genetic modification of the human germ line, but some explicitly allow adjustments for prevention, diagnosis, or treatment as long as the intent is not to modify the germ line. Such provisions arguably open the door to utilizing the potential beneficial somatic gene therapies arising out of applications of technologies, such as CRISPR-Cas9. An example of such legislation is the Lithuanian Law on Ethics in Biomedical Research (2002), which provides that human biomedical studies that modify the human genome may only be carried out for prevention, diagnosis, or treatment and only in the cases where they are not intended to alter the progeny genome (the germ line). Several jurisdictions address the modification of the human germ line, which is usually prohibited when mentioned. The majority of jurisdictions, however, do not appear to have a specific prohibition on germ-line alterations in humans.

The aim of this chapter is to provide an overview of the current legislation within the European Union (EU) member states, as well as other countries, such as the United States of America, Switzerland, Iceland, Norway, China, Singapore, and countries in the Middle East applying to genomics and personalized medicine, highlighting existing gaps, discrepancies, and potential controversies.

Applied Genomics and Public Health. DOI: https://doi.org/10.1016/B978-0-12-813695-9.00014-5

14.2 GENOMICS LEGISLATION IN THE UNITED STATES OF AMERICA

In the United States the most relevant law is the Genetic Information Nondiscrimination Act (GINA) 2008, which is an Act of Congress in the United States designed to prohibit some types of genetic discrimination. GINA disallows health insurers from segregation dependent on the hereditary data of enrollees. Accurately, health insurers may not utilize hereditary data to make qualification, inclusion, endorsing, or premium-setting choices.

Moreover, health insurers may not ask for or require people or their relatives to experience genetic testing or to give hereditary data. As characterized in the law, hereditary data incorporate family medicinal history, show sickness in relatives, and data with respect to people's and relatives' genetic tests. Title II of GINA is actualized by the Equal Employment Opportunity Commission and keeps businesses from utilizing hereditary data in work choices, for example, enlisting, terminating, advancements, pay, and occupation assignments. Furthermore, GINA prohibits bosses or other secured substances (work offices, work associations, joint work of the executives for preparing projects, and apprenticeship programs) from requiring or mentioning hereditary data and tests as a state of business. GINA incorporates a "look into exemption" to the general denial against health insurers or group health plans mentioning that an individual experiences a hereditary test. This exemption permits health insurers and group health plans occupied with research to ask (but not require) that an individual encounter a genetic analysis.

GINA has implications for individuals participating in research studies. The Office for Human Research Protections within US Department of Health and Human Services has issued a guidance on the GINA act,[4] including information on GINA's examination exclusion, considerations for institutional review boards (IRBs), and incorporating information on GINA into informed consent forms. The guidance in a matter of seconds is as portrayed next.

Given that GINA has implications regarding the actual or perceived risks of genetic research and a person's eagerness to take part in such research, investigators and IRBs ought to know about the securities provided by GINA as well as the limitations in the law's scope and effect. IRBs ought to consider the arrangements of GINA when assessing whether genetic research fulfills the criteria required for IRB's approval of research, especially whether the dangers are limited and sensible in connection to foreseen advantages and whether there are sufficient arrangements set up to ensure the protection of subjects and keep up the classification of their information. GINA is also relevant to informed consent. When investigators develop an IRB review, consent processes, and documents for genetic research, they should consider whether and how the protections provided by GINA ought to be reflected in the assent record's depiction of dangers and arrangements for guaranteeing the privacy of the information.

14.3 GENOMICS LEGISLATION IN THE EUROPEAN UNION

Contrary to the United States, the EU member states are characterized by a vast heterogeneity in the legislative framework that governs genetics and genomic medicine. In the following sections, we summarize the key legislation that relates to genetic services and genomic medicine in EU countries.

14.3.1 ESTONIA

Estonia has a comprehensive network of laws which relates to genetics. The most important instrument is the Human Gene Research Law, which specifically addressed genetic research and handling of genetic material and data. This law is one of the most comprehensive tools found in Europe, discussing genetic research.

The Human Gene Research Law of 2000[5] regulates genetic research and DNA samples, handling, and procedure. It establishes a gene database, specifies the rights of donors, and clarifies who has access to genetic information. Section 6 of the law provides that gene research and testing is allowed to explore and describe the relationship between genes, the environment, and people's lifestyles and to explore how to prevent or treat illnesses. DNA samples must be provided voluntarily. The law has extensive provisions specifying the rights of donors to know their genetic information, to order the destruction of their genetic information as well as obligations on persons processing the knowledge to handle it with respect and according to the donor's wishes. The law further specifies how the DNA is to be coded and deidentified for research purposes and prohibits any discrimination in employment relationships as a result, or based on, a person's genetic information. The law also prohibits any discrimination based on genetics in the insurance context, whereby insurance companies are prohibited from imposing different insurance conditions on people with varying risks of inheritance. The law also provides for the establishment of an ethics committee that scrutinizes research.

The Personal Data Protection Act of 2007[6] covers the area of protection and handling of personal data, including genetic data. Genetic data are expressly included as sensitive personal data and are thus subject to greater protections provided by most of the sections, 12–42, of the Act. There are strict limits on how sensitive personal data may be processed, and there is an obligation to register all processing of sensitive personal data with the relevant authority.

14.3.2 IRELAND

There seems to be a notable absence of legislation-governing gene editing and research on embryos (see this section for an article for more detail). The law relating to genetics covers environmental protection, patents, genetic testing, animal welfare, and DNA samples in the criminal context as well as data protection.

The Disability Act of 2005[7] governs genetic testing in persons and provides that testing shall not be carried out on a person unless it is prohibited by law and the consent of the data subject has been obtained following the Data Protection Act. The issue of genetic testing must be provided with all relevant information before the genetic sample being collected. The Data Protection Act of 2018[8] provides for the protection of personal data, which includes genetic data, and which is subject to a high protection level because it is specified as a particular category of data.

14.3.3 SWEDEN

Swedish legislation provides two statutes, which are directly relevant to genetic research: the Act on Biobanks in Healthcare etc. 2002:297 and the Act on Genetic Integrity etc. 2006:351.

The Act on Biobanks in Healthcare etc. 2002:297[9] does not explicitly refer to genes but its topic is closely related, and storage of tissues for research can have implications for genetic studies. The Act on Genetic Integrity etc. 2006:351[10] is the dominant legislation in Sweden relating to genetics. The Act prohibits insurance companies from requiring or requesting genetic information from the insured. Sections 3 and 4 prohibit genetic research, experiments, and treatment methods, which could lead to genetic changes that could be inherited. The Act also outlines the requirements for the analyses of gene information in the context of delivering health services or performing scientific research, as well as the conditions in which fetal genetic testing can be done. The Act also provides when gene therapies can be used and lists penalties for breaching any of the provisions of the Act. The Pharmaceuticals Act 2015:315[11] outlines how permissions for clinical trials are granted and provides extended decision periods for cases where the applicant wishes to use a gene therapy drug.

14.3.4 LATVIA

Latvia has one of the most directly targeted legislation on the topic, the Human Genome Research Law, which establishes a comprehensive legislative framework for the governance of genetics research in the country.

The Human Genome Research Law of 2003[12] is perhaps the most comprehensive legislation referred to in this chapter. Its purpose is to regulate the establishment and operation of a single-genome database of the Latvian population for genetic research and ensures the voluntary nature and confidentiality of gene donation. The Law establishes the Genome Research Board, which has the authority to examine projects and concepts related to genetic research and facilitate the provision of public information about such research. There are strict limits on which bodies are allowed to process genetic information contained in the database and register. Genetic analysis is, as per Section 8, permitted to study and describe the mutual connection between genes, individual state of health, and physical and social environment in order to discover disease diagnostic and treatment methods that will help to assess the health risks of individuals and prevent the causes of diseases.

As per the Human Genome Research Law of 2003, the Latvian Cabinet will publish regulations specifying the procedures for genetic research. All processing of information in the database shall be subject to the provisions for the processing of personal data. Donors of gene information must consent to the use of their data in a particular way and have the right to be informed of all the data produced as a result of the processing of their genetic data. It is only permitted to use the genome database for scientific research, research and treatment of the diseases of a gene donor, investigation of the health of society, and for statistical purposes. It is prohibited to use the genome database for any other purpose. The Law provides for how information is to be coded and deidentified during research. All research projects must comply with principles of ethics published by the Central Medical Ethics Committee, which shall evaluate compliance with those principles in all projects relating to genetics research.

14.3.5 ICELAND

Laws of Iceland provide comprehensive coverage of issues relating to genetics and cover topics, such as insurance, privacy law, inheritance, animal welfare, DNA police database, *in vitro* fertilisation (IVF) and use of embryos for stem-cell research, as well as genetically modified organisms (GMOs), patents, and food safety.

The Law on Insurance Contracts of 2004[13] prohibits companies from using results of genetic testing and any associated risks of developing certain diseases to determine the insurance policy of clients, nor are companies allowed to request such information. The Privacy Act of 2000[14] specifies genetic information as sensitive personal information, and hence all provisions governing the handling, storage, and access to confidential personal information govern genetic information, and the law affords this information the highest level of protection.

14.3.6 LITHUANIA

The Lithuanian laws on genomics are comprehensive and cover a wide variety of topics. The most important legislation is the Law on Ethics in Biomedical Research (2002). This Law is the most comprehensive legislation on genome research and gene technology used in the surveyed jurisdictions. The law provides that biomedical studies which modify the human genome may only be carried out for prevention, diagnosis, or treatment and only in the cases where they are not intended to alter the progeny genome (the germ line). According to the Law on Ethics in Biomedical Research (2002), per Article 4, it is prohibited to discriminate against a person or restrict his or her rights or legitimate interest based on results of the genetic studies. Article 5 provides the ethical requirements for biomedical research, which must be followed at all times. The Law specifies that consent is necessary for participation in research, and permission is also crucial for an individual's information to be stored in a biobank. The Law regulates the establishment, running, and obligations of biobanks, as well as the protection of all genetic information.

14.3.7 THE NETHERLANDS

Dutch laws relating to genomics and genomic medicine are comprehensive and cover topics of DNA in the context of crime, environmental protection, GMOs, patents, animal laws, and embryo research, as well as medical research on humans. The most significant piece of legislation is likely the Law on Medical Research on Humans of 26 February 1998,[15] which regulates medical research on humans. It specifies how gene therapies are to be researched, and it forbids experimentations on humans in a way that changes the germ line of humans.

14.3.8 NORWAY

Norwegian laws related to genetics are comprehensive with the most crucial legislative instrument to be the Act on Human Medicine Use of Biotechnology (Biotechnology Act) (2003), which regulates when genetic studies and gene therapy courses can be implemented. The Act also provides when and how genetic analysis and/or surveys can be conducted on individuals; and who is allowed access to genetic information.

14.3.9 FINLAND

Finland has various laws relating to genetics on many topics varying from biobanks, food safety, penal provisions, paternity testing, GMOs, and medical research. The Medical Research Act[16] specifies that an ethics committee must approve research proposals and outlines which factors the committee needs to consider in deciding on an application. The Act further specifies that the ethics committee has a more extended period for its decision when the request relates to gene therapies and/or GMOs.

14.3.10 LUXEMBOURG

The laws of Luxembourg on the genome are relatively comprehensive. The health code in Memorial A-343[17] establishes an ethics committee concerning gene therapy and genetic research, as well as any related clinical trials. The Code specifies how sensitive data, including genetic data, are to be handled, stored, and destroyed. Moreover, the Law of 2 August 2002 on the Protection of Individuals about the Processing of Personal Data provides for security protections of genetic data classified as health data, which is subject to higher protection than other kinds of data (Law of 2 August 2002 on the protection of individuals with regard to the processing of personal data).[18]

14.3.11 FRANCE

In France, most of the relevant provisions are contained in the Public Health Code, which provides for when medical genetic research can be carried out and what rules, namely, ethical considerations, must be considered and abided by. The Public Health Code is the most critical legislative instrument concerning genetics identified in the review (Code de la Sante Publique).[19] It contains provisions on genetic counseling, when such advice is to be provided and what qualifications genetic counselors must have. It further defines what genetic characteristics mean and what is genetic identification. The Code also regulates gene-therapy preparations and xenogeneic cell-therapy preparations, and importantly, Articles R1125-7 to R1125-13-1 regulate genetic research. The Code also regulates the examination of genetic characteristics by DNA fingerprinting for medical purposes.

14.3.12 PORTUGAL

The Portuguese laws relating to genetics are comprehensive and cover many topics. The most relevant and notable law is Law no. 12/2015 of 26 January on Personal Genetic Information and Health Information,[20] which provides an in-depth regulation of genetic research, storage of genetic information, as well as the limits of procedures, which can be utilized in working with genetic data. This appears to be one of the most comprehensive pieces of legislation focusing on genetics among the EU jurisdictions. The Law no. 12/2015 of 26 January on Personal Genetic Information and Health Information creates a comprehensive regulatory framework for genetic research, storage of genetic information, as well as the limits of procedures relating to genetic data. The Law protects the privacy of genetic information and places special protections on the data; for example, it dictates that genetic and health information be treated separately and only persons with the highest levels of access can have access. The Law prohibits the alteration of the human germ line and outlines what a genetic database is and what permissions must be obtained before setting up such a database. The Act has a chapter on genetic testing whereby it limits the availability of testing and provides that informed consent of the patient must be obtained before testing. The Law also prohibits discrimination of any person based on their genetic information or the fact that they have a specific genetic disease.

Similarly, insurance companies neither can request access to any genetic information, nor they can use such data in setting the insurance premiums. Further, genetic testing in employment is prohibited and cannot be used for recruitment purposes, unless there is some risk to a person with individual susceptibility. Human genome research is to follow the general rules on scientific research in the field of health, with additional confidentiality protections. Free access by the scientific community to emerging data on the human genome must be guaranteed. Research on the human genome is subject to approval by the ethics committees of the relevant hospital, university, or research institution. Such an analysis cannot be carried out without the informed consent of the subject.

The Law no. 67/1998 of 26 October on Personal Data Protection Act[21] regulates access to and handling of personal data (Personal Data Protection Act of Portugal); the personal data include genetic information that is classified as sensitive personal information, and special procedures must be employed when handling such data. Law no. 94/1999 of 16 July on Access to Documents[22] regulates access to official documents, and it specifies how documents containing genetic information can be accessed (Law on Access to Documents of Portugal).

14.3.13 SLOVENIA

In Slovenia the Medicines Law (2014) regulates medicines and clinical trials [Medicinal Products Act (ZZdr-2)]. It specifies that no clinical trials can be conducted where the drug/treatment would change the germ line of the patient. It also provides for proper authorizations necessary if a drug trial concerns a gene therapy treatment.

14.3.14 SPAIN

The Spanish Health Code specifies when genetic counseling is to be provided to individuals; and it mandates that personal data, including genetic data, and the identity of donors of organs must remain secret. The Code also and importantly includes the Law 14/2007 on Biomedical Research[23] whereby one of the purposes of the law is to regulate the procedures of genetic analysis and the storage and handling of genetic data. It is prohibited to discriminate against any person because of their genetic characteristics, or due to their refusal to undergo a genetic analysis or to participate in genetic research. The law specifies what information must be provided to subjects and what the scientist must acquire consent.

14.3.15 ITALY

The laws of Italy relating to genetics are reasonably varied, but only a few of the available instruments engage with the subject profoundly and mostly on the area of data-protection framework.

The Code for the Protection of Personal Data of 2003[24] includes genetics as personal data and provides how a person's consents must be informed and when notifications are necessary when the genetic data are being used.

14.3.16 CROATIA

No specific gene technology legislation is yet available for Croatia. The Law on the Protection of Patient's Rights of 2004[25] specifies that interventions aimed at changing the human genome can only be undertaken for preventative or therapeutic purposes, and no interventions are allowed with the view to improve the patient's germ line. In all cases the Central Ethics Committee must give its permission and opinion on the admissibility of clinical trials for gene therapy.

14.3.17 GREECE

In Greece the legislative framework relating to genetics is quite vague. The most significant laws are Law no. 2619/1998 (Chapter IV Articles 11–13), which is the ratification of "The Convention for the Protection of Human Rights and Dignity of the Human Being about the Application of Biology and Medicine" of the Council of Europe and Law no. 3418/2005. Both Laws refer to the human genome and prohibit any discrimination based on genetic characteristics in all contexts of life.

Moreover, genetic tests that could predict the appearance of genetic diseases or could be used to either identify the subject as a vector of a disease-responsible gene or to detect genetic predisposition that may only be carried out for health reasons or scientific research related to health. Also, any intervention aimed at modifying the human genome is only permissible for preventive, diagnostic, or therapeutic purposes, and only if it is not intended to introduce any modification in the progeny genome. Furthermore, genetic technology may not be used for civil or military purposes. With a new legislative measure, the specialties of clinical and laboratory geneticists have been established. The Plenary Session of the Greek Central Health Council at its 204th Meeting on 2006 set first, the organization and operation of Genetic Centers (terms, conditions, operating specifications, technical equipment, scientific interconnection with hospitals, etc.) and second the qualifications of Geneticists.[26]

14.3.18 CYPRUS

No specific legislation on gene technologies could be found in the laws of Cyprus; still, their legal system has extensive coverage of legislation relating to GMOs, IVF, environmental protection, personal data protection, and food safety. The Bioethics (Establishment and Functioning of the National Committee) Law of 2001 (150(I)/2001) establishes the National Bioethics Committee (Law 150(I)/2001).[27] The Committee analyses and evaluates projects relating to genetics and is also responsible for informing the public on its findings and monitoring implementation of the international obligations of Cyprus relating to genetics.

The Personal Data Processing Act (Protection of Individuals) Law of 2001 (138(I)/2001, arts 6, 8)[28] covers the data protection of sensitive personal data, which does not explicitly include genetic information but health information and information relating to ethnicity and race, which could presumably include some genetic information. The Act provides for stronger protections for sensitive data and limits the permitted uses of such data.

14.3.19 ROMANIA

The Romanian laws on genetics are comprehensive and cover topics ranging from DNA databases in criminal procedure, GMOs, animal laws, and extensive laws on embryo research, as well as the permitted biotechnological techniques. Civil Code of July 17, 2009 (Law 287/2009)[29] prohibits any medical intervention intended to modify the genetic information of offspring of an individual, except to prevent and/or treat genetic diseases. The Code further provides that genetic examination can only be undertaken for medical or scientific purposes and must be permitted by law. The identification of a person based on his/her genetic profile may be made only in the course of a civil or criminal judicial procedure or for medical or scientific research (permitted by law).

14.3.20 AUSTRIA

Austria has a fairly comprehensive legislation relating to genetics, which covers areas of data protection, patents, criminal prosecution, GMOs, as well as environmental protection. The most directly relevant legislative instruments are the Gene Technology Act and the Data Protection Act. The Gene Technology Act implements the Council Directive 98/91/EC among others and regulates the use of gene technologies and licensing. Importantly, Article 65 outlines the limited reasons for which genetic analysis for medical purposes can be carried out. The enumerated goals are to identify existing diseases and for the preparation to use gene therapy on the patient where the treatment is based on specific genetic markers. Genetic analysis can also be used to detect diseases, which are based on a germ-line mutation and to establish a predisposition for a disease, especially to evaluate the likelihood of a genetic disorder occurring in the future where there is scope for prevention. The Act also specifies when genetic analysis in humans can be used for scientific purposes and training (Gene Technology Act 510/1994, art 66)[30] and limits the use of gene data beyond the original intent of its collection. The Act further provides details of the qualifications and procedures while handling genetic data and situations, where genetic information can be revealed to family members. The Act also provides for when gene therapy is allowed and what must be satisfied before it can be administered to patients (Gene Technology Act 510/1994, arts 72, 74, 75).[31]

14.3.21 BULGARIA

Bulgaria's legislative coverage of topics relating to genetics is reasonably comprehensive, except for genetics in the criminal context. The most relevant legislation is the Health Law (2005) that touches on genetics in health care and medical research (Arts 127, 135, 138, 139, 144, 198; Health Law 2005).[32] This law further provides for genetic treatments to prevent hereditary diseases occurring in children. The legislation also covers some basics relating to genetic studies and the required licenses for any such research, as well as a provision to the effect that genetic laboratories may establish DNA banks for the collection and storage of genetic material for scientific and medical purposes. Importantly, Article 198 of the Law provides that medical and scientific research on humans shall not be carried out with genetic engineering products, which may lead to transmission of these changes to the offspring of the subjects. It appears, therefore, that permanent and inheritable changes in a person's genetic makeup are prohibited.

14.3.22 HUNGARY

The laws relating to genetics are reasonably extensive in Hungary. The most directly relevant legislation is the XXI Law on the Protection of Human Genetic Data, on the Rules of Human Genetic Testing and Research and Operation of Biobanks (2008).

The XXI Law on the Protection of Human Genetic Data, on the Rules of Human Genetic Testing and Research and Operation of Biobanks (2008) aims to protect people's genetic information (XXI Law on the Protection of Human Genetic Data, on the Rules of Human Genetic Testing and Research and the Operation of Biobanks of 2008). It provides how genetic data are to be handled, when it can be used for research, and how genetic analysis can be carried out and provides rules for biobanks. Genetic testing can be performed for prevention, diagnostic, therapeutic, or rehabilitation purposes and solely based on medical interest. People who have genetic data have a responsibility to protect the data. The law specifies how people are to be notified of genetic test results and what consent must be given before any testing is done. Genetic data are to be used mainly only for the purposes for which they were collected, but in limited circumstances, this can be extended. Genetic studies can use deidentified genetic data to determine the distribution of genetic variants between individuals.

14.3.23 CZECH REPUBLIC

There seems to be no specific legislation targeted at genetic technologies, the most closely related law appears to be the Penal Code, which prohibits individual acts with genetic data and health laws regulating genetic testing in people. Law N. 202/2017[33] relates to specific medical services. It regulates genetic testing in patients and specifies the circumstances in which genetic testing is permitted and what procedure must be followed. Law N. 66/2013,[34] which governs the provision of medical services, specifies when genetic information of a person can be used and in what way.

Law N. 101/2000[35] is the Czech Privacy Law, and in Article 4, it explicitly includes genetic data as sensitive personal information, thus subjecting genetic data to more critical measures of protection. This Law further specifies the obligations on those storing or processing confidential personal information.

14.3.24 GERMANY

Germany has a significant number of laws that touch on genetics, with the Genetic Testing Act (2009) No 50/2009 of 2009 that regulates genetic testing in humans[36] being the most relevant for present purposes. The Act provides when and how genetic testing can be performed and how any collected genetic information and samples are to be handled. It also prohibits discrimination based on genetic characteristics in all contexts of life as well as it bans discrimination based on the refusal to take a genetic test. The Act provides circumstances in which genetic counseling should be submitted to subjects and how results of tests are to be communicated, as well as how the quality of the test is to be assured. Furthermore, the Act provides that it is prohibited to require genetic data from a person in the health-insurance context, but it is allowed in the contact of functional disability and nursing-care insurance. The Family Law Procedural Law (2009) specifies when samples of DNA can be taken and compared in the context of determining familial relationships.

14.3.25 SWITZERLAND

Swiss laws relating to genetics are incredibly comprehensive and cover the whole variety of topics relating to the subject. Article 119 of the Federal Constitution[37] of the Swiss Federation gives people the right to be protected against the misuse of reproductive medicine and gene technology. The Confederation shall legislate on the use of human reproductive and genetic material. In doing so, it shall ensure the protection of human dignity, privacy, and the family and shall adhere in particular to the following principles—all forms of cloning and interference with the genetic material of human reproductive cells and embryos are unlawful; nonhuman reproductive and genetic materials may neither be introduced into nor combined with human reproductive material.

The Federal Act on Human Genetic Testing (2004)[38] stipulates the conditions under which human genetic testing may be performed in the medical, employment, insurance, and liability contexts. An overriding principle is given in Article 4, whereby no one can be discriminated against on the grounds of their genetic characteristics. All genetic testing is to be performed only with the consent of the subject of the test. Genetic tests may only be performed on individuals if they serve a medical purpose. The Act further outlines extensive privacy and data protection relating to genetic information and any results from genetic processing. The Act prohibits the use of genetic information in the context of insurance and employment. An employer may request an employee to take a genetic test for presymptomatic genetic diseases to prevent occupational diseases and accidents; this exception is narrow and highly regulated. The Federal Act on Research Involving Human Beings (2011) provides that biological material and genetic data in an uncoded form may be used for research projects if the subject has given informed consent for its use. The Act also outlines when data can be anonymized.

14.4 GENOMICS LEGISLATION IN ASIA

14.4.1 SINGAPORE

Singaporean law provides two statutes, such as Human Biomedical Research Act (2015) and Human Cloning and Other prohibited practices Act (2005 rev.), which are directly relevant to genetic research. Most relevant though is the Human Biomedical Research Act which in part 5 regulates that every person who has obtained individually identifiable information or human biological material for the purposes of social biomedical research must take all reasonable steps and safeguards as may be necessary, including rendering information or material nonidentifiable, to protect such information or materials against accidental or unlawful loss, modification or destruction, or unauthorized access, disclosure, copying, use or modification. Moreover, no research institution or person can conduct, supervise, or control any prohibited human biomedical research specified in the Act.

14.4.2 CHINA

In response to concerns about the rapidly growing field of individualized medicine, in February 2014 the China Food and Drug Administration (CFDA) and the National Health and Family Planning Commission (NHFPC) halted clinical assays and launched a series of regulatory requirements for next-generation sequencing (NGS). The CFDA requires that all components of an NGS commercial diagnostic system, including the reagents, instruments, and software that are packaged together and sold to laboratories, must be approved before being used in a clinical setting. All the clinical tests using NGS technology must be conducted under the corresponding regulations of the NHFPC document, "Medical Institution Clinical Laboratories Regulation for Amplification-Based Molecular Diagnostics."

14.5 GENOMICS LEGISLATION IN THE MIDDLE EAST

The issue of genetic testing is insufficiently addressed in regulations in the Arab world. Only Lebanon has written a law specially formulated to regulate it (No. 625 of 2004). Bahrain, Tunisia, and the United Arab Emirates (UAE) have referred to genetic testing in some stipulations, but the other 13 countries have ignored this issue altogether, or are working on draft guidelines, or have formulated simple guidelines that partially regulate genetic testing.[38]

14.5.1 UNITED ARAB EMIRATES

The UAE have no comprehensive guidelines for genetic testing and addresses the issue in the following draft law. Article 2/27 of Federal Law No. 28/2005, concerning personal status, mentions that "In order to proceed with a valid marriage contract, it is conditional on submitting a report from a competent medical committee formed by the Ministry of Health stating that the couples are free from diseases, prescribed by the same law according to which divorce may be requested."

14.5.2 LEBANON

Only Lebanon has directly and explicitly addressed the issue of genetic testing (Law No. 625 of 2004 on human genetic testing). Articles 1–6 of this law address that genetic variants, ethnic discrimination, genetic characteristics of the individual, prohibition of manipulations affecting human dignity, the confidentiality of tests, and prohibition of all commercial practices are related to genetics. Articles 7–20 concern genetic tests and their scientific or medical purposes, informed consent, delivery of results, paternity testing, the necessity of obtaining written permission from the guardians of minors and disabled persons, and the necessity of obtaining written permission from the Ministry of Health before testing groups or inhabitants of a particular region. Articles 21–26 concern the issue of DNA banks and medical confidentiality concerning the conservation or destruction of results or specimens. A draft law is currently being prepared to create a national database for DNA profiles. This database will be under the control and supervision of the Ministries of Justice and the Interior. Besides, all major institutes and universities in Lebanon have their IRBs to regulate all research related to the human genome and gene analysis in accordance with national and international guidelines on research bioethics.

14.5.3 QATAR

Qatar regulates issues related to the human genome and gene analysis through WHO's "Proposed International Guidelines on Ethical Issues in Medical Genetics and Genetic Services" under the title "Banked DNA," which are presented as an appendix to "Research Policies, Procedures, and Guidelines" by the Shafallah Genetics Medical Centre. According to these guidelines, existing stored specimens or samples, such as those in university or hospital departments or collections of blood spots, should not be subjected to new rules for consent or recontact that may be established in the future. An informed consent that would permit the use of a sample for genetic research in general, including unspecified future projects, appears to be the most efficient approach, which avoids costly recontact before each new research project. The consent should specify that family members may request and be given access to a sample to learn their genetic status. While spouses may not have such a right of access, their concerns should be considered (i.e., if the couple is planning to have children, it is the moral obligation of the party whose DNA has been banked to provide the spouse with any relevant information). All samples should be used with proper regard for confidentiality, except for forensic purposes. Qualified researchers may have access if identifying characteristics are removed. Particularly, between developing and developed countries, patenting has the potential to impede international collaboration to the detriment of service delivery to those with genetic disorders. Genetics differs from many areas of research in that essential new knowledge can come from a family or an ethnic group with a particular genetic variant. If this leads to the development of a diagnostic test or new therapies, equity requires that the donors or the community, in general, should receive some benefit.

14.6 DISCUSSION

In this chapter, we attempted to summarize the existing legislative frameworks regarding genomics and genomic medicine in several countries in Europe, Middle East, Asia, and the United States. The laws that were analyzed refer to the proper handling of the information produced, such as personal data protection, and DNA profiles in forensics, and try to embrace the new possibilities they offer, such as individualized therapy and genome editing.

While genomic technologies can be applied in several fields, the legislative frameworks across the world focus on specific activities, such as GMOs, embryo research, genetic testing services, and targeted gene-editing technologies. Clarity in the legal requirements across the world in these areas is primarily due to the guidelines implemented by international organizations (WHO, UNESCO, etc.) and, in this sense, the law in each country, where present, can be seen as a driving force in the harmonization of national legislation. However, maybe because of the absence of a specific international legal framework on genomics, there are currently few jurisdictions among the world which would explicitly allow the potential applications of new genomic technologies, such as CRISPR-Cas9, for gene therapies or severe genetic diseases, due to strict prohibitions on genetic modification of humans. Even in countries where there is no specific prohibition, the situation is often unclear as the relevant legislation does not appear to directly address specific topics, for example, nongerm-line genetic modifications of humans. Legislators should resolve the lack of clarity and, at the same time, introduce detailed legislation regarding those subjects.

In the United States of America, each state has its own legislative framework, which could be described in general as incomprehensive and vague in the subject of genetic testing. GINA, on the other hand, is a specific and relevant legislation.

Concerning genomic technologies, in particular, the regulatory framework appears very fragmented and heterogeneous. For example, many jurisdictions in Europe in particular, including the Czech Republic, Luxembourg, as well as Ireland, appear to have no specific legislation addressing them. It is possible that some regulations are in place beyond the first legislation frame, but lack of such law may prove problematic going forward. More specifically, as far as Europe is concerned, we can take Switzerland as the leading example of the legislative framework concerning genomics and personalized medicine. The Swiss Federal Act on Human Genetic Testing (2004) stipulates all the conditions under which human hereditary testing might be performed in the medicinal, business, protection, and obligation settings. Thus we can conclude that more countries in central Europe need to address legislation around the groundbreaking field of personalized medicine and genomics.

In Northern Europe, two countries, such as Estonia and Latvia, seem to have the most comprehensive legislation in this subject. Estonia has established the Human Gene Research Law (2000) that regulates genetic research, and Latvia has set up the Human Genome Research Law that establishes a far-reaching authoritative system for the administration of hereditary qualities explore in the nation. In Southern Europe, Portugal is the country with the most comprehensive legislation on Personal Genetic Information and Health Information. Genetic research, storage of genetic information, as well as the limits of procedures relating to genetic information, are well addressed in the Law no. 12/2015, which endorses the need for other countries in Southern Europe to follow the example of Portugal. Lastly, in Western Europe, France among the different countries has to show off the more appropriate legislative framework. Public Health Code of France provides for when medical genetic research can be carried out and what provisions, namely, ethical considerations, must be considered and abided by.

Finally, although the Arab World presents many accomplishments in the field of personalized medicine and genomics, yet many countries in the Middle East still have no regulations whatsoever concerning genetic testing. Bahrain, Tunisia, Lebanon, Qatar, and the UAE have referred to genetic testing in some stipulations, mostly though by following up draft guidelines of international organizations, which leads to the conclusion that much work needs to be done in the Arab World.

14.7 CONCLUSIONS

With a view to the future, countries should address how to regulate the use of genomic medicine and the applications of genomic technologies and place clear limits on when they should be allowed. Some apps could have considerable benefits, such as tailored treatment and prevention strategies to people's unique characteristics, and applications that could treat and potentially eradicate severe genetic diseases, such as Huntington's disease, or use of genomic DNA data in the field of criminal investigations. Even therapeutic uses of genomic information for gene-editing purposes give rise to serious ethical questions, which should be addressed before any laws are implemented. Consultation should include inputs from interested stakeholders, as well as the public. As the rules stand today, it is possible that in some countries, where there are no explicit prohibitions on areas, such as germ-line genetic modifications in humans, DNA modification in humans would be legal without much, if any, regulation in place.

REFERENCES

1. WHO definitions of genetics and genomics. <http://www.who.int/genomics/geneticsVSgenomics/en/>.
2. What is genomics? <https://www.ebi.ac.uk/training/online/course/genomics-introduction-ebi-resources/what-genomics>.
3. <https://www.ebi.ac.uk/training/online/course/genomics-introduction-ebi-resources/what-genomics>.
4. <https://www.hhs.gov/ohrp/regulations-and-policy/guidance/guidance-on-genetic-information-nondiscrimination-act/index.html>.
5. Human Gene Research Law of 2000, Estonia. <https://www.riigiteataja.ee/>.
6. Personal Data Protection Act of 2007, Estonia. <https://www.riigiteataja.ee/>.
7. Disability Act of 2005, Ireland. <http://www.irishstatutebook.ie/>.
8. Data Protection Act of 2018, Ireland. <http://www.irishstatutebook.ie/>.
9. Act on Biobanks in Healthcare etc. 2002:297, Sweden. <http://www.riksdagen.se/sv/dokument-lagar/>.
10. Act on Genetic Integrity etc. 2006:351, Sweden. <http://www.riksdagen.se/sv/dokument-lagar/>.
11. Pharmaceuticals Act 2015:315, Sweden. <http://www.riksdagen.se/sv/dokument-lagar/>.
12. Human Genome Research Law of 2003, Latvia. <https://likumi.lv/>.
13. Law on Insurance Contracts of 2004, Iceland. <http://www.althingi.is/lagas/148a/1990054.html>.
14. Privacy Act of 2000, Iceland. <http://www.althingi.is/lagas/148a/1990054.html>.
15. Law on Medical Research on Humans of 26 February 1998, Netherlands. <http://wetten.overheid.nl/>.
16. Medical Research Act of 9.4.1999/448, s10d (295/2004), Finland. <https://finlex.fi/fi/>.
17. Memorial A-343, of 2019, Luxembourg. <http://legilux.public.lu/>.
18. Law of 2 August 2002 on the protection of individuals with regard to the processing of personal data, Luxembourg. <http://legilux.public.lu/>.
19. Code de la Sante Publique, JORF, Décret no 2016-1537 of 2016, France. <https://www.legifrance.gouv.fr/affichTexte.do?cidTexte = JORFTEXT000033394083&categorieLien = id>.
20. Law no 12/2015 of 26 January on Personal Genetic Information and Health Information, Portugal. <http://www.pgdlisboa.pt/home.php>.
21. Law no 67/1998 of 26 October on Personal Data Protection Act Portugal. <http://www.pgdlisboa.pt/home.php>.
22. Law no 94/1999 of 16 July on Access to Documents Portugal. <http://www.pgdlisboa.pt/home.php>.
23. Law 14/2007 of 3 July on Biomedical Research, Spain. <http://www.isciii.es/ISCIII/es/contenidos/fd-investigacion/SpanishLawonBiomedicalResearchEnglish.pdf>.
24. The Code for the Protection of Personal Data of 2003, Italy. <http://www.normattiva.it/>.
25. Law on the Protection of Patients' Rights art 22 of 2004, Croatia.
26. <http://www.sige.gr/index.php/gr/genetics/laws/102-genetlaw1>.
27. Law (150(I)/2001) of 2001 on the Bioethics, Cyprus. <http://www.cylaw.org/>.
28. Law (138(I)/2001), arts 6, 8 of 2001 on Personal Data Processing Act (Protection of Individuals), Cyprus. <http://www.cylaw.org/>.
29. Law 287/2009 of 2009 on Civil Code, Romania. <http://legislatie.just.ro/>.

30. Act 510/1994, art 66 on Gene Technology, Austria. <https://www.ris.bka.gv.at/>.
31. Act 510/1994, arts 72, 74, 75 on Gene Technology, Austria. <https://www.ris.bka.gv.at/>.
32. Arts 127, 135, 138, 139, 144, 198 of 2005 on Health Law, Bulgaria. <https://www.lex.bg/>.
33. Law N. 202/2017 of 2017 on specific medical services, Czech Republic. <http://aplikace.mvcr.cz/sbirka-zakonu/start.aspx>; <http://zakony-online.cz/>.
34. Law N. 66/2013 of 2013 on the provision of medical services, Czech Republic. <http://aplikace.mvcr.cz/sbirka-zakonu/start.aspx>; <http://zakony-online.cz/>.
35. Law N. 101/2000, Article 4 of 2000 on Czech Privacy Law, Czech Republic. <http://aplikace.mvcr.cz/sbirka-zakonu/start.aspx>; <http://zakony-online.cz/>.
36. Genetic Testing Act (2009) No 50/2009 of 2009 on Genetic Testing, Germany. <http://www.buzer.de/>; <https://www.bgbl.de/>.
37. Federal Constitution of the Swiss Federation, Article 119, Switzerland. <https://www.admin.ch/>.
38. Federal Act on Human Genetic Testing of 2004, Switzerland. <https://www.admin.ch/>.

FURTHER READING

Ethics and Law in Biomedicine and Genetics: *An Overview of National Regulations in the Arab States*. United Nations Education, Scientific and Cultural Organization, Cairo, Egypt, 2011.

Overview of EU national legislation on genomics. In: *JRC Science for Policy Report, EUR 29404*. Publications Office of the European Union, 2018.

CHAPTER

GENOMICS, THE INTERNET OF THINGS, ARTIFICIAL INTELLIGENCE, AND SOCIETY

15

Vural Özdemir

Science Communication and Emerging Technology Governance, Toronto, ON, Canada

The Times They Are A-Changin

Bob Dylan.[1]

ABBREVIATIONS

AI artificial intelligence
IoT Internet of Things

15.1 INTRODUCTION

15.1.1 A NEW RELATIONSHIP FOR SCIENCE AND SOCIETY

Since the past 500 years, scientific knowledge has been understood as value free. Intellectuals of the Enlightenment era in the 16th century Europe argued that natural phenomena could be explained by the scientific method without being influenced by values of the scientists or the knowledge seeker in the laboratory.[2]

Yet, decades of research in social studies of science have shown that it is impossible to separate the "knowledge" from the "knower," and the "knower" from her/his "social context." Such insepa-rability of knowledge, politics, and human values applies to technical science as well as social sciences and humanities.[3–6]

With politics, I refer to the constitution and contestation of human power and unquestioned values and power hierarchies that shape innovation trajectories, from idea conception, research funding, and agenda setting to translational research, policymaking, promotion, and regulation of science.[7–10]

To the extent that science is designed, executed, and interpreted by humans, and all humans come with their values, assumptions, and prejudices (cognizant or not), lending a pure and value-free status to scientific practice remains a false and, ultimately, a self-destructive conviction.[5,11,12] This conviction, adopted widely in scientific communities, threatens

Applied Genomics and Public Health. DOI: https://doi.org/10.1016/B978-0-12-813695-9.00015-7

sustainable, responsible, and robust science by introducing, in addition to the vast technical risk factors, opaque and unchecked human values that in turn accrue as political risks on the science and innovation trajectory.[13]

For every technical risk, there are presumably dozens of political risks on the horizon for new ideas and their translation into veritable innovations—something that seasoned scientists instinctively know well but are often reluctant to acknowledge. For example, questions on politics of research practice such as "who and which ideas are accepted as priority, or excluded from funding? Why is researcher A extremely popular and powerful despite apparent lack of original ideas?" are not addressed by the scientific method and yet directly impact on the quotidian life of, and knowledge production by, scientists and social scientists[5,14,15].

In other words, our historical ignorance of the role played by human values and politics in coproduction of knowledge-based innovations has been omitted for the last 500 years since the Enlightenment era, falsely attributing to science and scientists, social scientists, and humanists a value-free apolitical status devoid of human values, power, and politics.

There are signs, however, that the times are changing. Chief among these is a broader take on science. The aim is to change toward making the human values and politics that are inherently embedded in scientific practice more transparent and accountable and thus reducing the potential of unchecked politics to serve as risks and liabilities to science.[16,17] Indeed, many funding agencies and science policies in industrialized societies now require scientists to reflect on such broader determinants (read: technology, human values, and politics) and outcomes of science and innovation.[18] There is also a move to reimagine the line between science, values, and society as being "porous," wherein both technical and societal knowledge are understood to be coproduced and thus shape each other.[19]

Human values, societal contexts, and politics shape scientific practices as much as new technologies do. A new two-pronged approach to science, for example, by study of (1) technical data and knowledge produced by a genome sequencer and (2) the societal context, values, and politics of the actors who generate such data and technical knowledge, not to mention the politics of the users of innovations, is becoming a mainstay practice in genomic sciences.

Importantly, rather than seeing the politics of science as being irrelevant to laboratory life, the politics of science is being understood as another crucial dimension to be addressed—instead of being swept under the carpet as it has been in the past 500 years.[20,21]

Not surprisingly, then, most books on genetics and genomics include a chapter on "technology and its societal aspects." The editor of this book has kindly asked me to write a similar chapter and yet, writing a chapter on the topic is increasingly difficult and challenging, but at the same time crucial in the current "postgenomic era"—for both the *types* of novel technologies entering the genomics innovation ecosystem and *how* we research the societal aspects of emerging technologies are rapidly changing. It is sure that the future will not be a linear extension of the past when it comes to genomics technology, society, and policy research.[22] This means that the research on the "societal context of technology" is not merely a "tick the box" exercise or a requirement to obtain competitive research funding, as it is occasionally perceived among the science communities.

This form of technology-related social research carries both instrumental or practical value to navigate the complex terrain of postgenomic technology uncertainties and principled normative value to achieve technological democracy and responsible innovation. "Technology and society" research is also about asking serious questions, such as in what kind of society and scientific practice we would like to live.

In the current climate of evolving postgenomic technology novelty and complexity, and the need for next-generation innovation policies on emerging technologies, the present chapter offers the readers an analysis of:

1. new postgenomic technologies that are both *anticipated* (e.g., proteomics, metabolomics, and other omics systems science technologies) and *unanticipated* such as artificial intelligence (AI) and the Internet of Things (IoT) that are converging with genomic science practices of the past three decades and
2. approaches to next-generation policy on emerging technologies such as AI and the IoT in a context of genomic sciences.

15.2 POSTGENOMIC TECHNOLOGIES AND SOCIETY

15.2.1 THE ANTICIPATED AND THE UNANTICIPATED

"Postgenomic technology" is a frequently used term in the literature. The term refers, however, to a diverse array of technologies and their equally diverse innovation actors and societal contexts.[23–26] The classic examples of postgenomic technologies include, in particular, those that enable functional genomics and multiomics data triangulation from genes to metabolites in the hierarchy of cell biology (e.g., proteomics, metabolomics, and others). These newer omics technologies not only share a focus on system sciences and systems thinking, as with genomics, but also bring about interesting and unprecedented social contexts.

Consider, for example, proteomics. Genotype-based diagnostics have tended to rest on a "single test-per-life-time" framework. Proteomics allows repeated measures testing for the same patient at different times to capture time-sensitive functional changes in cell biology and the whole organism. Proteomic diagnostics thus call for repeated reimbursement models that might prove to be more attractive (than single point genotype testing) for the diagnostics industry, and more comprehensive and function oriented for physicians and patients, in understanding the dynamic changes in patients' health longitudinally. No doubt, new actors with competing, cooperating, and contesting expectations, values, and priorities can be expected on the innovation trajectory as proteomic and other postgenomic technologies intersect and interact with genomics in the coming decade.

Emerging omics fields, such as proteomics, are not the only postgenomic technologies impacting on genomics.[27,28] Two other technologies, AI and the IoT, are noteworthy even though they are unrelated to the omics system sciences in their technology platforms, and previously unanticipated in terms of being understood among the postgenomic technologies (Fig. 15.1). Recent clinical applications of the IoT, discussed in the next section, point toward its enormous potential for genomics and multiomics science and innovation.

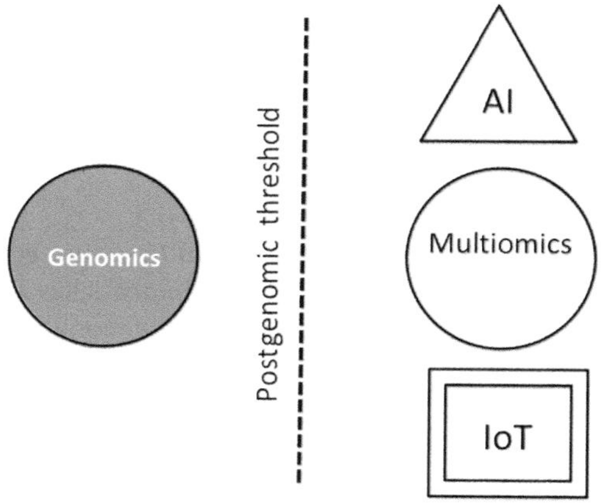

FIGURE 15.1 Transition from genomics to postgenomic technology ecosystems.

The dotted line represents the threshold for such transition where multiomics, AI, and the IoT or their various combinations offer a rich toolbox for scientific research while bringing about equally complex and rich societal corollaries. *AI*, Artificial intelligence; *IoT*, Internet of Things.

15.2.2 GENOMICS MEETS THE INTERNET OF THINGS AND ARTIFICIAL INTELLIGENCE—TOWARD A "QUANTIFIED PLANET"

Let us start with the following question for an introduction on the IoT: Can "network" be a new technology? Let us dig deeper into this question.

It is interesting to note that new technologies have often been imagined, especially in medicine and biology, as methods of measuring a chemical or aberrations in molecular mixtures such as plasma and blood. However, the "connectivity" itself, or information technologies and smart objects that create new types and speeds of connectivity for exchange, transformation, and diffusion of data, can be understood as a new technology as well.

For example, automated teller machines have been available since 1974 and represent one of the oldest smart devices, long before the invention of smartphones and genome sequencers. Gartner, a technology research firm, has observed that the number of wirelessly connected smart objects, excluding smartphones and computers, reached 8.4 billion worldwide by the end of 2017.[29] Put simply, this means that there are currently more smart objects than the number of humans on the planet. The Intel forecasts over 200 billion wirelessly connected objects by 2020.[30]

If this forecast proves correct, 2 years from now, each of us may carry more than two dozens of smart objects attached to our pockets, backpacks, bicycles, cars, and homes. Adding the capability of real-time data analysis and interpretation by AI and machine-learning algorithms, we have new technical tools based on extreme connectivity, and potentials for genomics, from real-time remote phenotypic data capture, genotype–phenotype association analyses, and multiomics data integration.[22]

The IoT refers to the abovementioned pervasive computing environment across the planet (not limited to the developed countries), and associated societal and human values in flux, which are collectively changing how scientific knowledge and innovation are produced and consumed.[27,31] Not only humans but virtually any object, animate, or inanimate are connected to the IoT and talk to each other through sensors and wireless connectivity, tracked in real-time, and in a state of constant learning from the Big Data they are generating and consuming at the same time. In this sense, we are moving toward a quantified planet that is much broader in scope and intensity than the quantified-self movement or personal genomics.[31]

Two recent scientific reports suggest that the IoT and AI are not far-fetched ideas and instead might soon be positioned for a "genomics turn." Consider, for example, the "digital drug" approved by the US Food and Drug Administration in November 2017 with a sensor that tracks, digitally, whether patients have ingested their medication.[32] The digital drug is approved for the treatment of certain mental health disorders, such as schizophrenia, where adherence to drug treatment needs breakthrough advances. Accordingly, it has been explained, "the sensor generates an electrical signal when it comes into contact with stomach fluid. This signal is then transmitted to a wearable patch on the patient's body, which then sends the information to the patient's smartphone. With the patient's consent, their doctor and up to four other people can be alerted when the drug is ingested."[32]

Another study with 408 people in the United Kingdom offers insights into the potential of Internet-enabled objects for dementia home care 24 hours a day, using a range of sensors, monitors, and trackers by a remote clinical monitoring team.[33] It is not difficult to envision an "AI turn" in medicine whereby such remote clinical phenotyping and monitoring can be complemented by incorporation of AI and real-time omics data analytics. As AI and the IoT move us toward a quantified planet, these examples signal how the future of health care and precision medicine might be shaped further by postgenomic technologies, whether those technical tools come in the form of yet another omics technology, such as proteomics, or new ways of building digital networks and connectivity.

To be sure, AI and the IoT depend on Big Data and extreme connectivity until everything is connected to everything else.[34] Such extreme connectivity has numerous imaginable and unimaginable implications for the quotidian life and how we relate to each other in a hyperconnected global society.[31]

Looking outside a narrow laboratory or technical vision, one could also argue that we are moving toward a "prediction-centered" society, living less in the present and more in a future imagined by predictive online diagnostics.

Extreme connectivity enabled by the IoT, Big Data, and AI might boost productivity and synchronize processes in smart hospitals. But these postgenomic tools also pose unprecedented threats, such as complete network collapse, when a component in such highly networked systems fails in a domino effect. The example of the cyberattack with the WannaCry malicious software that nearly collapsed the digital health networks in United Kingdom and the telecommunication sector in Spain in 2017 is an apt example in this context.[35]

Turning off our smartphone when we are out having dinner with friends and family might boost our creativity and ability to think outside the box. Absent such safe exits from extreme digital connectivity, constant connectivity might result in group thinking, "entrenchment" in the status quo, and echo chambers, not uncommon in social media groups nor in science and technology communities, and thus threaten the opportunities for highly novel or disruptive scientific innovations.[36]

This discussion brings us to a crucial question: How do we democratically weigh up the benefits and uncertainties of emerging technologies, such as AI, the IoT, proteomics, and other postgenomic technologies?

The next section provides a concise evolution of technological democracy and what we might seek out as elements of the next-generation technology policies in the current postgenomic era characterized in part by an emerging "AI turn."

15.3 TECHNOLOGY POLICY DESIGN

In this section, several prevailing approaches to technology policy design through societal research are highlighted, so as to better anticipate the uncertainties of emerging technologies and responsibly harness their innovative potentials. For this, we need to first return to the mid-20th century science policy and recall Vannevar Bush, the M.I.T. engineer, who wrote the July 1945 report "Science, The Endless Frontier" that still colors and shapes science policies in the early 21st century through a lens of technological determinism.[37]

The report by Bush emphasized "research in the purest realms of science" and in essence, the purity of science as a value-free activity detached from the concerns of society or the values of the scientists themselves.[38] As explained in Section 15.1, notions of science as being apolitical and purified from, and unaffected by, the values of the innovation actors, such as scientists, funders, or the society at large, are still "so deeply embedded in our cultural psyche that it seems like an echo of common sense."[38]

This chapter argues for a postgenomic technology policy design alternative to technological determinism and the self-destructive illusion of science as a value-free activity detached from the society. This proposal is consistent with decades of scholarship in social studies of science[39–44] and with arguments highlighted, for example, by Daniel Sarewitz:

> ...science will be made more reliable and more valuable for society today not by being protected from societal influences but instead by being brought, carefully and appropriately, into a direct, open, and intimate relationship with those influences.[38]

Postgenomic science falls under "Big Science" and has grown from an artisanal boutique-style activity in small laboratories to a large and arguably industrial-scale enterprise which is not equipped to answer the societal questions generated by and shaping science.[3,45]

Indeed, political philosophers from Plato onward suggested that questions about the social use, purpose, and relevance of science cannot be answered by science.[18] These questions pertain to social and political science research and the study of human values enacting on scientific practices.[3]

While the need to understand the societal dimensions of science has found more acceptance with new funding policies that require such social research, *how* such technology and society research should be conducted has not been debated extensively.[3,46] Many scientific communities tend to perceive such research as merely a requirement for funding applications, and as though it is being composed of a single brand of technology and society research and analysis.

One of the first attempts by scientists and other innovation actors to intervene and preempt the societal impacts of their research is exemplified by the invitation-only exclusive Asilomar conference held in 1975 on the Monterey Peninsula in California.

The aim of the Asilomar conference was for scientists to debate the ways to self-regulate their own trade and, specifically, the voluntary restrictions on recombinant DNA research. In addition to some 140 scientists, the Asilomar conference included a few journalists and policymakers, but not the public. Some 40 years on, entirely expert-led conferences such as Asilomar are no longer considered the best way to govern emerging technologies and address scientific uncertainties or controversies.[47]

In hindsight, Asilomar was an exercise by scientists to avoid government intervention and assuage public outcry and fears on recombinant DNA biotechnology. Technological democracy was not, however, one of the explicitly articulated aims of the Asilomar conference.[46]

Attempts by technology experts to guide the terms of public engagement in science have continued onward to the 1990s. The ethical, legal, and social implications (ELSI) project was launched in 1990 in the context of the Human Genome Project and has been institutionalized and funded by governments and various organizations.[48] The US ELSI program had a number of priority areas, such as fairness in the use of genetic information (e.g., by insurers and employers), privacy and confidentiality of genetic information, societal impacts and stigmatization due to human genetic differences, the adequacy of informed consent, clinical genetic testing, and commercialization of products. In Europe, a similar transdisciplinary program on the ethical, legal, and social aspects (ELSA) of science and technology was established by the European Commission in 1994. Unlike the US ELSI framework, the European ELSA program aimed to address not only genetics/genomics but also the broader field of life sciences and technologies.

As we move toward postgenomic research shaped by diverse technologies such as proteomics, AI, and extreme connectivity brought about by the IoT, not to mention new innovation actors from each emerging technology ecosystem and their values, we will no doubt face greater complexity, uncertainties, and opportunities on innovation trajectories. Next-generation technology policy design will be essential to imagine broader future(s) (*in plural*) for postgenomic scientific trajectories, rather than a preordained determinist singular innovation future. Policymaking approaches such as expert-led conferences (e.g., Asilomar) and the ELSI-type programs have led to a general awareness and acceptance (at times reluctantly by scientists) for societal research on technologies and in part aided by funding policies that demanded social research as a requisite for technology research support.

ELSI and similar programs have been a source of generous funding for social scientists and humanists as well but they have also been contested for shortcomings in design and conceptual framing, including technological determinism, compressed foresight, lack of sufficient independence from science and technology actors, regulatory capture, among others.[48–53]

These first-generation technology policy programs, while generating intensive public engagement activities on emerging technologies, risk and anticipated benefits, they ignored crucial questions about the politics of deliberation itself (constitution and contestation of power), for example, *how* and *by whom* public engagements were designed in the first place, who is included and who is left out in deliberations, what types of technology end points are selectively promoted through strategic use of argument, and other advantage seeking techniques.[3,9,54]

Another common theme among the first-generation technology policies was lack of debates on technology "opportunity costs" which is particularly important when scientific resources, be they research dollars, qualified personnel, or time, are limited. In addition, much of the first-generation technology policy framed the future narrowly, as a linear predictable extension of the past, in part to secure expectations for future-proof innovations. The latter promises and expectations generated could then be employed, by extension, to secure research investments from funders and other innovation investors and stakeholders.[55] The practice of imagining a technology and innovation future narrowly and as a predictable secure extension of the past is known as "compressed foresight,"[52] a practice in policymaking that is contested and clearly a divorce from the complex and uncertain realities of postgenomic innovation futures.

Taken together, and considering the earlier lessons learned from technology and society research from the past three decades before the postgenomic era, Table 15.1 articulates some of the envisioned attributes of the next-generation technology policy design for the readers.

Table 15.1 A Proposal on the Key Pillars for Next-Generation Innovation Policy Design, in Relation to the Current "Artificial Intelligence Turn" in Applied Genomics

Pillars for Technology Policy Design	Description
Purpose	Twofold: (1) A "science enabler" role by identifying, mapping, and responding to uncertainties on the innovation trajectory and (2) provides an independent critical analysis of science and emergent technology; importantly, questions the "opportunity costs" in technology assessment as well as the "frames on knowledge production" (i.e., *epistemology: how do we know what we know?)* in science, social science, and technology ethics
Intervention points	Upstream (*design*), midstream (*translational*), and downstream (*impacts*) on the innovation trajectory
Analytical distance between the analyst and the scientific practice	The innovation analyst operates at a credible and safe analytical distance from science, technology, and its actors and funders so as to preserve independence of the analysis and prevent the risk of cooption or entrenchment
Actors	Social scientists, philosophers, historians of science and technology, independent interdisciplinary scholars, activist and advocacy groups interested in rendering science more transparent and accountable, scientists schooled in social studies of science, feminist analyst of science, and technology with a view of making power embedded in science transparent and thus accountable
Disciplinary tools	History of science, social science, philosophy of technology and science, feminist analysis of technology, highly interdisciplinary
Analysis outcome	Normative (see earlier) and empirical
Time frame of analysis	Anticipatory, real time, and/or post hoc
Framing of innovation future(s)	Highly reflexive (opposite of technological determinism); futures are always in plural, in the making, contested: "uncertainty is not an accident of science but an integral part of it"

15.4 CONCLUSION AND OUTLOOK

Rather than seeing technology trajectories as being future proof, there is an increasing understanding (if not acceptance) by scientists that uncertainty is not an accident of science but rather an integral part of it. Thus next-generation technology policies would serve science and society well by adopting a broader imagination of the multiple possible technology futures (hence, remedying compressed foresight), taking a closer look at the politics of past societal research programs and public engagement initiatives (see also Chapter 12).[22]

As with science in the laboratory, societal research on new technologies is also political in that the process of knowing and knowledge coproduction is a result of technology, research activity (e.g., experiments, and social surveys), and researchers' own value systems. By bringing to the fore the human values attached to all researchers, social scientists and humanists included, we stand a much better chance to respond effectively to technology uncertainties, and with public deliberations that are well examined and deliberated for their design and politics.

In other words, creating unquestioned public engagement programs on innovations will not bring robust societal value, diversity, or social justice unless the design and the very politics of such social research are deliberated and questioned first.[3,46] Studying not only technology but also the embedded human values and politics would benefit science, society, and democracy in ways that stand the test of time and context.

ACKNOWLEDGMENTS

No funding was received in support of this innovation analysis. The views expressed reflect the personal opinions of the author only.

REFERENCES

1. Dylan B. *The Times They Are a-Changin*. New York: Columbia Records; 1964.
2. Harrison P. Curiosity, forbidden knowledge, and the reformation of natural philosophy in early modern England. *Isis*. 2001;92:265–290.
3. Özdemir V. Towards an "ethics-of-ethics" for responsible innovation. In: von Schomberg R, Hankins J, eds. *International Handbook on Responsible Innovation. A Global Resource*. Edward Elgar Publishing; 2019:70–82.
4. Özdemir V. Not all intelligence is artificial: data science, automation and AI meet HI. *OMICS*. 2019;23:67–69.
5. Özdemir V. The fly on the wall... *AGOS Newspaper*. Available from: <http://www.agos.com.tr/en/article/21777/the-fly-on-the-wall>; December 24, 2019c. Accessed 05.04.19.
6. Penders B. Marching for the myth of science: a self destructive celebration of scientific exceptionalism. *EMBO Rep*. 2017;18:1486–1489.
7. Haraway D. Situated knowledges: the science question in feminism and the privilege of partial perspectives. *Fem Stud*. 1988;14:575–599.
8. Özdemir V, Endrenyi L. Toward panvigilance for medicinal product regulation: clinical trial design using extremely discordant biomarkers. *OMICS*. 2019;23(3):131–133.

9. van Oudheusden M. Where are the politics in responsible innovation? European governance, technology assessments, and beyond. *J Responsible Innov*. 2014;1:67–86.
10. Winner L. *Autonomous Technology*. Cambridge, MA: The MIT Press; 1977.
11. Guston DH, Sarewitz D, Miller C. Scientists not immune to partisanship. *Science*. 2009;323:582.
12. Sarewitz D. CRISPR: science can't solve it. *Nature*. 2015;522(7557):413–414.
13. Slob M. Daniel Sarewitz on evidence-based policy. volTA magazine on science, technology and society in Europe. Available from: <http://volta.pacitaproject.eu/1-daniel-sarewitz-on-evidence-based-policy/>; 2012 Accessed 05.04.19.
14. Guston DH. Responsible innovation: who could be against that? *J Responsible Innov*. 2015;2:1–4.
15. von Schomberg R. A vision of responsible research and innovation. In: Owen R, Bessant J, Heintz M, eds. *Responsible Innovation*. Chichester, UK: Wiley; 2013:51–74.
16. Didier C, Duan W, Dupuy JP, et al. Acknowledging AI's dark side. *Science*. 2015;349(6252):1064–1065.
17. Long TB, Blok V. When the going gets tough, the tough get going: towards a new – more critical – engagement with responsible research and innovation in an age of Trump, Brexit, and wider populism. *J Responsible Innov*. 2017;4(1):64–70.
18. Fisher E. Interview with Prof. Erik Fisher, Arizona State University. Dawn of Responsible Innovation. *OMICS*. 2018;22(5):373–374.
19. Fisher E, O'Rourke M, Evans R, Kennedy EB, Gorman ME, Seager TP. Mapping the integrative field: taking stock of socio-technical collaborations. *J Responsible Innov*. 2015;2(1):39–61.
20. Feyerabend P. *Tyranny of Science*. Malden, MA: Polity; 2011.
21. Foucault M. In: Gordon C, ed. *Power/Knowledge: Selected Interviews and Other Writings, 1972–1977*. New York: Pantheon Books; 1980.
22. Özdemir V, Hekim N. Birth of industry 5.0: making sense of big data with artificial intelligence, "The Internet of Things" and next-generation technology policy. *OMICS*. 2018;22:65–76.
23. Özdemir V, Kolker E. Precision Nutrition 4.0: a big data and ethics foresight analysis—convergence of agrigenomics, nutrigenomics, nutriproteomics, and nutrimetabolomics. *OMICS*. 2016;20(2):69–75.
24. Özdemir V, Patrinos GP. David Bowie and the art of slow innovation: a fast-second winner strategy for biotechnology and precision medicine global development. *OMICS*. 2017;21(11):633–637.
25. Pavlidis C, Nebel JC, Katsila T, Patrinos GP. Nutrigenomics 2.0: the need for ongoing and independent evaluation and synthesis of commercial nutrigenomics tests' scientific knowledge base for responsible innovation. *OMICS*. 2016;20(2):65–68.
26. Pirih N, Kunej T. Toward a taxonomy for multiomics science? Terminology development for whole genome study approaches by omics technology and hierarchy. *OMICS*. 2017;21:1–16.
27. Gabbal A. Kevin Ashton describes "the internet of things." The innovator weighs in on what human life will be like a century from now. *Smithsonian Mag*. <www.smithsonianmag.com/innovation/kevin-ashton-describesthe-internet-of-things-180953749>; January 2015. Accessed 05.04.19.
28. Garvey C. Interview with Colin Garvey, Rensselaer Polytechnic Institute. Artificial Intelligence and Systems Medicine Convergence. *OMICS*. 2018;22(2):130–132.
29. Gartner. Gartner says 8.4 billion connected "things" will be in use in 2017, up 31 percent from 2016. <www.gartner.com/newsroom/id/3598917>; 2017 Accessed 05.04.19.
30. Intel. A guide to the internet of things. <www.intel.com/content/www/us/en/internet-of-things/info-graphics/guide-toiot.html>; 2017 Accessed 05.04.19.
31. Özdemir V. The Dark Side of the Moon: The Internet of Things, Industry 4.0, and The Quantified Planet. *OMICS*. 2018;22(10):637–641.
32. Gulland A. Sixty seconds on digital drugs. *BMJ*. 2017;359:j5365.
33. Robinson F. Household smart meters could be used to monitor our health. *BMJ*. 2018;361:k1855.

34. Burrus D. The internet of things is far bigger than anyone realizes. *Wired*. 2014;. <www.wired.com/insights/2014/11/theinternet-of-things-bigger/> Accessed 05.04.19.
35. Hern A, Gibbs S. What is WannaCry ransomware and why is it attacking global computers? *Guardian.*, May 12, 2017.
36. Pariser E. *The Filter Bubble: How the New Personalized Web Is Changing What We Read and How We Think*. London, UK: The Penguin Press; 2011.
37. Bush V. *Science: The Endless Frontier*. Washington, DC: U.S. Government Printing Office; 1945.
38. Sarewitz D. Saving science. *New Atlantis*. 2016;49:4–40. Spring/Summer.
39. Collins HM, Evans R. The third wave of science studies: studies of expertise and experience. *Soc Stud Sci*. 2002;32(2):235–296.
40. Funtowicz SO, Ravetz JR. Risk management as a postnormal science. *Risk Anal*. 1992;12:95–97.
41. Funtowicz SO, Ravetz JR. Science for the post-normal age. *Futures*. 1993;25:735–755.
42. Bijker WE, Hughes TP, Pinch T. *The social construction of technological systems. New Directions in the Sociology and History of Technology*. Cambridge, MA: MIT Press; 1987.
43. Callon M, Lascoumes P, Barthe Y, Burchell G. *Acting in an Uncertain World: An Essay on Technical Democracy*. Cambridge, MA: The MIT Press; 2011.
44. Collingridge D. *The Social Control of Technology*. New York: St. Martin's Press; 1980.
45. Alberts B. The end of "small science"? *Science*. 2012;337(6102):1583.
46. Özdemir V, Springer S. What does "Diversity" mean for public engagement in science? A new metric for innovation ecosystem diversity. *OMICS*. 2018;22(3):184–189.
47. Editorial (anonymous). After asilomar. *Nature*. 2015;526:293–294.
48. Fisher E. Lessons learned from the ethical, legal and social implications program (ELSI): planning societal implications research for the National Nanotechnology Program. *Technol Soc*. 2005;27(3):321–328.
49. Balmer AS, Calvert J, Marris C, et al. Taking roles in interdisciplinary collaborations: reflections on working in Post-ELSI spaces in the UK synthetic biology community. *Sci Technol Stud*. 2015;28 (3):3–25.
50. López JL, Lunau J. ELSIfication in Canada: legal modes of reasoning. *Sci Cult (Lond)*. 2012;21:77–99.
51. Nordmann A, Schwarz A. Lure of the "Yes": the seductive power of technoscience. In: Maasen S, Kaiser M, Kurath M, Rehmann-Sutter C, eds. *Governing Future Technologies: Nanotechnology and the Rise of an Assessment Regime*. Heidelberg: Springer; 2010:255–277.
52. Williams R. Compressed foresight and narrative bias: pitfalls in assessing high technology futures. *Sci Cult (Lond)*. 2006;4:327–348.
53. Chomsky N, Hutchison P, Nyks K, Scott JP. *Requiem for the American Dream: The 10 Principles of Concentration of Wealth & Power*. New York: Seven Stories Press; 2017.
54. Thoreau F, Delvenne P. Have STS fallen into a political void? Depoliticisation and engagement in the case of nanotechnologies. *Polit Soc*. 2012;11:205–226.
55. Borup M, Brown N, Konrad K, van Lente H. The sociology of expectations in science and technology. *Technol Anal Strateg Manage*. 2006;18:285–298.

FURTHER READING

Hekim N, Özdemir V. A general theory for "post" systems biology: iatromics and the environment. *OMICS*. 2017;21(7):359–360.

Özdemir V. Veracity Over Velocity in Digital Health. *OMICS*. 2019;23(6):295–296.

CHAPTER

16

ECONOMIC EVALUATION OF GENOMIC AND PERSONALIZED MEDICINE INTERVENTIONS: IMPLICATIONS IN PUBLIC HEALTH

Vassileios Fragoulakis[1], George P. Patrinos[2,3,4] and Christina Mitropoulou[1]

[1]The Golden Helix Foundation, London, United Kingdom [2]Department of Pharmacy, University of Patras School of Health Sciences, Patras, Greece [3]Department of Pathology, College of Medicine and Health Sciences, United Arab Emirates University, Al-Ain, United Arab Emirates [4]Zayed Center of Health Sciences, United Arab Emirates University, Al-Ain, United Arab Emirates

16.1 INTRODUCTION

The term "medical genetics" has been defined as the science of human biological variation as it relates to health and disease; the study of the etiology, pathogenesis, and natural history of diseases and disorders that are at least partially genetic in origin; and the application of genetics to medicine or to medical practice.[1] A milestone in the history of the discipline of genetics is the year of 1953 when James Watson and Francis Crick came up with the double helix model of DNA.[2]

After the double helix structure had been described, researchers spent a considerable amount of effort to find how the information, that is, contained in the DNA is translated into proteins.[3,4] With this knowledge the field of genetics evolved, mainly focusing in explaining the hereditary nature of certain diseases. Yet, genetic information was also proven to be important to guide therapeutics, a field that is also known as pharmacogenomics.[5,6] Initially, the term "pharmacogenetics" was introduced in 1956[7] to describe the relation between hereditary factors and drug metabolizing capacity. Later, the term "pharmacogenomics" was introduced, not only covering pharmacogenetics but also including acquired genomic variants and mRNA expression profiles affecting drug metabolism, while the term "personalized medicine" was introduced,[8] to indicate the combined knowledge of genetics to predict disease susceptibility, disease prognosis, or treatment response of a person to improve the person's health.

It is known that the side effects for the same drug vary from patient to patient and experimental evidence suggests that the variable phenotypic expression of drug treatment efficacy and toxicity is determined by a complex interplay of multiple genetic variants and environmental factors.[9] Pharmacogenomics is referred to as "... the delivery of the right drug to the right patient at the right dose," and several pharmacogenomic testing approaches currently exist to identify the underlying pharmacogenomic biomarkers.[10–12]

Applied Genomics and Public Health. DOI: https://doi.org/10.1016/B978-0-12-813695-9.00016-9

16.2 PHARMACOGENOMICS, PERSONALIZED MEDICINE, AND HEALTH ECONOMICS

Understanding the relative benefits and costs of alternative pharmacogenomics strategies is important in order to ensure that patients receive not only effective but also economically efficient care. The aforementioned progress made in the development of personalized medicine has coincided with health-care systems placing greater emphasis on evidence-based clinical practice,[13–18] particularly as they are operating within an increasingly budget-scarce environment. It is often argued that personalizing treatment will inevitably improve clinical outcomes for patients and help achieve more effective use of health-care resources. Hence, demand is increasing for demonstrable evidence of clinical utility and economic viability to support the use of personalized medicine in health care.[19]

Health economics is a branch of economics concerned with issues related to efficiency, effectiveness, value, and behavior in the production and consumption of health and health care.[20] There are three key features in health economics analyses as currently applied: (1) they are more focused on the benefits received by the health-care system and society as a whole rather than the individual/patient, namely, improvements in quality of life for the majority of patients, on average expansion of life expectancy and resources saved, to name a few; (2) the recipient of the medical intervention is, in the majority of cases, not the most informed medical decision-maker (the so-called "asymmetry of information"[21]) and, as such, does not have a complete picture of the potential benefits and harms of a given medical intervention or decision; and (3) most of the time, the recipients of medical interventions do not directly pay for these treatments; rather, payment is received from a third party that the recipient supports through taxes, medical premiums, or a shared model of employer/employee contributions. As such, health economics aims to better understand the value and costs of a certain medical intervention compared with another by taking into assumption all the factors that impact on patients, health-care providers, health-care system in general, and, ultimately, society.

16.3 ECONOMIC EVALUATION: TERMINOLOGY AND CONCEPT

Economic evaluation of health services is a branch of health economics that deals with the "systematic evaluation of the benefits and costs arising from the comparison of different health technologies." It represents a way of thinking and problem solving rather than a simple set of terms or methods used by health economists. The term "evaluation" refers to a process of comparing various choices to rank them, based on certain economic and clinical criteria, by order of attractiveness. The definition of economic evaluation includes also the word "systematic." It must be mentioned that analyses based on simplistic criteria of cost comparison between treatments (such as the price of one product compared with another), which include neither the entire financial burden nor the associated benefit for each treatment, are not systematic. When performing an economic evaluation, we seek an accurate comparison of the alternatives to judge their attractiveness to help the decision makers evaluate them. Therefore economic evaluation is useful to policymakers, who can so be informed about the available options (and the consequences of adopting them for the population's health and the country's budget). In the United States and elsewhere, this combination is referred to as "value."[22–25] Value can be thought of as a relationship between outcomes and cost, and this relationship is emerging as a central dogma for health-care reform.

Economic evaluation is utilized concurrently with evidence-based medicine (EBM). As mentioned earlier, EBM is usually synonymous with the search for knowledge (regarding effectiveness) through systematic review of the literature to identify and propose optimal practices for a health system. Its purpose is to inform the people doing clinical work about these practices and to change current practices if they are suboptimal.

In many cases, however, the distinction between "good clinical practice" and "available economic resources" is controversial and often provides results opposite to those initially expected. For example, the adoption of practices that offer little additional therapeutic benefit but considerable burden may undercut the overall ability of the health system to treat patients in the future, not to mention the next generation. In other cases, society elects—through its health-care system—to transfer resources to those less fortunate, regardless of whether this transfer will achieve a small increase in overall social welfare. Such social groups are usually unable to "obtain welfare" as easily as wealthier population groups. In this sense, help for such groups is not a behavior that leads to efficiency maximization; however, society might choose to uphold the concept of equity at the cost of maximizing efficiency, a fact that may not be taken into account by EBM. In this framework, economic evaluation attempts to link EBM, to the wishes of society, patients, and the state to better achieve multiple goals such as viability, societal fairness, and improved efficiency in the health system. In conclusion, economic evaluation combines objective data (prices of production factors, medical technology, etc.) with preference data to rule on which of the available options maximize the welfare of the society in general or of specific patient groups.

It should be also stated that economic evaluation is a very technical subject which at present is still at the stage of developing new quantitative approaches, whether these are entirely new or borrowed from other related fields such as statistics, mathematics, or econometrics. Of course, we have to bear in mind that despite the attractive veneer of objectivity given by the concise and elegant mathematical nomenclature, the actual subject of financial resource management is in practice fundamentally a political issue and the translation of "knowledge" into "political decision" involves other factors that are mostly outside the province of the academic community.[26] Nonetheless, scientists and technology-driven academic communities might play valuable roles in shaping the knowledge trajectories from lab to innovation-in society through greater transparency and a sociological read of the scientific laboratory and practices.

16.4 METHODS USED IN ECONOMIC EVALUATION

From a technical point of view, economic evaluation uses various tools to evaluate medical interventions, which will be briefly discussed next. The two more commonly used approaches in the field are cost-effectiveness analysis (CEA) and, most importantly, cost–utility analysis (CUA) (Table 16.1).

Table 16.1 Types of Economic Evaluation Analyses

Type of Analysis	Costs	Consequences	Result
CMA	Monetary units	Identical in all respects	Least cost alternative
CEA	Monetary units	Different magnitude of a common *measure*, e.g., LY gained and blood pressure reduction	Cost per unit of consequence, e.g., cost per LY gained
CUA	Monetary units	Single of multiple effects not necessarily common, *valued* as utility, e.g., QALY	Cost per unit of consequence, e.g., cost per QALY
CBA	Monetary units	As for CUA but *valued* in monetary units	Net cost:benefit ratio

CMA, Cost-minimization analysis; CFA, cost-effectiveness analysis; CUA, cost–utility analysis; CBA, cost–benefit analysis; QALY, quality-adjusted life-year; LY, life years.

16.4.1 COST-MINIMIZATION ANALYSIS

A cost-minimization analysis (CMA) can only be used to compare two or more health technologies that have proven to be fully equivalent in survival, quality of life, therapeutic effect, tolerability, safety, and compliance. In such a case the focus of the analysis shifts only to their treatment overall cost to choose the least costly as the preferable therapeutic option. It must be noted, however, that the equivalence of safety and efficacy is not very often in clinical practice, and in that sense, this approach is not so common in health economics literature.[27] It must be noted, however, that some cases were no substantial efficacy and safety differences among agents exist, and the decision makers have limited available resources for reimbursement, a CMA was considered the appropriate methodological approach to evaluate alternative therapies from an economic perspective.

16.4.2 COST-EFFECTIVENESS ANALYSIS

CEA is defined as an analytical technique intended for the systematic comparative evaluation of the overall cost and benefit generated by alternative therapeutic interventions for the management of a disease. This sort of analysis aims to determine whether a medical intervention for disease diagnosis, prevention, and/or treatment improves clinical outcomes enough to justify the additional costs compared with alternative approaches. At first, this method focuses on evaluating an intervention in physical units and also is interested in determining life expectancy (final outcome) and not the quality of life. In that sense, this very important method deals primarily with the patients' ultimate health outcomes instead of intermediate indicators. Such intermediate/clinical indicators are of interest only to the clinical scientist when making a diagnosis, proposing treatment, or deciding on future research but are not particularly useful in CEA except in the circumstances when an intermediate indicator is "translated" in life expectancy through a robust chain of evidence mainly via statistical modeling.

As an example, progression-free survival (PFS) in oncology has an important clinical interest because it defines the period, over which the patient is free from disease progression and could be used as a measure of effectiveness in an CEA, but it is not used indiscriminately in economic evaluation because in some type of cancers a different PFS is unable to demonstrate the physical equivalent of clinical observation, such as prolongation of life or improved life expectancy.[28–30]

In that sense, economic evaluation "ignores" a drug's mode of action, its route of administration, or its pharmacokinetic properties. Therefore provided a certain technology has been approved by the competent authorities for use in the general population, it is assessed by the single indicator that is of the most interest for insurance carriers and for individuals in charge of health-care budgets: patient survival. With this "flexible" (but objective) approach, it is possible to calculate the comparative cost and benefit generated by different health technologies targeting the same disease, as long as the outcomes are measured on the same scale (months or years of life), and the cost is presented in a common unit of measurement.

It should be noted that CEA is not a method to indicate which medical intervention or health-care technology reduces the cost, but rather it is used to inform which medical interventions and/or health-care technologies provide the greatest value for a given amount of health-care expenditure. Another feature of CEA is that it evaluates clinical events outcomes, such as cost per life-year gained, but does not allow for direct comparisons among technologies that are used in different therapeutic areas. For example, one cannot easily make a verdict in order to spend a certain amount of money to prevent a rare genetic disorder or a specific type of cancer within the cost-effectiveness context; for this, one should take into consideration other factors that will be discussed next.

When conducting a CEA, we face four different scenarios. Let us consider two different treatments, PGx (new/pharmacogenomic treatment) and S (standard treatment), each associated with a specific effectiveness (E) and cost (C) for the management of a disease. Comparative evaluation will give the following four possible scenarios (Fig. 16.1).

Scenario A: Innovative treatment has *greater* effectiveness but also *greater* total cost (upper right quadrant). This is the most common case. Usually, an increase in mean survival with the

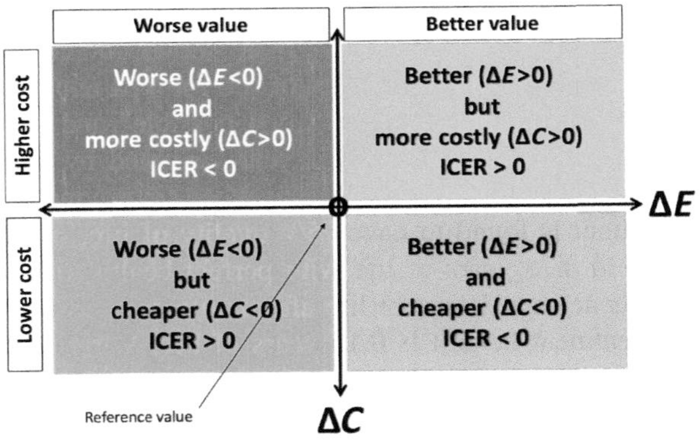

FIGURE 16.1

The cost-effectiveness plane.

innovative treatment and a corresponding increase in overall cost associated with its administration are observed. In recent years, new discoveries have been made continuously as new substances, new biological and gene therapies, new diagnostic methods, targeted treatments, or new procedures have been developed. Technology, in most cases, increases the cost of services because of the developing, purchasing, and operating expenses; the price of technology (which is a fraction of the overall cost of the intervention) tends to incorporate these costly processes. In such cases, there should be a criterion by which to assess whether the increased cost is justified by the additional effectiveness.

Scenario B: Innovative treatment has *greater* effectiveness but also *lesser* total cost (lower right quadrant). Here, the new treatment provides more effectiveness and is associated with lower cost than the standard treatment. This is not very often the case since health-care providers bear very high research and development costs that they wish to transfer to the end consumer or the public insurance funds while also making some profit because they are profit-seeking enterprises.

Scenario C: Innovative treatment has *lesser* effectiveness but also *greater* total cost (upper left quadrant). This unpromising scenario for the new technology includes increased cost but lower effectiveness from its use compared with the standard treatment. It is not very often the case in practice.

Scenario D: Innovative treatment has *lesser* effectiveness but also *lesser* total cost (lower left quadrant). Here, the difference in survival is for the standard treatment, even though the use of the new treatment is associated with resource savings. In scenario D (as in scenario A) the ultimate decision to adopt or reject the new technology will be based on weighing the savings (which is the desirable outcome) and the reduced effectiveness (which is a negative consequence) when comparing treatments. This scenario represents the generic health technologies.

Notably, CEA needs a quantitative criterion to judge when scenario A or D is the case to choose among health technologies. This is discussed in the following subsection.

16.4.3 COST-UTILITY ANALYSIS

CUA attempts to address the aforementioned CEA limitations by measuring outcomes through a metric called a quality-adjusted life-year (QALY)[31] that allows for comparisons across medical interventions and also taking into account the quality of life for a patient. Quality is often measured on a scale of 0, or of 100, where 0 represents the "worst possible" and 100 is the "highest or best possible" state of health. A QALY is a period of 1 year weighted by the quality of life that the patient is experiencing when suffering from a disease or when improving as a result of a treatment. For example, if a cancer patient is found to have 75% quality of life, then 1 year of life with this type of cancer is equivalent to 0.75 years of life with perfect health (0.75 QALYs). If the patient improves to 90% after treatment, then 1 year of life after treatment is equivalent to 0.9 years of life in perfect health, and the treatment benefit is 0.15 years of life. Various methodological tools are used to value a patient's health state and quality of life. Some of these are specialized for specific diseases, whereas others seek to evaluate a patient's general state of health. Some are based on simple indices, and others are more comprehensive but also more difficult to assess. The subjects in such studies are usually patients, but they may also be health professionals, such as nurses or physicians, or the general population. Examples of such efforts are the EuroQol EQ-5D and others.[32–34]

Because of the importance of the quality of life and because this type of analysis will (in theory) facilitate broad comparisons between different medical interventions by reducing them all to a common measure of value (the QALY), CUAs are becoming more and more common, and many organizations such as the UK National Institute for Health and Care Excellence (NICE) encourage their use.[35]

As mentioned earlier in economic evaluation, there is a "gray zone" where the additional benefit is associated with higher cost, and the result will be uncertain until the expense considered acceptable for an additional year of life is fully quantified. The mathematical formula that quantifies the ratio of differences among alternative treatments is called incremental cost-effectiveness ratio (ICER) and is described as follows:

$$\text{ICER} = \frac{\Delta C}{\Delta E} \quad \text{or} \quad \text{ICER} = \frac{(C_{\text{IN}} - C_{\text{STD}})}{(E_{\text{IN}} - E_{\text{STD}})},$$

where E_{IN}, E_{STD}, C_{IN}, and C_{STD} correspond to the mean effectiveness of the new (innovative) treatment, the mean effectiveness of the standard treatment, the mean cost of the new (innovative) treatment, and the mean cost of the standard treatment, whereas $\Delta E = E_{\text{IN}} - E_{\text{STD}}$ and $\Delta C = C_{\text{IN}} - C_{\text{STD}}$ are the definitions of the differences ("Δ," stands from Greek $\Delta\acute{\varepsilon}\lambda\tau\alpha$ which means "difference") in cost and effectiveness, respectively. ICER indicates the amount of money we have to spend in order to achieve one more unit of effectiveness. In this case, CUA indicates the amount we have to spend to achieve one more QALY. The amount a society is willing to pay to obtain 1 year of life is called willingness to pay (WTP) and is represented by the Greek letter lambda (λ). If λ is greater than ICER, then the new treatment is considered a cost-effective option. Let us consider the example next:

	Survival (year)	Quality of Life (%)	QALYs	Cost (€)
Genome-guided treatment	10	60	$10 \times 60\% = 6$	10,000
Standard treatment	8	50	$8 \times 50\% = 4$	6,000

ICER = (10,000 − 6000)/(6 − 4) = €2000 per QALY. If $\lambda = 5000$ per QALY, the genome-guided treatment might be considered a cost-effective option and must be reimbursed by the healthcare system. It must be noted that the determination of ICER is a technical manner, while the amount of λ represents a political decision. In many cases, such decisions regarding the distribution of resources for health care are contingent on each country's individual political and historical background and are not determined by economic models. The estimation of λ remains a subject of extensive debate,[36–44] and even large organizations, such as the UK NICE, have yet to announce a clear decision on its "correct" size. In various other countries, however, WTP values have been proposed for the "purchase" of 1 year of life to provide a transparent criterion for this difficult undertaking. According to the World Health Organization, the desired value for the indicator is approximately three times the average per capita income of the country, while for the United Kingdom, a value between £40,000 and £60,000 is the maximum accepted value in most cases. A value between $50,000 and $100,000 is considered cost-effective, a value less than $20,000 is considered particularly attractive, and values more than $100,000 are considered particularly costly and are rejected. It should be noted, however, that the determination of λ has direct effects on health system budgets and therefore should not be done independently of each economy's available funds.

16.4.4 COST-BENEFIT ANALYSIS

This analysis gives a monetary value to every aspect of health care and medical intervention, which can be very challenging in health care because health-care providers are often very reluctant to place a monetary value on health; as such, this prominent approach is truly difficult to perform accurately.

16.4.5 COST-THRESHOLD ANALYSIS

This analysis operates in reverse in that the analyst defines a threshold of cost-effectiveness (usually derived from the calculated cost-effectiveness of a given intervention). Once this is defined, the analyst asks the question, what are the necessary performance characteristics of a given test or intervention such that the cost of the test/intervention meets or exceeds the threshold? This approach is being used more frequently by developers to understand whether their test or intervention is ready for clinical use or needs additional development. In genomics, this has been used to assess when a genomic risk panel or pharmacogenomics test would have sufficient discriminatory power to justify its use in clinical care. All these approaches consider the cost of the medical intervention itself together with the accompanying costs, but they differ in how they measure the outcome or utility of an intervention.

16.5 ECONOMIC EVALUATION IN GENOMIC AND PERSONALIZED MEDICINE

The recent evolution in the field of genomics has been remarkable. The first human genome was sequenced recently, costing between US$500 million and US$1 billion.[45] After 5 years of research, this cost fell considerably, making more realistic the incorporation of this novel discipline in the clinical setting. The biggest advances in genome sequencing have been increasing speed and accuracy, resulting in reduction in manpower and cost. The speed is thanks to parallel analysis and high-throughput technology (next-generation sequencing, NGS), which permits either whole-genome sequencing (WGS) or parts of it to be sequenced in hours, at great depth and increasing sensitivity.[46]

This information might be used to assist practitioners concerning the diagnosis, prognosis, and clinical management for a variety of disorders, particularly cancer and rare diseases.[45] Over the past decade, genomic sequencing research studies have increased in size, and the same will be the case in the future.

Large-scale sequencing projects such as the 100,000 Genome Project in the United Kingdom and the All of Us Program in the United States are collecting an unprecedented amount of genomic, clinical, and health-care resource use data on individuals with cancer or rare diseases, as well as healthy individuals, but the health economic evidence base for whole-exome sequencing (WES) and WGS is very limited. A recent analysis identified just a few economic evaluations of either WGS or WES, but the majority of them do not fill the appropriate requirements for a full economic analysis.[47,48] This study has estimated the cost-effectiveness of generating information on incidental findings using NGS technologies but evaluated a population screening approach rather than estimating health outcomes for a particular disorder.[49] Methodological uncertainty among health economists may play its role for the lack of evidence on the health outcomes associated with genomic sequencing. Over the past decade, health economists have repeatedly questioned whether metrics such as the QALY, which focuses on clinical utility, can fully quantify the outcomes that are important to patients when they undergo genomic testing.[50–54]

Some applications of genomic sequencing generate information that may not improve quality of life or extend life expectancy, but this kind of applications may impact on patient wellbeing via nonclinical routes, generating "personal utility."

In evaluating the utility of human genome-wide assays the answer will differ depending on the interested parties. For the purposes of regulating medical tests a restrictive sense of clinical utility might be used, while for the purposes of using limited third party or public health resources, cost-effectiveness should be evaluated in a societal context. In taking account of personal utility, cost-effectiveness may be calculated on an individual and societal basis. Overall measures of utility may vary significantly between individuals depending on potential changes in lifestyle, health awareness and behaviors, family dynamics, and personal choice and interest as well as the psychological effects of disease risk perception.[51] For instance, this is the case for those suffering from rare diseases who often have lengthy diagnostic journeys but few treatment options. This could also be an issue if individuals without known health problems undergo genomic sequencing and find out that they have an elevated risk of a disease, but no preventive action can be taken to manage this risk.[45]

Conclusively, there is no consensus in the field on whether QALYs are sufficient to capture the clinical benefits of sequencing or on the best way to capture the nonclinical benefits of sequencing. In addition to that, health economists are also yet to agree on whether information on personal utility should feed into resource-allocation decisions at all in this context, given that the costs associated with genomic sequencing will be met by the health-care budget. In this light the majority of existing economic evaluations undertake via beyond "narrow" outcome measures such as diagnostic yield. Most existing studies evaluating the outcomes associated with WGS or WES are designed to inform "local" resource-allocation decisions (laboratories, individual hospitals, etc.), instead of contributing to within-country Health Technology Assessments.[55,56] In theory the large-scale sequencing projects can address all these issues involving experienced professionals and collecting a big amount of appropriate data in a structured manner, but this is not the case in practice, with the exception of the Cancer 2015 study in Australia which collected some EQ-5D data.[57]

Despite these challenges, evidence on the relative cost-effectiveness of WGS and WES will soon be required to inform the translation of these technologies into clinical practice and thus to be incorporated in daily routine aiming to improve population wellbeing. Among others, there are several steps that health economists can take to improve the evidence based on the clinical and nonclinical utilities of the available genome technologies such as WGS and WES that will underlie these implementation decisions.

Hence, it is crucial to generate the evidence on the clinical utility of genomic sequencing based on traditional methods used by the Health Technology Assessment Bodies around the world and in particular instruments such as EQ-5D questionnaires to generate utility weights that can be used to calculate QALYs. Probably these instruments will not be enough sensitive to differentiate the quality of life before and after undergoing genomic sequencing, but this assumption should be tested in empirical studies[58] to ensure that any evolution in methods is evidence based.

Priority must be given to the fact that large-scale sequencing studies will incorporate preference-based HRQoL instruments to the participants. If this is not feasible since the data collection process is already underway, smaller studies that collect HRQoL data from smaller subgroups of patients could still provide useful information on the clinical utility of WGS and WES and then could to be extrapolated to related clinical populations.

The big challenge for a health economist will be to explore the use of alternative health-state valuation techniques to generate utility weights within the QALY framework. Studies that link this evidence to patients' survival and quality of life could inform decision-making regarding the translation of these technologies into clinical practice. There are some attempts in the related literature to allow analysts to combine information on both clinical and personal utilities within a single metric in order to perform a cost–benefit analysis.[59] In general, health economists will have a vital role to play in the translation of genomic sequencing into clinical practice and to ensure that appropriate and timely decisions will be made regarding the allocation of scarce health-care resources to genomic sequencing.

16.6 EXAMPLES OF ECONOMIC EVALUATION IN GENOMIC AND PERSONALIZED MEDICINE

As has been previously mentioned, the number of published studies is still relatively small[60] (Simeonidis et al., 2019). Nonetheless, the studies that have been published represent a diverse approach to the application of economic analysis to genomic medicine instances that emphasize examination of the critical aspects of the analysis that can impact its validity.

16.6.1 USING PHARMACOGENOMICS TO PREVENT ADVERSE DRUG REACTIONS

Adverse drug events (ADEs) are a major contributor to morbidity, mortality, and costs of care. One of the most well-known examples involves the drug abacavir. Abacavir is a synthetic carbocyclic nucleoside analog with inhibitory activity against human immunodeficiency virus (HIV-1). In combination with other antiretroviral agents, it is indicated for the treatment of HIV-1 infection. Serious and sometimes fatal hypersensitivity reactions have been associated with abacavir. Studies of patients who experienced an abacavir-associated ADE identified an association between the ADE and a specific genetic variant in the HLA complex, HLA-B 57:01. Patients who carry the HLA-B 57:01 allele are at high risk for experiencing a hypersensitivity reaction to abacavir. Approximately 0.5% of patients who are HLA-B 57:01-negative will develop hypersensitivity, whereas more than 70% who are HLA-B 57:01 positive will develop hypersensitivity. This is a very straightforward case for the application of economic analysis. The first was performed in 2004.[61] The patient level data for abacavir ADE were obtained from a large HIV clinic, and the analysis included several types of cost (genetic cost, treatment of hypersensitivity, etc.) and the cost and selection of alternative antiretroviral regimens. The investigators concluded that based on the choice of comparators, the testing strategy ranged from dominant (less expensive and more beneficial compared with no testing) to an ICER of €22,811. Several subsequent analyses have been performed, all of which have determined testing prior to the use of abacavir as being cost-effective and potentially cost saving under some assumptions. A study conducted in 2008[62] used a simulated model of HIV disease based on the prospective randomized evaluation of DNA screening in a clinical trial study. The study modeled three different approaches: (1) triple therapy including abacavir; (2) genetic testing prior to triple therapy with tenofovir substituted for abacavir for patients who carry the HLA-B 57:01 allele; and (3) triple therapy with tenofovir substituted for abacavir for all patients. Abacavir and tenofovir were assumed to have equal efficacy, and the cost of the tenofovir treatment was $4 more than the abacavir treatment. Outcomes were QALYs and lifetime medical costs. The authors concluded that the genetic testing strategy was preferred and resulted in a cost-effectiveness ratio of $36,700/QALY compared with no testing. The authors highlighted that the model was subjected to the assumption that abacavir and tenofovir had equivalent efficacy and abacavir therapy was less expensive. Hence, the results have to be considered strictly in this specific setting and on the basis of resources and drug prices. If any of the underlying parameters change, so may the results and the conclusions of this analysis. Thus periodically updated analysis is required in order to reexamine the results of the question at hand.

Another economic study[63] was conducted to compare a pharmacogenomics versus a nonpharmacogenomics-guided clopidogrel treatment for coronary artery syndrome patients undergoing percutaneous coronary intervention (PCI) in the Spanish health-care setting (549 participants). In this study, patients were classified into two groups: the Retrospective group was treated with clopidogrel based on the clinical routine practice and the Prospective group was initially genotyped for the presence of *CYP2C19* variant alleles before treatment with those carrying more than one *CYP2C19* variant alleles given prasugrel treatment. The analysis predicted a survival of 0.9446 QALYs in the pharmacogenomics arm and 0.9379 QALYs in the nonpharmacogenomics arm within a 1-year horizon. The cumulative costs per patient were €2971 and €3205 for the Prospective and Retrospective groups, respectively. Data analysis showed that pharmacogenomics-guided clopidogrel treatment strategy may represent a cost-effective choice compared with nonpharmacogenomics-guided strategy for patients undergoing PCI.

In a very recent economic study that was conducted by our group to estimate the effectiveness of *DPYD* genotyping based on the cost of toxicity management and the clinical benefit per genotype group [noncarriers (Group A) vs carriers (Group B)], in a large group of patients, treated with a fluoropyrimidines (FL)-based chemotherapy, who suffered from various types of cancer within the Italian health-care setting.[64] The mean QALYs was 4.18 [95% Confidence Interval (CI): 3.16–5.55] in Group A, while it was 3.03 (95% CI: 1.94–4.25) in Group B, indicating a difference at −1.15 (95% CI: −2.90 to 0.46). The most frequent adverse reactions occurred in Group A was Grade IV Neutropenia at 5.71% (95% CI: 3.78%–7.81%), followed by febrile neutropenia at 0.94% (95% CI: 0.19%–1.89%). For Group B the percentage of those experiencing a Grade IV neutropenia was estimated at 7.35% (95% CI: 4.14%–17.07%) a difference of 1.63% (95% CI: 4.26%–10.85%) compared to the corresponding percentage of Group A. febrile neutropenia was 4.92% (95% CI: 3.37%–12.20%) in Group B, a difference of 3.97% (95% CI: 3.40%–11.44%) compared to Group A. Analysis showed that there was evidence for survival and cost difference between DPYD variants noncarriers and carriers in favor of the first group taking also into consideration the quality of life of patients.

16.6.2 BETWEEN ADVERSE DRUG REACTIONS AND EFFICACY

Thiopurine medications [including 6-mercaptopurine (6-MP), 6-thioguanine, and azathioprine] interfere with purine metabolism and are used for a variety of indications, including acute lymphoblastic leukemia (ALL) and inflammatory and autoimmune diseases, and as immunosuppressants in organ transplant recipients. Myelosuppression is generally considered an adverse event; however, in the case of ALL, it is the goal of the treatment, in so far as myelotoxicity is needed, to eliminate the malignant lymphoblasts and induce remission. The end point of the induction therapy in ALL is elimination of at least 99.9% of blasts from the blood and peripheral bone marrow, which necessitates a significant degree of generalized myelosuppression. In this context, myelosuppression can be looked at as both a beneficial and potentially harmful outcome. A previous study[65] examined an important outcome of ALL therapy, minimal residual disease (MRD) in a population of pediatric patients treated with standard therapies that included 6-MP who had undergone thiopurine methyltransferase (*TPMT*) genotyping. MRD is the strongest predictor of relapse of ALL, so it is a very important intermediate outcome of treatment. The genotype information collected was not used to adjust the dose of 6-MP. Analysis of the study data showed that patients with a genotype that predicted reduced TPMT activity had an MRD rate of 9.1% compared with 22.8% in the group predicted to have normal TPMT activity. This study emphasized the importance of considering outcomes that reflect both efficacy and harm.

In one of the earliest published CEA performed by the Institute for Prospective Technological Studies,[65] the cost-effectiveness of TMPT genotyping prior to thiopurine treatment in children with ALL was examined. Information for the cost-effectiveness model parameters was collected from literature surveys and interviews with experts from four European countries. The model indicated that *TPMT* gene testing in ALL patients has a favorable cost-effectiveness ratio. This conclusion was based on parameters collected for *TPMT* genotyping costs, estimates for frequency of TMPT deficiency, rates of thiopurine-mediated myelosuppression in TPMT-deficient individuals, and myelosuppression-related hospitalization costs in each of the four countries studied. The mean calculated cost per life-year gained by *TPMT* genotyping in ALL patients in the four study countries was €2100 (or €4800 after 3% discount) based on genotyping costs of €150 per patient. Based on their work, the most severe adverse event was myelosuppression. The estimates of the frequency were derived primarily from adult studies in which 6-MP was used for inflammatory bowel disease or other inflammatory conditions. A challenging parameter to determine is how many of the adverse events can be attributed to the presence of a *TPMT* gene variant. Although the presence of decreased TPMT activity is associated with increased risk of myelosuppression, the nature of the treatment means that even those with normal activity are at risk for a severe adverse event. In discussing the results the authors note that *TPMT* genotyping can reduce health-care costs through the avoidance of myelosuppression episodes compared with no genotyping. However, a different clinical approach could have an impact in efficacy was not taken into consideration. The authors assumed that the alterations in management would reduce ADEs with no impact on the treatment response for the primary disease; ALL but this assumption may not be the case for these patients.[66]

16.7 COST-EFFECTIVENESS ANALYSIS IN GENOMIC MEDICINE AND THE DEVELOPING WORLD

In an economic evaluation study involving elderly atrial fibrillation patients using warfarin treatment, it was shown that 97% of elderly Croatian patients with atrial fibrillation belonging to the pharmacogenomics-guided group did not have any major complications, compared with 89% in the control group, and, most importantly, the ICER of the pharmacogenomics-guided versus the control groups was calculated to be just €31,225/QALY.[67] These data suggest that pharmacogenomics-guided warfarin treatment may represent a cost-effective therapy option for the management of elderly patients with atrial fibrillation in Croatia, which may be the case for the same and other anticoagulation treatment modalities in neighboring countries.

In another study conducted in Serbia[68] the aim was to be assessed whether genotyping for the *CYP2C19**2 allele was cost-effective for myocardial infarction patients receiving clopidogrel treatment in the Serbian population compared with the nongenotype-guided treatment. Results shown that 59% of the *CYP2C19**1/*1 patients had a minor or major bleeding event versus 42.9% of the *CYP2C19**1/*2 and *CYP2C19**2/*2, while a reinfarction event occurred only in 2.3% of the *CYP2C19**1/*1 patients, compared with 11.2% of the *CYP2C19**1/*2 and *CYP2C19**2/*2 patients. Under the study's assumptions the analysis indicated that performing the genetic test prior to drug prescription represents a cost-saving option.

Apart from the differences in current drug prices and resource utilization in different countries, another important parameter to determine the cost-effectiveness of a certain medical intervention in different health-care systems is the variable frequencies of the pharmacogenomic biomarkers. As such, one should bear in mind that a pharmacogenomics-guided medical intervention that is not cost-effective in a certain country may be cost-effective in another country, even if no significant cost differences exist between these two countries because of the higher frequency of a pharmacogenomic biomarker in the general population. This suggests that economic evaluation studies in pharmacogenomics must be replicated in every country to inform policymakers prior to the implementation of a pharmacogenomic-guided medical intervention to evaluate its cost-effectiveness based on characteristics specific to each country.

16.8 MODELS FOR ECONOMIC EVALUATION IN GENOMIC MEDICINE

The primary question that health economics attempts to answer is how to distribute the available funds to the various public and private health-care providers in order to cover the existing needs in an economically viable way. In practice, when evaluating an innovative (genomic) treatment in comparison to an existing treatment, we determine the ICER.[69] This ratio is the difference between the overall costs of the two health technologies divided by the difference in benefit. The ICER indicates the additional amount of resources that must be expended in order to provide 1 additional year of life to society. In order to reach a final decision as to which of the two health technologies should be adopted by a country's health-care system, we should have a rule to determine whether the ICER is attractive or not. Thus we need to compare the amount of money needed in order to achieve higher effectiveness (the ICER) with the amount of money that the responsible agencies (budget holders) are willing to invest to obtain it. The latter amount is called "willingness to pay" and is denoted by "WTP" or by the Greek letter λ. λ represents the state's institutional representatives' willingness to invest additional resources in order to obtain more QALYs. When the ICER is lower than the λ, then a new health technology is considered to be advantageous for the society and is adopted by the system.

Even though this rule appears to be methodologically attractive, it has several issues that need to be addressed.[70–73] Among others, the major weaknesses of the analysis are that (1) it assumes that the available funds are limitless and can be directly adjusted to the cost of new technologies; (2) λ is determined arbitrarily; (3) the budget analysis does not depend on λ, something that is not true for any health-care system; (4) innovation—as defined by the difference in effectiveness between health technologies (difference in QALYs)—is not included in any way in the model but is assumed to be socially irrelevant; and (5) issues of ethical patient management can be examined using purely economic criteria.

For genomic medicine the development of a new a model is important because the newer personalized medicine treatments appear to be more advantageous compared to older standard treatments. In order to overcome the simplistic aforementioned assumptions, a new genome economic model (GEM) was developed.[26] The GEM model proposes that at least two limits should be applied in relation to the λ. One of these limits is defined by the budget and determines which options are cost-effective as well as cost-affordable, meaning that they can be adopted by the health-care system based on the available funds. The GEM model also addresses the issue as to whether there is also a lower limit for health technologies which can save resources. In other words, once a society has achieved a certain level of health through technology, it is reluctant to sacrifice that effectiveness based solely on economic criteria, an assumption that has been ignored by the classical analysis. Another special feature of this model is that it also handles the concept of innovation. In order to also incorporate this concept in the GEM model, the λ was modified (not fixed as in classical model) to correlate with the magnitude of the difference between the standard treatment and the new treatment.

If, for example, a new technology is marginally better than the standard treatment in terms of effectiveness (and also similar in terms of manufacture, active ingredient, etc.), we would expect to see a very small λ because the budget holders would consider the two technologies almost identical based on evidence from the final outcomes. If the new technology had a small difference in effectiveness, the λ would be lower than the one proposed by the classic model, with a tendency to increase, and if the new technology had a significant difference to the standard treatment then the λ would be greater than the one proposed by the classic model. As we approach the income restriction, of course, the additional amount of money that society is willing to invest in order to obtain a little more effectiveness would tend to become zero, since the value of money at that point would be more important. The highest effectiveness that the society is determined to incrementally reimburse should be specified in advance in this model.

Although the GEM addresses certain simplifying abovementioned assumptions, in the form previously presented, it cannot solve the problem of distribution of a given budget to different health technologies that treat different conditions, so as to achieve maximal societal utility. The generalization of the GEM model resolves the problem of distribution of a given budget to different health technologies that treat different conditions, so as to achieve maximal societal utility. This represents an important issue, and thus the presentation of this model is out of scope of the chapter and can be found by the interested reader elsewhere.[74]

16.9 CONCLUSIONS AND FUTURE CHALLENGES

Genomic CEA has the potential to inform assessments about the value of current and emerging technologies and prioritize value-based decisions about adoption and investment. Efforts to close evidence gaps can strategically target areas of greatest need and potential health and cost impacts. Economic evaluations from the perspective of the relevant stakeholder can provide information and guidance for decision-making and policy-making.

Results from these evaluations can include estimated rages and threshold levels for key outcome variables to achieve desirable real-world results. Most importantly, economic models must be developed for genomic medicine that is flexible and adaptable.[75] Ultimately, high-quality and robust models have to be developed that can be utilized by experienced stakeholders without high-level training in economics to encourage routine use of these models to assist in decision-making.

Lastly, it must be stressed that the ICER calculation by itself does not allow conclusions to be drawn about the cost-effectiveness of the various intervention options. Such conclusions require a quantitative criterion, below which an option is considered effective and above which the option is rejected. The estimation of this indicator remains a subject of extensive political debate rather than scientific analysis, and even large organizations such as the UK NICE have yet to announce a clear decision on its "correct" size. In other words the actual subject of financial resource management is in practice, fundamentally a political problem and, unfortunately, quantitative methods used by health economists represent a tool and not an integrated procedure, while also the translation of "economic knowledge" into "political decision" involves other factors that are mostly outside the scope of the academic/research community.

ACKNOWLEDGMENTS

This work was encouraged by the Genomic Medicine Alliance Health Economics working group.

REFERENCES

1. Epstein CJ. Medical genetics in the genomic medicine of the 21st century. *Am J Hum Genet*. 2006;79(3): 434–438.
2. Sussman I. 65 YEARS OF THE DOUBLE HELIX: Could Watson and Crick have envisioned the true impact of their discovery? *Endocr Relat Cancer*. 2018;25(8):E9–E11.
3. Siggers T, Gordan R. Protein-DNA binding: complexities and multi-protein codes. *Nucleic Acids Res*. 2014;42(4):2099–2111.
4. Agirrezabala X, Frank J. From DNA to proteins via the ribosome: structural insights into the workings of the translation machinery. *Hum Genomics*. 2010;4(4):226–237.
5. Wang JF, Wei DQ, Chou KC. Pharmacogenomics and personalized use of drugs. *Curr Top Med Chem*. 2008;8(18):1573–1579.

6. Aneesh TP, et al. Pharmacogenomics: the right drug to the right person. *J Clin Med Res.* 2009;1(4): 191–194.
7. Kalow W. Human pharmacogenomics: the development of a science. *Hum Genomics.* 2004;1(5): 375–380.
8. Eden C, et al. Medical student preparedness for an era of personalized medicine: findings from one US medical school. *Per Med.* 2016;13(2):129–141.
9. Shastry BS. Pharmacogenetics and the concept of individualized medicine. *Pharmacogenomics J.* 2006;6(1):16–21.
10. Lauschke VM, Milani L, Ingelman-Sundberg M. Pharmacogenomic biomarkers for improved drug therapy-recent progress and future developments. *AAPS J.* 2017;20(1):4.
11. Rodriguez-Antona C, Taron M. Pharmacogenomic biomarkers for personalized cancer treatment. *J Intern Med.* 2015;277(2):201–217.
12. Kalia M. Biomarkers for personalized oncology: recent advances and future challenges. *Metabolism.* 2015;64(3 suppl 1):S16–S21.
13. Barzkar F, Baradaran HR, Koohpayehzadeh J. Knowledge, attitudes and practice of physicians toward evidence-based medicine: a systematic review. *J Evid Based Med.* 2018;11(4):246–251.
14. Sadeghi-Bazargani H, Tabrizi JS, Azami-Aghdash S. Barriers to evidence-based medicine: a systematic review. *J Eval Clin Pract.* 2014;20(6):793–802.
15. Zwolsman S, et al. Barriers to GPs' use of evidence-based medicine: a systematic review. *Br J Gen Pract.* 2012;62(600):e511–e521.
16. van Dijk N, Hooft L, Wieringa-deWaard M. W*hat are the barriers to residents' practicing evidence-based medicine? A systematic review. Acad Med.* 2010;85(7):1163–1170.
17. Tenny S, Varacallo M. *Evidence Based Medicine (EBM).* Treasure Island, FL: StatPearls; 2019.
18. Reddy KR, Freeman AM, Esselstyn CB. An urgent need to incorporate evidence-based nutrition and lifestyle medicine into medical training. *Am J Lifestyle Med.* 2019;13(1):40–41.
19. Bertier G, et al. Integrating precision cancer medicine into healthcare-policy, practice, and research challenges. *Genome Med.* 2016;8(1):108.
20. Kernick DP. Introduction to health economics for the medical practitioner. *Postgrad Med J.* 2003;79(929): 147–150.
21. Blomqvist A. The doctor as double agent: information asymmetry, health insurance, and medical care. *J Health Econ.* 1991;10(4):411–432.
22. Lakdawalla DN, et al. Defining elements of value in health care—a health economics approach: an ISPOR Special Task Force Report [3]. *Value Health.* 2018;21(2):131–139.
23. Janamian T, Crossland L, Wells L. On the road to value co-creation in health care: the role of consumers in defining the destination, planning the journey and sharing the drive. *Med J Aust.* 2016;204(7 suppl): S12–S14.
24. Shaw LJ, Miller DD. Defining quality health care with outcomes assessment while achieving economic value. *Top Health Inf Manage.* 2000;20(3):44–54.
25. Gentry S, Badrinath P. Defining health in the era of value-based care: lessons from England of relevance to other health systems. *Cureus.* 2017;9(3):e1079.
26. Fragoulakis V, et al. An alternative methodological approach for cost-effectiveness analysis and decision making in genomic medicine. *OMICS.* 2016;20(5):274–282.
27. Briggs AH, O'Brien BJ. The death of cost-minimization analysis? *Health Econ.* 2001;10(2):179–184.
28. Hwang TJ, Gyawali B. Association between progression-free survival and patients' quality of life in cancer clinical trials. *Int J Cancer.* 2019;144(7):1746–1751.
29. Kovic B, et al. Evaluating progression-free survival as a surrogate outcome for health-related quality of life in oncology: a systematic review and quantitative analysis. *JAMA Intern Med.* 2018;178 (12):1586–1596.

30. Feyerabend S, et al. Survival benefit, disease progression and quality-of-life outcomes of abiraterone acetate plus prednisone versus docetaxel in metastatic hormone-sensitive prostate cancer: a network meta-analysis. *Eur J Cancer*. 2018;103:78–87.
31. Weinstein MC, Torrance G, McGuire A. QALYs: the basics. *Value Health*. 2009;12(suppl. 1):S5–S9.
32. Devlin NJ, Brooks R. EQ-5D and the EuroQol Group: past, present and future. *Appl Health Econ Health Policy*. 2017;15(2):127–137.
33. Hyland ME. A brief guide to the selection of quality of life instrument. *Health Qual Life Outcomes*. 2003;1:24.
34. Chen TH, Li L, Kochen MM. A systematic review: how to choose appropriate health-related quality of life (HRQOL) measures in routine general practice? *J Zhejiang Univ Sci B*. 2005;6(9):936–940.
35. Bamji AN. NICE behaviour: QALYs in the community. *BMJ*. 2007;335(7619):527–528.
36. Zhao FL, et al. Willingness to pay per quality-adjusted life year: is one threshold enough for decision-making?: results from a study in patients with chronic prostatitis. *Med Care*. 2011;49(3):267–272.
37. Lancsar E, et al. Deriving distributional weights for QALYs through discrete choice experiments. *J Health Econ*. 2011;30(2):466–478.
38. Gyrd-Hansen D, Kjaer T. Disentangling WTP per QALY Data: Different Analytical Approaches, Different Answers. *Health Econ*; 2012;21(3):222–237.
39. Donaldson C, et al. The social value of a QALY: raising the bar or barring the raise? *BMC Health Serv Res*. 2011;11:8.
40. Kirkdale R, et al. The cost of a QALY. *QJM*. 2010;103(9):715–720.
41. Towse A. Should NICE's threshold range for cost per QALY be raised? Yes. *BMJ*. 2009;338:b181.
42. Raftery J. Should NICE's threshold range for cost per QALY be raised? No. *BMJ*. 2009;338:b185.
43. McCabe C, Claxton K, Culyer AJ. The NICE cost-effectiveness threshold: what it is and what that means. *Pharmacoeconomics*. 2008;26(9):733–744.
44. Mason H, Baker R, Donaldson C. Willingness to pay for a QALY: past, present and future. *Expert Rev Pharmacoecon Outcomes Res*. 2008;8(6):575–582.
45. Buchanan J, Wordsworth S. Evaluating the outcomes associated with genomic sequencing: a roadmap for future research. *Pharmacoecon Open*. 2019;3(2):129–132.
46. Schwarze K, et al. Are whole-exome and whole-genome sequencing approaches cost-effective? A systematic review of the literature. *Genet Med*. 2018;20(10):1122–1130.
47. Schofield D, et al. Cost-effectiveness of massively parallel sequencing for diagnosis of paediatric muscle diseases. *NPJ Genom Med*. 2017;2.
48. Stark Z, et al. Prospective comparison of the cost-effectiveness of clinical whole-exome sequencing with that of usual care overwhelmingly supports early use and reimbursement. *Genet Med*. 2017;19(8):867–874.
49. Bennette CS, et al. The cost-effectiveness of returning incidental findings from next-generation genomic sequencing. *Genet Med*. 2015;17(7):587–595.
50. Buchanan J, Wordsworth S, Schuh A. Issues surrounding the health economic evaluation of genomic technologies. *Pharmacogenomics*. 2013;14(15):1833–1847.
51. Foster MW, Mulvihill JJ, Sharp RR. Evaluating the utility of personal genomic information. *Genet Med*. 2009;11(8):570–574.
52. Grosse SD, Wordsworth S, Payne K. Economic methods for valuing the outcomes of genetic testing: beyond cost-effectiveness analysis. *Genet Med*. 2008;10(9):648–654.
53. Pokorska-Bocci A, et al. Personalised medicine in the UK: challenges of implementation and impact on healthcare system. *Genome Med*. 2014;6(4):28.
54. Payne K, et al. Cost-effectiveness analyses of genetic and genomic diagnostic tests. *Nat Rev Genet*. 2018;19(4):235–246.
55. Valencia CA, et al. Clinical impact and cost-effectiveness of whole exome sequencing as a diagnostic tool: a pediatric center's experience. *Front Pediatr*. 2015;3:67.

56. Willig LK, et al. Whole-genome sequencing for identification of Mendelian disorders in critically ill infants: a retrospective analysis of diagnostic and clinical findings. *Lancet Respir Med.* 2015;3(5): 377–387.
57. Lorgelly PK, et al. Condition-specific or generic preference-based measures in oncology? A comparison of the EORTC-8D and the EQ-5D-3L. *Qual Life Res.* 2017;26(5):1163–1176.
58. Davison N, et al. Exploring the feasibility of delivering standardized genomic care using ophthalmology as an example. *Genet Med.* 2017;19(9):1032–1039.
59. Buchanan J, Wordsworth S, Schuh A. Patients' preferences for genomic diagnostic testing in chronic lymphocytic leukaemia: a discrete choice experiment. *Patient.* 2016;9(6):525–536.
60. Simeonidis S, Koutsilieri S, Vozikis A, Cooper DN, Mitropoulou C, Patrinos GP. Application of Economic Evaluation to Assess Feasibility for Reimbursement of Genomic Testing as Part of Personalized Medicine Interventions. *Front Pharmacol.* 2019;10:830.
61. Hughes DA, et al. Cost-effectiveness analysis of HLA B*5701 genotyping in preventing abacavir hypersensitivity. *Pharmacogenetics.* 2004;14(6):335–342.
62. Schackman BR, et al. The cost-effectiveness of HLA-B*5701 genetic screening to guide initial antiretroviral therapy for. *HIV AIDS.* 2008;22(15):2025–2033.
63. Fragoulakis V, et al. Cost-effectiveness analysis of pharmacogenomics-guided clopidogrel treatment in Spanish patients undergoing percutaneous coronary intervention. *Pharmacogenomics J.* 2019;. in press.
64. Fragoulakis V, et al. Estimating the effectiveness of DPYD genotyping in individuals of Italian origin suffering from cancer based on the cost of chemotherapy-induced toxicity. *Am J Hum Genet.* 2019;104(6):1158–1168.
65. Stanulla M, et al. Thiopurine methyltransferase (TPMT) genotype and early treatment response to mercaptopurine in childhood acute lymphoblastic leukemia. *JAMA.* 2005;293(12):1485–1489.
66. van den Akker-van Marle ME, et al. Cost-effectiveness of pharmacogenomics in clinical practice: a case study of thiopurine methyltransferase genotyping in acute lymphoblastic leukemia in Europe. *Pharmacogenomics.* 2006;7(5):783–792.
67. Mitropoulou C, et al. Economic evaluation of pharmacogenomic-guided warfarin treatment for elderly Croatian atrial fibrillation patients with ischemic stroke. *Pharmacogenomics.* 2015;16(2):137–148.
68. Mitropoulou C, et al. Economic analysis of pharmacogenomic-guided clopidogrel treatment in Serbian patients with myocardial infarction undergoing primary percutaneous coronary intervention. *Pharmacogenomics.* 2016;17(16):1775–1784.
69. O'Brien BJ, Briggs AH. Analysis of uncertainty in health care cost-effectiveness studies: an introduction to statistical issues and methods. *Stat Methods Med Res.* 2002;11(6):455–468.
70. Donaldson C, Birch S, Gafni A. The distribution problem in economic evaluation: income and the valuation of costs and consequences of health care programmes. *Health Econ.* 2002;11(1):55–70.
71. Donaldson C, Currie G, Mitton C. Cost effectiveness analysis in health care: contraindications. *BMJ.* 2002;325(7369):891–894.
72. Gafni A. Willingness to pay. What's in a name? *Pharmacoeconomics.* 1998;14(5):465–470.
73. Eckermann S, Briggs A, Willan AR. Health technology assessment in the cost-disutility plane. *Med Decis Making.* 2008;28(2):172–181.
74. Fragoulakis V, et al. Performance ratio based resource allocation decision-making in genomic medicine. *OMICS.* 2017;21(2):67–73.
75. Snyder SR, et al. Generic cost effectiveness models: a proof of concept of a tool for informed decision-making for public health precision medicine. Public Health Genomics 2018;21(5–6):217–227.

CHAPTER

PRICING, BUDGET ALLOCATION, AND REIMBURSEMENT OF PERSONALIZED MEDICINE INTERVENTIONS

17

Christina Mitropoulou[1], Vassileios Fragoulakis[1], Athanassios Vozikis[2] and George P. Patrinos[3,4,5]

[1]The Golden Helix Foundation, London, United Kingdom [2]Department of Economics, University of Piraeus, Piraeus, Greece [3]Department of Pharmacy, School of Health Sciences, University of Patras, Patras, Greece [4]Department of Pathology, College of Medicine and Health Sciences, United Arab Emirates University, Al Ain, United Arab Emirates [5]Zayed Center of Health Sciences, United Arab Emirates University, Al Ain, United Arab Emirates

17.1 INTRODUCTION

Personalized medicine (also termed as precision medicine) strategies hold promise to individualize drug treatment modalities and to detect the onset of disease, ideally at its earliest stages, either presymptomatically or through the determination of individual risk, in an effort to prevent disease progression. At present, this can be performed in a handful of monogenic disorders, such as the hemoglobinopathies and cystic fibrosis, cancers, and for over a hundred of the commonest drug treatment modalities.[1] As such, personalized medicine has the potential to shift the emphasis in medicine from clinical and/or therapeutic intervention to prevention, to inform the selection of optimal treatment modalities and reduce trial-and-error prescribing based on an individual's genome, to help avoiding adverse drug reactions and to improve quality of life. Moreover, it can reveal alternative or even additional uses of existing medications, also known as drug repositioning, and direct the selection and design of novel therapeutics. Overall, the interventions of personalized medicine can aid in controlling the overall cost of health care in medium to long term.[2]

While personalized medicine promises to improve the quality of clinical care, ideally lowering health-care costs, the economic issues and policy making accompanying this field still present a serious hurdle.[3] The decision to introduce personalized medicine into the public health-care system lies upon a range of stakeholders in society, such as innovators, research organizations and academic centers, corporate entities, funders and payers, patients and healthy individuals, regulators, policymakers, and legislators, requiring their combined efforts and concerted actions, most of whom are generally favorable to the implementation of personalized medicine.[4–6] As such, if all necessary policies and measures are properly implemented to foster the overall positive attitude of most stakeholders, ensuring that personalized medicine interventions are implemented in an efficient and effective manner, then the remaining stakeholders, who might hold a neutral-to-negative opinion, will also be supportive.[4]

Applied Genomics and Public Health. DOI: https://doi.org/10.1016/B978-0-12-813695-9.00017-0

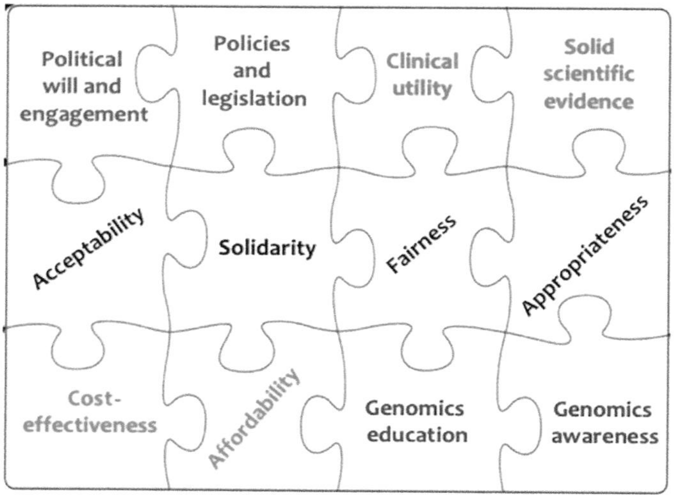

FIGURE 17.1

Conceptual depiction of the interdependencies between different political, societal, and economic factors affecting pricing and reimbursement of personalized medicine interventions as the interconnected pieces of a puzzle. Different colors in the pieces of the puzzle depict a different set of factors that affect the favorable decision to reimburse such innovative interventions. If one of the pieces of the puzzle is missing, most importantly cost-effectiveness and affordability, then the progress toward incorporation and implementation of personalized medicine interventions will be stalled (see also the text for details).

The most important factors required for the successful implementation of personalized medicine interventions into the public health-care system include:

1. political engagement and willingness to change existing health care;
2. solid scientific evidence and demonstration of proven and undisputed clinical efficacy;
3. acceptability, fairness, solidarity, and appropriateness;
4. appropriate policy measures and legislation;
5. affordability and demonstrated cost-effectiveness; and
6. appropriate knowledge and education for clinicians and patients.[7–9]

In other words, if payers are unwilling to reimburse the costs of personalized medicine interventions, then progress toward incorporation and implementation will be stalled (depicted illustratively in Fig. 17.1). Reimbursement decisions in relation to genomic testing are complicated and although genomic testing has been performed for more than 20 years, the respective decision-making process is still evolving[10,11] (see also Chapter 10).

17.2 INSTITUTIONS INVOLVED IN PRICING AND REIMBURSEMENT

All European Union (EU) countries follow the EU directive on medical device regulations. Within this directive, a medical device must have the Conformité Européenne (CE) mark to be sold within the respective member countries denoting some sort of EU harmonization.[12] Also, for a device to be sold in the EU, many member states have adopted complex reimbursement policies, requiring the device to be included in an approved reimbursement list.

In addition, in the EU the regulatory structure for the reimbursement of medical devices, including personalized medicine interventions, differs from country to country. For example, in Germany, it is administered by Der Gemeinsame Bundesausschuss, in France by La Haute Autorité de Santé, in the United Kingdom by the National Health Service, in Italy by Il Servizio Sanitario Nazionale, and in Spain by El Instituto Nacional de la Salud. Consequently, the health-care procedure reimbursement structure for European nations also differs from country to country due to the differences in health-care budgets, health-care policies, etc.[13]

Contrary to the EU, in the United States the Centers for Medicare and Medicaid Services manages the Clinical Laboratory Improvement Amendments (CLIA) program that regulates clinical laboratories, including those performing genomic testing. Several government agencies in the United States are currently working toward the development of regulatory standards for genomic testing laboratories and the comprehensive integration of genomic testing into clinical practice. Furthermore, the Centers for Disease Control and Prevention has the genetic testing policy group specifically focused on the CLIA regulations, as well as projects studying the validation of genomic tests and their integration into clinical practice. In addition, the Secretary's Advisory Committee on Genetics, Health and Society issued a comprehensive report summarizing the issues surrounding the reimbursement of genomic tests[9] (see the next section). On top of that the US National Human Genome Research Institute envisions to serve as a resource for advancing personalized medicine by assisting all types of payers in their efforts to evaluate emerging genomic tests for reimbursement and by promoting research into demonstrating health benefits and cost-effectiveness of personalized medicine interventions.

17.3 COVERAGE, PRICING, AND REIMBURSEMENT STRATEGIES FOR GENOMIC TESTS

The reimbursement structures in most European countries and the United States are faced with increasing health-care costs and patient-waiting times, which negatively impact on their financing their health-care expenditure. This has restricted the amounts that public payers are reimbursing, with respect to the innovative but also expensive molecular diagnostic tests.

While cost-effectiveness may or may not be considered by payers in their decision-making process, the importance of direct cost savings is crucial for some payers, particularly in relation to the use of these tests to avoid the occurrence of adverse drug reactions.[12,13] On top of that the European reimbursement environment is not uniform since each member state has its own policies, and reimbursement is being approved by either private or public insurance companies or, in some cases, a combination of the two. Approval for reimbursement from public health providers often requires lengthy negotiations. In this case, most of the European countries tend to review/authorize tests at the local level, which contrary to the national level reviews can be a significant barrier to consistent market access for genomic/pharmacogenomic tests and services.[12] Moreover, there are differences in health technology assessment (HTA) systems both between and within countries that would benefit from harmonization and also standardization.

17.4 COMPONENTS OF THE PROPOSED STRATEGY FOR PRICING AND REIMBURSEMENT IN PERSONALIZED MEDICINE

The ideal strategy for pricing and reimbursement for personalized medicine interventions should satisfy the following prerequisites:

1. universal access to essential genomic testing for all, and at acceptable prices for the health-care system;
2. sufficient regulation to ensure safety, efficacy, quality, fairness, and solidarity, while allowing space for innovation necessary to move the field forward;
3. appropriate use of genomic tests and information by physicians, according to patients' needs and clinical utility/actionability of testing outcomes; and
4. Investment in both human resources and capital for the research into the field of personalized medicine, evaluation of novel and existing diagnostic procedures, and monitoring of patient safety.

These objectives are discussed in detail in the following subsections.

17.4.1 UNIVERSAL ACCESS TO ESSENTIAL GENOMIC TESTING FOR ALL, AT ACCEPTABLE PRICES FOR THE HEALTH SYSTEM

Essential genomic tests are defined as those tests that satisfy the priority health-care needs of a given population and, like essential medicines, are selected with regard to disease prevalence, public health relevance, evidence of clinical efficacy and safety, and comparative costs and cost-effectiveness.

Ideally, genetic testing should be available for all citizens and should be made available in a timely manner, be distributed according to therapeutic need, and be accessible irrespective of locality, age, gender, or income. The latter will lead to greater health and patient safety, as well as greater efficiency and sustainability of personalized medicine. Currently, genomic tests are prescribed but not reimbursed; as such their prescription depends on the patients' ability to pay, rather than need, therapeutic rationale, or cost-effectiveness.[14]

In order to avoid these expensive out-of-pocket payments, it is first necessary to define and select the essential genomic tests that would be made available and reimbursed under the list of medical devices. This process would contribute to a more rational prescribing, less confusion, and greater familiarity with genomic testing among diagnostic laboratories and health-care providers, provided of course that this process will be regularly updated by clinical guidelines.[15] Classification of every genomic test should be documented by cost-effectiveness analyses with provisions to use HTAs in the decision-making process (see also the previous chapter). The key stakeholders tasked with ensuring the availability and affordability of genomic tests are the Ministries of Health, regulatory bodies, such as the National Medicines Authorities, payer organizations, medical device manufacturers, academic and other research institutes, pharmaceutical companies, pharmacies, hospitals and clinics, public and private diagnostic laboratories, and physicians.[4,15]

At the same time, affordable prices are crucial for ensuring that the population as a whole has access to essential genomic tests, such as pharmacogenomics tests. Increasing affordability should be a key commitment of government to guarantee that pecuniary issues do not create barriers to patients' access to genomic testing. To this end, only an effective, transparent, sustainable, and robust pricing system can guarantee affordable prices and hence access to genomic tests for the entire population. Under such an arrangement, pricing decisions take into account the comparative effectiveness of new genomic tests and their incremental effects relative to their incremental costs, as components of a cost-effectiveness analysis (see also the previous chapter). In addition, all stakeholders involved should have access to the entire pricing process and all the relevant information that was used to determine prices.

The criteria of genomic test classification should also be carefully controlled and regularly reviewed, especially in relation to the emergence of clinical, economic, fiscal, or other criteria, and how these different criteria should be weighed against each other in the decision-making process. In addition, a separate list of genomic tests that fail to meet certain scientific standards and/or lack clinical utility, such as genetic tests for athletic performance, nutrigenomic tests, intelligence, criminality, personality, and genomic identity testing, most of which are provided using the direct-to-consumer model (see also Chapters 10 and 19), should also be established.[16,17] For such tests, no reimbursement should be allowed, and out-of-pocket payments from the hopefully few interested individuals should be the only means to purchase them.

The first step would be the formation of a pricing committee within a country's HTA agency or regulatory authority that would oversee the pricing of all genomic tests. Members of the pricing committee should conduct, in a timely manner, regularly and robustly, background checks and maintain databases in order to advise appropriate prices for all new (or existing) genomic tests entering the market, aiming to allow reimbursement by public insurance funds. For certain genomic tests a tendering system might be developed in order to promote the price competition between manufacturers or, most importantly, distributors. However, such system should be fairly regularly and ultimately transparent.

17.4.2 SUFFICIENT REGULATION TO ENSURE SAFETY, EFFICACY, QUALITY, FAIRNESS, AND SOLIDARITY, WHILE ALLOWING SPACE FOR INNOVATION NECESSARY TO MOVE THE FIELD FORWARD

Robust regulation is a key parameter to ensure genomic test safety, efficacy, and quality and should be a result of strong cooperation among all relevant stakeholders. With this in mind, regulations to assure quality and safety, which can sometimes be lengthy and not so flexible to innovation, should be balanced with ensuring and preserving the opportunity for innovation.[18] Safety assurance is a key consideration for regulatory bodies and a safety process must be put in place for genomic testing services, to address issues pertaining to accuracy and/or testing reliability.

17.4.3 APPROPRIATE USE OF GENOMIC TESTS AND INFORMATION BY PHYSICIANS, ACCORDING TO PATIENTS' NEEDS AND CLINICAL UTILITY/ACTIONABILITY OF TESTING OUTCOMES

In personalized medicine physicians and patients act in concert to achieve the optimum outcome. Physicians have the responsibility for prescribing genomic tests rationally, ensuring appropriate, adequate, and cost-effective care, while well-informed and aware patients, pertaining to personalized medicine, should play an equally important role based on their personal circumstances and preferences. As such, development of an e-prescribing system would be really essential for more effective, efficient, and appropriate prescribing of genomic tests to all citizens. Such system would be designed such as to monitor physicians' prescribing, providers' implementation, and pharmacists' distribution records in order to rationalize the provision of genomic tests. A good example is the use of disease-specific "clinical utility gene cards" that establishes a priori peer-reviewed criteria for the indication of genomic testing.[19] e-Prescribing systems are gradually being adopted by several health-care systems aiming to reduce health-care costs and to maintain the quality of health care provided to patients, especially when coupled to electronic health records.[20]

Lastly, an important parameter in the prescription of genomic tests to citizens−patients is their provision according to international classification and existing guidelines, especially in the case of pharmacogenomic testing that depends upon standardized prescribing protocols across therapeutic categories. This underlines the need to provide personalized medicine interventions to patients at the level of health-care professionals, namely, the physician and the pharmacist, and not using the direct-to-consumer model (see also Chapter 19).

17.4.4 INVESTMENT IN HUMAN RESOURCES AND RESEARCH INTO THE FIELD OF PERSONALIZED MEDICINE, EVALUATION OF NOVEL AND EXISTING DIAGNOSTIC PROCEDURES, AND MONITORING OF PATIENT SAFETY

In the era of personalized medicine, generous investments are required in human capital and infrastructure to promote research in this emerging field, to evaluate novel and existing diagnostic procedures to ensure (1) patient's safety and effective patient care, (2) affordability of the health-care system, and (3) profitability for the industry. Research is a critical component for the personalized medicine sector, which is required for newer, innovative approaches and products. Thus research not only has the potential to make savings and curtail health-care expenditure but also to contribute to the economy as a whole through the development of innovative, possibly patentable technologies.

Investment in research and development is partly relying on the health-care system, on the national priorities for health-care research, and on a variety of factors, both financial and nonfinancial, which define a national policy. For example, a national policy promoting a more flexible and responsive regulatory framework to attract clinical trials might be a way forward to encourage investment in research and development in personalized medicine. In addition, HTA activities can contribute significantly in this direction, for example, by contributing to the making and updating of genomic tests list by the payers, by guiding public health funding from less cost-effective services toward diagnostic tests and pharmacotherapies with greater diagnostic, prognostic, and therapeutic values, respectively, by specifying clinical and prescribing protocols based on clinical cost-effectiveness criteria, and finally by advancing e-health systems.

17.5 PUBLIC HEALTH POLICY CONCERNS

Although lower prices may emerge from the development of increasingly efficient and widespread technologies and interventions, such as those of personalized medicine, it is crucial that reimbursement levels will not only ensure broad access to high quality interventions but also continue to encourage the development of a pipeline of more innovative ones, requiring substantial risk-based research. Many genomic tests, particularly those for predicting complex diseases and phenotypes, require prospective clinical trials and are currently not funded or reimbursed appropriately. Such lack of reimbursement restrict access to these tests for patients suffering from these conditions, as laboratories tend to offer those tests that are reimbursed contrary to those that are paid by the patients themselves and as such might be unprofitable. The latter requires better regulation measures to minimize the impact of the "free-market" shortcomings in the health-care sector, while incentivizing investments for research and development for genomic tests and companion diagnostics solutions.[15]

As such, given that molecular diagnostics have no direct health improvement effect but potentially impact on generating downstream health effects when used as indicated to inform therapeutic option, making the right policy decisions in this area is the single biggest challenge for all stakeholders. This would perhaps entail better coordination of the various HTA processes within a country specifically aimed for companion diagnostics solutions.

As mentioned earlier, HTA represents a rapidly growing area of government policy, especially over the last decade, which governments seek to contain costs. However, most of the paradigms for health-care appraisal and technology assessment have been developed for the purpose of comparing procedures or drug therapies. Today, tests are evaluated in the context of a much broader concept of their clinical utility, namely, cost-effectiveness, budget impact, and priorities (see also the previous chapter). Although most stakeholders would agree that (pharmaco)genomic tests are beneficial by having an impact both on patient management and the more accurate delivery of drug treatment modalities, many are still concerned that processes underlying "clinical utility" assessments are often not clear. To this end, more objective and reliable standards for these HTA and economic evaluation processes need to become broadly accepted. Since many argue that reimbursement should be tied to the value, the actual *value* that not only includes the economies for the payer but, most importantly, the utility for the patient needs to be acceptably defined, taking into consideration how difficult it has been to consider all the factors involved in *value* and to implement value-based pricing.[2]

As such, improving genomic testing quality reduces the rate of incorrect test, while test quality can be also ensured by laboratory accreditation, which constitutes the most extensive quality endorsement, covering genomic test quality inside the laboratory and at the interface with clinicians. To this end, all laboratories performing molecular diagnostic tests should be accredited and existing external quality-assessment schemes should be extended both at national and international levels. In fact, external quality-assessment schemes have been shown to increase the accuracy of genomic diagnostic testing.[21,22]

17.6 INCENTIVES FOR PERSONALIZED MEDICINE

In the case of personalized medicine, there are areas that could have a major impact on public health, but unfortunately traditional funding, pricing, or reimbursement systems fail to provide enough incentives. These areas include funding of the following:

1. the education of physicians and health-care providers, such as genomics education, and increasing literacy in genomics for patients and the general public[23,24];
2. allied professionals, such as genomic counselors (see also Chapter 18); and
3. rewarding and/or creating incentives to develop new tools, for example, translational tools (see Chapter 9: Translational Tools and Databases in Genomic Medicine) that could revolutionize some therapeutic areas.

It would be desirable to conduct surveys among clinicians to establish if there are any concerns with regard to companion diagnostics products, especially since many interviewed expert stakeholders disagreed about the role of clinicians.[4] Some stakeholders reported that clinicians are most reluctant to adopt companion diagnostics even when funding is guaranteed, whereas others reported that clinicians' adoption is not a barrier to access at all.[4] Whichever is true, close links between health-care professionals and clinicians help to increase clinicians' awareness about the kind of genomic tests available and how these are appropriately used and interpreted, especially when informal advice for patients is needed.

On the other hand patients rely traditionally on their doctors' professional opinion and generally follow their advice. In recent decades the clinician–patient relationship has evolved into a partnership model. Indeed, in a recent survey, roughly 75% of European patients, particularly those of younger age, wished to play a more active role in health-care decision-making.[25] For these patients, it is crucial to have easy access to reliable information about their own disease/condition, and treatment options in clear and lay language. This applies in particular to the complex field of personalized medicine.

To resolve these deficiencies, it is crucial not only to further investigate the knowledge and attitude of clinicians toward personalized medicine interventions but also to strengthen collaboration ties between clinicians and health-care professionals. It has been shown that US clinicians are not using pharmacogenomics due to lack of knowledge as to what tests are available, how to procure them, when to use them, how to interpret the results, and how to apply them in the context of an individual patient,[26] while the same situation is also evident for Europe.[6] Hence, increased collaboration leads to an exchange of knowledge, lowers the threshold to ask for advice, and generates awareness about the available testing options.

17.7 CONCLUSION AND FUTURE PERSPECTIVES

Personalized medicine presents special challenges to pricing and reimbursement because of the rapid evolution of new technologies and the lack of a clear understanding by all stakeholders of the requirements for inclusion in coverage policies. Coverage for genomic tests has been a complicated issue, as no consensus exists so far as to what should be covered; thus payers must consider options such as bundled payments or risk-sharing agreements, so that genomic tests are connected to specific drug treatment modalities and/or an entire continuum of health care. Hopefully this would positively impact implementation, which would subsequently help one to advance the field of genomic testing.

From the abovementioned information, it is evident that the pricing and reimbursement policy should have a central and pivotal role in the overall process of getting an innovative personalized

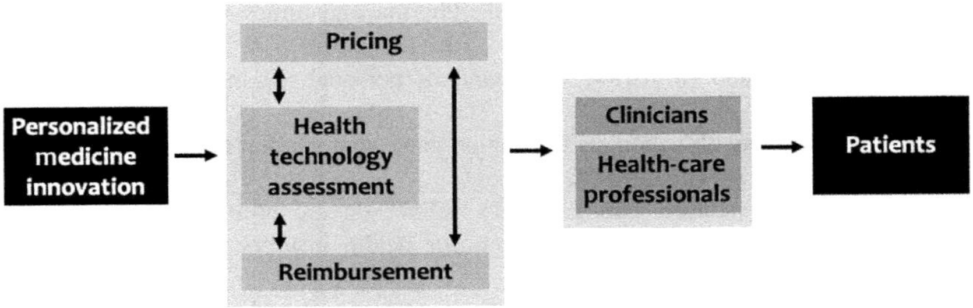

FIGURE 17.2

An overview of the sequence of stages a new personalized medicine intervention (e.g., a molecular diagnostics test or a companion diagnostic assay) has to pass before it becomes accessible to patients and society.

medicine intervention from conception to the end user that is the patient, through well-educated and thoroughly informed clinicians and health-care professionals (Fig. 17.2). To this process, HTA is an important parameter, which relies on the economic evaluation of personalized medicine interventions.[27]

To conclude, patient's outcomes and targeted therapies are decisive in improving patient care and are far superior to the currently practiced "one-size-fits-all" approaches. These, including the already approved system medicine approaches, tailored to the patient's genotypic profile, are expected to be the most cost-effective approach to clinical practice, coupled together with well-educated clinicians and health-care professionals. This would eventually become the standard of care, which could, in turn, assist to drive the costs of the health-care expenditure down.[27]

Lastly, adoption of an appropriate legal framework (see also Chapter 14) is deemed necessary in order to determine the appropriate conditions for reimbursement of clinically valid tests and to set the foundations of a stable, effective, and transparent pricing system.

REFERENCES

1. Manolio TA, Abramowicz M, Al-Mulla F, et al. Global implementation of genomic medicine: we are not alone. *Sci Transl Med*. 2015;7(290):290ps13.
2. Patrinos GP, Mitropoulou C. Measuring the value of pharmacogenomics evidence. *Clin Pharmacol Ther*. 2017;102(5):739–741.
3. Snyder SR, Mitropoulou C, Patrinos GP, Williams MS. Economic evaluation of pharmacogenomics: a value-based approach to pragmatic decision making in the face of complexity. *Public Health Genomics*. 2014;17(5–6):256–264.
4. Mitropoulou C, Mai Y, van Schaik RH, Vozikis A, Patrinos GP. Stakeholder analysis in pharmacogenomics and genomic medicine in Greece. *Public Health Genomics*. 2014;17(5–6):280–286.
5. Mai Y, Koromila T, Sagia A, et al. A critical view of the general public's awareness and physicians' opinion of the trends and potential pitfalls of genomic testing in Greece. *Per Med*. 2011;8(5):551–561.
6. Mai Y, Mitropoulou C, Papadopoulou XE, et al. Critical appraisal of the views of healthcare professionals with respect to pharmacogenomics and personalized medicine in Greece. *Per Med*. 2014;11(1):15–26.

7. Mette L, Mitropoulos K, Vozikis A, Patrinos GP. Pharmacogenomics and public health: implementing "populationalized" medicine. *Pharmacogenomics*. 2012;13(7):803–810.
8. Prainsack B. Personhood and solidarity: what kind of personalized medicine do we want? *Per Med.* 2014;11(7):651–657.
9. Williams MS. The public health genomics translation gap: what we don't have and why it matters. *Public Health Genomics*. 2012;15(3–4):132–138.
10. Williams MS. The genetic future: can genomics deliver on the promise of improved outcomes and reduced costs? Background and recommendations for health insurers. *Dis Manage Health Outcomes*. 2003;11(5):277–290.
11. Williams MS. Insurance coverage for pharmacogenomic testing. *Per Med.* 2007;4(4):479–487.
12. Merchant M. *Pricing and Reimbursement Strategies for Diagnostics: Overcoming Reimbursement Issues and Navigating the Regulatory Environment*. Warwick, UK: Business Insights Ltd; 2010.
13. Britnell M. *In Search of the Perfect Health System*. London, UK: Palgrave Macmillan; 2015.
14. Miller I, Ashton-Chess J, Spolders H, et al. Market access challenges in the EU for high medical value diagnostic tests. *Per Med.* 2011;8(2):137–148.
15. Vozikis A, Cooper DN, Mitropoulou C, et al. Test pricing and reimbursement in genomic medicine: towards a general strategy. *Public Health Genomics*. 2016;19(6):352–363.
16. Patrinos GP, Baker DJ, Al-Mulla F, Vasiliou V, Cooper DN. Genetic tests obtainable through pharmacies: the good, the bad and the ugly. *Hum Genomics*. 2013;7(1):17.
17. Pavlidis C, Lanara Z, Balasopoulou A, Nebel JC, Katsila T, Patrinos GP. Meta-analysis of nutrigenomic biomarkers denotes lack of association with dietary intake and nutrient-related pathologies. *OMICS*. 2015;19(9):512–520.
18. Evans JP, Watson MS. Genetic testing and FDA regulation: overregulation threatens the emergence of genomic medicine. *JAMA*. 2015;313(7):669–670.
19. Dierking A, Schmidtke J, Matthijs G, Cassiman J-J. The EuroGentest Clinical Utility Gene Cards continued. *Eur J Hum Genet*. 2013;21(1):1.
20. Kierkegaard P. E-prescription across Europe. *Health Technol*. 2013;3:205–219.
21. Berwouts S, Fanning K, Morris MA, Barton DE, Dequeker E. Quality assurance practices in Europe: a survey of molecular genetic testing laboratories. *Eur J Hum Genet*. 2012;20(11):1118–1126.
22. Berwouts S, Morris M, Dequeker E. Approaches to quality management and accreditation in a genetic testing laboratory. *Eur J Hum Genet*. 2010;18(suppl 1):S1–S19.
23. Kampourakis K, Vayena E, Mitropoulou C, et al. Key challenges for next generation pharmacogenomics. *EMBO Rep*. 2014;15(5):472–476.
24. Reydon TA, Kampourakis K, Patrinos GP. Genetics, genomics and society: the responsibilities of scientists for science communication and education. *Per Med.* 2012;9(6):633–643.
25. Coulter A, Jenkinson C. European patients' views on the responsiveness of health systems and healthcare providers. *Eur J Public Health*. 2005;15(4):355–360.
26. Crews KR, Hicks JK, Pui C-H, Relling MV, Evans WE. Pharmacogenomics and individualized medicine: translating science into practice. *Clin Pharmacol Ther*. 2012;92(4):467–475.
27. Simeonidis S, Koutsilieri S, Vozikis A, Cooper DN, Mitropoulou C, Patrinos GP. Application of economic evaluation to assess feasibility for reimbursement of genomic testing as part of personalized medicine interventions. *Front Pharmacol*. 2019;10:830.

CHAPTER

GENETIC COUNSELING

18

Janet L. Williams

Genomic Medicine Institute, Geisinger, Danville, PA, United States

18.1 INTRODUCTION AND BACKGROUND

Genetic counseling is currently defined as "the process of helping people understand and adapt to the medical, psychological, and familial implications of genetic contributions to disease."[1] Genetic counseling began in the purview of medical geneticists who viewed genetic counseling as a central tenet of disease prevention through avoidance of harmful or unfavorable findings from a population viewpoint and for the benefit of individuals and families.[2] In the 1950s, Sheldon Reed proposed the idea of a medical social worker and promoted the term genetic counseling.[3] While Reed saw physicians, psychologists, or other PhDs as the only adequately qualified professionals to provide genetic counseling, Melissa Richter et al. at Sarah Lawrence College envisioned a master's level professional with education involving laboratory science, statistical expertise, and an understanding of the psychological aspects of human genetic disease.[4] The first program for graduate education in genetic counseling to establish a new health-care provider, the genetic counselor, was initiated at Sarah Lawrence College in Bronxville, New York, in 1969. Although it grew out of the practice of medical genetics by physicians, genetic counseling has never been primarily geared toward diagnosing genetic disease. Genetic counseling focuses on the communication process around genetic conditions and the attendant psychosocial adjustment that accompanies the recognition of genetic disease in families.

Today, genetic counseling in many countries is provided by trained, certified, registered, and/or licensed professionals, who are genetic counselors. Genetic counselors across the world tend to have similar educational training within structured masters level graduate programs. There are also nurses who have gone through additional genetics training and who are considered genetics nurses. Since the inauguration of the Sarah Lawrence graduate program, many more programs have been established. There are currently 45 accredited graduate programs for genetic counseling in North America.[5] Patch and Middleton[6] estimate that there are approximately 7000 genetic counselors across the globe in at least 28 countries. Information about international genetic counseling graduate programs can be found at the website of the *Transnational Alliance for Genetic Counseling*[7] (Table 18.1).

Applied Genomics and Public Health. DOI: https://doi.org/10.1016/B978-0-12-813695-9.00018-2

Table 18.1 Global State of the Genetic Counseling Profession[8]

Region	Countries Where Genetic Counseling Exists as a Profession[a]	2018 Estimated Number of GCs	Year of First Established Master's Training Program (Total No. of Programs)
North American	United States, Canada	4400	1969 (42, with 5 under review 1/2018)
Europe	Denmark, France, Ireland, Netherlands, Norway, Portugal, Romania, Spain, Sweden, Switzerland, United Kingdom	900	1992 (8)
Middle East	Israel, Saudi Arabia	<100	1997 (2)
Oceana	Australia, New Zealand	~300	1995 (1 graduate diploma); 2008 (Masters) (2)
Africa	South Africa	~25	1988 (2)
Asia	India, Indonesia, Japan, Malaysia, Philippines, Singapore, South Korea, Taiwan	350	2003 (5)
Central/South America	Cuba	~900	1999 (1)

[a]*The existence of the profession does not imply governmental acknowledgment of the profession or a regulatory process, rather the other profession exists separately from physicians or other health-care providers offering genetic counseling services. Other countries not listed have small numbers of genetic counselors trained in other countries who may be offering both clinical services or consulting services through corporate or academic laboratories.*
From Ormond KE, Laurino MY, Barlow-Stewart K, et al. Genetic counseling globally: where are we now? Am J Med Genet C Semin Med Genet. *2018;178C:98–107. Available from: https://doi.org/10.1002/ajmg.c.31607.*

As genetic and genomic information becomes more broadly incorporated into our overall understanding of health and illness, it is important to increase general understanding of the process of genetic counseling. The complete definition of genetic counseling presented by Resta et al. in 2006 states

Genetic counseling is the process of helping people understand and adapt to the medical psychological and familial implications of genetic contributions to disease. This process integrates the following:

- interpretation of family and medical histories to assess the chance of disease occurrence or recurrence;
- education about inheritance, testing, management, prevention, resources and research; and
- counseling to promote informed choices and adaptation to the risk or condition.

The definition does not specify who should provide genetic counseling, rather it describes a process. While genetic counselors are uniquely trained health-care providers, they will not be the only purveyors of genetic and genomic information, particularly as sequencing is employed within a population health setting. All medical providers will need familiarity with genetic and genomic sequencing information and will need to hone their ability to communicate this information and its implications with their patients. This chapter will review fundamentals of the genetic counseling encounter and propose applications of genetic counseling processes in population health initiatives.

18.2 FUNDAMENTALS OF GENETIC COUNSELING

This section of the chapter will review the assumptions and elements of genetic counseling that support the communication process characteristic of genetic counseling visits. Genetic counseling is a skill-based process that facilitates communication of genetic and genomic information in a variety of health-care settings as well as in wellness venues that seek to incorporate genomic information in the assessment of personalized or precision health.

18.2.1 ACCESS TO GENETIC COUNSELING

There is an historical assumption that genetic counseling should be voluntary, that is, the information should be offered when indicated, but that individuals and families may decide whether to participate in counseling or in testing. This likely arose primarily in response to concerns about the potential to impact decisions about genetic testing and reproductive choices. Individuals should be free to make the decisions they wish without being deemed irresponsible if they choose *not* to prevent genetic disease. The issue of voluntary access can also represent a barrier to genetic counseling if an individual is unaware that such a resource exists, or if they are lacking the financial or social resources to receive it. There is an obligation to provide education to individuals and families about genetic counseling and the potential to benefit from advances in genomic science. In addition, there is an obligation to advocate for policies that enable equal access for any and all who might benefit from the information.

18.2.2 TOOLS OF PRACTICE

Collecting personal medical and family medical history is fundamental in the genetic counseling process. The collected data inform the differential diagnosis (with a physician or in a targeted disease clinic), risk formulation, medical management recommendations, and educational and psychosocial support requirements. Delineation of family history, often in pedigree format, is a key tool in the genetic counseling process. While the use of a pedigree format is not necessary for risk assessment, as illustrated in the later example, one can clearly see the transmission from generation to generation in an autosomal dominant inheritance pattern, suggestive of hereditary breast and ovarian cancer syndrome (Fig. 18.1).

Formal instructions for standardized family history collection and analysis can be found in "Recommendations for Standardized Human Pedigree Nomenclature."[9] Other options include electronic family history collection with analysis built in relative to certain specialty areas. Examples of online software tools include CRA Health (https://www.crahealth.com/) and Progeny (https://progenygenetics.com). There are many such online tools and this chapter does not endorse one tool over any other. The landscape of online family history and risk assessment software continues to grow with new features including analysis and clinical decision support added frequently.

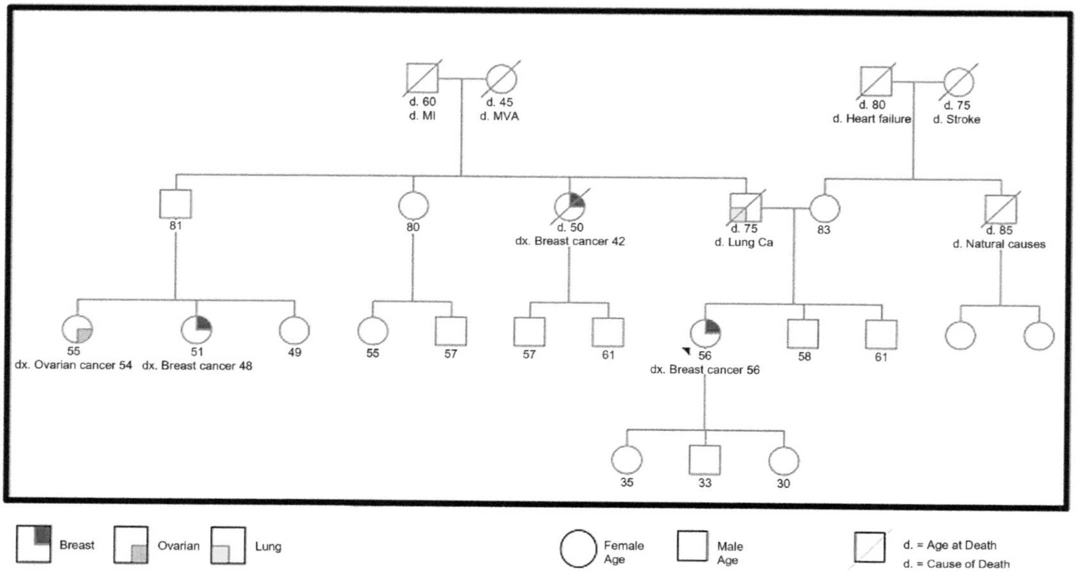

FIGURE 18.1

Example: Breast cancer risk assessment pedigree.

18.2.3 PATIENT EDUCATION

Genetic counseling provides information relative to the condition in question, the genetics of the condition, diagnosis and management, recurrence risk, family members at risk, possible psychosocial impacts of the condition, resources for the patient and family members, strategies for prevention and relevant research directed to improved understanding of the condition, and potential treatment options. Principles of patient-centered shared decision-making and ensuring informed choice rely on the use of multiple methods of patient education including literacy assessment to ensure that the patient understands the information being provided (see also Chapter 11: Assessing the Stakeholder Landscape and Stance Point on Genomic and Personalized Medicine). Many resources for patient educational materials can be found in online resources, such as the Genetic Alliance (www.geneticalliance.org), European Rare Diseases Organization (www.eurordis.org), the Genetics Home Reference (https://ghr.nlm.nih.gov/), and at disease-specific organization or foundation websites.

18.2.4 COMPLETE DISCLOSURE OF INFORMATION

A primary principle of genetic counseling holds that all relevant information belongs to the individual and should be disclosed. Selectively withholding certain information is seen as disrespectful and paternalistic. For example, an early dilemma that was frequently discussed involved whether to inform a phenotypic female with androgen insensitivity, of the cytogenetic finding of her XY karyotype. Within the genetic counseling world, the female individual owns that information for her own life planning and medical management. Another dilemma arose in the context of discovered but undisclosed nonpaternity with laboratory testing. Here, the response regarding disclosure has been more nuanced. Some have questioned the "relevance" of the information to the reason for testing and asked whether the social impact to the family unit would be greater and potentially more deleterious than the value to knowing the finding of nonpaternity. This scenario often plays out on a case-by-case basis with different decisions made given the specific circumstances of the individual, the family, and the provider.

A different challenge is presented in the new reality of genomic sequence information with the tremendous amount of information that can be obtained and the potential for secondary findings. What is considered complete and transparent disclosure of the information? This remains a conundrum as while there are recommendations of what secondary findings should be disclosed by laboratories, there are no practice guidelines yet to direct the disclosure content and process.[10]

18.2.5 SHARED DECISION-MAKING

The patient-centered focus of genetic counseling relies on provision of the information that is necessary to support decision-making, adjustment to a diagnosis, embarking on altered health management, or any other action prompted by the information disclosed during genetic counseling. For many professionals, the practice of genetic counseling has been portrayed as *nondirective* counseling. Historically, this meant refraining from offering advice especially regarding the topic of reproductive decision-making. More recently, there is a recognition that with the increase in knowledge regarding genetic disorders, there are many more interventions that are disease preventing or risk reducing, and genetic counseling will be accountable for health outcomes related to the information provided. The role of nondirective counseling has diminished in favor of the model of shared decision-making. In this model, both parties involved share information toward the goal of consensus about the preferred treatment and agreement on how to proceed with the treatment. The process seeks to balance the provider and patient contribution to and responsibility in any decision-making process.

18.2.6 PSYCHOSOCIAL ASSESSMENT

Genetic counseling seeks to empower the patient and family to cope with their genetic condition. This may include assessment of social, cultural, economic, emotional, and lived experiences. Having individuals tell their story of what the condition means for them, and how they currently cope with the issues associated, has been shown to encourage patients to see themselves as competent. It allows patients and their families to project the best potential outcomes they envision for themselves and may help to anticipate options for adjustment. For some patients, the adjustment may feel overwhelming. In such a situation, referral for more in-depth resources to support the individual and/or family may be very helpful. Recognizing the psychosocial strengths and challenges can inform the support that the patient and family may require to be successful in their efforts to respond to increased risk and to make the necessary changes in their health-care management and ultimately to lead to healthy adjustment to what may be a new norm for defining quality of life.

Resources that offer support for individuals and families can often be found on the same websites as those that offer patient education.

18.2.7 CONFIDENTIALITY, PRIVACY, AND DATA SHARING

In many ways, genetic information is not different from other laboratory results or physical exam findings. Such health information is protected as the individual's personal right to privacy. The main difference is that genetic information has implications beyond the patient, that is, for their family, which has prompted many countries to enact policies which specifically apply to the use and sharing of genetic information. Genomic sequence information can pinpoint an individual precisely and identify relatives according to their degree of relatedness. This realization has become more evident to the general public with the highly publicized use of ancestry testing data in the United States to identify serial killers and other criminals.[11] Such genetic information also holds much more information about an individual than the reason that it might have been acquired. Individuals providing genetic counseling will need to comply with their national, regional, and local policies regarding data storage and data sharing as well as policies regarding familial communication and disclosure of medical information.

18.3 GENETIC COUNSELING IN POPULATION HEALTH INITIATIVES

The focus of public health in the developed world, and increasingly in the developing world, has changed from recognizing and preventing the spread of infectious diseases to one of promoting preventive health and wellness through the recognition of the contribution of a variety of social and lifestyle factors and greater emphasis on the contribution of genetic factors to overall health outcomes. Options for prevention have become available through the widespread adoption of testing for genetic disorders. Newborn screening is perhaps the most successful and widely utilized population application. Outside of newborn screening, the past 50 years of attention has been directed at single-gene disorders identified through single-gene testing. Recent advances in genomic sequencing technology have ushered in a new approach to population health and some would argue opens the door for greater inclusion in the public health agenda. There are several global population health genomics initiatives including efforts, such as Genome England, the Belgian Medical Genomics Initiative, the Estonian Pilot Program for Personal Medicine, and the Canadian Genomics and Personalized Health Competition.[12] These initiatives aim to harness the contribution that genomic sequencing information may provide in addressing public health. Specifically, they are focused on identifying the implementation strategies that will enable broad-scale applications of genomic sequencing information to more precisely differentiate those with high-risk genetic contributions to disease. This has often been labeled as "precision public health" and described as delivering the right intervention to the right people at the right time.[13] As genomic information becomes integrated into the public health agenda, the role that genetics plays in health and wellness will need to be translated at the individual level. Genetic counseling is well suited for this task. Genetic counseling is already being employed in several population health initiatives. In the next section, three examples of genetic counseling will be discussed to demonstrate the role it can play in population health initiatives that employ genomic information in their program.

18.3.1 CLINICAL GENOME RESOURCE

The Clinical Genome (ClinGen) Resource (www.clinicalgenome.org) is a resource funded by the National Institutes of Health in the United States that was developed to provide an authoritative central resource where clinicians, researchers, laboratory specialists, or the public can find definitive information regarding the clinical relevance of genes and variants. The ClinGen website provides many resources and tools focused on gene and variant curation, actionability, variant pathogenicity, registries, options for data sharing, and links to other resources. ClinGen funding also included the formulation of an expert working group, the Consent and Disclosure Recommendations (CADRe) workgroup, charged with developing "guidance to facilitate communication about genetic testing and improve the patient experience."[14] The workgroup set out to establish whether a rubric could be used to inform nongenetics providers in planning the genetic counseling strategy for discussing genetic testing results. The CADRe group provides a conceptual framework that can be individualized to patient characteristics and the contextual setting and uses three communication approaches (Fig. 18.2).

The first involves "traditional genetic counseling" consisting of detailed discussion with a clinician who has genetics expertise, collection of family history, review of genetic testing options, and provision of appropriate patient educational and support materials often lasting more than 30 minutes. The second communication approach is characterized as a "targeted" conversation with the ordering clinician in the context of a brief overview of the condition, the genetic test, and assessment of the need for more in-depth traditional genetics counseling based on indicated psychosocial concerns. The final approach proposed involved a "brief communication" with the clinician and could include educational materials, assessment of patient response, and the potential need for a more involved communication approach. Their work has highlighted the need for clinicians to take the initiative to better understand basic elements of genetic testing and counseling, and awareness that the discussion of the results at a minimum include such aspects as why the test is important for the health-care management of the individual, the potential results, what a test can and cannot tell you, and what results mean for family members. In addition, patients and clinicians may benefit from having genetics professionals a part of team in approaching the return of genetic and genomic results. The CADRe rubric may provide an informative resource for clinicians evaluating the best options for genetic counseling discussions with their patients relative to the genetic or genomic results.

18.3.2 GENOMIC SEQUENCING IN HEALTHY POPULATIONS

Some genomic sequencing programs in healthy populations have included plans for communication of results. In particular, the UK 100,000 Genome Project has embarked on an ambitious process not only to train genetic counselors as part of the investment in genomic medicine but also to have genetic counselors contribute to a plan for increased education in genomics competency for all health-care providers.[6] A national Genomics Education Programme has been instituted with the goal of "upskilling existing staff so that they can make the most of genomic technologies in their

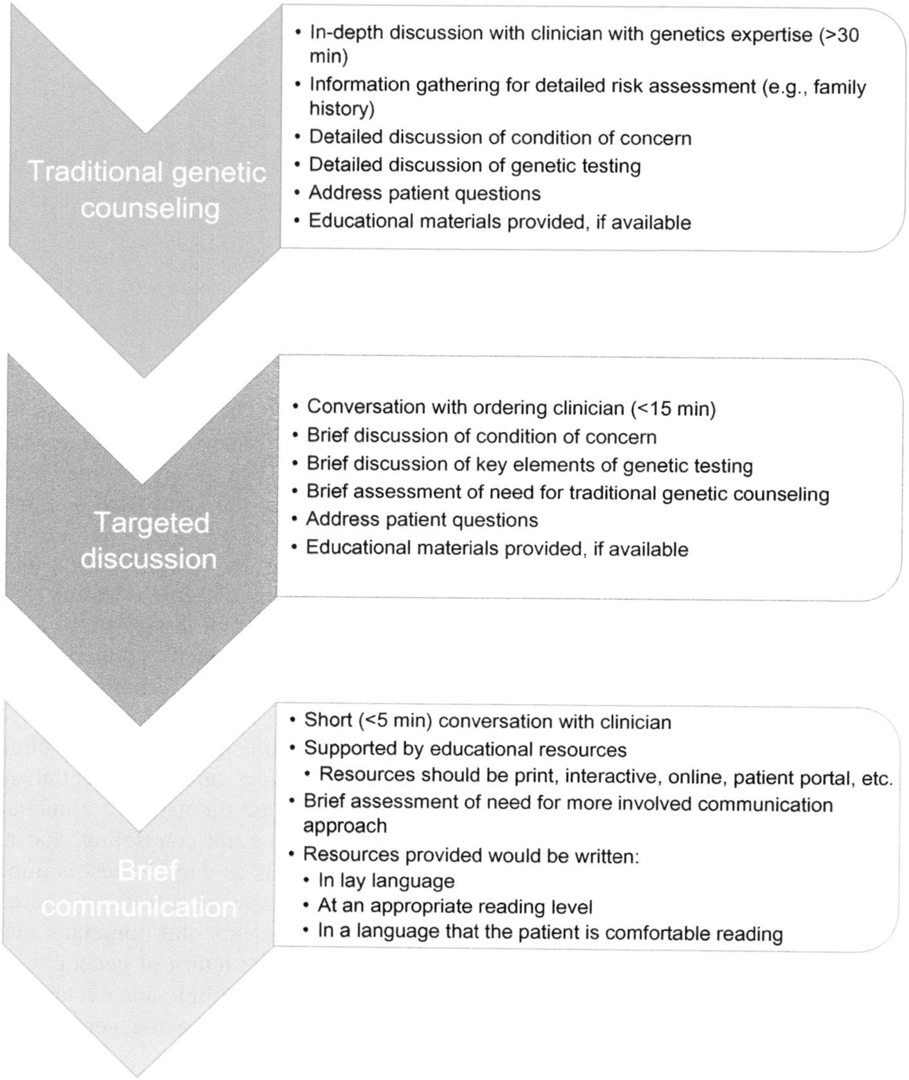
Traditional genetic counseling
• In-depth discussion with clinician with genetics expertise (>30 min)
• Information gathering for detailed risk assessment (e.g., family history)
• Detailed discussion of condition of concern
• Detailed discussion of genetic testing
• Address patient questions
• Educational materials provided, if available
Targeted discussion
• Conversation with ordering clinician (<15 min)
• Brief discussion of condition of concern
• Brief discussion of key elements of genetic testing
• Brief assessment of need for traditional genetic counseling
• Address patient questions
• Educational materials provided, if available
Brief communication
• Short (<5 min) conversation with clinician
• Supported by educational resources
• Resources should be print, interactive, online, patient portal, etc.
• Brief assessment of need for more involved communication approach
• Resources provided would be written:
• In lay language
• At an appropriate reading level
• In a language that the patient is comfortable reading

FIGURE 18.2

Description of levels of communication as utilized in the ClinGen Resource CADRe.[14] *CADRe*, Consent and Disclosure Recommendations; *ClinGen*, Clinical Genome.

work."[15] The program offers online modules and general education courses as well as formal degree-granting programs with MSc in genomic medicine. In addition, they have identified the educational needs of primary-care providers and developed a strategy in response to address those needs. Genetic counselors play a critical role as a resource however this program seeks to equip all health-care providers, especially primary-care providers, with the knowledge skills to understand the implications of genomic information and to provide appropriate management.

A second example of genetic counseling in the setting of genomic sequencing in a general population is the Genomic Screening and Counseling (GSC) program initiated at Geisinger, a large integrated health-care delivery system in rural central Pennsylvania and southern New Jersey in the United States. A strategic partnership between Geisinger, a health-care system, and Regeneron Pharmaceuticals, Inc., a pharmaceutical company, led to the large-scale enrollment of patient-participants into a research program that collected biospecimens, performed exome sequencing, and used clinical data from electronic health record systems for the purposes of discovery research.[16] This investment led to the development of a program to return medically actionable genomic sequencing results to participants in the MyCode Community Health Initiative. The participant population is recruited in outpatient clinics across the system and is unrelated to health indication or presence or absence of disease. To date, over 225,000 individuals have consented to participate and approximately 92,000 individuals have had research grade exome sequencing. As of December 2018, just over 1044 participants have had results returned. A listing of the results returned and information about the program can be found at https://www.geisinger.org/mycode. The interesting juxtaposition that this effort has revealed is the reversal of the traditional genetics visit process which often starts with evaluation of the physical characteristics of an individual, the phenotype, and proceeds to a decision about genetic testing. In population genomic screening a visit may start with the genotype of the individual as the indication for a visit.[17] In this setting, genetic counseling requires adjustment to accommodate the individual who may have the genotype finding that implies risk for a genetic condition but no phenotype.[18] The GSC program is based on dual roles in the communication process including genomics personnel and the patient's primary-care provider. While many primary-care providers disclose the genomic results, they also rely on the genomics professionals to elaborate on the implications of the results. The genetic counseling process in this setting is focused on the response and adaptation of the patient-participant, answering questions and providing educational materials, facilitating appropriate follow-up management such as enhanced cancer surveillance and coordinating communication with at-risk family members. The foundational elements of genetic counseling as presented previously are still relevant; it is the order and context that has changed.

The MyCode Community Health Initiative offers a view to the public health potential of utilizing population genomic sequencing to offer preventive health management for high-risk familial conditions. Examples of such high-risk conditions include three disorders recognized by the US Centers for Disease Control Public Health Genomics program as having sufficient evidence to implement identification and management of individuals at risk.[17] The conditions include Hereditary Breast and Ovarian Cancer syndrome, Lynch syndrome, and Familial Hypercholesterolemia.[19] The key to realizing the public health benefit is in appropriate design and management of the screening program as well as appropriate communication of the genomic sequencing information. Application of this model to hereditary breast and ovarian cancer syndrome has identified the value of such an approach in recognizing that clinical care identified only 20% of the at-risk individuals as defined by the presence of a pathogenic/likely pathogenic variant in either *BRCA1* or *BRCA2*.[20]

18.3.3 CASCADE TESTING: REACHING OUT TO AT-RISK FAMILY MEMBERS

There are many ways in which genetic information is similar to other laboratory and medical information; however, one of the ways it is significantly different is that it involves implications beyond the individual. More than any other medical or health information, genetic information is likely to pose uncertainty and potentially significant risk for family members. It is this recognition of the implied significantly increased risk for family members that has prompted discussion of considering cascade testing, an issue of public health concern.

Cascade testing, or screening, is the term used for the process to identify and sequentially test the at-risk relatives of a patient for whom a genetic finding is known.[21] Studies evaluating the cost-effectiveness of genomic medicine have pointed to cascade testing as the tipping point at which each additional relative identified increases the economic value that is realized (e.g., Lynch syndrome).[22] However, beyond the recognized added economic value, cascade testing promises to enable greater impact on the public health opportunities for prevention of genetic disease. Using population genomic screening to find the first person in a family and then employing cascade testing to identify additional at-risk individuals will further the preventive potential of genomic information. An important element in the process is genetic counseling to ensure adequate understanding of the implications of genomic information, appropriate contextualization of the information within the personal and family medical setting of the recipient, communication of the importance of follow-up to realize the full preventive potential, and facilitation of cascade testing to broaden the impact of the result.

A recent review of barriers to cascade testing noted that current methods have not been particularly effective in accomplishing cascade testing. Historically, cascade education efforts directed to at-risk relatives have been left to the patient for completion. Many patients report feeling unable and unwilling to communicate this information to family members for a wide variety of reasons including feeling unprepared and uninformed, concerned about responses from their relatives, feeling ostracized for having the condition, as well as no longer having contact with certain relatives. Support of the importance of the information by health-care providers in the context of genetic counseling about results is one way to support communication with relatives. Others are looking to online tools, such as the resource My46 (www.my46.org) or KinTalk (www.kintalk.org), to aid communication of results with relatives. Cascade screening is seen as a critical component in genetic counseling that elevates the activity to public health relevancy.

18.3.4 LESSONS LEARNED

In the first example, a national resource in the United States is described that facilitated development of a rubric to support the disclosure and communication process involved with genetic and genomic testing in clinical practice. The working group anticipated that not all genetic counseling will be provided by genetics professionals. The rubric establishes an approach to evaluating when professional genetics expertise might be most beneficial to the patient. It also defines the elements of genetic counseling expected regardless of the provider. Although different in approach, the United Kingdom has also developed a national resource to facilitate genetic counseling as a critical component of the national public health genomics initiative in the United Kingdom. The efforts are focused on raising the genomics literacy of all health-care providers through online and traditional educational modalities. In addition, the effort has increased training of genetic counselors to accomplish the goal of broader access to accurate genomics information.

The second example presents the importance of genetic counseling in facilitating communication of genomic information when the result is the primary presenting indication for a health-care visit. In this setting, genetic counselors support primary-care providers in disclosing genomic sequencing results. The paired relationship is welcomed by patient-participants who not only prefer to hear medical information from their own provider but who also want access to the most thorough explanation of the genomic result. The relationship is also welcomed by providers, who as they have more familiarity with genomic results, begin to build confidence in providing genetic counseling relative to those results.

The third example utilizes cascade screening as a well-recognized high value activity that has significant potential as an overarching public health genomics initiative. The importance of genetic counseling is clearly evident in the expectation that identification of at-risk relatives and encouragement to share genomic results with relatives occurs within the genetic counseling process. The population benefit of cascade testing and the potential for economic value are dependent on successful approaches to genetic counseling.

18.4 IMPLICATIONS FOR PUBLIC HEALTH GENOMICS

Genetic counseling (and likely genetic counselors) is essential to realizing the full potential of public health genetics and genomics. In the past, genetic counselors have been viewed as the referral resource for genetic counseling for rare diseases and prenatal screening. With the introduction of national and/or population sequencing programs, there is increasing appreciation for the process of genetic counseling and the realization that not all genomics-related patient encounters will involve a trained genetic counselor. The examples in this chapter discuss ways in which genetic counselors can support, and in many instances, direct population efforts to integrate genomic information into regular health-care efforts.

With the increased attention to the possibilities of genomic sequencing technologies in population (and public) health activities, there is the concomitant need to increase research focus on best practice approaches to genetic counseling for genomic sequence information. Identifying outcomes of significance for patients will help inform practical guidance for disclosure processes for genetics and nongenetics professionals.

This chapter points to the key role played by genetic counselors in the translation of the genomic sequencing technology. While genetic counseling is not required to be performed by genetic counselors, they are the specialist uniquely trained to understand and communicate this information to patients and providers. The current number of graduate education programs across the world is inadequate to support the role that genetic counselors could play to promote the benefit of public health genomics. There is need to recognize the value that genetic counselors bring in the form of increased opportunities for graduate education in genetic counseling.

REFERENCES

1. Resta R, Biesecker BB, Bennett RL, et al. A new definition of genetic counseling: national society of genetic counselor's task force report. *J Genet Couns*. 2006;15(2):77–83. Available from: https://doi.org/10.1007/s10897-005-9014-3.
2. Paul D. From eugenics to medical genetics. *J Policy Hist*. 1997;9(1):96–116.
3. Reed SC. *Counseling in Medical Genetics*. Philadelphia, PA: WB Saunders; 1955.
4. Stern AM. A quiet revolution: the birth of the genetic counsellor at Sarah Lawrence College, 1969. *J Genet Couns*. 2009;18(1):1–11. Available from: https://doi.org/10.1007/s10897-008-9186-8.
5. Accreditation Council for Genetic Counseling. Program Directory. <https://www.gceducation.org> Accessed 12.01.18 [and again 9.18.2019].
6. Patch C, Middleton A. Genetic counselling in the era of genomic medicine. *Br Med Bull*. 2018;126(1):27–36. Available from: https://doi.org/10.1093/bmb/ldy008.
7. University of South Carolina School of Medicine. Transnational Alliance for Genetic Counseling. <https://tagc.med.sc.edu/about.asp> Accessed 12.01.18 [and again 9.18.2019].
8. Ormond KE, Laurino MY, Barlow-Stewart K, et al. Genetic counseling globally: where are we now? *Am J Med Genet C Semin Med Genet*. 2018;178C:98–107. Available from: https://doi.org/10.1002/ajmg.c.31607.
9. Bennett RL, French KS, Resta RG, et al. Standardized human pedigree nomenclature: update and assessment of the recommendations of the National Society of Genetic Counselors. *J Genet Couns*. 2008;17(5):424–433. Available from: https://doi.org/10.1007/s10897-008-9169-9.
10. Green RC, Berg JS, Grody WW, et al. ACMG recommendations for reporting of incidental findings in clinical exome and genome sequencing. *Genet Med*. 2013;15(7):565–574. Available from: https://doi.org/10.1038/gim.2013.73. Erratum in: Genet Med. 2017 May;19(5):606.
11. Berkman BE, Miller WK, Grady C. Is it ethical to use DNA to solve crimes? *Annu Intern Med*. 2018;169(5):333–334. Available from: https://doi.org/10.7326/M18-1348.
12. Manolio TA, Abramowicz M, Al-Mulla F, et al. Global implementation of genomic medicine: we are not alone. *Sci Transl Med*. 2015;7(290):290ps13. Available from: https://doi.org/10.1126/scitranslmed.aab0194.
13. Khouty MJ, Bowen MS, Clyne M, et al. From public health genomics to precision public health: a 20-year journey. *Genet Med*. 2018;20(6):574–582. Available from: https://doi.org/10.1038/gim.2017.211.
14. Ormond KE, Hallquist MLG, Buchanan AH, et al. Developing a conceptual, reproducible, rubric based approach to consent and result disclosure for genetic testing by clinicians with minimal genetics background. *Genet Med*. 2019;21:727–735. Available from: https://doi.org/10.1038/s41436-018-0093-6.
15. Genomics Education Programme. NHS Health Education England Genomics Education Programme. <https://www.genomicseducation.hee.nhs.uk/> Accessed 12.15.18 [and again 9.18.2019].
16. Carey DJ, Fetterolf SN, Davis FD, et al. The Geisinger MyCode community health initiative: an electronic health record–linked biobank for precision medicine research. *Genet Med*. 2016;18(9):906–913. Available from: https://doi.org/10.1038/gim.2015.187.
17. Williams MS, Buchanan AH, Davis FD, et al. Patient-centered precision health in a learning health care system: Geisinger's genomic medicine experience. *Health Aff*. 2018;37(5):757–764. Available from: https://doi.org/10.1377/hlthaff.2017.1557.
18. Schwartz MLB, McCormick CZ, Lazzeri AL, et al. A model for genome-first care: returning secondary genomic findings to participants and their healthcare providers in a large research cohort. *Am J Hum Genet*. 2018;103(3):328–337. Available from: https://doi.org/10.1016/j.ajhg.2018.07.009. Epub2018 Aug 9.
19. Centers for Disease Control and Prevention. Public Health Genomics. Genomic Application Toolkit. <https://www.cdc.gov/genomics/implementation/toolkit/tier1.htm> Accessed 12.13.18 [and again 9.18.2019].

20. Manickam K, Buchanan AH, Schwartz MLB, et al. Exome sequencing-based screening for BRCA1/2 expected pathogenic variants among adult biobank participants. *J Am Med Assoc*. 2018;1(5):e182140. Available from: https://doi.org/10.1001/jamanetworkopen.2018.2140.
21. Schwiter R, Rahm AK, Williams JL, Sturm AC. How can we reach at-risk relatives? Efforts to enhance communication and cascade testing uptake: a mini-review. *Curr Genet Med Rep*. 2018;6(2):21–27. Available from: https://doi.org/10.1007/s40142-018-0134-0.
22. Mvundura M, Grosse SD, Hampel H, Palomaki GE. The cost-effectiveness of genetic testing strategies for Lynch syndrome among newly diagnosed patients with colorectal cancer. *Genet Med*. 2010;12(2):93–104. Available from: https://doi.org/10.1097/GIM.0b013e3181cd666.

CHAPTER

DEFINING GENETIC-TESTING DELIVERY AND PROMOTIONAL STRATEGIES FOR PERSONALIZED MEDICINE 19

Christina Mitropoulou[1], Despina Giannouri[2], Kariofyllis Karamperis[3], Sam Wadge[3] and George P. Patrinos[2,4,5]

[1]The Golden Helix Foundation, London, United Kingdom [2]Department of Pharmacy, School of Health Sciences, University of Patras, Patras, Greece [3]Department of Twin Research and Genetic Epidemiology, King's College London, London, United Kingdom [4]Department of Pathology, College of Medicine and Health Sciences, United Arab Emirates University, Al-Ain, United Arab Emirates [5]Zayed Center of Health Sciences, United Arab Emirates University, Al-Ain, United Arab Emirates

19.1 INTRODUCTION

In recent years, significant advances have been made in our understanding of the genetic basis of inherited disorders and the correlations between mutant genotype and clinical phenotype, both for monogenic and multifactorial conditions.[1,2] These advances, in conjunction with the advent of high-throughput genetic analysis and deep resequencing, have served to reshape the field of modern medical practice[3] and are reflected in the rapid development of the genetic-testing industry.[4,5] Nowadays, there are a wide variety of public entities and private companies that offer a broad range of antenatal and postnatal molecular genetic-testing services for monogenic and multigene disorders, classical and molecular cytogenetic analysis for chromosomal rearrangements, pharmacogenomic testing, and even predictive genomics for genetic disorders (see also Chapter 10). In addition, many laboratories also offer molecular genetic-testing services in microbiology and virology. At the same time, genetic-testing services are becoming more affordable so that we can already envisage genome resequencing for as little as $1000.

However, the rapid expansion of the genetic-testing industry has not come without problems. In particular, some laboratories still offer genetic-analysis services using in-house (home-brew) kits rather than quality-controlled and certified assays. In addition, test results are not invariably interpreted by a qualified professional (e.g., a genetic counselor), whereas other laboratories are not yet accredited for the provision of genetic-testing services.[6] Moreover, it transpires that in several cases, genetic analysis is routinely conducted without obtaining informed consent from those persons requesting the test. This raises serious ethical concerns in relation to the preservation of the

Applied Genomics and Public Health. DOI: https://doi.org/10.1016/B978-0-12-813695-9.00019-4

anonymity of the individuals tested[7] (see also Chapter 13), the fate of their genetic material, and, most importantly, the safeguarding of test results in order to avoid genetic stigmatization.[8]

There are different models for the provision of genetic-testing services, based on which all the existing genetic-testing laboratories currently operate. In this chapter, we provide an overview of the existing models for the delivery of genetic-testing services and we analyze the various means to promote these services to the end users, that is, patients and their families. We also allude to efforts undertaken in various countries to map the genetic-testing services' environment.

19.2 GENETIC-SERVICE DELIVERY MODELS

Delivery models for the provision of genetic-testing services can be classified into five distinct categories according to which a health-care professional plays the most prominent role in patient pathways to care:

1. Genetic services provided by geneticists
2. Genetic services as part of primary care
3. Genetic services provided by the medical specialist
4. Genetic services integrated into large-scale population-screening programs
5. Genetic services provided using the direct-to-consumer (DTC) model

In all of these models, there are four key players that participate in different order, depending on the model: (1) the patient, who wishes to undertake the genetic testing, (2) the general practitioner (GP) or medical specialist who refers the patient to the genetic laboratory, (3) the genetic laboratory that performs the genetic test, and (4) the genetic counselor or clinical geneticist, who explains the results to the patient. A detailed description of these models is provided in the following subsections.

19.2.1 GENETIC SERVICES PROVIDED BY GENETICISTS

In this model the professional team that requests the genetic services may include clinical and/or laboratory geneticists, genetic counselors, and other health-care professionals (e.g., genetic nurses). The professional team is responsible for risk assessment, counseling, and genetic testing of individuals or families affected or at risk of genetic disorders. Depending on the case, the genetic team collaborates with other medical specialists (e.g., oncologists, cardiologists, neurologists, and psychiatrists) who could also be part of the genetic service. Classical examples of this model are genetic services for rare diseases.

According to this model, patients could access genetic services through two different paths:

1. Patient—GP/medical specialist—genetic counselor/clinical or laboratory geneticist > genetic laboratory
2. Patient—genetic counselor/clinical or laboratory geneticist—genetic laboratory

The first path involves a patient seeking medical assistance from a GP or any medical specialist who then makes a referral to the genetic service, where a genetic counselor or a clinical or laboratory geneticist can perform a risk assessment. If a genetic test is relevant and available, they may suggest genetic testing to the patient; then, samples are collected, and tests are performed in the genetic laboratory. Based on the results of the test, genetic counselors or medical geneticists recommend surveillance and/or intervention. Clinical management of genetic conditions may involve various medical specialists, other than geneticists (e.g., oncologists, cardiologists, neurologists, and psychiatrists).

The second path involves a patient who without a medical referral, inquires about a genetic service directly, where a genetic counselor or a medical geneticist can perform a risk assessment. Subsequently, the latter path is identical to the former path from this point onward.

The majority of the genetic tests offered under this model include newborn screening.

19.2.2 GENETIC SERVICES AS A PART OF PRIMARY CARE

In this model, primary-care units play a prominent role, in which GPs must have specific genetic skills and can undertake an initial risk assessment using standardized referral guidelines. In some cases, GPs refer patients, who are categorized as "high risk," to genetic services, while in other cases, they can provide genetic counseling, request genetic testing, and interpret the results from the genetic tests. Again, in this model, there are three possible paths:

1. Patient—GP—genetic counselor—genetic laboratory
2. Patient—GP—genetic laboratory—genetic counselor
3. Patient—GP—genetic laboratory

The first path involves a patient who contacts a GP, who first undertakes the initial risk assessment and subsequently makes referrals to a genetic service, where a genetic counselor or a clinical and/or laboratory geneticist performs counseling and suggests genetic testing to the patient.

The second path is similar to the first one, but instead, genetic counseling is offered to patients after the delivery of the genetic results. Thus patients were seen by the genetic counselor only after the genetic test.

The third path involves a patient who contacts a GP capable of performing the risk assessment, undertaking genetic counseling, and suggesting genetic testing.

19.2.3 GENETIC SERVICES PROVIDED BY THE MEDICAL SPECIALIST

In this model, genetic tests can be requested directly by medical specialists (e.g., oncologists, cardiologists, neurologists, and psychiatrists) who are capable of managing patients suffering from genetic disorders without requiring consultation from medical geneticists. In this case, medical specialists can request genetic testing, communicate genetic test results to their patients and families, and set up treatment with or without consultation from a clinical or laboratory geneticist.

Again, there are two main paths for this model:

1. Patient—medical specialist—genetic laboratory
2. Patient—medical specialist—genetic counselor—genetic laboratory

The first path involves a patient who contacts (with or without a GP referral) a medical specialist who is capable of performing a risk assessment, undertaking genetic counseling, and suggesting genetic testing. A variation of this model includes the involvement of a genetic nurse instead of a medical specialist.

The second path involves a patient who contacts a medical specialist, who will undertake the initial risk assessment, and then requests genetic counseling, collaborating with the medical geneticist or genetic counselor for the management of the patient.

19.2.4 GENETIC SERVICES INTEGRATED INTO LARGE-SCALE POPULATION-SCREENING PROGRAMS

In this model, genetic services are provided as part of well-organized population-screening programs such as newborn screening, cancer screening, and Ashkenazi Jewish genetic screening. There are three possible paths for this model:

1. Patient—GP/medical specialist—genetic counselor/clinical geneticist—genetic laboratory
2. Patient—GP/medical specialist—genetic laboratory
3. Patient—genetic counselor/clinical geneticist—genetic laboratory

The first path involves a patient who takes part in a large-scale population−based screening program; a health-care professional, who is involved in the screening program, can perform an initial risk assessment and refer the patient for genetic counseling. The genetic counselor or clinical geneticist can then undertake genetic counseling, suggest genetic testing, and, based on the results of the test, can recommend the appropriate intervention. A variation of this theme is when genetic counseling is offered to patients after the genetic analysis. The second path involves a patient who takes part in a population-based screening program. In this case a duly-qualified physician involved in the screening program can perform risk assessment, undertake genetic counseling, and suggest genetic testing. Based on the results of the genetic test, the physician can recommend surveillance and/or the necessary intervention. Lastly, the third path involves a patient who first contacts a genetic counselor or a medical geneticist, who can undertake genetic counseling, suggest genetic testing, and, based on the results of the test, can suggest surveillance through available population-based screening programs and/or the necessary intervention.

This model is more common in countries with large-scale screening programs such as the United States and the United Kingdom.

19.2.5 GENETIC SERVICES PROVIDED USING THE DIRECT-TO-CONSUMER MODEL

In this model, private genetic-testing laboratories or, worse, companies that outsource these genetic-testing services offer genetic-testing services typically through websites. Health-care professionals are not involved in the process and medical referrals are not required for genetic testing through DTC companies. In other words, patients are self-referred. Furthermore, the private companies usually do not offer risk assessment and genetic counseling. In this model, patients purchase the genetic test, usually online, take their own sample at home using a buccal swap kit that is delivered to their home or purchased from a pharmacy, send it to the genetic laboratory, and receive the results directly by electronic or regular mail. Some DTC companies offer genetic counseling but only post testing.[9] This model is quite popular in the United Kingdom and the United States, among other countries[9,10].

19.3 MARKETING IN PUBLIC HEALTH

Public health has always been called upon to address major health problems at local, national, regional, and even global level, such as the control of transmittable diseases, the improvement of the physical environment, the quality of valuable goods such as water and food supplies, the provision of medical care, and the relief of disability and destitution. In order to achieve what is mentioned earlier, social marketing plays a decisive role, as an effective approach for developing programs to promote healthy behaviors.[11] In recent years, social marketing is being applied to an even wider range of public-health activities and programs, from the safe drinking water campaign in Madagascar to the promotion of mosquito nets in Nigeria.[12] Social marketing practice and successful social-marketing campaigns can be found all over the world. Countries active in applying social-marketing techniques to public health vary at the levels of economic and technological developments and differ in social, cultural, and regulatory environments.[12]

19.4 MARKETING IN GENETIC-TESTING SERVICES

In the post-genomic era, a new public-health challenge is that of genetic testing. The completion of the Human Genome Project created a completely new field of genome research that was then included in clinical practice to promote the health of the population. In 2005, by unanimous decision, a group of 18 specialists created an organization to integrate genome-based knowledge and technologies into health services and public policy for the benefit of population health called the Public Health Genomics. As is obvious, the multiple possibilities offered by genetic tests have attracted many pharmaceutical companies and industries.[13] Despite widespread discussions on the reliability and use of genetic tests (see also Chapter 10), they are rapidly growing, attracting more and more companies and consumers alike. It seems that genetic-testing practices are rapidly increasing in rare-disease diagnostics and for personalized medicines, which in turn, are fueling the growth of this market. According to recent statistics, the total number of genetic tests that are actively marketed by Clinical Laboratory Improvement Amendments–certified laboratories in the United States was 74,448, representing approximately 10,000 unique test types for the year 2018. On the other hand, there are 135,000 inactive genetic tests either because the lab stopped offering them, or because the test changed and assigned a new one. It is estimated that 10 new genetic tests are added daily. Based on the above statistical data, 86% of genetic tests were single-gene tests, while the remaining tests are panel-based including 9311 multianalyte assays (via algorithmic analyses), 85 noninvasive prenatal tests, 122 whole exome–sequencing tests, and 873 whole genome–analysis tests, which included whole genome–sequencing tests.[14] Prenatal tests accounted for the highest percentage of spending on genetic tests and spending on hereditary cancer tests accounted for the second highest. Apart from single tests, the market for multipanel gene panels has continued to grow. The frequency of these testing products in the United States represents more than 8% of the total number (9488) on the market during the year 2018. The significant delay between the introduction of tests and insurance coverage is often a major problem for proper promotion and production of some products, significantly affecting its sales[14].

The clinical sequencing market is growing at a compound annual growth rate of 28% and is forecasted to be worth $7.7 billion worldwide by 2020. This growth is due to the growing demand for tests with better performance characteristics and clinical validity compared to other applications such as prenatal screening and monitoring cancer recurrence. In addition the dramatic reduction in the cost of sequencing, development of informatic tools and the ability to analyze complex data, and of new methods, such as circulating cell-free DNA techniques, and improving data in the interpretation of results are considered to be some of the reasons for their unexpected growth in worldwide markets.[14] So far, it seems that whole exome sequencing is not only the predominant force in the market compared to other next-generation technologies but also to different types of genetic tests (Fig. 19.1). More precisely, clinicians seem to be interested in whole exome sequencing due to its more comprehensive diagnostic approach than single-gene, panel, or some microarray testing—yet providing more focused and medically actionable results than whole genome–sequencing.[15]

19.5 DEFINING THE MARKETING STRATEGY FOR GENETIC-TESTING SERVICES

Briefly, marketing is defined as the management process, which is responsible for identifying how to anticipate and satisfy the customer requirements. As a more in-depth opinion, Kotler defines marketing as "... a process by which individuals and groups acquire what they need and desire, through the creation exchange of products and values with others"[16]. According to Janice Denegri-Knott, Mike Molesworth, and Richard Scullion (University of Bournemouth, Dorset, United Kingdom), "... in some ways, marketing is as old as the ancient civilization of Greece." Ultimately, the marketing orientation of businesses began in the 1960s with the production and sales orientations. According to the new philosophy, businesses should first identify what the consumer wants and then try to produce it rather than produce what they can and then try to change needs of consumers. It is understood that marketing began within an introverted environment focusing on what a business produced. Today, modern marketing operates within an extrovert environment with multipurpose tools such as those in social media, which provide valuable insights into the business for the prevailing demand in markets and consumer preferences. The goal is therefore to achieve more effective customer satisfaction through an integrated marketing-strategy program.[17]

19.5.1 TARGET AUDIENCE

The target audience of the genetic-testing services is considered to be health-care professionals, namely, physicians and pharmacists, public bodies, for example, hospitals, private clinics, and laboratories, and the state, for example, the Ministry of Health. The indirect target audience is considered to be the end users of genetic testing, since we believe that the DTC model (see earlier) does

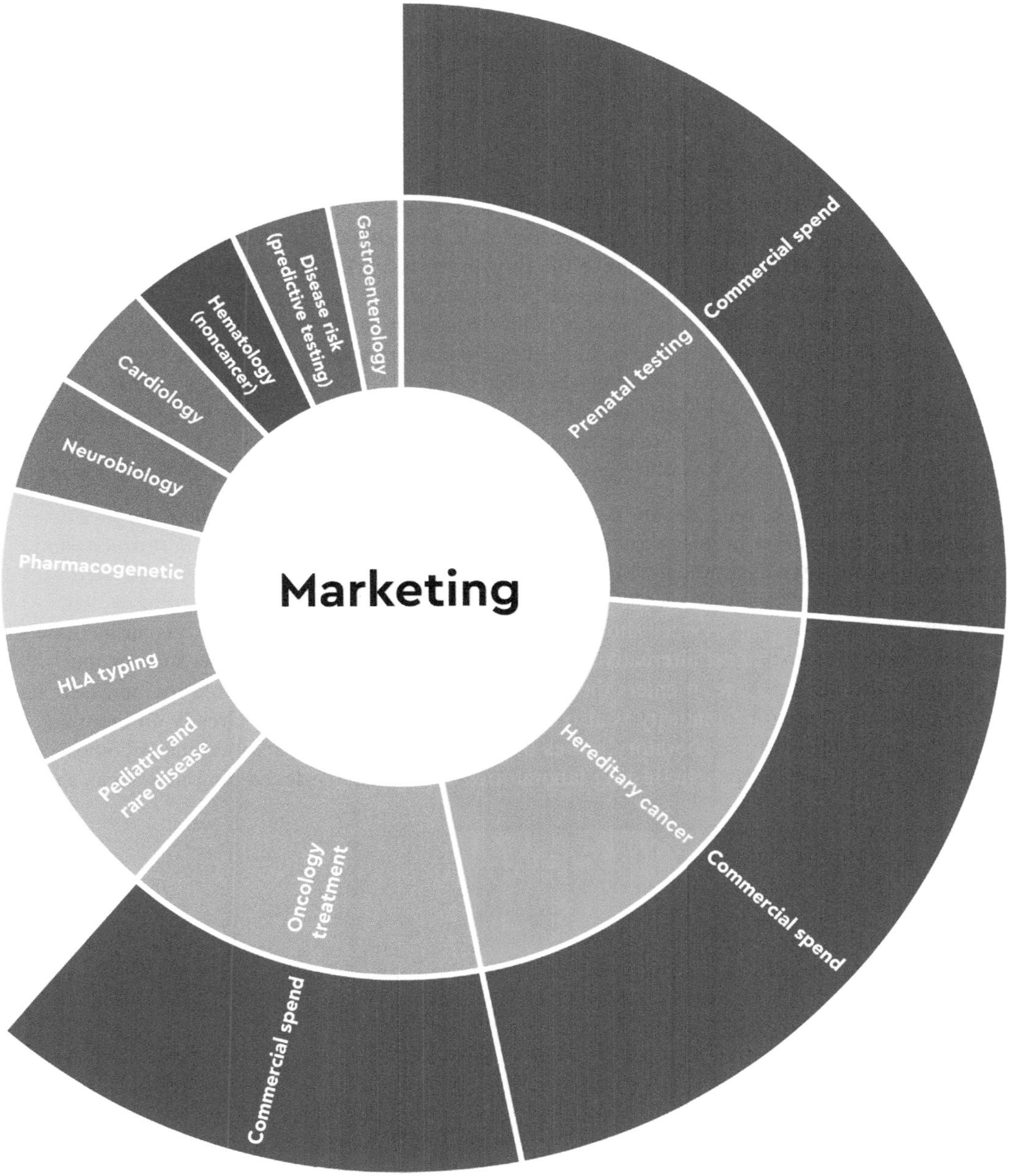

FIGURE 19.1

A catalog of genetic testing services in the market.

not meet the ethical requirement for a health-care service. Therefore genetic-testing services should be promoted to the health-care professional community in order to properly inform the interested parties.

19.5.2 THE MARKETING MIX

The marketing mix is the combination of human and material resources needed to fulfill the programs and objectives of the business in one market.[17] The strategy of the marketing mix manages the aforementioned resources to maximize the growth in sales. A well-designed set of strategies not only includes an attractive pricing and the development of new products, but additionally must be promoted in the best possible way to the public based on integrated communication programs. Proper communication is the key factor in creating and maintaining good customer relationships of trust. At this point, it should be noted that the value of the factors and the improvement of the company's development strategy varies according to the product and creates particular parameters that can vary according to the final product being distributed on the market.[17]

The marketing mix, which is often referred to as "4Ps" (Fig. 19.2), deals with the following:

- *Product*: In this case, products are defined as the genetic-testing services. Once the product comes into the market, it goes through four stages, namely, *import*, *growth*, *maturation*, and *decline*, also known as a product life cycle. At the *import* stage, the product is promoted to make it public and available on the market. As this phase is unlikely to be profitable, the products/genetic-testing services should be monitored continuously until there is an increase in sales. Otherwise, the safest alternative would be to withdraw the product. As soon as the product sales begin to rise, it enters the growth phase, which is characterized by increased sales and profits. The stage of maturity is also the most popular for products already on the market where the genetic-testing laboratory scores the most profits and the competition becomes stronger. In this phase, genetic-testing laboratories are constantly developing new ideas and

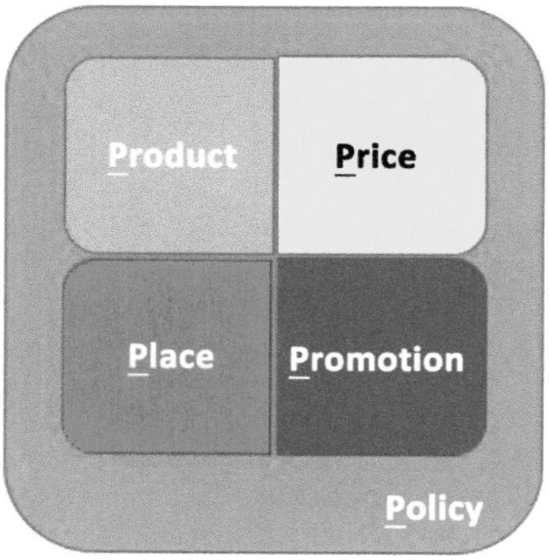

FIGURE 19.2

Outline of the 4Ps concept of the marketing mix.

investing money in product research and development, in order to compete with other assets, stay on the market, and succeed. During the final stage of decline, as profits are no longer high for that product, it may be withdrawn from the market.[18] Genetic tests, particularly those related to pharmacogenomics or genetic predisposition to an inherited disease, are considered to belong to the second stage of development as more and more people are sensitized, informed, and interested to look for a timely diagnosis, which will improve the quality of their lives.

- *Price*: It is the monetary value at which the consumer purchases the product or service—in this case, genetic-testing services. The price range of a genetic-testing service could start from as low as 50€ or $ to up to thousands of € or $. In the case of whole genome, whole exome, or targeted resequencing, data interpretation is often provided as an additional service. Apart from the monetary value of the genetic testing service, the emotional and psychological cost and time constraints the target group bears to overcome barriers and successfully adopt the desired behavior are equally important issues. The amount that a customer has to pay for a genetic test plays a crucial role for the development of the product. Based on the fact that health is a key priority for most people, the price must be affordable to be distributed and used in multiple groups of interest. Furthermore, this has a big impact on the entire marketing strategy as well as a great effect on the sales and demand of the product. Nowadays, genetic and pharmacogenomic-testing services cost less than 300€ in Europe and the United States, allowing them to grow rapidly and use more and more.
- *Place*: This refers to the way the product is distributed to the customer. In the case of the genetic testing service, this is provided through accredited (or not) genetic-testing laboratories, specifically where the target group will be exposed to the product, where it will interact and think about the subject, and perform the desired behavior. Globally, genetic disorders and the development of genetic diseases vary as environmental factors are known to play an important role in genetics. Genetic and pharmacogenomic tests must be provided based on the needs of each population and their afflicting condition. For that reason the development of each genetic test depends on that, while the way the product is distributed to the customer is also equally important.
- *Promotion*: Promotion is probably the most important stage in the development of each product and refers to the way it reaches the end user/customer and how the user is ultimately prompted to purchase it. In this case, special emphasis should be given to the promotional strategy and message for the target audience and the marketing medium. In the first case the promotional message should clearly state the importance of analyzing the results of a genetic test by a specialist while prompting physicians to suggest them to their patients when they consider the need for such an examination. As far as the marketing medium is concerned, genetic-testing services are unique and, as such, require particular attention for their advertising and promotion. Contrary to the majority of consumer products, the most appropriate and scientifically sound promotional means would be scientific journals and websites that appeal to health-care professionals as well as informational days for doctors to raise awareness about the usefulness and relevance of these tests for their patients. Direct marketing of genetic tests to the general public, using promotional media, such as ads in television and newspapers, cold calls or mass postal or e-mail, is considered to be inappropriate for such a specialized type of service. Prior to the promotion of a genetic test, it is considered necessary to inform and recruit appropriate medical staff as they should provide the customer with all the necessary information and guide.

Proper education is highly recommended for health-care professionals and scientists, such as geneticists, physicians, genetic counselors, especially, given their lack of genomics knowledge (see also Chapter 12). The main approaches, such as branding, advertising, public relations, corporate identity, social media outreach, sales management, special offers, and exhibitions, are integral parts of the promotion strategy.

- *Policy*: In the case of public-health marketing, policy is the fifth broad level of marketing decision (hence a fifth P in the 4Ps marketing mix), which has been added to the basic framework according to the Center for Disease Control and Prevention. Policy refers to the creation or usually the modification of existing legislation and regulations trying to encourage or require the desired behavior change.

19.5.3 DIVERSITY

Genetic-testing services differ greatly from a hematological or biochemical test. Genetic testing can be offered in a wide variety of types, such as preclinical, prenatal, or even preimplantation genetic diagnosis, single variant or panel-based multivariant genetic screening, targeted resequencing, whole exome and whole genome sequencing, and pharmacogenomic testing.

19.5.4 CONFIDENCE

As previously mentioned, the target audience for genetic tests is physicians and patients. For both of these groups, different approaches must be sought.

When it comes to physicians, it is necessary to adequately inform them about the benefits from using these tests for the early diagnosis of genetic diseases, rationalizing drug use, and hence, minimizing the risk of developing adverse drug reactions for their patients. This would assist them in more targeted drug prescriptions. This is particularly needed for the older generation of physicians whose genetics education level is low.

Similarly, patients are also required to get informed about the benefits of personalized treatment based on their genetic profile and the benefits of undertaking a genetic test, which provides an enormous amount of information about the health outcomes. Also, being able to diagnose a genetic disease will lead the patient to early prevention and to avoid future treatment costs and, most importantly, to improve the quality of his/her life.

19.5.5 STRENGTHS, WEAKNESSES, OPPORTUNITIES, AND THREATS ANALYSIS

SWOT analysis is very important for launching a new product and takes into consideration the strengths, weaknesses, opportunities, and threats from launching this product or service. Strengths and weaknesses are part of internal analysis, while opportunities and threats are elements of external analysis. In the case of the genetic-testing services the following elements are part of the SWOT analysis.

19.5.5.1 Strengths

- Analysis of the human genome has led to the early prognosis and diagnosis of diseases
- Testing the general public at relatively very low prices compared to the benefits they get

- Improving the quality of life of the patients
- Reducing the health-care costs from the reciprocal reduction of adverse reactions and informed health decisions

19.5.5.2 Weaknesses

- Inability to interpret and understand the results of genetic tests by consumers without the assistance of a qualified health-care professional
- A time-consuming process of exporting the results
- Pitfalls in the technology used for genetic testing

19.5.5.3 Opportunities

- Continuous technology and industry development and scientific progress
- Implementation of next-generation sequencing technology in molecular diagnostics
- Raising general public and end-user awareness
- Ability to treat diseases at a personal level and individualize treatment with fewer adverse reactions
- Lower costs for drug treatment modalities
- Development of new diagnostic applications, such as array-on-demand, for the diagnosis of the genetic basis of unknown genetic diseases.

19.5.5.4 Threats

- Use of genetic tests at home using, for example, the DTC model
- Performing genetic testing without the informed consent from the interested party
- Ensuring patients' anonymity and personal data protection assurance
- Lack of sufficient funding
- Lack of the necessary legal framework governing the genetic testing—service environment
- Improper promotion of genetic testing by doctors, mostly those with incomplete genetic knowledge
- Lack of genomics awareness of the general public
- Lack of a stable health-care environment
- Lack of a well-defined and consistent national strategy on genomic medicine
- Insufficient support of genetic-testing services by pharmaceutical companies and other stakeholders due to conflict of interests
- Lack of accreditation of genetic-testing laboratories
- Lack of reimbursement of genetic tests by payers and insurance companies
- Denial of insurance based on genetic profile and genetic stigmatization

19.5.6 POLITICAL, ECONOMIC, AND SOCIAL POLICIES, TECHNOLOGICAL DEVELOPMENTS, LEGISLATION, AND THE ENVIRONMENT ANALYSIS

The PESTLE analysis focuses on the analysis of the external environment and presents the prevailing conditions, challenges, and changes that may be encountered by an enterprise, relating to the political (P), economic (E), and social (S) policies, technological developments (T), legislation (L), and the environment (E).

As far as the implementation of the PESTLE analysis for genetic-testing services is concerned, *political* stability is necessary for the legislation of the necessary bills such as the establishment of the genetic specialties or the taxation environment for the genetic-testing services. In terms of *economy* the slow pace of the economy may result in the improper promotion of the genetic-testing services through the Internet, while the high unemployment rate and the low wages leave little room for choice for such a high specialized type of testing, such as genetic testing, as compared to the classic annual hematological and biochemical tests. In other words, in an uncertain economic environment with an even more uncertain future, priorities are set to ensure basic needs, such as food and housing, and unfortunately not something as specialized as the genetic tests.

The most important parameter of the PESTLE analysis for the genetic testing services is the *social* circumstances. In particular, lack of genetic awareness from the general public (see also Chapter 12) can negatively impact on the understanding of the significance of the genetic tests. In fact a large percentage of the general public would be interested to carry out a genetic test, although not all of them are duly informed about, for example, the role of genetic material and the means for DNA isolation. Of equal importance are the *ethical* considerations in the PESTLE analysis. Genetic tests are not just consumer products, which can be directly promoted to the consumer and as such, it is of utmost importance to ensure that genetic-testing services are promoted to the general public in the most appropriate way to safeguard public health and at the same time the interests of the patients. Often, however, promotion of these genetic-testing services is not adequately performed, since the end user is either misled or misinformed about the usefulness of these tests. For example, certain genetic laboratories promote these tests via brochures sent to prospective consumers by mass e-mail campaigns or misinform the end user about the usefulness of a genetic test such as those tests that are aimed to predict athletic performance or dietary intake. Also, very often, the genetic test results are not received directly from the genetic laboratories, but they are mailed to the end users using either electronic or regular mail. This is totally unacceptable, since this approach not only bypasses the genetic counselor but also creates an issue with patient's anonymity. Also, the availability of some genetic tests online may create more ethical issues such as in the case of the paternity test that can be ordered without the consent of both parties. The legal requirements are also important to be clearly set, especially those safeguarding the person being tested. The most well-known legislation is the GINA Act in the United States, while similar legislation measures are also in place in other countries (see also Chapter 14). Also, the DTC genetic-testing provision model is currently poorly covered by the existing legal framework, while there is huge heterogeneity in the respective legislation framework among different countries, mostly in Europe (Mai et al., 2011). Similar legislation measures are also in place for the required accreditation of the genetic-testing laboratories requiring international external quality audits and certain accreditations such as the ISO 15189:2007 in the EU or the College of American Pathologists certification in the United States. Lastly, the analysis of the various *technological conditions* refers to the genetic-testing technologies, which are constantly developing, not only positively impacting on the science itself but also having a direct impact on the genetic-testing costs.

19.6 DEFINING THE LANDSCAPE OF GENETIC TESTING IN VARIOUS COUNTRIES

The landscape of private genetic-testing services is still poorly developed in many parts of the world. Even though genetic-testing services are mostly well regulated in the United States and Western European countries, such as The Netherlands, the United Kingdom, and Switzerland, in other European countries, Asia, and the Middle East, genetic-testing services often face several issues, mostly resulting from the lack of proper regulatory framework, which consequently leads to a number of different ethical issues. EuroGentest (http://www.eurogentest.org) has attempted to rectify this deficiency by initiating a drive to harmonize genetic-testing services in Europe. In parallel, OrphaNet (http://www.orpha.net) has attempted to database the plethora of genetic-testing laboratories in Europe but these efforts have often been hampered by the willingness of some laboratories to communicate the requested details of their operations. It is therefore clear that in the emerging era of personalized genomics, the task of "fine-mapping" genetic-testing services in Europe is assuming ever greater urgency.

The detailed description of the genetic-testing environment lies outside the scope of this chapter. Instead, we opted to summarize in the following subsections, the current landscape of genetic-testing services in two health-care environments, that of Greece and Malaysia, in an effort to highlight the need of harmonizing and introducing proper legislative measures for better regulation of the genetic testing service environment to the benefit of the patients.

19.6.1 OVERVIEW OF GENETIC TESTING SERVICES IN MALAYSIA

The Malaysian population is characterized by a multitude of genetic disorders due to its vast ethnic diversity. The landscape of genomic testing and genetic counseling services in Malaysia is quite diverse. There are 20 Malaysian, mostly private, genetic-testing laboratories, all located in Peninsular Malaysia, offering genomic services including genomic testing and counseling. Such services include paternity testing, family tree analysis, ancestry and prenatal testing, clinical tests, animal testing as well as lifestyle genomic tests (wellness and fitness). Interestingly enough, clinical and lifestyle (wellness and fitness) genomic tests are the most common genomic services pursued in Malaysia (40% of the genetic-testing laboratories), followed by prenatal testing services (35%).[19] In contrast, various relationship/ancestry tests are only provided by very few genetic testing laboratories (15%), while an equal number of genetic laboratories offer pharmacogenomic testing services.

Interestingly, two genetic-testing laboratories offer several DTC testing services (paternity, relationship, and infidelity tests), both of them offering "discreet delivering options." Statements such as "discreet testing services" and "discreet samples" are also included in the online material of other genetic-testing laboratories. Also, genetic counseling services appear to be extremely limited, since most genetic-testing laboratories solely offer "DNA test reports" that provide the test results accompanied by ways to interpret them. Some paternity tests include conclusive results that "confirm" or "exclude" the alleged father. Only one private molecular diagnostics company offers complete genetic counseling support for physicians and patients. This may perhaps be explained by the fact that there is still no formal recognition of the role of genetic counselors in the country and there are no legal requirements for the provision of genetic-counseling services. In particular, there are few genetic counselors, who often lack the knowledge, expertise and skills to properly inform interested parties.

19.6.2 GENETIC-TESTING SERVICES IN GREECE

In Greece the first genetic-testing services appeared in Athens in the early 1960s, the first being Cytogenetics Laboratories. Since the early 2000s the number of genetic-testing laboratories has increased significantly. However, the country still lacks formal genetics centers organized within a national genetic-testing network such as in the case of the United Kingdom. Also, the number of physicians trained in clinical and/or laboratory genetics is very extremely limited, especially since the clinical and laboratory genetics specialties were only established in Greece in 2018.

Sagia et al.[20] have previously surveyed genetic-testing laboratories in Greece. This study indicated that the majority of them are involved in the provision of molecular genetic analysis for inherited disorders, followed closely by classical and/or molecular cytogenetic testing and molecular genetic testing for microbiology and predictive genomics. Physicians appear to be the main target group for the genetic laboratories, followed by the general public and other interested parties such as other genetic laboratories, diagnostic centers, hospitals, and pharmaceutical companies. The main specialties that the diagnostic laboratories mainly address are obstetricians/gynecologists, followed by pathologists, cardiologists, psychiatrists, and other specialties, namely, oncologists, pediatricians, hematologists, and neurologists. This may not be unexpected since obstetricians and gynecologists usually order molecular genetic and, particularly, cytogenetic tests to screen for fetal malformations, particularly in cases with a family history. In addition, psychiatrists, cardiologists, and oncologists are the physicians who order pharmacogenomic tests more frequently, since these are the disciplines in which pharmacogenomic testing has been most widely adopted.

In Greece, genetic-testing services are also provided using the DTC model, where 26 laboratories were previously identified offering DTC genetic tests.[21] In particular, these private genetic-testing laboratories offer genetic tests for predisposition to complex diseases, paternity testing, pharmacogenomics, nutrigenomics, antiaging, and genetics of athletic performance. From these, two genetic-testing laboratories are partnering with major pharmacy groups to promote their services using the over-the-counter business model.[22] Interestingly, these companies do not operate as medical entities but rather as genetic-analysis laboratories, indicating that the interpretation of the results from the genetic tests is the sole responsibility of the person(s) who requested the genetic test. This is particularly the case for the vast majority of the genetic-testing laboratories that only "sell" genetic tests online and in some cases outsource the actual genotyping to other genetic laboratories abroad. It is therefore not surprising that such private genetic-testing laboratories often offer genetic-testing services for pharmacogenomics and/or paternity testing together with genetic services for inbred animals (e.g., race horses) and/or genetically modified organisms.

19.7 CONCLUSION AND PERSPECTIVES

The provision of genetic services, along with research in the fields of genomics and genetics, has evolved in recent years to meet the increasing demand of consumers interested in the prediction of genetic diseases and various inherited traits. In this chapter, we attempted to summarize and classify delivery models for the provision of genetic testing, presented the steps to be undertaken when defining the optimal promotional strategy for genetic testing services while we also presented the landscape of provision of genetic-testing services in two countries, Europe and Asia.

New models of genetic-service delivery are currently under development worldwide for addressing the increasing demand for accessible and affordable genetic-testing services. These models require the seamless integration of genetics into all medical specialties, collaboration among different health-care professionals and stakeholders, and the redistribution of professional roles. We believe that there is no generic model for the provision of genetic-testing services. On the contrary, an appropriate model for genetic-service provision must be ideally defined according to the type of health-care system, the genetic test provided within a genetic program, and the cost-effectiveness of the intervention, once this is determined (see also Chapter 16), as only genetic applications with proven efficacy, solid scientific evidence, and cost-effectiveness should be implemented in health-care systems and made available to all citizens.

ACKNOWLEDGMENT

Part of this work was supported by the European Commission grant COST CHIPME (IS3101) and is part of the GoGreece initiative.

REFERENCES

1. Chen JM, Férec C, Cooper DN. Revealing the human mutome. *Clin Genet*. 2010;78(4):310–320.
2. Cooper DN, Chen JM, Ball EV, et al. Genes, mutations, and human inherited disease at the dawn of the age of personalized genomics. *Hum Mutat*. 2010;31(6):631–655.
3. Metzker ML. Sequencing technologies – the next generation. *Nat Rev Genet*. 2010;11(1):31–46.
4. Ginsburg GS, Willard HF. Genomic and personalized medicine: foundations and applications. *Transl Res*. 2009;154(6):277–287.
5. Caulfield T, Ries NM, Ray PN, Shuman C, Wilson B. Direct-to-consumer genetic testing: good, bad or benign? *Clin Genet*. 2010;77(2):101–105.
6. Burnett D. Quality control in the genetic laboratory. In: Patrinos GP, Ansorge W, eds. *Molecular Diagnostics*. 2nd ed. Burlington, MA: Elsevier/Academic Press; 2009. pg.
7. Gurwitz D, Bregman-Eschet Y. Personal genomics services: whose genomes? *Eur J Hum Genet*. 2009;17 (7):883–889.
8. Guttmacher AE, McGuire AL, Ponder B, Stefánsson K. Personalized genomic information: preparing for the future of genetic medicine. *Nat Rev Genet*. 2010;11(2):161–165.
9. Kaye J. The regulation of direct-to-consumer genetic tests. *Hum Mol Genet*. 2008;17(R2):180–183.
10. Mai Y, Koromila T, Sagia A, Cooper DN, Vlachopoulos G, Lagoumintzis G, et al. A critical view of the general public's awareness and physicians' opinion of the trends and potential pitfalls of genetic testing in Greece. *Per. Med*. 2011;8(5):551–561.
11. Cleeren E, van der Heyden J, Brand A, van Oyen H. Public health in the genomic era: will Public Health Genomics contribute to major changes in the prevention of common diseases? *Arch Public Health*. 2011;69(1):8.

12. Grier S, Bryant CA. Social marketing in public health. *Annu Rev Public Health*. 2005;26(1):319–339.
13. Avard D, Bucci LM, Burgess MM, et al. Public health genomics (PHG) and public participation: points to consider. *J Public Deliber*. 2009;5(1):7.
14. Phillips KA, Deverka PA, Hooker GW, Douglas MP. Genetic test availability and spending: where are we now? Where are we going? *Health Affairs (Project Hope)*. 2018;37(5):710–716.
15. Katsanis SH, Katsanis N. Molecular genetic testing and the future of clinical genomics. *Nat Rev Genet*. 2013;14(6):415–426.
16. Kotler PT, Armstrong G. Principles of Marketing. 17th ed. Pearson. 2017.
17. Morgan AN, Whitler AK, Feng H, Chari S. Research in marketing strategy. *J Acad Mark Sci*. 2018;. Available from: https://doi.org/10.1007/s11747-018-0598-1.
18. Armstrong G, Kotler P. *Marketing: An Introduction*. 10th ed. Pearson Prentice Hall; 2011. ISBN-13: 9780136102434.
19. Balasopoulou A, Mooy FM, Baker DJ, et al. Advancing global precision medicine: an overview of genomic testing and counseling services in Malaysia. *OMICS*. 2017;21(12):733–740.
20. Sagia A, Cooper DN, Poulas K, Stathakopoulos V, Patrinos GP. A critical appraisal of the private genetic and pharmacogenomic testing environment in Greece. *Per Med*. 2011;8(4):413–420.
21. Kechagia S, Yuan M, Vidalis T, Patrinos GP, Vayena E. Personal genomics in Greece: an overview of available direct-to-consumer genomic services and the relevant legal framework. *Public Health Genomics*. 2014;17(5–6):299–305.
22. Patrinos GP, Baker DJ, Al-Mulla F, Vasiliou V, Cooper DN. Genetic tests obtainable through pharmacies: the good, the bad and the ugly. *Hum Genomics*. 2013;7(1):17.

FURTHER READING

Dequeker E, Ramsden S, Grody WW, Stenzel TT, Barton DE. Quality control in molecular genetic testing. *Nat Rev Genet*. 2001;2(9):717–723.

Kricka LJ, Fortina P, Mai Y, Patrinos GP. Direct-access genetic testing: the view from Europe. *Nat Rev Genet*. 2011;12(10):670.

CHAPTER

REGULATORY ASPECTS OF GENOMIC MEDICINE AND PHARMACOGENOMICS

20

Konstantinos Ghirtis[1,2]

[1]Human Medicines Assessment Section, Product Assessment Division, National Organization for Medicines, Athens, Greece [2]Department of Pharmaceutical Chemistry, Faculty of Pharmacy, National and Kapodistrian University of Athens, Zografou, Greece

20.1 INTRODUCTION: PUBLIC-HEALTH SYSTEM AND REGULATION

20.1.1 PUBLIC HEALTH AND THE NEED FOR REGULATION THEREOF; ISSUES OF AGENCY AND CONFIDENCE

The entire public-health system is based on the constant buildup, elaboration, reiteration, consolidation, and assurance of trust and confidence among the various key players with the individual, patient or healthy, naturally at its center.[1] Only then one will give one's consent and comply with the proposed health interventions and advice. Patient compliance, adherence, and failure, thereof, are among the biggest challenges to health systems, leading to costly implications for the person's own health and, consequently, the health system as well as the society as a whole.[2] There is a limit as to what a person can know or even be/become aware of about his/her health condition, as a lot of health problems are far from being clinically manifested and evident and only come to one's attention later in the course of the disease with ramifications for the timely seeking of and achieving access to adequate health care. Furthermore, a person's own health-status appraisal is dubious and subjective. In addition, for one to seek health assistance, he/she has to be convinced that he/she indeed has a problem or imminent deterioration of his/her health and that seeking assistance is better than doing nothing, waiting for the condition's resolution, and eventual recovery on its own, instead.[3] The asymmetry of information[4] erodes the necessary trust further limiting timely patient access to adequate and effective health care.

Hence, specialized, knowledgeable, and experienced professionals, the health professionals, help direct his/her health-related decisions. There needs to be an assurance about their competence and qualifications; therefore these professionals need to be licensed to practice. In turn, these professionals need to utilize proper diagnostic methodologies to facilitate their personal judgment and act as the layperson's best agent[5] and help him/her reach effective decisions on interventions, of preventative, curative, or palliative intent, if needed.

Applied Genomics and Public Health. DOI: https://doi.org/10.1016/B978-0-12-813695-9.00020-0

20.1.2 HEALTH INTERVENTIONS AND NEED FOR REGULATION THEREOF

As far as these interventions are concerned, they need to be of proven and substantiated effectiveness and safety. Their regulatory assessment is largely split in three parts, administrative, quality, and clinical.[6] The administrative part concerns the proof of establishment and traceability of all parties involved in the life cycle management of the intervention, starting from its efficient production, placing in the market, and distribution including (pharmaco)vigilance and final disposal in an environmentally acceptable manner. In addition, they need to be of proven adequate quality of manufacture and distribution all the way till they reach the intended recipient, healthy or sick, according to the state-of-the-art. Finally, they should be qualified for the targeted health condition and their benefit should be hoped for. These are examined and established by the regulatory authorities, who set the conditions (indications, dosing, contraindications, warnings, etc.) under which their use may be beneficial, that is, showing a positive benefit/risk ratio. Then, the health professionals can apply them with confidence as deemed suitable for the particular patient. In essence, they implement and tailor the regulatory conclusions to match the individual's needs.

20.1.3 BRIDGING REGULATION WITH MEDICAL PRACTICE

The regulatory decisions are taken on the basis of population-based clinical-study results. These studies are strictly controlled to erase any source of confounding events, as much as possible, which may disrupt and perturb the essential randomization.[7] Only, whatever data can be randomized can also be analyzed with scientific rigor and are thus rendered internally valid for the regulatory purposes.[8] Then, the health professional can draw and extrapolate, from the clinical trial population data to the particular individual patient's setting and circumstances. In this way, he/she applies and implements the regulatory decisions to the specific patient he/she faces in the real-world clinical setting. In this way, he/she bridges the tightly regulated clinical-trial setting to the real-world realities, providing and enhancing the external validity of the use of the intervention according to the regulatory conclusions. Any in-use problem, ideally, is referred back through (pharmaco)vigilance to the health authorities and closes the circuitry of information flow between the key players in health-care goods' provision and distribution.[9] The authorities, in turn, update the conditions under which the health intervention maintains its positive benefit/risk ratio and accordingly notify the health-product manufacturers and practitioners alike of any meaningful to the clinical application modification.

Ideally, the right patient will take the right medicine/health intervention at and for the right time.[10] Apparently, there are a lot of variables and uncertainties, which can arise in every "right" of this ideal, for the health system, situation. The regulatory effort tries to narrow down all those potential caveats to allow the health professional practice confidently and knowledgeably for the patient's utmost benefit. The latter will then comply and seek medical assistance, if and when needed, and do so in a highly desired, timely, and adequate fashion.

20.2 REGULATION OF IN VITRO DIAGNOSTIC MEDICAL DEVICES

20.2.1 GENERAL REGULATORY REQUIREMENTS

Regulatory requirements are driven and evolve by the awareness of medical accidents and the progress of science.[11] The ideal would be to limit this evolution to the ones based on knowledge/

awareness gained/raised by the former and establish conditions to anticipate and prevent/limit, as much as possible, the latter. Different government entities have established different structures for the registration of in vitro diagnostic medical devices that are used to generate the genomic and not only information. These requirements are quite recent since such products became part of standard medical practice later than, for example, the drugs whose regulation, in turn, commenced as adulteration preventive measures with no proof of efficacy and safety. Similar to other medical interventions' regulation, they are assessed on the grounds of their quality of manufacture, clinical utility when used for their intended purpose and administrative traceability[12]

20.2.2 IN VITRO DIAGNOSTIC MEDICAL-DEVICE REGULATION IN THE UNITED STATES AND EUROPEAN UNION

Two of the major regulatory efforts are established in the United States (US) and the European Union (EU). Their different forms of governance are reflected in the manner of organization of these efforts. They both participate in the International Medical Device Regulators Forum[13] with the aim of, eventually, globally harmonizing requirements for medical devices.

In the EU the national member-state (MS) governments maintain their authority and responsibility for the health of their citizens and residents of their countries/MSs.[14] However, with the aim of ensuring an as-much-as-possible free circulation of goods and services throughout the EU, a lot of coordination, harmonization, and mutual-recognition efforts have legislatively taken place.[15] In vitro diagnostic medical devices are regulated separately than the other medical devices. Each in vitro diagnostic that is placed in the EU market needs to bear a "CE" mark meaning European Conformity to a set of rules established in the respective Directive 98/79/EC and requirements stemming from therein.[16] The national legislations have to follow this directive. Each MS has to designate a national competent authority, which oversees one or more notified bodies (NBs). The NBs award the "CE" marking based on the conformity declared and postulated by the manufacturer. These NBs are assessed for their competence in performing their regulatory duties per particular device category.

With time, experience, and technological progress, a lot of inadequacies of the previous legislative framework and a need for revision became apparent. This was undertaken and materialized in EU Regulation 2017/746.[12] The explicit aim was to ensure a high level of safety and health promotion while supporting innovation. The key elements are the rules-based risk classification, taking into account the intended purpose of the devices and their inherent risks, performance evaluation, both technical and clinical, vigilance and market surveillance proportional to the risk classification along with traceability. Also, the function and transparency of the NBs is strengthened by mutual inspections and the addition of a centralized expert committee, the Medical Devices Coordination Group (MDCG), where every MS is represented.

In the United States, there is a centralized entity, the Food and Drug Administration (FDA), under the US federal law,[17] licensing the medical devices, in general, via its dedicated Center for Devices and Radiological Health (CDRH) throughout the states. No formal separation of in vitro diagnostics is made, and all are treated as medical devices, which they of course are. However, they are acknowledged in specific guidance further downstream. They also follow a risk-based approach; the FDA governing law (Title 21)[18] specifies detailed medical device categories each with its own risk classification as opposed to the rule-based classification of EU. In the EU legislation, it is explicitly mentioned that the concept of clinical benefit for in vitro diagnostic medical devices is different than the therapeutic ones. It concerns providing patient's accurate medical information juxtaposed against information obtained through the use of other diagnostic methodologies, while the final clinical outcome for the patient is dependent on further diagnostic and/or therapeutic options available.

20.2.3 COMPANION DIAGNOSTIC IN VITRO MEDICAL DEVICE REGULATION

Companion diagnostics are in vitro diagnostic medical devices eventually placed in the medium–high risk class in both legislatures, since the information they provide may have a critical effect on one's health. The EU legislation[12] specifies and postulates that

> *(in-vitro diagnostic)*... Devices are classified as class C if they are intended: ... (f) to be used as companion diagnostics; (g) to be used for disease staging, where there is a risk that an erroneous result would lead to a patient management decision resulting in a life-threatening situation for the patient or for the patient's offspring; (h) to be used in screening, diagnosis, or staging of cancer; (i) for human genetic testing; (j) for monitoring of levels of medicinal products, substances or biological components, when there is a risk that an erroneous result will lead to a patient management decision resulting in a life-threatening situation for the patient or for the patient's offspring

So, they are assessed for their technical competence regarding accuracy, precision, and reproducibility of generated results. Diagnostic sensitivity and specificity and positive/negative predictive value are of paramount importance. Beyond those, their manufacture should follow Good Manufacture Practice (GMP) principles and be placed under a quality-management system. As part of their quality-management system, manufacturers should also systematically and actively collect information from postmarket experience with their devices. Then, they can update their technical documentation from their market-surveillance activities, as needed. A comprehensive postmarket surveillance system is envisioned within the quality-management system. The traceability of devices, by a so-called unique device identification (UDI) system, in the EU, for example, should significantly enhance the effectiveness of the post-market safety–related activities and minimize errors, while also assisting in the detection of falsified devices. The proper device distribution and disposal will also benefit from such a system.

At the same time, companion diagnostic devices are essential for defining patients who are most likely to benefit from a specific treatment with a medicinal product as well as patients who run a risk of developing adverse reactions during treatment. So, they become essential for the safe and effective use of those medications. Then, the applicant has to submit clinical performance data from suitable clinical studies. The demonstration of the clinical evidence is based on scientific validity, analytical performance, and clinical performance of the device. This performance demonstration is not a static, one-time one, but one that should be updated throughout the life cycle of the product, reflecting the progress of science and change in the art of medicine. A continuous performance evaluation process is thus established for the benefit of patients.

20.2.4 COMPANION DIAGNOSTICS AND BIOMARKERS

It should be further clarified that companion diagnostics are useful for defining patients' eligibility for a specific treatment with a medicinal product through the quantitative or qualitative determination of specific markers, the biomarkers.[12] These can be present in healthy subjects and/or in patients and be prognostic for a disease or predictive for a treatment. In both regulatory authorities a close-collaboration and codecision process with the drug regulatory body, such as the FDA Center for Drug Evaluation and Research (CDER) in the United States and National Medicine Agencies (NMAs) or European Medicines Agency (EMA), Committee for Medicinal Products for Human Use (CHMP) in the EU, is required for the final benefit/risk of the medication/companion diagnostic combination assessment, in its totality. Also, both FDA and EMA have biomarker-qualification programs, which timely assist the applicants in the early development of a suitable biomarker for their products. There,

> *a clear understanding of the relationship between a biomarker and the clinical outcome, a defined use for the biomarker in drug development and an identified biomarker measurement assay, preferably analytically validated*

are sought, discussed and proven. Since 2014 the two agencies have aligned their demands for biomarker qualification to enable better collaboration and harmonization of regulatory decisions).[19] A rough comparison between the two regulatory authorities follows in Table 20.1:

A large number of biomarkers have been developed for early detection (Fig. 20.1) (Rolfo, 2017). Nevertheless, relatively few are supported by evidence sufficient for regulatory approval as

Table 20.1 Comparison Between In Vitro Diagnostic Device Regulatory Legislative Entities in the United States (US) and the European Union (EU).

Regulatory Area	US	EU
Member-state sovereignty	No	Yes
Regulatory authority	FDA, CDRH	Notified bodies
Risk-based approach	Yes	Yes
In vitro diagnostic classification	Within-device category	Rules
Companion diagnostic classification	Medium–high	Medium–high
Biomarker qualification	Yes	Yes
Assessment along medication	FDA, CDER	EMA, CHMP/NMAs

CDER, Center for Drug Evaluation and Research; *CDRH*, Center for Devices and Radiological Health; *CHMP*, Committee for Medicinal Products for Human Use; *EMA*, European Medicines Agency; *FDA*, Food and Drug Administration; *NMAs*, National Medicine Agencies.

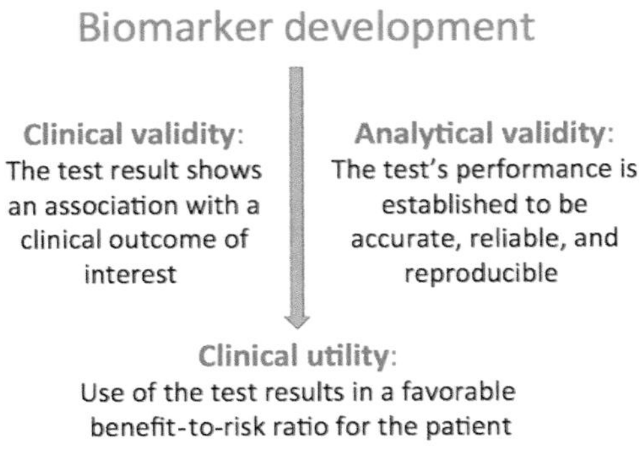

Biomarker development
Clinical validity:
The test result shows an association with a clinical outcome of interest
Analytical validity:
The test's performance is established to be accurate, reliable, and reproducible
Clinical utility:
Use of the test results in a favorable benefit-to-risk ratio for the patient

FIGURE 20.1

A general biomarker-development strategy and regulatory requirements.[21]

Preclinical exploratory	*Phase I*	*Promising directions identified*
Clinical assay and validation	*Phase II*	*Clinical assay detects established disease*
Retrospective longitudinal	*Phase III*	*Biomarker detects disease early before it becomes clinical and a "screen positive" rule is defined*
Prospective screening	*Phase IV*	*Extent and characteristics of disease detected by the test and the false referral rate are identified*
Cancer control	*Phase V*	*Impact of screening on reducing the burden of disease on the population is quantified*

FIGURE 20.2

A timeline of new in vitro diagnostic development in the case, for example, of cancer control.[20]

the standards for validation of clinical relevance in appropriate populations are very rigorous. An example of a potential development pathway follows Fig. 20.2.[20]

20.3 GENOMIC INFORMATION AND REGULATION

20.3.1 GENERAL ASPECTS

With the genomic-medicine technological explosion, the health sector is experiencing another paradigm shift. As with all novelties, its proper use will determine whether its full potential to the patients', health systems', and societies' benefit will be harnessed at its fullest and whether more intended and/or unintended medical tragedies, as in the past, will follow or not. With regard to the regulation of medicinal products for human use, there are a lot of ways in which the genomic information can add on, expand, explain, and complement the existing, phenotypically/clinically/empirically derived one from the pre- and clinical-trial setting and, if properly generated and applied, enhance both the internal and external validity of the regulatory conclusions. The genetic information itself must be generated according to the principles of good science, using validated methods and instrumentation in accredited facilities performed by qualified personnel.[12] The methodology evolves so the requirements need to be accordingly updated. The legislation needs to contain elements that allow flexibility, adjustments, and adaptation as regards to the ever-advancing state-of-the-art.

Issues of data privacy and confidentiality need to be tackled beforehand.[22] The health systems need to be reorganized for making the genetic-analysis technology available to the general

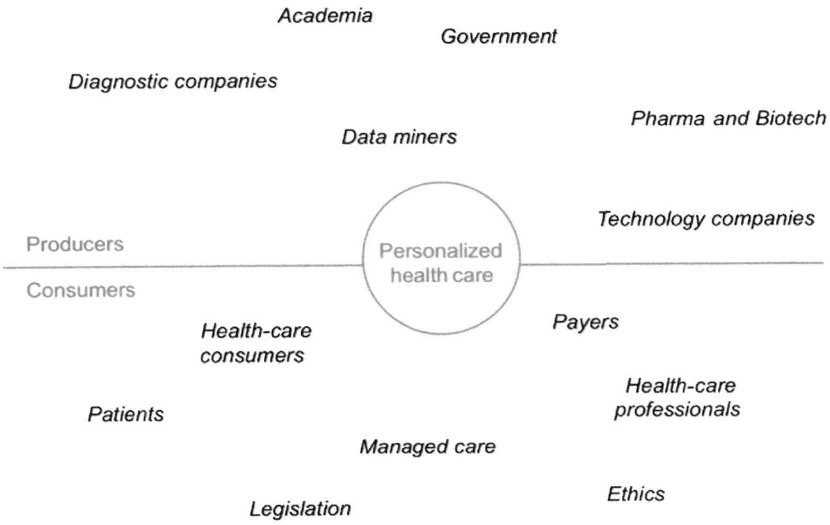

FIGURE 20.3

Personalized health care. Stakeholder space.[23]

population in an as equitable, accessible, and standardized manner as possible. All this new technology will then help in containing the ever-increasing health-care costs by enabling more accurate diagnoses and treatment courses, avoiding failures and inadequacies thereof, which increase the disease burden and costs (Fig. 20.3).[23]

20.3.2 ASPECTS AND APPLICATIONS OF INDIVIDUALIZED EVIDENCE-BASED PATIENT BENEFIT

Every person carries his/her own unique genetic material and genomic medicine exploits and relates it with the disease characteristics a patient is presented with, enabling proper diagnosis. Not only this source of interindividual, germline, inherited variability[24] but also the intraindividual, somatic, acquired genetic information over one's own course of life course varies in space/time.[25] The epigenetically, variably controlled expression also adds to the complexity of the task. The disciplines of epigenomics, as well as the rest of *-omics*, greatly assist in deciphering the relationship between genomic and phenotypic variability.[26] As stated earlier, the proper diagnosis is of paramount importance for the adequately informed clinical decisions. Genetic variants may be the underlying cause for phenotypic deviations from the established model of disease. In this way, they allow for proper target validation/qualification, which leads to better tailored therapeutics.[27] This is of paramount importance in the era of personalized and evidence-based medicine, because it is now possible to investigate and prove the value of an intervention.

There have already been numerous examples where the genomic information is employed for the benefit of patients. In pharmacogenomics the pharmacokinetic individual variability has been traced up to genomic variability. An example is the fast and slow alcohol metabolizers.[28] Another one is the detection of lower activity of the active irinotecan metabolite SN-38 metabolizing (inactivating) enzyme UGT1A1 in homozygous for *UGT1A1*28*-allele individuals.[29] The prescribing and adverse event monitoring are accordingly modified in susceptible individuals. The pharmacodynamic individual variability has also found numerous applications. For example, only carriers of Epidermal Growth Factor Receptor (EGFR) -activating variants will be amenable for the successful treatment of non−small cell lung cancer patients with erlotinib.[30]

Stratification of clinical-trial populations by a genetic feature may reveal a particularly sensitive or resistant patient subpopulation, which was not possible before. The target patient population may, thus, be better defined and reflected in the prescribing information. Alternatively, a companion diagnostic with a similar aim might be developed and included in the treatment package. On the other hand, the number of patients in the subpopulations is thinning out to the point of loss of power for the trial to detect statistically significant and clinically meaningful differences. Further stratification may make single-cohort studies inevitable, which is undesired for the regulatory purposes' situation.[31,32] The example of cystic fibrosis with its +350 predisposing genetic defects is one (extreme but not unusual in the area of medicines for indications of diseases with orphan designation) such example.[33]

20.3.3 THE CASE FOR CANCER AND FURTHER RAMIFICATIONS OF GENOMIC MEDICINE

Of course, nowhere else is the utility of genomic medicine more manifested than in cancer, a disease whose hallmark is the genetic instability, both in space and time (spatial and temporal tumor heterogeneity),[34−36] which remains largely incurable. The number of identified genetic defects, both germ-line and somatic, has greatly increased, and the newly discovered medicines have led to substantially longer survivals, unimagined in many cases before.[37] Overall survival remains the gold-standard end point for an anticancer agent's clinical trials and medicinal product licensure.[38] The extended survival times, now experienced with these new agents, would mean that the patients have to be followed-up for much longer times, which, among others, will lead to subject loss due to attrition or crossover, etc. because of other causes disrupting the rigor of randomized controlled trials' settings. Hence, these trials would require surrogate end points, and genomic medicine may help identify earlier events in the course of cancer, which may be associated with increased overall survival, shortening the required clinical-trial times and other, otherwise, inevitable losses to attrition.[39] This is not to say that molecular biology is set to compete with the pathologist but rather assist and complement his/her work, identifying a tumor and its stage as it changes over time.[40] Pathologists play important roles in evaluating histology and biomarker results and establishing detection methods. Pathologists are expected to act as "genetic interpreters" or "genetic translators" and build a link between the molecular subtypes with the histological features of tumors. Subsequently, by using their findings, oncologists will carry out targeted therapy based on molecular classification.

Currently, biomarker-based patient selection and clinical relevance remain a challenge.[41] Although a number of genomic signatures[42,43] have been discovered along with diagnostic, prognostic, and predictive genomic drivers that can predict risk of cancer relapse, therapy monitoring and detect resistance variants, the question, when does a driver become a target or remains a mere, so-called passenger, stays unresolved and a case-by-case justification is required .[44–48]

Type I errors, that is, failure to detect a patient eligible for treatment when he/she in fact is, control is of major importance from the regulators' perspective.[7] Genomic information, alone, can take a candidate treatment up to Phase II of its development only. However, the translational research is greatly accelerated both because of proper target identification and more tailored subsequent dosing after the recommended Phase II dose studies. Simple broad-blanket indications based on histology of tumor origin *alone* are no more the case. In addition, genomic information has enabled more adaptive clinical trial designs such as prognostic enrichment and basket and umbrella trials.[49–53] A cautionary approach is recommended as the issues of multiplicity loom above regulatory decision-making as the independence of such studies seems to be missing. Also, the drug sensitivity does not seem to be homogeneous across histologies.[54–57]

In addition, contrary to histologically dependent indications in which a trend toward limiting the broadness of indication becomes all the more manifested, the genetic information has led, in some cases and under conditions, to broadening of an indication to tumors that bear the genetic defect regardless of their tissue of origin, the so-called tissue-agnostic or "histology-independent" ones.[58,59] Thus far this has been limited to final-line therapies utilizing agents of known safety profile after justification where there is no other alternative. The issue of their clinical relevance remains to be seen. Confirmatory basket-like trial designs have recently appeared to bridge the confirmatory with the exploratory setting, especially if they may lead to niche indications in cases of unmet medical need.[60] The question of tumor type has evolved to biomarker within tumor type and potentially biomarker of tumor of any type.

One can imagine in vitro, in vivo correlation and modeling and simulation studies with extrapolations to patients where it is unethical or just too limited a patient pool exists to perform proper Phase III confirmatory trials such as in rare cancers or pediatric patients.[61] In these cases the employment of a biomarker can add scientific rigor to otherwise futile settings. A case-by-case strong scientific/biological rationale, compelling, even if only exploratory, clinical data, comparability of responses as with the "regularly" approved indications can be a helpful exercise.[50] Absolute certainty regarding drug effect will not exist for every biomarker–tumor–drug combination even though the initial data generation can achieve higher r^2 coefficients of determinations for in vitro diagnostic–acquired information.[62–64] The end decision shall remain a clinical one where lower r^2 are, lacking a better alternative, acceptable at least for the time being.

Essentially, with the focus on a biomarker, the tip of the iceberg of tumor complexity is barely touched upon, in most cases, for what a biomarker is attempting to do is introduce a binary output as a condition for intervening in a whole biological continuum of concurrent events, which renders the question of any threshold an intractable one (Fig. 20.4).[21] There is already criticism that all these are attempts to reinstate failed trials, for there is heterogeneity of genetic alterations of tumor, immune tumor contextures, and in host metabolic profiles along with overtime variability.[59] The level of actionable variants and their clinical relevance is a never-ending discussion.[64–66]

Precision medicine in oncology, for the time being, involves more or less indiscriminate sequencing of bioptically acquired genetic material that is hardly of any representativeness of the heterogeneous in space and time mix of tumor cells, as explained earlier. Recent findings challenge the very notion of "driver variants" as despite their seemingly ubiquitous presence, only some of the potential totality of candidate tumors is undergoing carcinogenic transformation. Mostly

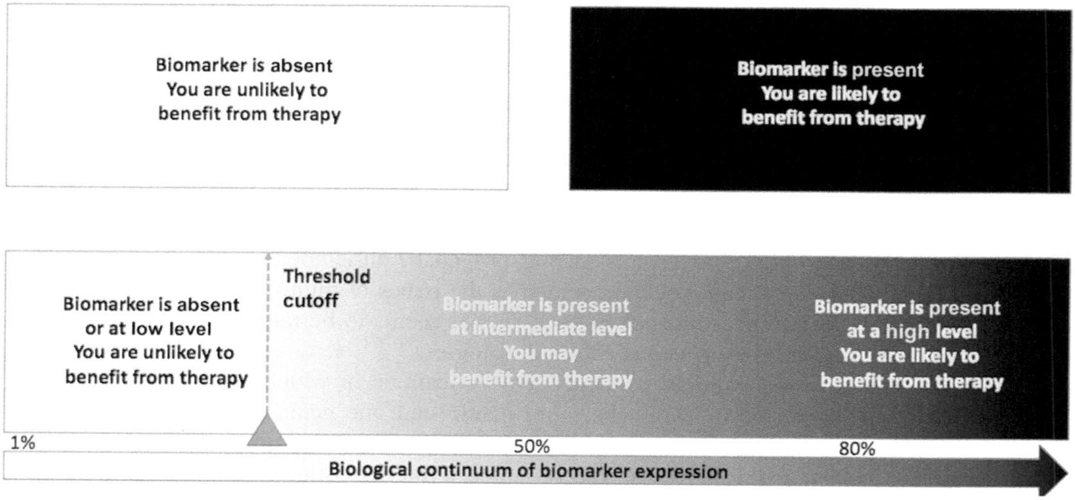

FIGURE 20.4

Binary versus biological continuum of biomarker expression and clinical decision-making.[21]

disturbing is the discovery that on surviving cytotoxic treatment, cancer cells express progenitor, stem cell–like phenotypes, which are exactly the effectors of the highly undesired recurrence, casting further doubt on the real benefit of such targeting. All these point to the absence of an often-assumed prescriptive linear relationship between tumor genotype and phenotype.[67]

20.3.4 GENERAL CONSIDERATIONS FOR REGULATORY EVOLUTION FOLLOWING GENOMIC MEDICINE

In addition, lacking a better alternative and due to the abundance of unmet medical need, there is extensive off-label use of already-approved medications.[68–70] These are frequently backed up by exploratory Phase II-based genomic data. Genomic medicine may permit early access to potentially useful indications where there is an unmet medical need. The example of histology-independent applications and circumstances of acceptability mentioned earlier, point to the more rigorous use of such treatments, even in the off-label setting. On the other hand, extrapolation to patient subpopulations, such as the pediatric one, is not always successful.

Proper incentives should be given as the entities to be treated in these settings tend to be rarer but not rare enough to be treated as a designated orphan indication.[71,72] Also, conditional marketing authorizations and other conditional approval measures may be used. An appropriate mixture of incentives and sanctions should be used to force companies to further pursue potential new uses of their approved products along with the evolution of genomic medicine.[73] In EU pharmaceutical legislation, for example, there are no sanctions for lack of postapproval efficacy studies even though they are stated as a postapproval obligation in the initial marketing authorization.[74] This is not the case for the postapproval safety studies for which there is an abundance of regulatory tools to impose upon the marketing authorization holders.[9] The framework is more rigorous and demanding, awarding conditional-only approvals, asking for postapproval safety studies, and incorporating data from pharmacovigilance. No such enforcement exists for postapproval efficacy studies with the evolution of the state-of-the-art in the respective, say, auxiliary in vitro companion diagnostic device. A commitment to it will then become a business decision whether to keep pursuing a particular genomic medicine–refinement pathway or not.[75]

20.4 CONCLUSION AND A LOOK AHEAD

The possibilities of genomic medicine seem ever expanding and the challenges to its proper use are at the same time mounting. Caution should be exercised because timely communication looks prerequisite for the integrity of confidence elaboration of all stakeholders involved. There needs to be a standardization of laboratorial methodologies from a public-health perspective to enable the generation of interpretable data across the health system.[12] From the data scarcity of the previous century, the enabling technology allows the generation of lots of data, rightly termed "big data".[76] The abundance of data, along with artificial-intelligence tools, enables, at the same time, a more guided precision medicine decision-making (Fig. 20.5).[77]

At the same time the feasibility and importance of translational and confirmatory clinical research are increasing. A sincere and in-depth exploration of conflicting findings indicating nongenetic and complex tumor phenotype dynamics will have to take place if the genomic medicine is to ultimately lead to undisputed clinical benefits.[64] Risk-sharing collaborative research programs, pooling companies' resources may be a way forward.[76,78] The competitive model of the pharmaceutical industry might have to be revisited. The interested companies should work closely with the

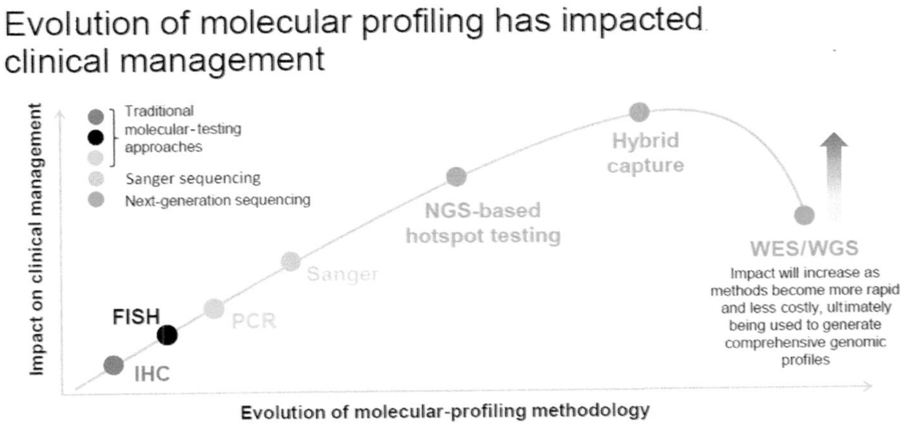

FIGURE 20.5

Evolution of molecular profile and its impact on clinical management. *FISH*, Fluorescence in situ hybridization; *IHC*, immunohistochemistry; *NGS*, next-generation sequencing; *PCR*, polymerase chain reaction; *WES*, whole-exome sequencing; *WGS*, whole-genome sequencing.[77]

regulatory authorities to discuss their plans to utilize genetic information in their development programs as early as possible. This qualification strategy mentioned earlier may save a lot of time and resources in the process. The regulators should also adjust to the new reality and positively react by issuing as-clear-as-possible guidelines, keeping up with the evolutions and fostering innovation. Societies of Physician and health professional should issue updated clinical-use guidelines for proper application of the knowledge gained. In a lot of cases, a change in the disease-treatment paradigm is foreseeable and physician discretion might be further limited by genetic information. Consequently, genomic information can help diminish the agency problems due to the asymmetry of information between patient and health-care provider. The patient will know something about him/herself and will feel more empowered in seeking appropriate health-care information with confidence. Health-care professionals should prepare to share their dominant position in this asymmetric information environment and welcome the seemingly inexorable introduction of evidence-based new approach in their everyday practice as the shift from histology to a molecular classification takes place. In addition, the ever-pervasive small-area variations in medical practice and the problems with a properly functioning health system will be decreased with the advent of genomic medicine.[79–81]

Only then, a truly and meaningfully informed consent can be granted (see also Chapter 18). Special consideration should be given to the suitable communication of the genetic test results to the patients and their caregivers. Miscommunication may lead to opposite results and mistrust and join the series of medical tragedies that have accompanied medical innovations in the past.

Clinical experientialism will remain the pillar of medical practice for the foreseeable time being. However, "absence of evidence is not evidence of absence";[42] so, the enabling and ever-evolving technology will keep producing clues for the understanding of life biology. A lot more paradigm shifts may follow, and the civil society has to be alert to respond, suitably recognize, and utilize the innovation(s') potential and ascribe tangible value for the benefit of most fellow humans.

REFERENCES

1. Institute of Medicine. *Crossing the quality chasm: A new health system for the 21st Century*. Washington, DC: National academies press; 2001.
2. World health organization. *Adherence to long-term therapies: evidence for action*. Geneva (Switzerland): WHO; 2003.
3. Taber JM, Leyva B, Persoskie A. Why do people avoid Medical Care? A qualitative study using national data. *J Gen Intern Med*. 2015;30:290–297.
4. Stiglitz J.E. Information and the change in the paradigm in economics, Nobel prize lecture, 2001. Available from: http://www.nobelprize.org/uploads/2018/06/stiglitz-lecture.pdf
5. Tofan G, Bodolica V, Spraggon M. Governance mechanisms in the physician–patient relationship: a literature review and conceptual framework. *Health Expect*. 2013;16:14–31.
6. DIRECTIVE 2001/83/EC OF THE EUROPEAN PARLIAMENT AND OF THE COUNCIL on the community code relating to medicinal products for human use. OJ L 2001;311:67-128.
7. Chow S-C, Liu J-P. *Design and analysis of clinical trials: Concepts and methodologies*. 3rd Ed Wiley; 2013.
8. Steyerberg EW. *Clinical prediction models: A practical approach to development, validation, and updating*. New York: Springer; 2009.

9. DIRECTIVE 2010/84/EU OF THE EUROPEAN PARLIAMENT AND OF THE COUNCIL amending, as regards pharmacovigilance, Directive 2001/83/EC on the community code relating to medicinal products for human use. OJ L 2010;348:74-99.
10. Crews KR, Hicks JK, Pui C-H, Relling MV, Evans WE. Pharmacogenomics and individualized medicine: Translating science into practice. *Clin Pharmacol Ther*. 2012;92:467–475.
11. Rägo L, Santoso B. Drug Regulation: History, Present and Future. In: Van Boxtel CJ, Santoso B, Edwards IR, eds. *Drugs benefits and risks: International textbook of clinical pharmacology. Rev*. 2nd Ed Amsterdam: IOS press; 2008:65–76.
12. REGULATION (EU) 2017/746 OF THE EUROPEAN PARLIAMENT AND OF THE COUNCIL on in vitro diagnostic medical devices and repealing Directive 98/79/EC and Commission Decision 2010/227/EU. OJ L 2017;117:176-332.
13. International medical device regulators forum. Available from: http://www.imdrf.org/.
14. European commission. *State of health in the EU: Companion report 2017*. Luxembourg: Publications office of the European Union; 2017. Available from: https://ec.europa.eu/health/sites/health/files/state/docs/2017_companion_en.pdf.
15. European commission. Free movement in harmonized and non-harmonized sectors. Available from: http://ec.europa.eu/growth/single-market/goods/free-movement-sectors_en.
16. DIRECTIVE 98/79/EC OF THE EUROPEAN PARLIAMENT AND OF THE COUNCIL on in vitro diagnostic medical devices. OJ L 1998; 331:1-37.
17. Federal food, drug, and cosmetic act. Pub L No. 75-717, 52Stat 1040 (1938).
18. US Government publishing office. CFR Title 21:GPO 2017.
19. Manolis E, Koch A, Deforce D, Vamvakas S. European medicines agency experience with biomarker qualification. *Methods Mol Biol*. 2015;1243:255–272.
20. Pepe MS, Etzioni R, Feng Z, Potter JD, Thompson ML, Thornquist M, et al. Phases of biomarker development for early detection of cancer. *J Natl Cancer Inst*. 2001;93:1054–1061.
21. Rolfo C. Definition of a driver. Cellular/tissular mechanisms supporting that a driver becomes a target. Multiple drivers, mechanisms of resistance. *Workshop on site and histology - independent indications in oncology, European medicines agency (London)*. 2017;14:12. Available from: http://www.ema.europa.eu/docs/en_GB/document_library/Presentation/2018/02/WC500243466.pdf.
22. Brothers KB, Rothstein MA. Ethical, legal and social implications of incorporating personalized medicine into healthcare. *Per Med*. 2015;12:43–51.
23. Jakka S, Rossbach M. An economic perspective on personalized medicine. *HUGO J*. 2013;7:1–6.
24. Lupski JR, Stankiewitz PT. *Genomic disorders: The genomic basis of disease*. Humana press; 2007.
25. Erickson PR. Somatic gene mutation and human disease other than cancer: an update. *Mutat Res*. 2010;705:96–106.
26. Zhang X, Kuivenhoven JA, Groen AK. Forward individualized medicine from personal genomes to interactomes. *Front Physiol*. 2015;6: Art.364. Available from: https://www.frontiersin.org/articles/10.3389/fphys.2015.00364/full.
27. Turan N, Katari S, Coutifaris C, Sapienza C. Explaining inter-individual variability in phenotype: Is epigenetics up to the challenge? *Epigenetics*. 2010;5:16–19.
28. Jones JD, Comer SD, Kranzler HR. The Pharmacogenetics of Alcohol Use Disorder. *Alcohol Clin Exp Res*. 2015;39:391–402.
29. Innocenti F, Undevia SD, Iyer L, Chen PX, Das S, Kocherginsky M, et al. Genetic variants in the UDP-glucuronosyltransferase 1A1 gene predict the risk of severe neutropenia of irinotecan. *J Clin Oncol*. 2004;22:1382–1388.
30. Kobayashi K, Hagiwara K. Epidermal growth factor receptor (EGFR) mutation and personalized therapy in advanced nonsmall cell lung cancer (NSCLC). *Target Oncol*. 2013;8:27–33.

31. Hatswell AJ, Baio G, Berlin JA, Alar I, Freemantle N. Regulatory approval of pharmaceuticals without a randomised controlled study: analysis of EMA and FDA approvals 1999–2014. *BMJ Open*. 2016. Available from: http://bmjopen.bmj.com/content/bmjopen/6/6/e011666.full.pdf.
32. Gedeborg R, Cline C, Zethelius B, Salmonson T. Pragmatic clinical trials in the context of regulation of medicines. *Ups J Med Sci*. 2019;124:37–41.
33. Clinical and Functional Translation of Cystic Fibrosis Cystic fibrosis transmembrane conductance regulator. CFTR2 Variant List History. Available from: https://www.cftr2.org/mutations_history
34. Salk JJ, Fox EJ, Loeb LA. Mutational heterogeneity in human cancers: origin and consequences. *Ann Rev Pathol*. 2010;5:51–75.
35. Weinstein JN, Collisson EA, Mills GB, Mills Shaw KR, Ozenberger BA, Ellrott K, et al. The cancer genome atlas pan-cancer analysis, project. *Nat Genet*. 2013;45:1113–1120.
36. Rogozin IB, Pavlov YI, Goncearenco A, De S, Lada AG, Poliakov E, Panchenko AR, Cooper DN. Mutational signatures and mutable motifs in cancer genomes. *Brief Bioinform*. 2018;19:1085–1101.
37. American Society of Clinical Oncology. The State of Cancer Care in America, 2017: A Report by the American Society of Clinical Oncology. *J Oncol Pract*. 2017;13:e353–e394.
38. European Medicines Agency. Guideline on the evaluation of anticancer medicinal products in man (Rev 5, 2.9.2017). Available from: http://www.ema.europa.eu/docs/en_GB/document_library/Scientific_guideline/2017/11/WC500238764.pdf
39. Zhao F. Surrogate End Points and Their Validation in Oncology Clinical Trials. *J Clin Oncol*. 2016;34:1436–1437.
40. Yu Y. Molecular classification and precision therapy of cancer: immune checkpoint inhibitors. *Front Med*. 2018;12:229–235.
41. Mayer B, Heinzel A, Lukas A, Perco P. Predictive Biomarkers for Linking Disease Pathology and Drug Effect. *Curr Pharm Des*. 2017;23:29–54.
42. Ciriello G, Miller ML, Aksoy BA, Senbabaoglu Y, Schultz N, Sander C. Emerging landscape on oncogenic signatures across human cancers. *Nat Genet*. 2013;45:1127–1133.
43. Sanchez-Vega F, Mina M, Armenia J, Chatila WK, Luna A, La KC, et al. Oncogenic Signaling Pathways in The Cancer Genome Atlas. *Cell*. 2018;173:321–337.
44. Burrell RA, Swanton C. Tumour heterogeneity and the evolution of polyclonal drug resistance. *Mol Oncol*. 2014;8:1095–1111.
45. Beckman RA, Schemmann S, Yeang C-H. Impact of genetic dynamics and single-cell heterogeneity on development of nonstandard personalized medicine strategies for cancer. *PNAS*. 2012;109:14586–14591.
46. Beckman RA, Clark J, Chen C. Integrating predictive biomarkers and classifiers into oncology clinical development programmes. *Nat Rev Drug Discov*. 2011;10:735–748.
47. Hyman DM, Taylor BS, Baselga J. Implementing Genome-Driven Oncology. *Cell*. 2017;168:584–599.
48. Thompson LL, Jeusset LM, Lepage CC, McManus KJ. Evolving Therapeutic Strategies to Exploit Chromosome Instability in Cancer. *Cancers (Basel)*. 2017;9:151.
49. Collignon O. Statistical considerations for the development of diagnostic tests Workshop on site and histology - Independent indications in oncology. *European medicines agency (London)*. 2017. Available from: http://www.ema.europa.eu/en/documents/presentation/presentation-session-4-statistical-considerations-development-diagnostic-tests-olivier-collignon_en.pdf.
50. Woodcock J, LaVange LM. Master protocols to study multiple therapies, multiple diseases, or both. *N Engl J Med*. 2017;377:62–70.
51. Ondra T, Jobjörnsson S, Beckman RA, Burman CF, König F, Stallard N, et al. Optimized adaptive enrichment designs. *Stat Methods Med Res*. 2019;28:2096–2111.

52. Beckman RA, Antonijevic Z, Kalamegham R, Chen C. Adaptive Design for a Confirmatory Basket Trial in Multiple Tumor Types Based on a Putative Predictive Biomarker. *Clin Pharmacol Ther.* 2016;100:617–625.
53. Beckman RA, Chen C. Efficient, Adaptive Clinical Validation of Predictive Biomarkers in Cancer Therapeutic Development. *Adv Exp Med Biol.* 2015;867:81–90.
54. Chiu YD, Koenig F, Posch M, Jaki T. Design and estimation in clinical trials with subpopulation selection. *Stat Med.* 2018;37:4335–4352.
55. Chen C, Li X, Shuai Y, Antonijevic Z, Kamegham R, Beckman RA. Statistical Design and Considerations of a Phase 3 Basket Trial for Simultaneous Investigation of Multiple Tumor Types in One Study. *Statistics in Biopharmaceutical Research.* 2016;8:248–257.
56. Renfro LA, Sargent DJ. Statistical controversies in clinical research: basket trials, umbrella trials, and other master protocols: a review and examples. *Ann Oncology.* 2017;28:34–43.
57. Simon R, Geyer S, Subramanian J, Roychowdhury S. The Bayesian basket design for genomic variant-driven phase II trials. *Semin Oncol.* 2016;43:13–18.
58. Pavlidis N, Pentheroudakis G. Cancer of unknown primary site. *Lancet.* 2012;379:1428–1435.
59. Lacombe D, Burock S, Bogaerts J, Schoeffski P, Golfinopoulos V, Stupp R. The dream and reality of histology agnostic cancer clinical trials. *Mol Oncol.* 2014;8:1057–1063.
60. Flaherty KT, Le DT, Lemery S. *Tissue-agnostic drug development. Am Soc Clin Oncol Educ Book.* 2017;37:222–230.
61. Barretina J, Caponigro G, Stransky N, Venkatesan K, Margolin AA, Kim S, et al. The Cancer Cell Line Encyclopedia enables predictive modelling of anticancer drug sensitivity. *Nature.* 2012;483:603–607.
62. Dudley JC, Lin MT, Le DT, Eshleman JR. Microsatellite Instability as a Biomarker for PD-1 Blockade. *Clin Cancer Res.* 2016;22:813–820.
63. Bhatt DL, Mehta C. Adaptive Designs for Clinical Trials. *N Engl J Med.* 2016;375:65–74.
64. Hyman DM, Puzanov I, Subbiah V, Faris JE, Chau I, Blay J-Y, et al. Vemurafenib in Multiple Nonmelanoma Cancers with BRAF V600 Mutations. *N Engl J Med.* 2015;373:726–736.
65. Simon R. Genomic Alteration-Driven Clinical Trial Designs in Oncology. *Ann Intern Med.* 2016;165:270–278.
66. Simon R, Blumenthal G, Rothenberg M, Sommer J, Roberts S, Armstrong D, et al. The role of nonrandomized trials in the evaluation of oncology drugs. *Clin Pharmacol Ther.* 2015;97:502–507.
67. Brock A, Huang S. Precision Oncology: Between Vaguely Right and Precisely Wrong. *Cancer Res.* 2017;77:6473–6479.
68. Wittich CM, Burkle CM, Lanierb WL. Ten Common Questions (and Their Answers) About Off-label Drug Use. *Mayo Clin Proc.* 2012;87:982–990.
69. Sutherland A, Waldek S. It is time to review how unlicensed medicines are used. *Eur J Clin Pharmacol.* 2015;71:1029–1035.
70. Christian L, Duttge G. Ethical and legal framework and regulation for off-label use: European perspective. *Ther Clin Risk Manag.* 2014;10:537–546.
71. Tsigkos S, Llinares J, Mariz S, Aarum S, Fregonese L, Dembowska-Baginska B, et al. Use of biomarkers in the context of orphan medicines designation in the European Union. *Orphanet J Rare Dis.* 2014;9:13. Available from: https://ojrd.biomedcentral.com/articles/10.1186/1750-1172-9-13.
72. Hunter NL, Rao GR, Sherman RE. Flexibility in the FDA approach to orphan drug development. *Nat Rev Drug Discov.* 2017;16:737–738.
73. Nayroles G, Frybourg S, Gabriel S, Kornfeld Å, Antoñanzas-Villar F, Espín J, et al. Unlocking the potential of established products: toward new incentives rewarding innovation in Europe. *J Mark Access Health Policy.* 2017;5:1298190. Available from: https://www.ncbi.nlm.nih.gov/pmc/articles/PMC5508393/pdf/zjma-5-1298190.pdf.

74. Commission Delegated Regulation (EU) No 357/2014 of 3 February 2014 supplementing Directive 2001/83/EC of the European Parliament and of the Council and Regulation (EC) No 726/2004 of the European Parliament and of the Council as regards situations in which post-authorisation efficacy studies may be required. OJ L 2014;107:1–4.
75. Woloshin S, Schwartz LM, White B, Moore TJ. The Fate of FDA Postapproval Studies. *N Engl J Med.* 2017;377:1114–1117.
76. Oktay Y, Gottwald M, Schüler P, Martin CM. Opportunities and Challenges for Drug Development: Public–Private Partnerships, Adaptive Designs and Big Data. *Front Pharmacol.* 2016;7: Art.461. Available from: https://www.frontiersin.org/articles/10.3389/fphar.2016.00461/full.
77. Thomas M. How could molecular profiling impact histology independent labels in the future? *European medicines agency (London).* 2017. Available from: http://www.ema.europa.eu/en/documents/presentation/presentation-session-3-how-could-molecular-profiling-impact-histology-independent-labels-future_en.pdf.
78. Allen CE, Laetsch TW, Mody R, Irwin MS, Lim MS, Adamson PC, et al. Target and Agent Prioritization for the Children's Oncology Group—National Cancer Institute Pediatric MATCH Trial. *J Natl Cancer Inst.* 2017;109. Available from: https://doi.org/10.1093/jnci/djw274.
79. Willard HF, Angrist M, Ginsburg GS. Genomic medicine: genetic variation and its impact on the future of health care. *Philos Trans R Soc Lond B Biol Sci.* 2005;360:1543–1550.
80. Schork NJ, Nazor K. Integrated Genomic Medicine: A Paradigm for Rare Diseases and Beyond. *Adv Genet.* 2017;97:81–113.
81. Bilkey GA, Burns BL, Coles EP, Mahede T, Baynam G, Nowak KJ. Optimizing Precision Medicine for Public Health. *Front Public Health.* 2019;7:42.

CHAPTER

GENOMIC MEDICINE IN EMERGING ECONOMIES

21

Catalina Lopez Correa
Genome British Columbia, Vancouver, BC, Canada

21.1 INTRODUCTION

Advances in genomics, and all other "omic"-related technologies, are revolutionizing the practice of medicine around the world. Genomic analysis is being used to provide a much more precise diagnosis to help guide treatments and to indicate the prognosis of several human diseases, in particular in rare diseases, cancer, and infectious diseases.[1] Genomic analysis is gradually being adopted by diagnostic laboratories and hospitals in the United States, Canada, and Europe, and guidelines from US Food and Drug Administration (http://www.fda.gov) and the European Medicines Agency (http://www.ema.europa.eu) are being developed to provide a regulated framework for the clinical implementation of these new technologies.[2]

However, the adoption of genomic technologies in clinical practice in resource-limited countries and regions has been much slower. In this chapter, we will describe the advances in genomic technologies, the challenges emerging economies are facing in implementing these technologies, and also the opportunities that should be explored in order to democratize the use of genomics.

21.2 FROM SANGER SEQUENCING TO NEXT-GENERATION SEQUENCING AND NATION-WIDE GENOMIC PROGRAMS

Genomic analysis, and genome sequencing in particular, has been the key innovation that has allowed scientist to analyze the genetic information contained in the DNA of every living organism. Genome sequencing technologies are the key drivers of precision (personalized) medicine approaches. The first generation of genome sequencing was introduced in the 1970s[3] and was distributed more widely and industrialized during the 1980s. The first generation of sequencing technologies was recognized by its high accuracy but was expensive and had a very low throughput, which limited its use at a large scale. Even with its limitations, Sanger sequencing was the technology used to sequence the first human genome. In fact, the cost of sequencing the first human genome was about $3 billion, and it took several international institutes, hundreds of researchers, and 13 years to complete.[4] Given the high cost of the first sequencing machines, only research institutions in developed countries could afford to adopt this technology when it was introduced in the market.

Applied Genomics and Public Health. DOI: https://doi.org/10.1016/B978-0-12-813695-9.00021-2

Luckily, in the past few years the cost of sequencing has declined exponentially thanks to a series of next-generation sequencing (NGS) technologies that have been developed and marked a new era in genomics, because they allowed to analyze very large amounts of DNA at a much more reasonable cost.[5,6] Following the development of NGS technologies, in 2014 the US government launched a unique program to drive the cost of genome sequencing down to $1000 per genome. This dramatic decrease in the cost of sequencing has been often compared to Moore's law who observed that computing power tends to double—and that its price therefore halves—every 2 years, but it is now clear that sequencing technologies have outpaced Moore's law. As personal computers changed the world, the great advances in genome sequencing technologies are now revolutionizing the biomedical field.[7] The genomic revolution is allowing scientist to sequence genomes much faster and at a much lower cost, leading to a more decentralized and democratized access to genomic data.[8]

All these advances in sequencing and genomic technologies together with an increasing number of results that are having an impact in patient care, going from disease prevention to more accurate diagnosis and personalized treatment decisions, have led to the development of a series of cutting-edge genomic medicine or precision medicine programs in various countries. Most of these programs are using local expertise and are aligning their efforts according to their cultural, political, and social backgrounds. On one hand, these programs are helping advance the use and application of genomics in clinical practice, and on the other the challenge remains that most of these programs are being developed in isolation. Organizations such as the National Human Genome Research Institute and the Global Genomic Medicine Collaborative (G2MC) are now fostering connections between different global initiatives and are also working on capturing and disseminating research outcomes, best practices, policies, and health economic analysis to help develop international standards in order to advance the global implementation of genomics medicine.[9]

However, the pace of implementation of genomic medicine practices is not always equally met in developing and resource-limited countries or emerging economies, where significant barriers exist, often related to lack of resources, lack of technology and knowledge transfer, and lack of training. Some of the barriers that these regions are encountering when trying to implement clinical genomic initiative are described more in detailed here.

21.3 CAPACITY BUILDING AND COST OF SETTING UP SEQUENCING CENTERS

Even with the dramatic increase in sequencing speed and the decrease in sequencing cost, establishing a genome sequencing facility with NGS remains a very costly endeavor. It has been estimated that the cost of establishing a sequencing facility varies from $100,000 to $700,000. The costs are higher when establishing a sequencing facility in developing countries (shipment, customs, reagents, maintenance, etc.), and the challenge is to maintain these facilities up a running at a cost-effective manner.[10] Also, laboratories in developing countries have great difficulty to cover contract services and to buy consumables and reagents that will ensure a fully operational high-throughput sequencing facility, at a cost that is affordable and that can compete with sequencing facilities in developed countries. The cost needed to develop and maintain these types of sequencing facilities far exceeds the funds that scientist in developing countries can access through local grants. Given these limitations, in many cases, scientists in developing countries are outsourcing genome sequencing services to private companies (e.g., Genotypic, Macrogen, Eurofins, and BGI) located in different regions of the world. However, it is important to invest in building capacity around genomic technologies in low- and middle-income countries to ensure the training of local scientists and the development of local initiatives. Investing in infrastructure not only for sequencing but more importantly in bioinformatics is one of the crucial steps to ensure emerging economies benefit from the genomics revolution.[11]

21.4 LACK OF DIVERSITY ON INTERNATIONAL DATABASES

Many thousands of whole genomes have been sequenced after the first human genome sequence was published in 2001. However, most of these genomes have been sequenced in Europe or in the United States and belong, in the majority of the cases, to individuals of Caucasian origin. Despite the general agreement regarding the need to increase diversity on the human genomic data that are being generated and the need for a more democratized and equitable access to genomic technologies, genomics research remains largely focused on populations of European descent. A 2009 study indicated that 96% of all genomics studies were done on populations of European descent.[12] An updated analysis done in 2016 indicated that non-European participants represented 19% of individuals studied in Genome-Wide Association Studies analysis.[13,14] Most of this increase was associated to genomic data generated from Asian populations with other ethnic groups showing only a minimal increase[15] even though over three-fourths of the world population live in Africa.[16]

21.5 PARACHUTE RESEARCH

Lots of the research work around genomics in developing countries and emerging economies has been done in collaboration with northern and more developed countries. In recent years the term "parachute research" has emerged to describe initiatives where the results of these collaborations only, or mostly, benefit one side of the group. There is an increasing resistance against parachute researchers: the ones who go to another country (particularly low- and middle-income countries), use the local infrastructure, take samples of patients and populations, and then go back home to write an academic paper for a high-ranking journal without even acknowledging the participation of the local scientists and without ever returning results or data to the patients or populations used in the study.[17]

It is clear that North–South collaborations are of great value, but these collaborations should be performed on equal and respectful bases where all contributions are fully recognized, and most importantly, these collaborations should be based on needs and priorities defined by the low- and middle-income countries. This mutual beneficial collaboration model has been suggested in 2013 by the Genomic Medicine Alliance (www.genomicmedicinealliance.org), and several collaborations have emerged based on this model.[18] Some groups have initiated pilot studies that analyze the

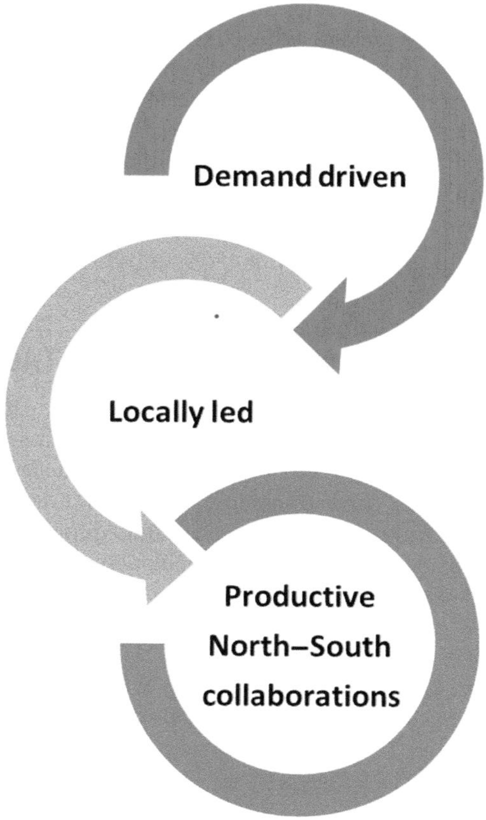

FIGURE 21.1

Model for productive North–South collaborations.
This figure illustrates the need for a more interactive collaboration where the exchange is initiated based on local needs, and the results will benefit emerging economies directly.

development of demand-driven and locally led research (Fig. 21.1) where countries from the South fully benefit from collaborations with northern countries.[19]

21.6 EDUCATION AND CAPACITY BUILDING

The public health benefits from genetic and genomic innovations can only be realized when we fully understand how to best implement innovations in clinical practice. Several groups have studied the barriers for clinical implementation in Europe and the United States and have determined that clinician's knowledge and understanding of the uses and applications of genomics in medicine is one of the key barriers that have slow down the adoption of genomic technologies in clinical practice.[20] Several programs are now being set up in England and United States to help educate health-care providers on the use of genomics.

In low- and middle-income countries the problem is even larger. The challenge is not only that the health-care provider are not informed and not trained to be able to fully understand and benefit from genomic innovations, the largest challenge is the lack of bioinformatics and computational expertise, and also the lack of a critical mass of scientists that are trained in cutting-edge genomic technologies.[21]

If we think about the patient's journey, from the moment that a patient visits a doctor and the doctor decides to prescribe a test to the moment the patient gets the report with the results of the genomic or genetic test (Fig. 21.2), there are many professionals involved. All these professionals (doctors, nurses, pharmacists, lab technicians, bioinformaticians, etc.) need to be trained in order to fully implement clinical genomics to help improve patient's health and outcomes. Biomedical scientists and health-care professionals in low- and middle-income countries frequently fail to fully appreciate the potential that this new technologies offer to improve medical diagnosis and treatment. Investing in training genomics scientist and informing health-care providers and patients about genomics is highly recommended to ensure these new technologies are adopted.

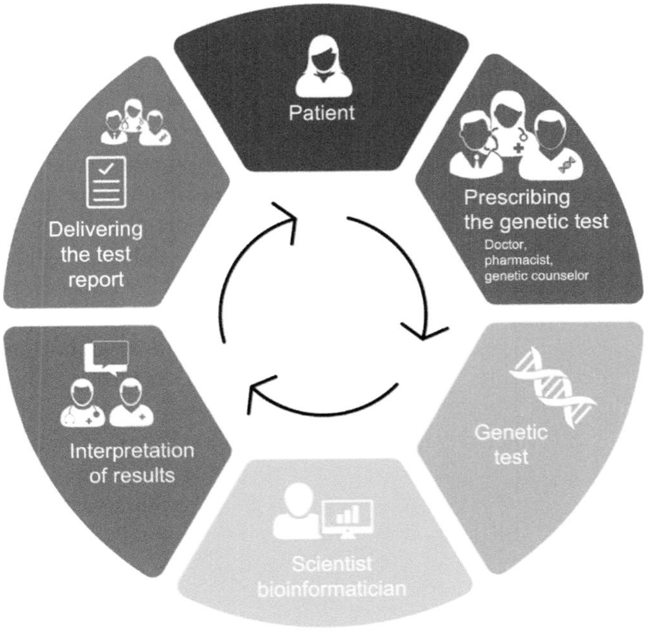

FIGURE 21.2

The patient journey.
This figure illustrates the different professionals that are involved in ordering, performing, analyzing, and providing the results of a genomic test.

Aside from these barriers, there are also growing opportunities to advance the implementation of genomics in emerging economy settings. Developing countries may be resource-limited but are also potentially rich in producing data in various genomic medicine-related disciplines, from the perspective of public health genomics. Even though the pace of implementation of genomic medicine varies from country to country, depending on a number of different parameters, in recent years, we have observed a series of success stories that are indicating the great potential genomic medicine has to help solve some of the most pressing health challenges present in low- and middle-income countries. Some of the opportunities for the implementation of genomic medicine in these countries are described next.

21.7 FAST-SECOND WINNER MODEL

Most of the developed countries follow a very linear path to innovation. This path includes large investments in fundamental research to advance some initiatives into translation and application that could led to concrete return on investment and commercial opportunities.[22] Instead of this linear innovation model used in developed countries, which goes from discovery science toward application and translational, a new innovation model for emerging economies has been proposed.[23] The new approach has been called "Fast-Second Winner" model of innovation. This model offers multiple entry points into the global genomics innovation ecosystem for developing countries, whether or not extensive discovery projects and infrastructure is already in place. Some low- and middle-income countries have already been using this model of innovation and have advanced the implementation of genomic medicine at a public health level. One of these examples is the implementation of pharmacogenomics testing in some Asian countries. Whereas some European countries are heavily investing on large whole-genome sequencing initiatives similar to the one developed by Genomics England,[24] scientists from Thailand proposed a simple pharmacogenomics test that could be used to avoid severe adverse drug events such as Stevens–Johnson syndrome (SJS) and toxic epidermal necrolysis (TEN). Adverse drug events such as SJS/TEN can occur in patients who take carbamazepine and have a particular gene variant; even though the events are rare, the burden on the patients and the cost to the health-care system are high due to the clinical severity. The scientist from Thailand then developed a simple pharmacogenomics ID card that is cost-effective and easy to use and that can indicate when an individual has the genetic variants associated with the severe adverse drug event. This pharmacogenomics ID card is now being covered by the public health-care system in Thailand and is being implemented in other Asian countries such as Taiwan.[25] Other countries have used different strategies to implement pharmacogenomics in clinical practice and have successfully aligned the local expertise with specific regional needs to advance the use and implementation of genomics.[26]

21.8 HEALTH BIOTECHNOLOGY IN LATIN AMERICA

As described before, emerging economies as well as low- and medium-income countries are starting to use genomics and health biotechnology in many different ways. A study published in 2018 analyzed the patterns of health biotechnology publications in six Latin American countries (Argentina, Brazil, Chile, Colombia, Cuba, and Mexico) from 2001 to 2015. They found that many of these countries had a publication growth patter higher than some industrialized nations, but the visibility of their research (measured by the number of citations) did not reach the world average, with the only exception of Colombia.[27] Some of the examples cited in the paper are the development of an immune-enzymatic assay for the detection of serum antibodies against *Trypanosoma cruzi* (Chagas disease) in Chile and a diagnostic assay that uses biomarkers for tuberculosis detection in Colombia. The article also indicates that Latin American countries are active in international research collaboration with Colombia being the most active (64% of papers coauthored internationally), whereas Brazil was the least active (35% of papers).

Data published in this and other papers indicate that emerging economies are embracing genomics and biotechnology and are also developing international collaborations on areas of key priorities for the regions such as infectious diseases.

21.9 GENOMICS IN AFRICA

One of the key barriers and bottlenecks to advance the use of genomics in emerging economies is the lack of bioinformatics and computational biology expertise and the lack of local datasets. In most of low- and middle-income countries the use and analysis of large-scale genomics datasets have been limited by the availability of scientific expertise, the lack of infrastructure, and the lack of data generated from local populations. The African continent, with funding support from the National Institute of Health and the Welcome Trust, developed a new initiative called the Human Heredity and Health in Africa (H3Africa; https://h3africa.org). This initiative was created to help advance research of the genomic and environmental aspects of diseases prevalent in Africa, to build capacity around bioinformatics and around genomics research, and to generate reference datasets arising from African populations.[28] As part of the H3Africa initiative, a specific bioinformatics network (H3ABioNet) was created to enable scientist to analyze their own data.[29] The network has significantly contributed to create new infrastructure and to develop human capacity that will allow the continent to be more self-sufficient without having to always rely on the expertise and infrastructure available in developed countries.

21.10 GLOBAL INITIATIVES

There are several global initiatives around genomics that have emerged in the last few years. Two worth noting are the Global Alliance for Genomics and Health (GA4GH; https://www.ga4gh.org) and the G2MC (https://g2mc.org). On the one hand the GA4GH has focused mainly on the research aspects of genomics by developing a policy framework and technical standards to enable responsible sharing and aggregation of genomic data. One of the initial GA4GH driver projects is the Clinical Genome Resource initiative (ClinGen) aiming at developing a knowledge base to support the understanding of genes and variants for use in precision medicine.[30] ClinGen has been used as a reference dataset to help advance phenotype genotype correlations that can have an impact on patient diagnosis.

On the other hand, the G2MC group is focusing on the clinical application and implementation of genomic medicine by fostering collaboration and engaging multiple stakeholders across the globe. The G2MC group is focusing on enabling the demonstration of value and supporting the effective use of genomics in medicine. The G2MC started as a discussion group where participants from many different regions (mostly developed countries) joined forces to develop standards for clinical implementation of genomics.[9] The organization is now an independent entity that is embracing and attracting several members of low- and middle-income countries.

It is clear that many advances have been made in the last 5–6 years and that we can now whiteness several success stories on the use and implementation of clinical genomics in emerging economies. These success stories will continue to grow, but the speed and impact will fully depend of the capacity emerging economies will have to invest in science and technology in order to build local capacity and expertise.

REFERENCES

1. McCarthy JJ, McLeod HL, Ginsburg GS. Genomic medicine: a decade of successes, challenges, and opportunities. *Sci Transl Med. 12*. 2013;5(189):189sr4.
2. Knowles L, Luth W, Bubela T. Paving the road to personalized medicine: recommendations on regulatory, intellectual property and reimbursement challenges. *J Law Biosci*. 2017;4(3):453–506.
3. Sanger F, Nicklen S, Coulson AR. DNA sequencing with chain-terminating inhibitors. *Proc Natl Acad Sci USA*. 1977;74(12):5463–5467.
4. Sboner A, Mu XJ, Greenbaum D, Auerbach RK, Gerstein MK. The real cost of sequencing: higher than you think!. *Genome Biol*. 2011;12(8):125.
5. Shendure J, Ji H. Next-generation DNA sequencing. *Nat Biotechnol*. 2008;26(10):1135–1145.
6. Mardis ER. A decade's perspective on DNA sequencing technology. *Nature*. 2011;470(7333):198–203.
7. Check HE. Technology: the $1,000 genome. *Nature*. 2014;507(7492):294–295.
8. Li PE, Lo CC, Anderson JJ, et al. Enabling the democratization of the genomics revolution with a fully integrated web-based bioinformatics platform. *Nucleic Acids Res. 9*. 2017;45(1):67–80.
9. Manolio TA, Abramowicz M, Al-Mulla F, et al. Global implementation of genomic medicine: we are not alone. *Sci Transl Med*. 2015;7(290):290ps13.
10. Helmy M, Awad M, Mosab KA. Limited resources of genome sequencing in developing countries: challenges and solutions. *Appl Transl Genom*. 2016;9:15–19.
11. Forero DA, Wonkam A, Wang W, et al. Current needs for human and medical genomics research infrastructure in low and middle income countries. *J Med Genet*. 2016;53(7):438–440. Available from: https://doi.org/10.1136/jmedgenet-2015-103631.
12. Bentley AR, Callier S, Rotimi CN. Diversity and inclusion in genomic research: why the uneven progress? *J Community Genet*. 2017;8(4):255–266.
13. Hindorff LA, Bonham VL, Brody LC, et al. Prioritizing diversity in human genomics research. *Nat Rev Genet*. 2018;19(3):175–185. Available from: https://doi.org/10.1038/nrg.2017.89. Epub 2017 Nov 20.
14. Hindorff LA, Bonham VL, Ohno-Machado L. Enhancing diversity to reduce health information disparities and build an evidence base for genomic medicine. *Per Med*. 2018;15(5):403–412. Available from: https://doi.org/10.2217/pme-2018-0037. Epub 2018 Sep 13.
15. Popejoy AB, Fullerton SM. Genomics is failing on diversity. *Nature*. 2016;538(7624):161–164.

16. Nordling L. How the genomics revolution could finally help Africa. *Nature*. 2017;544(7648):20–22.
17. Bockarie M, Machingaidze S, Nyirenda T, Olesen OF, Makanga M. Parasitic and parachute research in global health. *Lancet Glob Health*. 2018;6(9):e964.
18. Cooper DN, Brand A, Dolzan V, et al. Bridging genomics research between developed and developing countries: the Genomic Medicine Alliance. *Per Med*. 2014;11(7):615–623.
19. Kok MO, Gyapong JO, Wolffers I, Ofori-Adjei D, Ruitenberg EJ. Towards fair and effective North-South collaboration: realising a programme for demand-driven and locally led research. *Health Res Policy Syst*. 2017;15(1):96.
20. Sperber NR, Carpenter JS, Cavallari LH, et al. Challenges and strategies for implementing genomic services in diverse settings: experiences from the Implementing GeNomics In pracTicE (IGNITE) network. *BMC Med Genomics*. 2017;10(1):35. Available from: https://doi.org/10.1186/s12920-017-0273-2.
21. Isaacson Barash C. Translating translational medicine into global health equity: what is needed? *Appl Transl Genom*. 2016;9:37–39. Available from: https://doi.org/10.1016/j.atg.2016.03.004. Published2016 Mar 10.
22. Wiechers IR, Perin NC, Cook-Deegan R. The emergence of commercial genomics: analysis of the rise of a biotechnology subsector during the Human Genome Project, 1990 to 2004. *Genome Med*. 2013;5(9):83. Available from: https://doi.org/10.1186/gm487. Published 2013 Sep 20.
23. Mitropoulos K, Cooper DN, Mitropoulou C, et al. Genomic medicine without borders: which strategies should developing countries employ to invest in precision medicine? A new "Fast-Second Winner" strategy. *OMICS*. 2017;21(11):647–657. Available from: https://doi.org/10.1089/omi.2017.0141. PubMed PMID: 29140767.
24. Turnbull C, Scott RH, Thomas E, et al. 100 000 Genomes Project. The 100 000 Genomes Project: bringing whole genome sequencing to the NHS. *BMJ*. 2018;361:k1687. Available from: https://doi.org/10.1136/bmj.k1687. Erratum in: BMJ. 2018 May 2;361:k1952. PubMed PMID:29691228.
25. Sukasem C, Chantratita W. A success story in pharmacogenomics: genetic ID card for SJS/TEN. *Pharmacogenomics*. 2016;17(5):455–458. Available from: https://doi.org/10.2217/pgs-2015-0009. Epub 2016 Mar 30. PubMed PMID: 27027537.
26. Mitropoulos K, Al Jaibeji H, Forero DA, et al. Success stories in genomic medicine from resource-limited countries. *Hum Genomics*. 2015;9:11. Available from: https://doi.org/10.1186/s40246-015-0033-3. PubMed PMID: 26081768.
27. León-de la ODI, Thorsteinsdóttir H, Calderón-Salinas JV. The rise of health biotechnology research in Latin America: a scientometric analysis of health biotechnology production and impact in Argentina, Brazil, Chile, Colombia, Cuba and Mexico. *PLoS One*. 2018;13(2):e0191267. Available from: https://doi.org/10.1371/journal.pone.0191267.
28. Mulder N, Abimiku A, Adebamowo SN, et al. H3Africa: current perspectives. *Pharmgenomics Pers Med*. 2018;11:59–66. Available from: https://doi.org/10.2147/PGPM.S141546. Published 2018 Apr 10.
29. Mulder NJ, Adebiyi E, Adebiyi M, et al. H3ABioNet Consortium, as members of the H3Africa Consortium Development of bioinformatics infrastructure for genomics research. *Global Heart*. 2017;12(2):91–98.
30. Dolman L, Page A, Babb L, et al. ClinGen advancing genomic data-sharing standards as a GA4GH driver project. *Hum Mutat*. 2018;39(11):1686–1689. Available from: https://doi.org/10.1002/humu.23625. PubMed PMID: 30311379.

Index

Note: Page numbers followed by "*f*" and "*t*" refer to figures and tables, respectively.

H

I

J

K

L

M

N

O

P

Q

R

S

9780128136959